Lecture Notes in Electrical Engineering

Volume 240

For further volumes:
http://www.springer.com/series/7818

James J. (Jong Hyuk) Park
Joseph Kee-Yin Ng · Hwa Young Jeong
Borgy Waluyo
Editors

Multimedia and Ubiquitous Engineering

MUE 2013

Volume II

Springer

Editors

James J. (Jong Hyuk) Park
Department of Computer Science
Seoul University of Science
and Technology (SeoulTech)
Seoul
Republic of South Korea

Joseph Kee-Yin Ng
Department of Computer Science
Hong Kong Baptist University
Kowloon Tong
Hong Kong SAR

Hwa Young Jeong
Humanitas College
Kyung Hee University
Seoul
Republic of South Korea

Borgy Waluyo
School of Computer Science
Monash University
Clayton, VIC
Australia

ISSN 1876-1100
ISBN 978-94-007-6737-9
DOI 10.1007/978-94-007-6738-6
Springer Dordrecht Heidelberg New York London

ISSN 1876-1119 (electronic)
ISBN 978-94-007-6738-6 (eBook)

Library of Congress Control Number: 2013936524

© Springer Science+Business Media Dordrecht(Outside the USA) 2013
This work is subject to copyright. All rights are reserved by the Publisher, whether the whole or part of the material is concerned, specifically the rights of translation, reprinting, reuse of illustrations, recitation, broadcasting, reproduction on microfilms or in any other physical way, and transmission or information storage and retrieval, electronic adaptation, computer software, or by similar or dissimilar methodology now known or hereafter developed. Exempted from this legal reservation are brief excerpts in connection with reviews or scholarly analysis or material supplied specifically for the purpose of being entered and executed on a computer system, for exclusive use by the purchaser of the work. Duplication of this publication or parts thereof is permitted only under the provisions of the Copyright Law of the Publisher's location, in its current version, and permission for use must always be obtained from Springer. Permissions for use may be obtained through RightsLink at the Copyright Clearance Center. Violations are liable to prosecution under the respective Copyright Law. The use of general descriptive names, registered names, trademarks, service marks, etc. in this publication does not imply, even in the absence of a specific statement, that such names are exempt from the relevant protective laws and regulations and therefore free for general use.
While the advice and information in this book are believed to be true and accurate at the date of publication, neither the authors nor the editors nor the publisher can accept any legal responsibility for any errors or omissions that may be made. The publisher makes no warranty, express or implied, with respect to the material contained herein.

Printed on acid-free paper

Springer is part of Springer Science+Business Media (www.springer.com)

Message and Organization Committee

Message from the MUE 2013 General Chairs

MUE 2013 is the FTRA 7th event of the series of international scientific conferences. This conference will take place in May 9–11, 2013, in Seoul Korea. The aim of the MUE 2013 is to provide an international forum for scientific research in the technologies and application of Multimedia and Ubiquitous Engineering. It is organized by the Korea Information Technology Convergence Society in cooperation with Korea Information Processing Society. MUE2013 is the next event in a series of highly successful international conferences on Multimedia and Ubiquitous Engineering, MUE-12 (Madrid, Spain, July 2012), MUE-11 (Loutraki, Greece, June 2011), MUE-10 (Cebu, Philippines, August 2010), MUE-09 (Qingdao, China, June 2009), MUE-08 (Busan, Korea, April 2008), and MUE-07 (Seoul, Korea, April 2007).

The papers included in the proceedings cover the following topics: *Multimedia Modeling and Processing*, *Ubiquitous and Pervasive Computing*, *Ubiquitous Networks and Mobile Communications*, *Intelligent Computing*, *Multimedia and Ubiquitous Computing Security*, *Multimedia and Ubiquitous Services*, *Multimedia Entertainment*, *IT and Multimedia Applications*. Accepted and presented papers highlight new trends and challenges of Multimedia and Ubiquitous Engineering. The presenters showed how new research could lead to novel and innovative applications. We hope you will find these results useful and inspiring for your future research.

We would like to express our sincere thanks to Steering Chairs: James J. (Jong Hyuk) Park (SeoulTech, Korea), Martin Sang-Soo Yeo (Mokwon University, Korea). Our special thanks go to the Program Chairs: Eunyoung Lee (Dongduk Women's University, Korea), Cho-Li Wang (University of Hong Kong, Hong Kong), Borgy Waluyo (Monash University, Australia), Al-Sakib Khan Pathan (IIUM, Malaysia), SangHyun Seo (University of Lyon 1, France), all Program Committee members and all the additional reviewers for their valuable efforts in the review process, which helped us to guarantee the highest quality of the selected papers for the conference.

We cordially thank all the authors for their valuable contributions and the other participants of this conference. The conference would not have been possible without their support. Thanks are also due to the many experts who contributed to making the event a success.

May 2013

Young-Sik Jeong
Leonard Barolli
Joseph Kee-Yin Ng
C. S. Raghavendra
MUE 2013 General Chairs

Message from the MUE 2013 Program Chairs

Welcome to the FTRA 7th International Conference on Multimedia and Ubiquitous Engineering (MUE 2013), to be held in Seoul, Korea on May 9–11, 2013. MUE 2013 will be the most comprehensive conference focused on the various aspects of multimedia and ubiquitous engineering. MUE 2013 will provide an opportunity for academic and industry professionals to discuss recent progress in the area of multimedia and ubiquitous environment. In addition, the conference will publish high quality papers which are closely related to the various theories and practical applications in multimedia and ubiquitous engineering. Furthermore, we expect that the conference and its publications will be a trigger for further related research and technology improvements in these important subjects.

For MUE 2013, we received many paper submissions; after a rigorous peer review process, we accepted 62 articles with high quality for the MUE 2013 proceedings, published by Springer. All submitted papers have undergone blind reviews by at least two reviewers from the technical program committee, which consists of leading researchers around the globe. Without their hard work, achieving such a high-quality proceeding would not have been possible. We take this opportunity to thank them for their great support and cooperation. We would like to sincerely thank the following invited speaker who kindly accepted our invitations, and, in this way, helped to meet the objectives of the conference: Prof. Hui-Huang Hsu, Tamkang University, Taiwan. Finally, we would like to thank all of you for your participation in our conference, and also thank all the authors, reviewers, and organizing committee members. Thank you and enjoy the conference!

Eunyoung Lee, Korea
Cho-Li Wang, Hong Kong
Borgy Waluyo, Australia
Al-Sakib Khan Pathan, Malaysia
SangHyun Seo, France
MUE 2013 Program Chairs

Organization

Honorary Chair:	Makoto Takizawa, Seikei University, Japan
Steering Chairs:	James J. Park, SeoulTech, Korea
	Martin Sang-Soo Yeo, Mokwon University, Korea
General Chairs:	Young-Sik Jeong, Wonkwang University, Korea
	Leonard Barolli, Fukuoka Institute of Technology, Japan
	Joseph Kee-Yin Ng, Hong Kong Baptist University, Hong Kong
	C. S. Raghavendra, University of Southern California, USA
Program Chairs:	Eunyoung Lee, Dongduk Women's University, Korea
	Cho-Li Wang, University of Hong Kong, Hong Kong
	Borgy Waluyo, Monash University, Australia
	Al-Sakib Khan Pathan, IIUM, Malaysia
	SangHyun Seo, University of Lyon 1, France
Workshop Chairs:	Young-Gab Kim, Korea University, Korea
	Lei Ye, University of Wollongong, Australia
	Hiroaki Nishino, Oita University, Japan
	Neil Y. Yen, The University of Aizu, Japan
Publication Chair:	Hwa Young Jeong, Kyung Hee University, Korea
International Advisiory Committee:	Seok Cheon Park, Gachon University, Korea
	Borko Furht, Florida Atlantic University, USA
	Thomas Plagemann, University of Oslo, Norway
	Roger Zimmermann, National University of Singapore, Singapore
	Han-Chieh Chao, National Ilan University, Taiwan
	Hai Jin, HUST, China
	Weijia Jia, City University of Hong Kong, Hong Kong
	Jianhua Ma, Hosei University, Japan
	Shu-Ching Chen, Florida International University, USA

viii Message and Organization Committee

	Hamid R. Arabnia, The University of Georgia, USA
	Stephan Olariu, Old Dominion University, USA
	Albert Zomaya, University of Sydney, Australia
	Bin Hu, Lanzhou University, China
	Yi Pan, Georgia State University USA
	Doo-soon Park, SoonChunHyang University, Korea
	Richard P. Brent, Australian National University, Australia
	Koji Nakano, University of Hiroshima, Japan
	J. Daniel Garcia, University Carlos III of Madrid, Spain
	Qun Jin, Waseda University, Japan
	Kyung-Hyune Rhee, Pukyong National University, Korea
Publicity Chairs:	Chengcui Zhang, The University of Alabama at Birmingham, USA
	Michele Ruta, Politecnico di Bari, Italy
	Bessam Abdulrazak, Sherbrooke University, Canada
	Junaid Chaudhry, Universiti Teknologi Malaysia, Malaysia
	Bong-Hwa Hong, Kyung Hee Cyber University, Korea
	Won-Joo Hwang, Inje University, Korea
Local Arrangement Chairs:	HyunsungKim, Kyungil University, Korea
	Eun-Jun Yoon, Kyungil University, Korea
Invited Speaker:	Hui-Huang Hsu, Tamkang University, Taiwan
Program Committee:	A Ra Khil, Soongsil University, Korea
	Afrand Agah, West Chester University of Pennsylvania, USA
	Akihiro Sugimoto, National Institute of Informatics, Japan
	Akimitsu Kanzaki, Osaka University, Japan
	Angel D. Sappa, Universitat Autonoma de Barcelona, Spain
	Bartosz Ziolko, AGH University of Science and Technology, Poland
	Bin Lu, West Chester University, USA

Brent Lagesse, BBN Technologies, USA

Ch. Z. Patrikakis, Technological Education Institute of Piraeus, Greece

Chang-Sun Shin, Sunchon National University, Korea

Chantana Chantrapornchai, Silpakorn University, Thailand

Chih Cheng Hung, Southern Polytechnic State University, USA

Chulung Lee, Korea University, Korea

Dakshina Ranjan Kisku, Asansol Engineering College, India

Dalton Lin, National Taipei University, Taiwan

Dariusz Frejlichowski, West Pomeranian University of Technology, Poland

Debzani Deb, Winston-Salem State University, USA

Deqing Zou, Huazhong University of Science and Technology, China

Ezendu Ariwa, London Metropolitan University, United Kingdom

Farid Meziane, University of Salford, UK

Florian Stegmaier, University of Passau, Germany

Francisco Jose Monaco, University of Sao Paulo, Brazil

Guillermo Camara Chavez, Universidade Federal de Minas Gerais, Brazil

Hae-Young Lee, ETRI, Korea

Hai Jin, Huazhong University of Science and Techn, China

Hangzai Luo, East China Normal University, China

Harald Kosch, University of Passau, Germany

Hari Om, Indian School of Mines University, India

Helen Huang, The University of Queensland, Australia

Hermann Hellwagner, Klagenfurt University, Austria

Hong Lu, Fudan University, China

Jeong-Joon Lee, Korea Polytechnic University, Korea
Jin Kwak, Soonchunhyang University, Korea
Jinye Peng, Northwest University, China
Joel Rodrigue, University of Beira Interior, Portugal
Jong-Kook Kim, Korea University, Korea
Joyce El Haddad, Universite Paris-Dauphine, France
Jungong Han, Civolution Technology, the Netherlands
Jun-Won Ho, Seoul Women's University, Korea
Kilhung Lee, Seoul National University of Science and Technology, Korea
Klaus Schoffmann, Klagenfurt University, Austria
Ko Eung Nam, Baekseok University, Korea
Lidan Shou, Zhejiang University, China
Lukas Ruf, CEO Consecom AG, Switzerland
Marco Cremonini, University of Milan, Italy
Maria Vargas-Vera, Universidad Adolfo Ibanez, Chile
Mario Doeller, University of applied science, Germany
Maytham Safar, Kuwait University, Kuwait
Mehran Asadi, Lincoln University of Penssylvania, USA
Min Choi, Chungbuk National University, Korea
Ming Li, California State University, USA
Muhammad Younas, Oxford Brookes University, UK
Namje Park, Jeju National University, Korea
Neungsoo Park, Konkuk University, Korea
Ning Zhou, University of North Carolina, USA
Oliver Amft, TU Eindhoven, Netherlands
Paisarn Muneesawang, Naresuan University, Thailand
Pascal Lorenz, University of Haute Alsace
Quanqing Xu, Quanqing Xu, Data Storage Institute, A*STAR, Singapore
Rachid Anane, Coventry University, UK

Rainer Unland, University of Duisburg-Essen, Germany
Rajkumar Kannan, Affiliation Bishop Heber College, India
Ralf Klamma, RWTH Aachen University, Germany
Ramanathan Subramanian, Advanced Digital Sciences Center, Singapore
Reinhard Klette, The University of Auckland, New Zealand
Rene Hansen, Aalborg University, Denmark
Sae-Hak Chun, Seoul National University of Science and Technology, Korea
Sagarmay Deb, University of Southern Queensland, Australia
Savvas Chatzichristofis, Democritus University of Thrace, Greece
Seung-Ho Lim, Hankuk University of Foreign Studies, Korea
Shingo Ichii, University of Tokyo, Japan
Sokratis Katsikas, University of Piraeus, Greece
SoonSeok Kim, Halla Universty, Korea
Teng Li, Baidu Inc., China
Thomas Grill, University of Salzburg, Austria
Tingxin Yan, University of arkansas, USA
Toshihiro Yamauchi, Okayama University, Japan
Waleed Farag, Indiana University of Pennsylvania, USA
Wee Siong Ng, Institute for Infocomm Research, Singapore
Weifeng Chen, California University of Pennsylvania, USA
Weifeng Zhang, Nanjing University of Posts and Telecommunication, China
Wesley De Neve, Ghent University iMinds and KAIST,
Won Woo Ro, Yonsei University, Korea
Wookho Son, ETRI, Korea
Wei Wei, Xi'an University of Technology, China

Xubo Song, Oregon Health and Science University, USA
Yan Liu, The Hong Kong Polytechnic University, Hong Kong
Yijuan Lu, Texas State University, USA
Yingchun Yang, Zhejiang University, China
Yong-Yoon Cho, Sunchon University, Korea
Yo-Sung Ho, GIST, Korea
Young-Hee Kim, Korea Copyright Commission, Korea
Young-Ho Park, Sookmyung Women's University, Korea
Zheng-Jun Zha, National University of Singapore, Singapore
Zhu Li, Samsung Telecom America, USA

Message from ATACS-2013 Workshop Chair

Welcome to the Advanced Technologies and Applications for Cloud Computing and Sensor Networks (ATACS-2013), which will be held from May 9 to 11, 2013 in Seoul, Korea.

The main objective of this workshop is to share information on new and innovative research related to advanced technologies and applications in the areas of cloud computing and sensor networks. Many advanced techniques and applications in these two areas have been developed in the past few years. Sensor networks are becoming increasingly large and produce vast amounts of raw sensing data, which cannot be easily processed, analyzed, or stored using conventional computing systems. Cloud computing is one promising technique of efficiently processing these sensing data to create useful services and applications. The convergence of cloud computing and sensor networks requires new and innovative infrastructure, middleware, designs, protocols, services, and applications. ATACS-2013 will bring together researchers and practitioners interested in both the technical and applied aspects of Advanced Techniques and Application for Cloud computing and Sensor networks. Furthermore, we expect that the ATACS-2013 and its publications will be a trigger for further related research and technology improvements in this important subject.

ATACS-2013 contains high quality research papers submitted by researchers from all over the world. Each submitted paper was peer-reviewed by reviewers who are experts in the subject area of the paper. Based on the review results, the Program Committee accepted eight papers.

Message and Organization Committee   xiii

I hope that you will enjoy the technical programs as well as the social activities during ATACS-2013. I would like to send our sincere appreciation to all of the Organizing and Program Committees who contributed directly to ATACS-2013. Finally, we special thank all the authors and participants for their contributions to make this workshop a grand success.

Joon-Min Gil
Catholic University of Daegu
ATACS-2013 Workshop Chair

ATACS-2013 Organization

General Chair: Joon-Min Gil, Catholic University of Daegu, Korea
Program Chair: Jaehwa Chung, Korea National Open University, Korea
Publicity Chair: Dae Won Lee, Seokyeong University, Korea
Program Committee: Byeongchang Kim, Catholic University of Daegu, Korea
                     Hansung Lee, Electronics and Telecommunications Research Institute (ETRI), Korea
                     HeonChang Yu, Korea University, Korea
                     Jeong-Hyon Hwang, State University of New York at Albany, USA
                     JongHyuk Lee, Samsung Electronics, Korea
                     Ki-Sik Kong, Namseoul University, Korea
                     KwangHee Choi, LG Uplus, Korea
                     Kwang Sik Chung, Korea National Open University, Korea
                     Mi-Hye Kim, Catholic University of Daegu, Korea
                     Shanmugasundaram Hariharan, TRP Engineering College (SRM Group), India
                     Sung-Hwa Hong, Mokpo National Maritime University, Korea
                     Sung Suk Kim, Seokyeong University, Korea
                     Tae-Gyu Lee, Korea Institute of Industrial Technology (KITECH), Korea
                     Tae-Young Byun, Catholic University of Daegu, Korea
                     Ui-Sung Song, Busan National University of Education, Korea
                     Yong-Hee Jeon, Catholic University of Daegu, Korea
                     Yunhee Kang, Baekseok University, Korea
                     Zhefu Shi, University of Missouri, USA

Message from PSSI-2013 Workshop Chair

The organizing committee of *the FTRA International Workshop on Pervasive Services, Systems and Intelligence* (*PSSI 2013*) would like to welcome all of you to join the workshop as well as the FTRA MUE 2013. Advances in information and communications technology (ICT) have presented a dramatic growth in merging the boundaries between physical space and cyberspace, and go further to improve mankind's daily life. One typical instance is the use of smartphones. The modern smartphone is equipped with a variety of sensors that are used to collect activities, locations, and situations of its user continuously and provide immediate help accordingly. Some commercial products (e.g., smart house, etc.) also demonstrate the feasibility of comprehensive supports by deploying a rapidly growing number of sensors (or intelligent objects) into our living environments. These developments are collectively best characterized as ubiquitous service that promises to enhance awareness of the cyber, physical, and social contexts. As such, researchers (and companies as well) tend to provide tailored and precise solutions (e.g., services, supports, etc.) wherever and whenever human beings are active according to individuals' contexts. Making technology usable by and useful to, via the ubiquitous services and correlated techniques, humans in ways that were previously unimaginable has become a challenging issue to explore the picture of technology in the next era.

This workshop aims at providing a forum to discuss problems, studies, practices, and issues regarding the emerging trend of pervasive computing. Researchers are encouraged to share achievements, experiments, and ideas with international participants, and furthermore, look forward to map out the research directions and collaboration in the future.

With an amount of submissions (13 in exact), the organizing chairs decided to accept six of them based on the paper quality and the relevancy (acceptance rate at 46 %). These papers are from Canada, China, and Taiwan. Each paper was reviewed by at least three program committee members and discussed by the organizing chairs before acceptance.

We would like to thank three FTRA Workshop Chair, Young-Gab Kim from Korea University, Korea for the support and coordination. We thank all authors for submitting their works to the workshop. We also appreciate the program committee members for their efforts in reviewing the papers. Finally, we sincerely welcome all participants to join the discussion during the workshop.

James J. Park
Neil Y. Yen
Workshop Co-Chairs

FTRA International Workshop on Pervasive Services, Systems and Intelligence (PSSI-13)

Workshop Organization

Workshop Chairs:: James J. Park (Seoul National University of Science and Technology, Korea)
Neil Y. Yen (The University of Aizu, Japan)

Program Committee:: Christopher Watson, Durham University, United Kingdom
Chengjiu Yin, Kyushu University, Japan
David Taniar, Monash University, Australia
Jui-Hong Chen, Tamkang University, Taiwan
Junbo Wang, the University of Aizu, Japan
Lei Jing, the University of Aizu, Japan
Marc Spaniol, Max-Planck-Institute for Informatic, Germany
Martin M. Weng, Tamkang University, Taiwan
Nigel Lin, Microsoft Research, United States
Ralf Klamma, RWTH Aachen University, Germany
Vitaly Klyuev, the University of Aizu, Japan
Xaver Y. R. Chen, National Central University, Taiwan
Wallapak Tavanapong, Iowa State University, United States
Renato Ishii, Federal University of Mato Grosso do Sul, Brazil
Nicoletta Sala, U. of Lugano, Switzerland and Università dell'Insubria Varese, Italy
Yuanchun Shi, Tsinghua University, China
Robert Simon, George Mason University, USA

Contents

Part I Multimedia Modeling and Processing

Multiwedgelets in Image Denoising 3
Agnieszka Lisowska

**A Novel Video Compression Method Based on Underdetermined
Blind Source Separation** 13
Jing Liu, Fei Qiao, Qi Wei and Huazhong Yang

Grid Service Matching Process Based on Ontology Semantic 21
Ganglei Zhang and Man Li

Enhancements on the Loss of Beacon Frames in LR-WPANs 27
Ji-Hoon Park and Byung-Seo Kim

**Case Studies on Distribution Environmental Monitoring
and Quality Measurement of Exporting Agricultural Products** 35
Yoonsik Kwak, Jeongsam Lee, Sangmun Byun, Jeongbin Lem,
Miae Choi, Jeongyong Lee and Seokil Song

**Vision Based Approach for Driver Drowsiness Detection
Based on 3DHead Orientation** 43
Belhassen Akrout and Walid Mahdi

Potentiality for Executing Hadoop Map Tasks on GPGPU via JNI ... 51
Bongen Gu, Dojin Choi and Yoonsik Kwak

**An Adaptive Intelligent Recommendation Scheme for Smart
Learning Contents Management Systems** 57
Do-Eun Cho, Sang-Soo Yeo and Si Jung Kim

Part II Ubiquitous and Pervasive Computing

An Evolutionary Path-Based Analysis of Social Experience Design . . . 69
Toshihiko Yamakami

Block IO Request Handlingfor DRAM-SSD in Linux Systems 77
Kyungkoo Jun

**Implementation of the Closed Plant Factory System Based
on Crop Growth Model** 83
Myeong-Bae Lee, Taehyung Kim, HongGeun Kim, Nam-Jin Bae,
Miran Baek, Chang-Woo Park, Yong-Yun Cho and Chang-Sun Shin

Part III Ubiquitous Networks and Mobile Communications

**An Energy Efficient Layer for Event-Based Communications
in Web-of-Things Frameworks** 93
Gérôme Bovet and Jean Hennebert

A Secure Registration Scheme for Femtocell Embedded Networks . . . 103
Ikram Syed and Hoon Kim

Part IV Intelligent Computing

**Unsupervised Keyphrase Extraction Based Ranking Algorithm
for Opinion Articles** 113
Heungmo Ryang and Unil Yun

**A Frequent Pattern Mining Technique for Ranking Webpages
Based on Topics** 121
Gwangbum Pyun and Unil Yun

**Trimming Prototypes of Handwritten Digit Images with Subset
Infinite Relational Model** 129
Tomonari Masada and Atsuhiro Takasu

Ranking Book Reviews Based on User Influence 135
Unil Yun and Heungmo Ryang

**Speaker Verification System Using LLR-Based Multiple
Kernel Learning** 143
Yi-Hsiang Chao

Contents   xix

Edit Distance Comparison Confidence Measure for Speech Recognition 151
Dawid Skurzok and Bartosz Ziółko

Weighted Pooling of Image Code with Saliency Map for Object Recognition 157
Dong-Hyun Kim, Kwanyong Lee and Hyeyoung Park

Calibration of Urine Biomarkers for Ovarian Cancer Diagnosis 163
Yu-Seop Kim, Eun-Suk Yang, Kyoung-Min Nam, Chan-Young Park, Hye-Jung Song and Jong-Dae Kim

An Iterative Algorithm for Selecting the Parameters in Kernel Methods 169
Tan Zhiying, She Kun and Song Xiaobo

A Fast Self-Organizing Map Algorithm for Handwritten Digit Recognition 177
Yimu Wang, Alexander Peyls, Yun Pan, Luc Claesen and Xiaolang Yan

Frequent Graph Pattern Mining with Length-Decreasing Support Constraints 185
Gangin Lee and Unil Yun

An Improved Ranking Aggregation Method for Meta-Search Engine 193
Junliang Feng, Junzhong Gu and Zili Zhou

Part V   Multimedia and Ubiquitous Computing Security

Identity-Based Privacy Preservation Framework over u-Healthcare System 203
Kambombo Mtonga, Haomiao Yang, Eun-Jun Yoon and Hyunsung Kim

A Webmail Reconstructing Method from Windows XP Memory Dumps 211
Fei Kong, Ming Xu, Yizhi Ren, Jian Xu, Haiping Zhang and Ning Zheng

On Privacy Preserving Encrypted Data Stores 219
Tracey Raybourn, Jong Kwan Lee and Ray Kresman

Mobile User Authentication Scheme Based on Minesweeper Game ... 227
Taejin Kim, Siwan Kim, Hyunyi Yi, Gunil Ma and Jeong Hyun Yi

Design and Evaluation of a Diffusion Tracing Function for Classified Information Among Multiple Computers 235
Nobuto Otsubo, Shinichiro Uemura, Toshihiro Yamauchi and Hideo Taniguchi

DroidTrack: Tracking Information Diffusion and Preventing Information Leakage on Android 243
Syunya Sakamoto, Kenji Okuda, Ryo Nakatsuka and Toshihiro Yamauchi

Three Factor Authentication Protocol Based on Bilinear Pairing 253
Thokozani Felix Vallent and Hyunsung Kim

A LBP-Based Method for DetectingCopy-Move Forgery with Rotation 261
Ning Zheng, Yixing Wang and Ming Xu

Attack on Recent Homomorphic Encryption Scheme over Integers ... 269
Haomiao Yang, Hyunsung Kim and Dianhua Tang

A New Sensitive Data Aggregation Scheme for Protecting Data Integrity in Wireless Sensor Network 277
Min Yoon, Miyoung Jang, Hyoung-il Kim and Jae-woo Chang

Reversible Image Watermarking Based on Neural Network and Parity Property 285
Rongrong Ni, H. D. Cheng, Yao Zhao, Zhitong Zhang and Rui Liu

A Based on Single Image Authentication System in Aviation Security 293
Deok Gyu Lee and Jong Wook Han

Part VI Multimedia and Ubiquitous Services

A Development of Android Based Debate-Learning System for Cultivating Divergent Thinking 305
SungWan Kim, EunGil Kim and JongHoon Kim

Development of a Lever Learning Webapp for an HTML5-BasedCross-Platform 313
TaeHun Kim, ByeongSu Kim and JongHoon Kim

Contents

Looking for Better Combination of Biomarker Selection and Classification Algorithm for Early Screening of Ovarian Cancer 321
Yu-Seop Kim, Jong-Dae Kim, Min-Ki Jang, Chan-Young Park and Hye-Jeong Song

A Remote Control and Media Sharing System Based on DLNA/UPnP Technology for Smart Home 329
Ti-Hsin Yu and Shou-Chih Lo

A New Distributed Grid Structure for k-NN Query Processing Algorithm Based on Incremental Cell Expansion in LBSs 337
Seungtae Hong, Hyunjo Lee and Jaewoo Chang

A New Grid-Based Cloaking Scheme for Continuous Queries in Centralized LBS Systems 345
Hyeong-Il Kim, Mi-Young Jang, Min Yoon and Jae-Woo Chang

New Database Mapping Schema for XML Document in Electronic Commerce 353
Eun-Young Kim and Se-Hak Chun

A Study on the Location-Based Reservation Management Service Model Using a Smart Phone 359
Nam-Jin Bae, Seong Ryoung Park, Tae Hyung Kim, Myeong Bae Lee, Hong Gean Kim, Mi Ran Baek, Jang Woo Park, Chang-Sun Shin and Yong-Yun Cho

A Real-time Object Detection System Using Selected Principal Components 367
Jong-Ho Kim, Byoung-Doo Kang, Sang-Ho Ahn, Heung-Shik Kim and Sang-Kyoon Kim

Trajectory Calculation Based on Position and Speed for Effective Air Traffic Flow Management 377
Yong-Kyun Kim, Deok Gyu Lee and Jong Wook Han

Part VII Multimedia Entertainment

Design and Implementation of a Geometric Origami Edutainment Application 387
ByeongSu Kim, TaeHun Kim and JongHoon Kim

Gamification Literacy: Emerging Needs for Identifying Bad Gamification 395
Toshihiko Yamakami

Automatic Fixing of Foot Skating of Human Motions from Depth Sensor 405
Mankyu Sung

Part VIII IT and Multimedia Applications

A Study on the Development and Application of Programming Language Education for Creativity Enhancement: Based on LOGO and Scratch 415
YoungHoon Yang, DongLim Hyun, EunGil Kim, JongJin Kim and JongHoon Kim

Design and Implementation of Learning Content Authoring Framework for Android-Based Three-Dimensional Shape 423
EunGil Kim, DongLim Hyun and JongHoon Kim

A Study on GUI Development of Memo Function for the E-Book: A Comparative Study Using iBooks 431
Jeong Ah Kim and Jun Kyo Kim

Relaxed Stability Technology Approach in Organization Management: Implications from Configured-Control Vehicle Technology 439
Toshihiko Yamakami

Mapping and Optimizing 2-D Scientific Applications on a Stream Processor 449
Ying Zhang, Gen Li, Hongwei Zhou, Pingjing Lu, Caixia Sun and Qiang Dou

Development of an Android Field Trip Support Application Using Augmented Reality and Google Maps 459
DongLim Hyun, EunGil Kim and JongHoon Kim

Implementation of Automotive Media Streaming Service Adapted to Vehicular Environment 467
Sang Yub Lee, Sang Hyun Park and Hyo Sub Choi

Contents xxiii

**The Evaluation of the Transmission Power Consumption
Laxity-Based (TPCLB) Algorithm** 477
Tomoya Enokido, Ailixier Aikebaier and Makoto Takizawa

**The Methodology for Hardening SCADA Security Using
Countermeasure Ordering** 485
Sung-Hwan Kim, Min-Woo Park, Jung-Ho Eom and Tai-Myoung Chung

**Development and Application of STEAM Based Education
Program Using Scratch: Focus on 6th Graders' Science
in Elementary School** 493
JungCheol Oh, JiHwon Lee and JongHoon Kim

**Part IX Advanced Technologies and Applications for Cloud
Computing and Sensor Networks**

**Performance Evaluation of Zigbee Sensor Network for Smart
Grid AMI** 505
Yong-Hee Jeon

**P2P-Based Home Monitoring System Architecture Using
a Vacuum Robot with an IP Camera** 511
KwangHee Choi, Ki-Sik Kong and Joon-Min Gil

**Design and Simulation of Access Router Discovery
Process in Mobile Environments** 521
DaeWon Lee, James J. Park and Joon-Min Gil

**Integrated SDN and Non-SDN Network Management Approaches
for Future Internet Environment** 529
Dongkyun Kim, Joon-Min Gil, Gicheol Wang and Seung-Hae Kim

**Analysis and Design of a Half Hypercube Interconnection
Network** 537
Jong-Seok Kim, Mi-Hye Kim and Hyeong-Ok Lee

**Aperiodic Event Communication Process for Wearable
P2P Computing** 545
Tae-Gyu Lee and Gi-Soo Chung

**Broadcasting and Embedding Algorithms for a Half
Hypercube Interconnection Network** 553
Mi-Hye Kim, Jong-Seok Kim and Hyeong-Ok Lee

Obstacle Searching Method Using a Simultaneous Ultrasound Emission for Autonomous Wheelchairs 561

Byung-Seop Song and Chang-Geol Kim

Part X Future Technology and its Application

A Study on Smart Traffic Analysis and Smart Device Speed Measurement Platform 569

Haejong Joo, Bonghwa Hong and Sangsoo Kim

Analysis and Study on RFID Tag Failure Phenomenon 575

Seongsoo Cho, Son Kwang Chul, Jong-Hyun Park and Bonghwa Hong

Administration Management System Design for Smart Phone Applications in use of QR Code 585

So-Min Won, Mi-Hye Kim and Jin-Mook Kim

Use of Genetic Algorithm for Robot-Posture 593

Dong W. Kim, Sung-Wook Park and Jong-Wook Park

Use of Flexible Network Framework for Various Service Components of Network Based Robot 597

Dong W. Kim, Ho-Dong Lee, Sung-Wook Park and Jong-Wook Park

China's Shift in Culture Policy and Cultural Awareness 601

KyooSeob Lim

China's Cultural Industry Policy 611

WonBong Lee and KyooSeob Lim

Development of Mobile Games for Rehabilitation Training for the Hearing Impaired 621

Seongsoo Cho, Son Kwang Chul, Chung Hyeok Kim and Yunho Lee

A Study to Prediction Modeling of the Number of Traffic Accidents 627

Young-Suk Chung, Jin-Mook Kim, Dong-Hyun Kim and Koo-Rock Park

Part XI Pervasive Services, Systems and Intelligence

A Wiki-Based Assessment System Towards Social-Empowered Collaborative Learning Environment 633
Bruce C. Kao and Yung Hui Chen

Universal User Pattern Discovery for Social Games: An Instance on Facebook 641
Martin M. Weng and Bruce C. Kao

Ubiquitous Geography Learning Smartphone System for 1st Year Junior High Students in Taiwan 649
Wen-Chih Chang, Hsuan-Che Yang, Ming-Ren Jheng and Shih-Wei Wu

Housing Learning Game Using Web-Based Map Service 657
Te-Hua Wang

Digital Publication Converter: From SCORM to EPUB 665
Hsuan-pu Chang

An Intelligent Recommender System for Real-Time Information Navigation 673
Victoria Hsu

Part XII Advanced Mechanical and Industrial Engineering, and Control I

Modal Characteristics Analysis on Rotating Flexible Beam Considering the Effect from Rotation 683
Haibin Yin, Wei Xu, Jinli Xu and Fengyun Huang

The Simulation Study on Harvested Power in Synchronized Switch Harvesting on Inductor 691
Jang Woo Park, Honggeun Kim, Chang-Sun Shin, Kyungryong Cho, Yong-Yun Cho and Kisuk Kim

An Approach for a Self-Growing Agricultural Knowledge Cloud in Smart Agriculture 699
TaeHyung Kim, Nam-Jin Bae, Chang-Sun Shin, Jang Woo Park, DongGook Park and Yong-Yun Cho

Determination of Water-Miscible Fluids Properties 707
Zajac Jozef, Cuma Matus and Hatala Michal

Influence of Technological Factors of Die Casting on Mechanical
Properties of Castings from Silumin 713
Stefan Gaspar and Jan Pasko

Active Ranging Sensors Based on Structured Light Image
for Mobile Robot 723
Jin Shin and Soo-Yeong Yi

Improved Composite Order Bilinear Pairing
on Graphics Hardware 731
Hao Xiong, Xiaoqi Yu, Yi-Jun He and Siu Ming Yiu

Deployment and Management of Multimedia Contents
Distribution Networks Using an Autonomous Agent Service 739
Kilhung Lee

Part XIII Advanced Mechanical and Industrial Engineering, and Control II

Design Optimization of the Assembly Process Structure
Based on Complexity Criterion 747
Vladimir Modrak, Slavomir Bednar and David Marton

Kinematics Modelling for Omnidirectional Rolling Robot 755
Soo-Yeong Yi

Design of Device Sociality Database for Zero-Configured
Device Interaction 763
Jinyoung Moon, Dong-oh Kang and Changseok Bae

Image Processing Based a Wireless Charging System
with Two Mobile Robots 769
Jae-O Kim, Chan-Woo Moon and Hyun-Sik Ahn

Design of a Reliable In-Vehicle Network Using
ZigBee Communication 777
Sunny Ro, Kyung-Jung Lee and Hyun-Sik Ahn

Wireless Positioning Techniques and Location-Based Services:
A Literature Review 785
Pantea Keikhosrokiani, Norlia Mustaffa, Nasriah Zakaria
and Muhammad Imran Sarwar

Part XIV Green and Human Information Technology

Performance Analysis of Digital Retrodirective Array Antenna System in Presence of Frequency Offset 801
Junyeong Bok and Heung-Gyoon Ryu

A Novel Low Profile Multi-Band Antenna for LTE Handset 809
Bao Ngoc Nguyen, Dinh Uyen Nguyen, Tran Van Su,
Binh Duong Nguyen and Mai Linh

Digital Signature Schemes from Two Hard Problems 817
Binh V. Do, Minh H. Nguyen and Nikolay A. Moldovyan

**Performance Improvements Using Upgrading Precedences
in MIL-STD-188-220 Standard** 827
Sewon Han and Byung-Seo Kim

**Blind Beamforming Using the MCMA and SAG-MCMA
Algorithm with MUSIC Algorithm**. 835
Yongguk Kim and Heung-Gyoon Ryu

**Performance Evaluation of EPON-Based Communication
Network Architectures for Large-Scale Offshore
Wind Power Farms**. 841
Mohamed A. Ahmed, Won-Hyuk Yang and Young-Chon Kim

A User-Data Division Multiple Access Scheme 849
P. Niroopan, K. Bandara and Yeon-ho Chung

**On Channel Capacity of Two-Way Multiple-hop MIMO
Relay System with Specific Access Control**. 857
Pham Thanh Hiep, Nguyen Huy Hoang and Ryuji Kohno

**Single-Feed Wideband Circularly Polarized Antenna
for UHF RFID Reader** 863
Pham HuuTo, B. D. Nguyen, Van-Su Tran, Tram Van
and Kien T. Pham

**Experimental Evaluation of WBAN Antenna Performance
for FCC Common Frequency Band with Human Body**. 871
Musleemin Noitubtim, Chairak Deepunya and Sathaporn Promwong

Performance Evaluation of UWB-BAN with Friis's Formula and CLEAN Algorithm 879
Krisada Koonchiang, Dissakan Arpasilp and Sathaporn Promwong

A Study of Algorithm Comparison Simulator for Energy Consumption Prediction in Indoor Space 887
Do-Hyeun Kim and Nan Chen

Energy Efficient Wireless Sensor Network Design and Simulation for Water Environment Monitoring 895
Nguyen Thi Hong Doanh and Nguyen Tuan Duc

An Energy Efficient Reliability Scheme for Event Driven Service in Wireless Sensor Actuator Networks 903
Seungcheon Kim

Efficient and Reliable GPS-Based Wireless Ad Hoc for Marine Search Rescue System 911
Ta Duc-Tuyen, Tran Duc-Tan and Do Duc Dung

Improved Relay Selection for MIMO-SDM Cooperative Communications 919
Duc Hiep Vu, Quoc Trinh Do, Xuan Nam Tran and Vo Nguyen Quoc Bao

Freshness Preserving Hierarchical Key Agreement Protocol Over Hierarchical MANETs 927
Hyunsung Kim

A Deployment of RFID for Manufacturing and Logistic 935
Patcharaporn Choeysuwan and Somsak Choomchuay

Real Time Video Implementation on FPGA 943
Pham Minh Luan Nguyen and Sang Bock Cho

Recovery Algorithm for Compressive Image Sensing with Adaptive Hard Thresholding 949
Viet Anh Nguyen and Byeungwoo Jeon

Estimation Value for Three Dimension Reconstruction 957
Tae-Eun Kim

Gesture Recognition Algorithm using Morphological Analysis 967
Tae-Eun Kim

Omnidirectional Object Recognition Based Mobile Robot Localization 975
Sungho Kim and In So Kweon

Gender Classification Using Faces and Gaits 983
Hong Quan Dang, Intaek Kim and YoungSung Soh

Implementation of Improved Census Transform Stereo Matching on a Multicore Processor 989
Jae Chang Kwak, Tae Ryong Park, Yong Seo Koo and Kwang Yeob Lee

A Filter Selection Method in Hard Thresholding Recovery for Compressed Image Sensing 997
Phuong Minh Pham, Khanh Quoc Dinh and Byeungwoo Jeon

Facial Expression Recognition Using Extended Local Binary Patterns of 3D Curvature 1005
Soon-Yong Chun, Chan-Su Lee and Sang-Heon Lee

Overview of Three and Four-Dimensional GIS Data Models 1013
Tuan Anh Nguyen Gia, Phuoc Vinh Tran and Duy Huynh Khac

Modeling and Simulation of an Intelligent Traffic Light System Using Multiagent Technology 1021
Tuyen T. T. Truong and Cuong H. Phan

A Numerical Approach to Solve Point Kinetic Equations Using Taylor-Lie Series and the Adomian Decomposition Method 1031
Hag-Tae Kim, Ganduulga, Dong Pyo Hong and Kil To Chong

Regional CRL Distribution Based on the LBS for Vehicular Networks 1039
HyunGon Kim, MinSoo Kim, SeokWon Jung and JaeHyun Seo

Study of Reinforcement Learning Based Dynamic Traffic Control Mechanism 1047
Zheng Zhang, Seung Jun Baek, Duck Jin Lee and Kil To Chong

Understanding and Extending AUTOSAR BSW for Custom Functionality Implementation 1057
Taeho Kim, Ji Chan Maeng, Hyunmin Yoon and Minsoo Ryu

A Hybrid Intelligent Control Method in Application of Battery Management System 1065
T. T. Ngoc Nguyen and Franklin Bien

Interpretation and Modeling of Change Patterns of Concentration Based on EEG Signals 1073
JungEun Lim, Soon-Yong Chun and BoHyeok Seo

Design of Autonomic Nerve Measuring System Using Pulse Signal 1081
Un-Ho Ji and Soon-Yong Chun

Semiconductor Monitoring System for Etching Process 1091
Sang-Chul Kim

Enhancing the Robustness of Fault Isolation Estimator for Fault Diagnosis in Robotic Systems 1099
Ngoc-Bach Hoang and Hee-Jun Kang

Software-Based Fault Detection and Recovery for Cyber-Physical Systems 1107
Jooyi Lee, Ji Chan Maeng, Byeonghun Song, Hyunmin Yoon, Taeho Kim, Won-Tae Kim and Minsoo Ryu

Sample Adaptive Offset Parallelism in HEVC 1113
Eun-kyung Ryu, Jung-hak Nam, Seon-oh Lee, Hyun-ho Jo and Dong-gyu Sim

Comparison Between SVM and Back Propagation Neural Network in Building IDS 1121
Nguyen Dai Hai and Nguyen Linh Giang

Anomaly Detection with Multinomial Logistic Regression and Naïve Bayesian 1129
Nguyen Dai Hai and Nguyen Linh Giang

Implementation of Miniaturized Automotive Media Platform with Vehicle Data Processing 1137
Sang Yub Lee, Sang Hyun Park, Duck Keun Park, Jae Kyu Lee and Hyo Sub Choi

Design of Software-Based Receiver and Analyzer System for DVB-T2 Broadcast System 1147
M. G. Kang, Y. J. Woo, K. T. Lee, I. K. Kim, J. S. Lee and J. S. Lee

Age-Group Classification for Family Members Using
Multi-Layered Bayesian Classifier with Gaussian Mixture Model ... 1153
Chuho Yi, Seungdo Jeong, Kyeong-Soo Han and Hankyu Lee

Enhancing Utilization of Integer Functional Units
for High-Throughput Floating Point Operations
on Coarse-Grained Reconfigurable Architecture................ 1161
Manhwee Jo, Kyuseung Han and Kiyoung Choi

An Improved Double Delta Correlator for BOC Signal
Tracking in GNSS Receivers 1169
Pham-Viet Hung, Dao-Ngoc Chien and Nguyen-Van Khang

Implementation of Automatic Failure Diagnosis
for Wind Turbine Monitoring System Based on Neural Network ... 1181
Ming-Shou An, Sang-June Park, Jin-Sup Shin,
Hye-Youn Lim and Dae-Seong Kang

Development of Compact Microphone Array
for Direction-of-Arrival Estimation 1189
Trình Quốc Võ and Udo Klein

Design and Implementation of a SoPC System
for Speech Recognition 1197
Tran Van Hoang, Nguyen Ly Thien Truong,
Hoang Trang and Xuan-Tu Tran

Index 1205

Part IX
Advanced Technologies and Applications for Cloud Computing and Sensor Networks

Performance Evaluation of Zigbee Sensor Network for Smart Grid AMI

Yong-Hee Jeon

Abstract Smart Grid is a convergent system of the existing power grid and Information Technology (IT). In Smart Grid system, AMI (Advanced Metering Infrastructure) is a system with various sensors to evaluate data related with the usage of various utility resources such as electricity, gas, and water. In this paper, IEEE 802.15.4 MAC protocol for the AMI sensor networks is analyzed. Based on the analysis, OPNET simulator was implemented for the performance evaluation. Particularly, this paper focuses on the effect of protocol parameters to the performance. For these parameters, the simulation results are presented to achieve the most efficient usage of network under different traffic loads. Based on the performance evaluation results, it is revealed that real-time data transmission is possible if the total offered load is restricted under 50 % with 100 nodes and the offered load period from 20 to 50 % makes the best trade-off in terms of network throughput, average delay, etc.

Keywords Smart grid · AMI sensor network · Performance · Simulation · OPNET

1 Introduction

In Smart Grid system, it is required to monitor and control in real-time the electric power grid for the efficient operation. Therefore the Smart Grid communication networks have a strict latency requirement such as the maximum message transmission time. As an important component of the Smart Grid system, AMI refers to the collection of systems to evaluate data related with the usage of various utility resources such as electricity, gas, and water.

Y.-H. Jeon (✉)
Department of IT Engineering, Catholic University of Daegu, Gyeongsan,
Republic of Korea
e-mail: yhjeon@cu.ac.kr

J. J. (Jong Hyuk) Park et al. (eds.), *Multimedia and Ubiquitous Engineering*,
Lecture Notes in Electrical Engineering 240, DOI: 10.1007/978-94-007-6738-6_61,
© Springer Science+Business Media Dordrecht(Outside the USA) 2013

Among the communication types of Smart Grid AMI, IEEE 802.15.4 Zigbee is one of the mostly considered types due to the superior performance characteristics. Zigbee uses the IEEE 802.15.4 as its MAC layer protocol. In the MAC protocol, there are numerous protocol parameters that affect the performance of Zigbee communication, such as Beacon Order (BO), Superframe Order (SO), Number of Backoffs (NB), Backoff Exponent (BE), etc. This paper intends to examine the effect of those parameters to the delay performance of AMI sensor networks, including security aspects [1].

2 Background

In the star topology of IEEE 802.15.4, communication types may be categorized into two types such as beacon-enabled mode and non beacon-enabled mode based on data communication types between coordinator and end device. In this paper, beacon-enabled mode with single hop star-topology is assumed. According to the IEEE 802.15.4 superframe structure, it consists of beacon interval for the synchronization of the superframe, active period for the actual data transmission, and inactive (sleep) period for low power consumption. Beacon interval and active period are determined by Beacon Order (BO) and Superframe Order (SO) values respectively. Active period is further divided into Contention Access Period (CAP) and Contention Free Period (CFP) depending on whether each node trying to transmit data contends with other nodes or not. In the CAP, Carrier Sense Multiple Access/Collision Avoidance (CSMA/CA) protocol is used for data transmission. In order to get Quality of Service (QoS) guarantee and for the successive data transmission, end device may request CFP to its coordinator [2, 3].

The slotted CSMA/CA algorithm starts by setting the appropriate initial values. Number of Backoffs (NB) refers to how many times the device may backoff due to the unavailability of the medium and is set to zero. Contention Window (CW) refers to the number of backoff periods that need to be clear of channel activity before the packet is allowed to transmit and is set to 2. Backoff Exponent (BE) determines the number of backoff periods for a device to wait prior to attempting the channel access. The default value of macMinBE is set to 3. Then, the boundary of the next backoff period is determined and a random number is generated in the range of 0 to $2^{BE} - 1$.

The procedure then counts down for this number of backoff periods, which is so called Random Backoff Countdown (RBC). Then the algorithm performs Clear Channel Assessment (CCA) and checks whether the channel is busy or not. If the channel is idle, the value of CW is decremented by one. After the CCA is performed one more time and if the channel is idle, the corresponding node occupies the medium. If the channel is busy, the value of CW goes back to 2, the number of NB is incremented by one, and the value of BE is chosen again. When the value of NB is greater than the limit of macMaxCSMABackoffs, the transmission is assumed to fail and the algorithm is terminated [3].

Table 1 Simulation parameters

Parameter	Value or types
Traffic model	CBR or Poisson
Number of nodes	[10 ... 100]
BO = SO	[0 ... 14]
CW	2
NB	5
macMinBE	[1 ... 5]
Packet size	103 byte
Offered loads	[20 % ... 300 %]

3 OPNET Simulation and Results

There are various protocol parameters in IEEE 802.15.4 such as BO, SO, NB, BE, etc. The number of combination for these parameters may be more than 9,000, considering 16 values of the SO parameter, 16 values of the BO parameter, 6 values of the NB parameter, and 6 values of the BE parameter. Thus parameter tuning is required for the efficient usage of Smart Grid AMI sensor networks. In this paper, OPNET simulator was implemented for the performance evaluation of Zigbee sensor networks. The performance test bed of Zigbee sensor networks was used with 100 end devices and 1 coordinator [4].

Table 1 shows simulation parameters. In general, Poisson traffic model is used due to the convenience of mathematical analysis. In the simulation, there was no significant difference between Poisson and Constant Bit Rate (CBR) models. Since AMI data is sampled periodically, CBR traffic model is assumed in this paper. In the simulation, the performance of uplink traffic from AMI node to coordinator is only considered in CSMA/CA beacon-enable mode [5, 6].

The maximum data rate of physical layer is assumed as 250 kbps and acknowledgement is not considered. OPNET model is broadly categorized into Node model and Process model. Node model is used to model each network device and process model is used to simulate a practical operational behavior of network device.

Figure 1 shows the values of throughput under different traffic loads by SO and BO parameters. For low SO (BO) values, especially when the SO value equals to zero, slotted CSMA/CA mechanism results in more than 50 % of overhead under all traffic loads. This is due to the fact that many backoff periods are wasted by CCA deference and beacon frames are generated more frequently. CCA deference occurs when the remaining backoff period in the superframe is smaller than the required number for the transmission of total frame. It is shown that the effect of SO to the network throughput is decreasing when the value of SO is larger than 4. It is analyzed that the probability of CCA deference becomes lower for higher SO values and thus the throughput may be increased by the reduction of collision. The figure shows that the corresponding throughput is 80 % for SO values greater than or equal to 4 and falls in the range of 40–76 % for lower than 4, both for lower than 50 % of traffic loads.

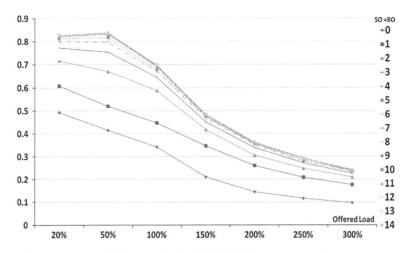

Fig. 1 Throughput under different traffic loads by SO, BO parameter values

Figure 2 shows average delay under different traffic loads by SO, BO parameter values. Similar to the effects on the throughput and link utilization, the average delay becomes saturated for high traffic loads and the effect is decreasing for the SO values greater than or equal to 4. However the delay is increasing as the SO values becomes larger. This is because the backoff delay is not increasing for lower SO values while the delay is increasing for higher SO values [5, 6].

Figure 3 shows that how the delay varies as times go by. In this figure, the large delay appears periodically with a constant interval. This phenomenon is analyzed to occur by the reason that many nodes try to access the medium right after beacon signal and thus increased collision.

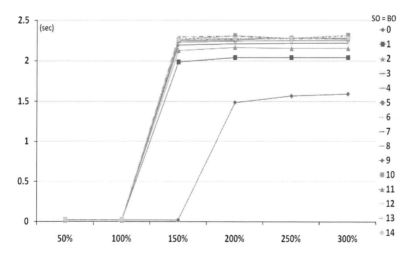

Fig. 2 Average delay under different traffic loads by SO, BO parameter values

Performance Evaluation of Zigbee Sensor Network 509

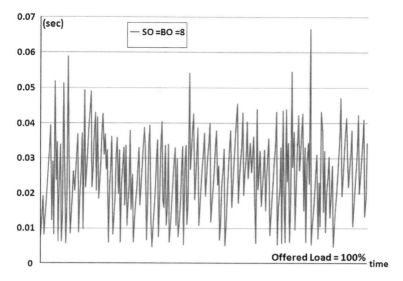

Fig. 3 Delay variations by time when SO = BO = 8 and offered traffic load = 100 %

Figure 4 shows the average delay by BE values.

BE is an important parameter in the backoff algorithm of slotted CSMA/CA. This parameter affects to the backoff delay prior to the channel access trial. In IEEE 802.15.4, this parameter may have the values in the range of 0–5. However, the range of 1–5 is used in the simulation because the value 0 disables the collision avoidance in the first iteration. As expected, if the macMinBE increases, the average delay increases as well. This is due to the fact that the probability of success becomes higher for the increased backoff delay by the increased macBinBE values.

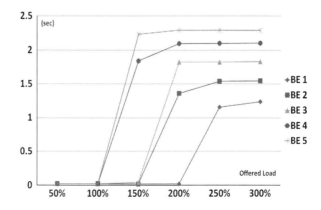

Fig. 4 Average delay under different traffic loads by BE values

The utility may be defined as {(throughput × reference delay)/delay}. It can be used to estimate the level of performance by the measured delay. The value was maximized in the range of 20–50 % period of offered traffic load. In the range of 50–100 % of offered traffic load, the utility value is decreasing slowly. When the offered traffic load is ≥150 %, the network performance becomes severely deteriorated, resulting in the deadlock. Based on the utility graph, it was revealed that the offered traffic load to achieve an optimal efficient usage of the Zigbee sensor network falls between the range of 20 and 50 %.

In addition to the delay performance of Zigbee sensor network as discussed so far, the security function for the network is also important. In the AMI system design and implementation, the confidentiality and integrity of sensor network data must be guaranteed. However, due to the page limitation, simulation results with regard to the security function are not described in this paper.

4 Conclusion

This paper deals with the performance issues of Smart Grid AMI Zigbee sensor network. In particular, the effect of protocol parameters in IEEE 802.15.4 to the network performance was examined. Based on the simulation results, it is revealed that the network may have an optimal operation with regards to network throughput, average delay, etc. in the range from 20 to 50 % of offered traffic loads. To deliver AMI sensor network data in real-time, it is analyzed that the offered traffic load is required to be restricted below 50 %.

References

1. Zigbee Alliance (2009) Zigbee smart energy profile 2.0 technical requirements document
2. IEEE 802.15.4 Standard-2003, Part 15.4 (2003) Wireless medium access control (MAC) and physical (PHY) layer specifications for low-rate wireless personal area networks (LR-WPANS), IEEE-SA standards board
3. Misic J, Misic VB (2008) Wireless personal area networks—performance, interconnections and security with IEEE 802.15.4. Wiley Series on Wireless Communications and Mobile Computing, Wiley, Chichester
4. OPNET Technologies Inc. OPNET Modeler Wireless Suite—ver. 11.5A. http://www.opnet.com
5. Rohm D, Goyal M, Hosseini H, Divjak A, Bashir Y (2009) A simulation based analysis of the impact of IEEE 802.15.4 MAC parameters on the performance under different traffic loads. Mob Inf Syst 5:81–99
6. Koubaa A, Alves M, Nefzi B, Tovar E (2006) A comprehensive simulation study of slotted CSMA/CA for IEEE 802.15.4 wireless sensor networks. Paper presented at IEEE International Workshop on Factory Communication Systems, pp 183–192

P2P-Based Home Monitoring System Architecture Using a Vacuum Robot with an IP Camera

KwangHee Choi, Ki-Sik Kong and Joon-Min Gil

Abstract We propose Peer-to-Peer-based (P2P-based) home monitoring system architecture to exploit a vacuum robot with an IP camera, a movable IP camera, without requiring a number of cameras for the whole monitoring at home. The key implementation issues for the proposed home monitoring system are (1) the easy configuration of a vacuum robot to connect to Wi-Fi networks, (2) the session management between a vacuum robot and a home monitoring server, and (3) the support of the Network Address Translator (NAT) traversal between a vacuum robot and a user terminal. In order to solve these issues, we use and extend the Wi-Fi Protected Setup (WPS), the Session Initiation Protocol (SIP), and the UDP hole punching. For easy configuration of a vacuum robot, we also propose the GENERATE method which is an extension to the SIP that allows for the generation of a SIP URI. In order to verify the feasibility of the proposed system, we have implemented the prototype and conducted the performance test using the authoritative call generator.

Keywords Home monitoring · Vacuum robot · Home networks · SIP

K. Choi
Service Development Unit, LG Uplus, 34 Gajeong-Dong, Yuseong-Gu,
Daejeon 305-350, Korea
e-mail: theidea@lguplus.co.kr

K.-S. Kong
Department of Multimedia, Namseoul University, 21 Maeju-Ri, Seonghwan-Eup,
Seobuk-Gu, Cheonan, Chungnam 331-707, Korea
e-mail: kskong@nsu.ac.kr

J.-M. Gil (✉)
School of Information Technology Engineering, Catholic University of Daegu, 13-13
Hayang-Ro, Hayang-Eup, Gyeongsan, Gyeongbuk 712-702, Korea
e-mail: jmgil@cu.ac.kr

J. J. (Jong Hyuk) Park et al. (eds.), *Multimedia and Ubiquitous Engineering*,
Lecture Notes in Electrical Engineering 240, DOI: 10.1007/978-94-007-6738-6_62,
© Springer Science+Business Media Dordrecht(Outside the USA) 2013

1 Introduction

According to the recent trend of ubiquitous computing, we can access information and services anywhere and at any time via any device. A home network is used for communication between the digital devices typically deployed in the home, usually a small number of personal computers (PCs) and consumer electronics such as a vacuum robot or an IP camera. An important function of the home network is the sharing of Internet access, that is, often a broadband service through a fiber-to-the-home (FTTH), cable TV, Digital Subscriber Line (DSL) or mobile broadband Internet service provider (ISP). If the ISP only provides one IP address, a router including a Network Address Translator (NAT) allows several computers and consumer electronics to share the IP address. Today, the deployment of a dedicated hardware router including a wireless access point (AP) is common, often providing Wi-Fi access. Recently, there are many products in the monitoring system [1, 2]. However, most of them only considered immovable cameras such as an IP camera or a web camera which is connected to the wired Internet. In this paper, in order to overcome such the limitation, we design and implement the home monitoring system exploiting a movable camera of a vacuum robot, which is the first work that has never been reported for the home monitoring system. In our home monitoring system, for easy configuration of a vacuum robot, the Wi-Fi Protected Setup (WPS) [3] is adopted, and the GENERATE method, which is an extension to the Session Initiation Protocol (SIP) [4], is proposed. Also, for signaling and P2P-based NAT traversal, the SIP and the UDP hole punching [5] are mainly used, respectively. In order to verify the capacity of the proposed system, the performance test is conducted by the authoritative call generator.

2 P2P-Based Home Monitoring Service

2.1 Overall Architecture

We describe the overall architecture for P2P-based home monitoring service, as depicted in Fig. 1. In the home network, one or more home routers, working as NATs, such as an AP or a home gateway, are connected to the main Internet.

The home device (HD) such as a vacuum robot is connected to the AP through Wi-Fi. Therefore, a vacuum robot is always located behind one or more home NATs such as an AP or a home gateway. The user terminal (UT) such as a smart phone or a PC is connected to the Internet via wired or wireless networks. Smart phones connected to mobile networks such as 3G or 4G are often located behind the NAT since mobile network operators have widely deployed the NATs to cope with IPv4 address pool exhaustion [5]. The home monitoring server (HMS) and the media relay server (MRS) are connected to the Internet, respectively.

P2P-Based Home Monitoring System Architecture 513

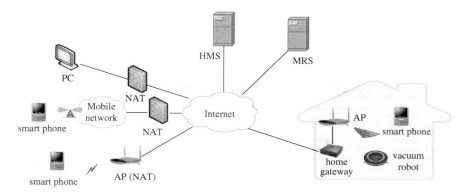

Fig. 1 System architecture for P2P-based home monitoring service

2.2 Components of the Proposed Architecture

In Fig. 2, the architecture for P2P-based home monitoring service consists of four main components: a HMS which controls session establishment based on SIP, a MRS which supports the media relay in the case that a HD or a UT is behind a symmetric NAT, a HD agent which is the embedded software in a HD such as a vacuum robot, and a UT agent which is the embedded software in a UT such as a smart phone.

The HMS contains the functionalities of SIP, UDP hole punching, and NAT type check. Firstly, the HMS acts as a SIP registrar, a back-to-back user agent (B2BUA), and a location server which provide user registration, user authentication, session routing, user location, and information management for HDs and UTs. Secondly, the HMS acts as a rendezvous server which helps two clients to set up direct P2P UDP session using UDP hole punching. Finally, the HMS has the NAT type check server which differentiates between cone NATs and symmetric NATs. According to

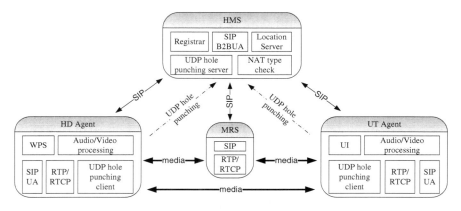

Fig. 2 Components for P2P-based home monitoring service

the Simple Traversal of UDP over NATs (STUN) protocol, NAT implementation is classified as a full cone NAT, an address restricted cone NAT, a port restricted cone NAT, or a symmetric NAT [6]. The UDP hole punching does not work with symmetric NAT even if this mechanism enables two clients to set up a direct P2P UDP session. In order to overcome this drawback, the MRS is activated to relay all the traffics between the peers in the case of symmetric NAT. The HD agent consists of WPS pairing, SIP user agent, UDP hole punching client, audio/video processing, and RTP/RTCP stack. The HD agent supports PBC-based WPS pairing which is familiar to most consumers to configure a network and enable security. It also contains SIP UA functionality for SIP session control, and performs UDP hole punching. Considering the low bandwidth of mobile networks, Speex and motion JPEG are used for audio codec and video codec, respectively. The UT agent consists of SIP user agent, UDP hole punching client, audio/video processing, RTP/RTCP stack, and user interface control for the user.

2.3 Service Flow

In this subsection, we describe the signaling flows of the home monitoring service. First of all, a user should configure a HD such as a vacuum robot to use home monitoring service. Figure 3 shows the message flow for the configuration of a HD. The flow consists of the WPS pairing step and the SIP Uniform Resource Identifier (SIP URI) generation step.

We assume that a user notifies his or her HD of the unique identify (e.g., MAC address) to the HMS through the on/off line subscription. In order to enable for a HD to access Wi-Fi networks easily and securely, the PBC-based WPS, as already mentioned, is used.

The AP and the HD have a physical or software-based button. During the setup period (e.g., 2 min) which follows the push of the AP's button, the HD can join the network by pushing its button if it is in range (**Step 1**). Due to the individual privacy issue, it is not desirable that the MAC address of the HD is exposed. In order to more easily provide a SIP URI for the HD, we propose the GENERATE method which is an extension to the SIP that allows for the generation of a SIP URI.

Figure 4 shows the example of the GENERATE method. In the GENERATE method, the hashing value of the MAC address of the HD is inserted using 'key' parameter, newly defined, into the 'from' header field. The user name values in the 'from' and 'to' fields are assigned to 'generate' since the HD has no SIP user name until it receives 200 OK. The HMS responds 200 OK including the SIP URI for the HD using 'val' parameter, newly defined, in the 'from' header field (**Step 2**).

The HD should maintain a UDP session with the HMS to receive a home monitoring request. Figure 5 shows the message flow for the establishment of a UDP session with the HMS. For the SIP registration, the HD agent sends REGISTER message to the SIP registrar of the HMS (S:5060). The AP (NAT) receives the REGISTER message and allocates a temporary IP address and a port for

P2P-Based Home Monitoring System Architecture

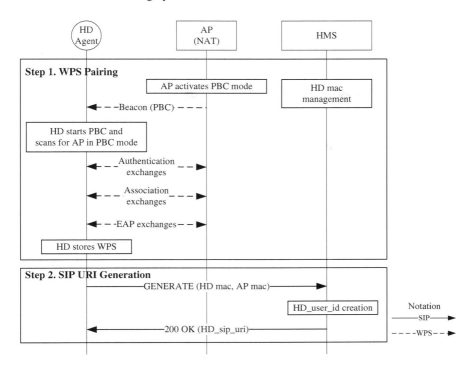

Fig. 3 Message flow for configuration of a HD

Fig. 4 The example of the GENERATE method

```
GENERATE sip:domain.com SIP/2.0
From:sip:generate@domain.com;tag=aaa50ab3d;key=3
8e4fcbe049be2fc4aba6537e78664f2
To: sip:generate@domain.com
    ...
SIP/2.0 200 OK
From:sip:generate@domain.com;tag=aaa50ab3d;val=h
omedevice1@domain.com
To: sip:generate@domain.com;tag=s694272
    ...
```

outgoing connection. Then, it translates the source IP address and the port for the message from A:a to A′:a′, and sends the message to the HMS. The HMS compares the IP address and the port in 'via' filed of the REGISTER message with the IP address and the port received from the network and the transport layers. If they are different, the HMS sends 200 OK message with received = A′ and rport = a′ parameters in the topmost 'via' header field to the UA (**Step 3**). After receiving 200 OK, the HD agent sends the UDP message, defined as CHECKNAT, to the NAT type check server of the HMS (S:5061). The HMS can obtain the port, a″, received from the transport layer. The HMS can know if the HD is behind the symmetric NAT by comparing a′ with a″. This NAT type status is used by the home

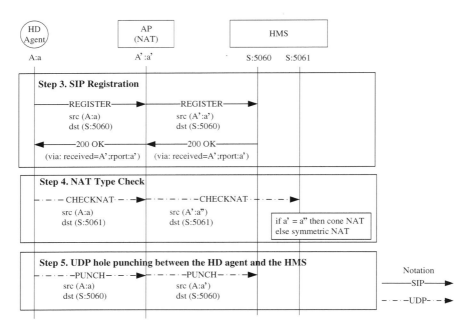

Fig. 5 Message flow for establishment of a UDP session between the HD agent and the HMS

monitoring flow, mentioning later (**Step 4**). The HD agent uses the UDP hole punching mechanism for NAT traversal, and usually transmits UDP message, defined as PUNCH, to the HMS periodically within the hole punching timer value of the NAT (**Step 5**). Similarly, the UT agent can establish a UDP session with the HMS.

Suppose the UT (e.g., smart phone) wants to establish a UDP session directly with the HD (e.g., vacuum robot). The procedure for the home monitoring service is as follows:

- The UT agent does not know how to reach the HD agent, so the UT agent sends the INVITE message, including its private IP address and port, to the HMS.
- The HMS checks if both the UT and the HD are not behind symmetric NATs.
- If both are not behind symmetric NATs, the procedure shown in Fig. 6a is performed. Otherwise, the procedure in Fig. 6b is performed.

Figure 6a shows the message flow for the home monitoring service in the P2P-based communication case. After receiving the INVITE from the UT agent, the HMS sends the INVITE message, including the UT's private and public IP addresses and ports, to the HD agent. The HD agent responds 200 OK, including its private IP address and port, to the HMS. Similarly, the HMS sends the response message, which is including the HD's private and public IP addresses and ports, to the UT agent (**Step 6**). The HD agent and the UT agent start sending the UDP

P2P-Based Home Monitoring System Architecture

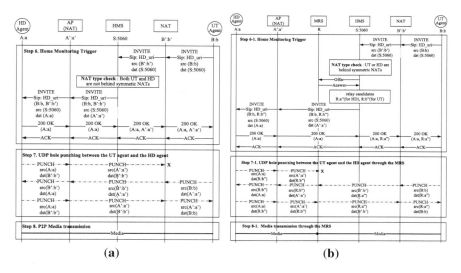

Fig. 6 Message flow for home monitoring service. **a** P2P case. **b** Relay case

messages, PUNCH, to establish a P2P session (**Step 7**). Once the session is established, each agent can send not only video media but also control data (**Step 8**).

Figure 6b shows the message flow for the home monitoring service in the relay-based communication case. The NAT traversal with the UDP hole punching does not work with symmetric NAT. To overcome this drawback, the HMS uses the MRS in the case of symmetric NAT. The HMS obtains the relay pairs (e.g., R:a″ and R:b″) for the HD and the UT from the MRS. The HMS works in the same way as the Step 6, except replacing each agent's public IP address and port with relaying IP address and port (**Step 6-1**). The HD agent and the UT agent start sending the UDP messages, PUNCH, to establish a session through the MRS (**Step 7-1**). Once the session is established, each agent can send not only video media but also control data through the MRS (**Step 8-1**).

Suppose that a user controls the vacuum robot using his or her smart phone in home. The UT and the HD are behind the common NAT. Since 78 % of the commercial NATs do not support hairpin translation, each agent starts to establish a UDP session directly using opponent's private IP address and port gained from Step 6 or Step 6-1 [5]. Each agent can send not only video media but also control data directly.

3 Performance Test

In order to verify the capacity and the stability of the proposed system, the performance test is essential. In this section, we focus on conducting the performance test of the HMS to find out tolerable load, aspects of SIP signaling traffic in the

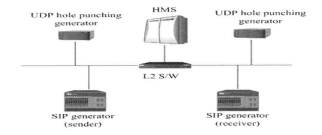

Fig. 7 Experimental test environment

proposed system. Figure 7 shows the experimental test environment. We use a Spirent's Abacus-5000 [7], which is one of the most popular SIP traffic generators, and the UDP hole punching generators, which can send simple UDP packets, respectively.

For performance evaluation, test conditions are listed in Table 1. The number of SIP subscribers, which consist of UTs and HDs, are set to 16,000. The SIP registration interval is set to 3,600 s, which is the default expiration for SIP registrations [8]. The UDP hole punching interval is set to 20 s [5]. Call holding time is set to 180 s, which is commonly used in the telecommunication simulation [9].

The performance test of the HMS proceeds as follows:

- The SIP subscribers, which represent UTs and HDs, register with the HMS by sending SIP REGISTER method every 3,600 s.
- Currently, they send the UDP packet, PUNCH, to the HMS every 20 s.
- The 30 Call attempts Per Second (CPS) are generated by senders which represent UTs. These calls are accepted by receivers which represent HDs.

We evaluated the performance of the HMS with increasing Call attempts Per Second (CPS) values by 5, starting at 10 CPS. The measurements and statistics reported by Abacus-5000 in Fig. 8 have shown that the performance of the HMS is 54.40 CPS with tolerable delay values in the test conditions. The test elapsed time is 2 h 12 min 21 s. The total call completion ratio and the total registration success ratio are 99.99 % (431,918 completions of 431,930 call attempts) and 100 % (210,644 successes of 210,644 registration attempts), respectively. This means that the HMS outperforms 95 % which is the call completion rate criteria for Internet telephony in Korea [10]. In this test, the performance factor is only SIP signaling traffic since media packets are transmitted by the P2P way, as mentioned in

Table 1 Test conditions

Item	Value	Remarks
SIP subscriber	16,000	UTs (8,000), HDs (8,000)
Registration interval	3,600	Second
Hole punching interval	20	Second
Call holding time	180	Second

Measurements Summary

Delay Type	Count	Minimum	Average	Maximum
Call length terminate (s)	215959	179.919	180.118	180.395
Call length originate (s)	215959	180.014	180.126	180.407
Response Time (msec)	215959	2	9	227
Call Setup (msec)	215959	64	80	387
Tear Down (msec)	215959	2	9	242
Post dial delay (msec)	215959	77	108	458
Reg 4XX response time	210644	2	10	243
Reg 200 response time	210644	2	9	222
Reg success time	210644	6	19	354

Test Status

Test status	DONE
Start time	07:59:12 PM 08/01/2011
End time	10:20:38 PM 08/01/2011
Elapsed time	02:12:21

Statistics

SIP Subscriber

	Script Attempts	Script Compl.	% Script Compl.	Call Attempts	Call Compl.	% Call Compl.	Call Attempts per Sec	Reg. Successes	Reg. Failures	Reg. Retry Attempts	Reg. In Progress	Errors
Total	431918	431918	100.00	431930	431918	99.99	54.40	210644	0	210644	0	12
Orig	215959	215959	100.00	215971	215959	99.99	27.20	106644	0	106644	0	12
Term	215959	215959	100.00	215959	215959	100.00	27.20	104000	0	104000	0	0

Fig. 8 The performance test result for the HMS reported by Abacus-5000

Fig. 6a. From the performance test of the HMS shown above, we can find that the HMS can support more than 50 CPS signaling load stably in given conditions.

4 Conclusion

In this paper, we proposed the P2P-based home monitoring service exploiting a vacuum robot with an IP camera, without requiring a number of cameras for the whole monitoring at home, which is the first trial in the literature. In order to provide the easy configuration of a vacuum robot, the session management, and the NAT traversal, we used the WPS, the SIP, and the UDP hole punching, respectively. Also, in order to configure a vacuum robot easily, we proposed the GENERATE method which is an extension to the SIP that allows for the generation of a SIP URI. In addition, we implemented the HMS, the MRS, the HD agent, and the UT agent to validate the effectiveness and feasibility of the proposed system. The performance test has shown that the HMS supports 99.99 % call completion ratio. These results indicate that the HMS outperforms 95 % which is the call completion rate criteria for Internet telephony in Korea.

Acknowledgments This research was supported by Basic Science Research Program through the National Research Foundation of Korea (NRF) funded by the Ministry of Education, Science and Technology (2012R1A1A4A01015777).

References

1. Zhuang H, Wang Z (2006) IP-based real time video monitoring system with controllable platform. In: Proceedings of 2nd IEEE/ASME international conference on mechatronic and embedded systems and applications, pp 1–4
2. Lam K, Chiu C (2003) Mobile video stream monitoring system. In: Proceedings of 11th ACM international conference on multimedia, pp 96–97

3. Alliance W (2007) Wi-Fi protected setup specification. WiFi Alliance Document
4. Rosenberg J, Schulzrinne H, Camarillo G, Johnston A, Peterson J, Sparks R, Handley M, Schooler E (2002) SIP: Session Initiation Protocol. IETF RFC 3261
5. Ford B, Srisuresh P (2005) Peer-to-peer communication across network address translators. In: Proceedings of USENIX annual technical conference, pp 139–140
6. Rosenberg J, Weinberger J, Huitema C, Mahy R (2003): STUN—Simple traversal of user datagram protocol (UDP) through network address translators (NATs). IETF RFC 3489
7. Ji L, Yin X, Wang X (2007) Conversational model based VoIP traffic generation. In: Proceedings of 3rd international conference on networking and services, pp 14–19
8. Rosenberg J (2004) A session initiation protocol (SIP) event package for registrations. IETF RFC 3680
9. Carothers C, Fujimoto R, Lin Y, England P (1994) Distributed simulation of large-scale PCS networks. In: Proceedings of 2nd international workshop on modeling, analysis, and simulation of computer and telecommunication systems, pp 2–6
10. Telecommunications Technology Associations (2005) TTAS.KO-01.0077. Voice quality criteria of internet telephony

Design and Simulation of Access Router Discovery Process in Mobile Environments

DaeWon Lee, James J. Park and Joon-Min Gil

Abstract With the development of mobile communications and Internet technologies, smart phones have become a necessity of life. To have better connection status, power consumption, and faster transmission speed, the most of mobile users want to access 802.11 wireless networks that are well known as a Wi-Fi. When entering a new area, a mobile host (MH) decides to use one of access routers (ARs) on available networks. However, since previous works are focused on layer 2 handoff for faster connection at 802.11, only the subsystem identification (SSID) and signal strength are considered to choose its new connection. This can fail to provide the MH with a suitable AR. Therefore, more information needs to be used to determine the suitable AR and seamless connectivity. In this paper, we extend a prefix information option in the router advertisement message to include the status information of an AR, such as status of the MH, capacity of the router, current load of the router, and depth of the network hierarchy. Also, we propose a decision engine by which the MH can analyze the status information of ARs and determine a suitable AR automatically based on the information. By analyzing our simulation results, we found that our AR discovery process has several advantages. For the MH, the packet loss can be reduced with the increase of wireless connection period. Additionally, load balancing was achieved for the AR and router, and the network topology was also able to become more efficient.

D. Lee
Division of General Education, SeoKyeong University, Jeongneung 4-dong, Sungbuk-gu, Seoul 136–704, Korea
e-mail: daelee@skuniv.ac.kr

J. J. Park
Department of Computer Science and Engineering, SeoulTech, 172 Gongreung 2-dong, Nowon-gu, Seoul 139–743, Korea
e-mail: parkjonghyuk1@hotmail.com

J.-M. Gil (✉)
School of Information Technology Engineering, Catholic University of Daegu, 13-13 Hayang-ro, Hayang-eup, Gyeongsan-si, Gyeongbuk 712–702, Korea
e-mail: jmgil@cu.ac.kr

J. J. (Jong Hyuk) Park et al. (eds.), *Multimedia and Ubiquitous Engineering*, Lecture Notes in Electrical Engineering 240, DOI: 10.1007/978-94-007-6738-6_63, © Springer Science+Business Media Dordrecht(Outside the USA) 2013

Keywords Fast scanning · Seamless connectivity · Neighbor discovery protocol · 802.11 networks

1 Introduction

Today, 802.11 wireless networks are the most popular access networks as demand for mobile access continues to increase. To have better connection status, power consumption, and faster transmission speed, most of mobile users want to access the 802.11 wireless networks that are well known as a Wi-Fi. The Mobile Hosts (MHs) access to 802.11 networks in one of two different modes. An MH can form spontaneous networks as a Mobile Router (MR; ad hoc mode) or it can get connected to an Access Router (AR), which is directly connected to a backbone network (infrastructure mode) [1]. When the MH changes its point of attachment, it should quickly discover and attach to a new point of attachment to reconnect to the network. This is known as a handoff. Most previous studies are focused on layer 2 handoffs in 802.11 networks [2–6]. When an MH starts up or enters a new cell, it needs to discover its environment including radio frequencies, neighbor points of attachment, and available services. Generally, the MH may have several available ARs. The user of the MH decides to use one of the ARs, based only on the SubSystem IDentification (SSID) and signal strength of the user's Internet connection [1]. In our previous work [7], we presented AR selection algorithm and compared its performance with that of pure 802.11 scanning algorithm.

In this paper, we focus on utilizing the layer 3 information to determine a suitable AR of available ones. The information includes status, capacity, current load, and depth of network hierarchy. In order to correctly describe the status information of ARs, we extend a prefix information option in the router advertisement message [8]. We also design a Decision Engine (DE) to analyze reachable ARs and determine a suitable AR. In the proposed protocol, the handoff MH selects the suitable AR automatically, and thus the proposed protocol provides load balance of ARs and efficiency of network topology.

This paper is organized as follows. In Sect. 2, we briefly describe related work. Section 3 describes the need for AR status information and presents the extended prefix information option. AR discovery algorithm is given in Sect. 4. Section 5 presents the performance evaluation with simulation results. Finally, Sect. 6 concludes the paper.

2 Related Work

Most related studies on the 802.11 discovery process have focused on reducing the scanning latency during a layer 2 handover, when an MH roams from one AP to another. One simple method to reduce the full scanning latency is to use selective scanning, which allows the scanning of a subnet of channels instead of probing each of them. The other methods have focused on reducing the value of the scanning timers (MinCT and MaxCT) [2, 3] by fixing the potential best values for both timers, presenting theoretical considerations and simulation results. The smooth handover [4] and the periodic scanning [5] are based on splitting the discovery phase into multiple sub phases. The objective of this division is to allow an MH to alternate between data packet exchange and the scanning process. An MH builds a list of target APs, maintaining information only on channel and SSID. These methods focus only on reducing latency to minimize the disconnected time of the MH. However, they do not ensure that the MH connects to a suitable AP. Instead, more information needs to be used to connect to the suitable AR. This information cannot be provided by a probe signal at layer 2. To collect this information, we instead focus on a neighbor discovery protocol at layer 3. The router advertisement message is one message format from the neighbor discovery protocol in IPv6 [8].

The neighbor discovery protocol corresponds to a combination of the IPv4 protocol ARP, ICMP, RDISC, and ICMPv4. The router advertisement messages contain prefixes that are used to determine whether another address shares the same link and/or address configuration, a suggested hop limit value, and so on. To collect the information to find a suitable AP, we focus on the router advertisement message [8], as shown in Fig. 1. However, the router advertisement message has not enough reserved fields to provide more information. Thus, we focus on the prefix information that must be used in the hierarchy architecture in IPv6. Figure 2 shows the prefix information option [8].

3 Extended Prefix Information Option

To access a new AR in an L2 handoff, the AR provides only SSID and authentication. However, the information provided by L2 is insufficient to determine a suitable AR. In this paper, we focus on an L3 handoff for AR selection. There are four information elements needed to determine the suitable AR. First, the status of the AR is necessary. Second, the maximum capacity of the AR is necessary. Third, the expected AR load is necessary; because several MHs use the same AR in a cell,

Fig. 1 Router advertisement message format

```
 0                   1                   2                   3
 0 1 2 3 4 5 6 7 8 9 0 1 2 3 4 5 6 7 8 9 0 1 2 3 4 5 6 7 8 9 0 1
+-+-+-+-+-+-+-+-+-+-+-+-+-+-+-+-+-+-+-+-+-+-+-+-+-+-+-+-+-+-+-+-+
|     Type      |     Code      |          Checksum             |
+-+-+-+-+-+-+-+-+-+-+-+-+-+-+-+-+-+-+-+-+-+-+-+-+-+-+-+-+-+-+-+-+
| Cur Hop Limit |M|O|  Reserved |        Router Lifetime        |
+-+-+-+-+-+-+-+-+-+-+-+-+-+-+-+-+-+-+-+-+-+-+-+-+-+-+-+-+-+-+-+-+
|                         Reachable Time                        |
+-+-+-+-+-+-+-+-+-+-+-+-+-+-+-+-+-+-+-+-+-+-+-+-+-+-+-+-+-+-+-+-+
|                          Retrans Timer                        |
+-+-+-+-+-+-+-+-+-+-+-+-+-+-+-+-+-+-+-+-+-+-+-+-+-+-+-+-+-+-+-+-+
|   Options ...
+-+-+-+-+-+-+-+-+-
```

Fig. 2 Prefix information option format

the network bandwidth is limited. The MH should find a free AR. The last is the depth of the network hierarch, which represents the logical location with respect to a border gateway in a subnet. To prevent frequent handoffs and provide fast transmission in a subnet, the MH should connect with a higher-level AP in a subnet hierarchy. The additional attributes in the extended router advertisement message are as follows: (1) Status: status of AR (stationary/portable), (2) Cap: maximum capacity of AR, (3) Load: current load of AR, (4) Depth: depth of network hierarchy.

Figure 3 shows a format of the extended prefix information option. However, we use the minimum bits in the reserved field because the signaling overhead on wired/wireless links is an important issue.

4 AR Discovery Algorithm

In this section, we propose a decision engine (DE) that analyzes the router advertisement messages and determines a suitable AR. Table 1 shows our AR discovery algorithm with the DE.

The proposed AR discovery algorithm consists of three parts. First, information is received from the router advertisement message. Second, it is used for decision-making, which is divided into two parts: an active state for an MH that moves frequently and an idle state for an MH that moves rarely. The last part of the

Fig. 3 Extended prefix information option format

Design and Simulation of Access Router Discovery Process

Table 1 AR discovery algorithm

---

1 If Power up or entering new subnet
   1.1 Send router solicitation message to ARs
   1.2 Wait for router advertisement messages of ARs
2 For all elements in each router advertisement messages
   2.1 Compare status
   2.2 Compare bandwidth
   2.3 Compare signaling strength
   2.4 Compare hierarchy
   2.5 Compare loadratio
   2.6 Decide candidate AR_list
/* the priority of active MH: status > bandwidth > signaling
   strength > hierarchy > loadratio */
/* the priority of idle MH : status > hierarchy > signaling
   strength > loadratio > bandwidth */
3 If Connection is needed
   3.1 Scanning candidate ARs
   3.2 Verify AR states
   3.3 Select state best AR
   3.4 Send binding update to new AR

---

algorithm covers what happens when the MH loses its connection. The MH broadcasts a neighbor-solicitation message and then connects to a new AR, following decision-making.

5 Performance Evaluation

5.1 Simulation Environment

We tested the performance of the proposed discovery process with AR discovery algorithm and compared it with that of 802.11 discovery process. To this end, we developed a simulator in JAVA and incorporated the DE into it. Figure 4 shows the simulation environment used to evaluate the proposed discovery process. Our simulations are conducted with three subnets, each of which has a border router (BR) to connect to the Internet. Subnet A is composed of three hierarchies. BR_A consists of an AR and six routers. Each router on subnet A has six ARs. Also, subnet B is composed of three hierarchies. BR_B consists of an AR and two routers. Each router on subnet B has three ARs. Subnet C is composed of two hierarchies. BR_C consists of an AR and a router. A router on subnet C has three ARs. Two kinds of mobile devices are generated: one is an MR that changes its point of attachment by random movement and the other is an MH. And, 70 MRs and 150 MHs were generated. Both of them were randomly located in the initial state and each of them had random mobility.

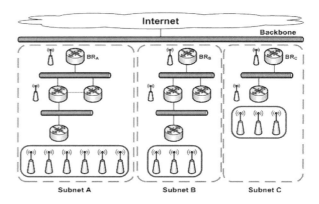

Fig. 4 Simulation environment

5.2 Simulation Results and Analysis

In our first experiment, we compared the packet loss of the proposed discovery process with that of previous works (periodic scanning, scanning timer with MinCT, and scanning timer with MaxCT). We measured the packet loss of each method in 20 min. Figure 5 shows the comparison of the packet loss. The results of this figure show that the proposed discovery process has an improvement of 20 (periodic scanning), 20 (MinCT), and 15 % (MaxCT) in the reduction of the packet loss.

Figure 6 shows the average ratio of router bandwidth in 20 min. The average ratio of router bandwidth means the average utilization of all routers in a domain. The proposed discovery process outperformed the other discovery processes and evenly distributed except in the initial state. The average ratio of the proposed discovery process is 32.9 %. The periodic scanning, MinCT, and MaxCT have the average ratio of 34.3, 33.7 and 33.2 %, respectively. The proposed discovery process showed an improvement of 1.0–1.5 %, as compared to the other three methods. From the results of Fig. 6, we can see that the proposed discovery process can achieve the load balancing of ARs and topological stability.

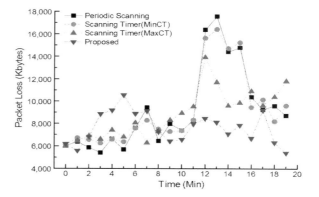

Fig. 5 Comparison of packet loss

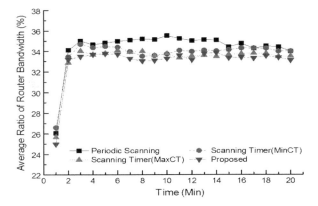

Fig. 6 Comparison of average ratio of router bandwidth

6 Conclusion

In this paper, we addressed the decision making process for determining a suitable AR when an MH enters a new cell. We proposed a DE to find the suitable AR on the 802.11 based MANET. An MH may have several available networks when powered up or moving into a new area. The user of the MH decides to use one of the ARs on an AR list. However, a decision based only on the AR's name and signal strength for the user's Internet connection has a limitation to provide the suitable AR for the MH. To determine the suitable AR, we focused on the neighbor discovery protocol in L3. We extended the prefix information option in the neighbor discovery protocol to include the AR's status information, such as status, capacity, current load, and depth of network hierarchy. The simulation results showed that the proposed AR discovery process has the following advantages: for the MH, the packet loss can be reduced with the increase of wireless connection period. Additionally, load balancing was achieved for the AR and router, and the network topology was also able to become more efficient.

Acknowledgments This research was supported by Basic Science Research Program through the National Research Foundation of Korea (NRF) funded by the Ministry of Education, Science and Technology (2012R1A1A4A01015777).

References

1. RFC 5416: Control and provisioning of wireless access points (CAPWAP) protocol binding for IEEE 802.11. http://www.rfc-editor.org/rfc/rfc5416.txt
2. Shin S, Forte AS, Rawat AS, Schulzrinne H (2004) Reducing MAC layer handoff latency in IEEE 802.11 wireless LANs. In: Proceedings of 2nd international workshop on mobile management and wireless access protocols, pp 19–26
3. Velayos H, Karlsson G (2004) Techniques to reduce the IEEE 802.11b handoff time. In: Proceedings of 2004 IEEE international conference on communications, pp 3844–3846

4. Liao Y, Gao L (2006) Practical schemes for smooth MAC layer handoff in 802.11 wireless networks. In: Proceedings of 2006 international symposium on world of wireless, mobile and multimedia Networks, pp 181–199
5. Montavont J, Montavont N, Noel T (2005) Enhanced schemes for L2 handover in IEEE 802.11 networks and their evaluations. In: Proceedings of IEEE 16th international symposium on personal, indoor and mobile radio communications, pp 1429–1434
6. Koutsopoulos I, Tassiulas L (2007) Joint optimal access point selection and channel assignment in wireless networks. IEEE/ACM Trans Network 15(3):521–532
7. Lee D, Kim Y, Lee H (2012) A study of an optimal discovery process in mobile ad hoc network. In: Proceeding of The 2012 FTRA international conference on advanced IT, engineering, and management
8. RFC 4861: Neighbor discovery for IP version 6 (IPv6). http://www.rfc-editor.org/rfc/rfc4861.txt

Integrated SDN and Non-SDN Network Management Approaches for Future Internet Environment

Dongkyun Kim, Joon-Min Gil, Gicheol Wang and Seung-Hae Kim

Abstract For years, computer scientists have been dreaming of innovating Internet in terms of performance, reliability, energy efficiency, security, and so on. However, it is nearly impossible to carry out practical large-scale experiments and verification, since new software and programs are hardly evaluated on the current Internet environment where routers and switches are totally closed. In this context, Software Defined Networking (SDN) concept has been introduced to deploy software-oriented and open programmable network coupled tightly with traditional Internet (non-SDN) environment. This tight integration results in easy and efficient deployment of SDN domains into non-SDN infrastructure, but it is also required that integrated management methods for SDN and non-SDN be developed. This paper proposes how SDN and non-SDN can be managed in an integrated manner in terms of two aspects: (1) non-SDN driven management approach and (2) SDN oriented management approach, in order to achieve reliable Future Internet environment.

Keywords SDN · Non-SDN · Network management · Future internet

---

D. Kim · G. Wang · S.-H. Kim
Korea Institute of Science and Technology Information, 52-11 Eoeun-dong,
Yuseong-gu, Daejeon, South Korea
e-mail: mirr@kisti.re.kr

G. Wang
e-mail: gcwang@kisti.re.kr

S.-H. Kim
e-mail: shkim@kisti.re.kr

J.-M. Gil (✉)
School of IT Engineering, Catholic University of Daegu, 13-13 Hayang-ro, Hayang-eup,
Gyeongsan-si, Gyeongbuk, South Korea
e-mail: jmgil@cu.ac.kr

J. J. (Jong Hyuk) Park et al. (eds.), *Multimedia and Ubiquitous Engineering*,
Lecture Notes in Electrical Engineering 240, DOI: 10.1007/978-94-007-6738-6_64,
© Springer Science+Business Media Dordrecht(Outside the USA) 2013

1 Introduction

There have been drastic demands over the past years in the Internet, compared to the earlier Internet designs [1]. One of the challenges is the requirement of open networking that makes it possible for users to innovate Internet by experimenting and applying newly developed technologies over practical large-scale networks. Internet innovation known as Future Internet, currently, is deeply involved with network performance, reliability, energy efficiency, security, etc., while Internet is entirely closed not to allow any new user-oriented software to be installed and tested on. Only can network vendors access and control the network devices on Internet regarding programmability, reconfiguration, and any innovative efforts.

Therefore, various researches have been performed to cope with the new user demands, and one of them is Software Defined Networking (SDN) with OpenFlow protocols [2] that has invoked significant interest in reconsidering traditional aspects of Internet architecture and design. Two important features of SDN are (1) implementation of network control plane decoupled from data plane (network forwarding hardware), and (2) relocation of control plane from hardware switch equipped with a typical low performance CPU.

Another interesting feature of SDN is that it can easily be integrated into the current Internet environment generally at layer 2 local networks. For example, OpenFlow devices are mostly Ethernet switches to support packet forwarding, which indicates that SDN switches are basically to be incorporated with other non-SDN Ethernet switches for layer-2 communications. Based on SDN's inherent integration with classical local networks (i.e., non-SDN), remote SDN domains communicate with other SDN domains via Internet (i.e., non-SDN) [3]. So, in order to keep the overall SDN and non-SDN environment reliable, it is inevitable that SDN and non-SDN need to interoperate in a tightly coupled way, and should be managed constantly and reliably through well-designed network management model.

In this context, this paper proposes combined network management models for both SDN and non-SDN environments adopting Distributed virtual Network Operations Center (DvNOC) [4] and several other related works, which will be introduced more in detail in Sect. 2. The proposed models are induced based on two approaches: (1) non-SDN adaptation into SDN in an SDN-oriented way and (2) SDN adjustment into non-SDN in a non-SDN-oriented way. The main difference of SDN and non-SDN management is that SDN management is interactive while non-SDN managenet is not two-way. SDN controller(s) and devices communicate with each other by exchanging (and analyzing) a variety of management information (e.g., status, topology), which results in "super-active" network management. Non-SDN management is generally not interactive though: there is no centralized controller, but one or more management servers which contact network devices and gather management information from them, whereas network devices cannot do the same jobs. Therefore, the suggested management models in this paper principally deal with how contrastive management methods can be integrated as the combined and efficient network management model.

The remainder of this paper is organized as follows. In Sect. 2, we introduce related works for the proposed network management approaches. Section 3 describes DvNOC architecture as a non-SDN management environment. The SDN and non-SDN integrated management models based on Future Internet are explained in Sect. 4. Finally, Sect. 5 concludes the paper.

2 Related Work

There are several open source SDN controllers that have been developed and being improved so far (in Table 1). Many experimental SDN applications exploit the open source controllers in the field of network virtualization, energy efficiency, mobility, traffic engineering, wireless mobile video streaming, and so on. Each of the controllers can be used for the proposed network management model as well in terms of Disaster Manager (DM) implementation and East/Westbound Interface/API designs with non-SDN manager.

Slice Around the World (SATW) initiative [5] is influenced by various Future Internet testbed activities in Europe, Asia, and elsewhere outside USA, and it consists of three components such as application/service, programmable SDN infrastructure, and international experimental facility. K-GENI [6] has participated in SATW initiative since 2011. Both K-GENI and SATW initiative are playgrounds on which SDN and non-SDN integrated network management model can be applicable.

IEEE 802.1ag [7] is a standard specifying Ethernet OAM for connectivity fault management (CFM) of paths through 802.1 bridges and local area networks. OpenFlow/SDN-based Ethernet OAM [8] adopts IEEE 802.1ag standard to exchange 802.1ag Ethernet frame based on NOX in Table 1, and to manage Open

Table 1 Open source SDN controllers in public domain

Name	Language	Platform(s)	License	Developer(s)	Notes
OpenFlow Ref. [11]	C	Linux	OpenFlow License	Stanford U., Nicira	Extensibility necessary
NOX [12]	Python, C++	Linux	GPL	Nicira	Actively developed
Beacon [13]	Java	Win, Mac, Linux, Android	GPL(core), FOSS L.	Stanford U.	Web UI framework
Maestro [14]	Java	Win, Mac, Linux	LGPL	Rice U.	–
Trema [15]	Ruby, C	Linux	GPL	NEC	Emulator included
RouteFlow [16]	Python, C++	Linux	Apache L.	CPqD	Virtual IP routing

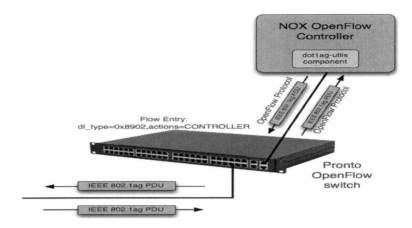

Fig. 1 IEEE 802.1ag frame monitoring using NOX controller

vSwitch based on CCM (Continuity Check Message). This research is very relevant to the integration of SDN and non-SDN management in the Ethernet layer as shown in Fig. 1, which describes how to operate and manage SDN-based Ethernet environment using IEEE 802.1ag PDU exchanges, interacting with NOX OpenFlow Controller. In turn, this SDN management can be extended to other non-SDN Ethernet devices using the IEEE 802.1ag standard.

Disaster Manager (DM) [9] is a more active management system than IEEE 802.1ag based OAM, devised to store and analyze the disaster-related information of network nodes by embedding a DM firmware into OpenFlow devices. An embedded DM actively acquires more than thousands of monitoring data from SysLog, for example, generated everyday in OpenFlow switches, and the DM analyzes the numerous datasets so that network devices can perform self-directed network disaster isolation and autonomous intelligent network management. The proposed management model incorporates DM as a premier building block to achieve "super-active" network management over combined SDN and non-SDN ecology.

3 Distributed Virtual Network Operations Center

DvNOC [4] was designed as a virtual network management framework for hybrid research network (that is a combination of circuit-oriented network and packet-switching network for providing both end-to-end dedicated lightpath and layer-3 routed path [10]). Virtual network resources can be managed by researchers and network engineers through DvNOC framework. Moreover, since recent advanced applications require global end-to-end network environment for collaborative researches over many individual network domains and countries, Network

Operations Center (NOC) to NOC cooperation is getting more and more important between multi-domain networks. In this regard, DvNOC provides the following functionalities to support collaborative efforts on multi-domain NOCs: *Multi-domain Network Awareness, Efficient NOC-to-NOC Cooperation, and User-oriented Virtual Network Management.*

DvNOC incorporates new features (e.g., resource repository) for special end-users (researchers and experimenters) as well as conventional users (network operators and engineers) in addition to the traditional NOC facilities so that the DvNOC manages resource information collected from local networks inside a network domain, ultimately in order to share the resource information with other NOCs on DvNOC domain. Each associated dNOC (distributed NOC) exchanges the operational dataset with others, interfacing with virtual NOC (vNOC).

Since DvNOC supports several specific functionalities that classical Internet management system hardly provides, we consider DvNOC a non-SDN management framework to be combined with SDN. Non-SDN oriented DvNOC architecture basically includes data acquisition, federation engine, and user interface, all of which are interacting with resource repository. Among those capabilities, federation engine comprises data ownership and policy management as well to coordinate multi-domain data exchanges over Autonomous Distributed Networks (ADNs) correctly. Users and applications can acquire the datasets stored in resource repository (by federation engine and data acquisition modules) in a very similar way as northbound interface/API in SDN provides network resource information for applications and users. In order to achieve northbound interface in DvNOC, it is capable of following four OpenAPIs (equivalent with Northbound Interface/APIs in SDN): *getListDataSet(), getSpecificDataSet(), getWholeDataSet(), setDataSet().*

4 Proposed SDN and Non-SDN Integrated Management Approaches Based on DvNOC Framework

SDN is designed, by nature, to embrace classical non-SDN networks including local and wide area networks. In addition, targeted users of both SDN and non-SDN are anything but dissimilar regarding services, applications, and users, such as end-to-end virtual network services, high performance data transmission, and experimental and research uses, etc. Therefore, it is inferred that integrated management of SDN and non-SDN infrastructure may lead to considerable synergistic effect for diverse future Internet demands.

In this paper, two network management models are proposed for integrated SDN and non-SDN environment based on DvNOC and related works introduced in Sect. 2. Basically, these two models aim to the same direction, but the detailed configurations are revised. Regarding SDN, a controller communicates with network hardware using OpenFlow protocol, while it interacts with applications and

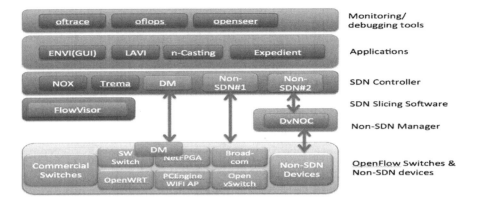

Fig. 2 The first model: SDN-oriented non-SDN management integration

users through northbound interface and APIs. Considering this structure, the simple way of managing combined SDN and non-SDN infrastructure is that an SDN controller acquires resource information from non-SDN hardware. That is, an open source SDN controller like NOX needs to be designed and implemented including some modules to interface with non-SDN devices. The first model is proposed under this consideration using DvNOC as a non-SDN management framework. In this sense, an SDN controller communicates with DvNOC in different layers, SDN in upper layer and DvNOC in lower layer as shown in Fig. 4, where an SDN controller can carry out supervisory management for integrated SDN and non-SDN network environment efficiently.

In Fig. 2, three main functionalities are included in the controller layer. First, DM is equipped with OpenFlow protocol in SDN controllers for more dynamic and interactive network management as well as network disaster isolation. Second, controllers embrace non-SDN communication module (non-SDN#1), adopting SNMP, Netconf, TL1, CLI, etc., which are classical network management protocols on Internet. Third, DvNOC is combined with controllers (non-SDN#2), achieving multi-domain network management based on hybrid research networks in order to meet some application and user demands in Table 2. Nonetheless, there is a trade-off between non-SDN#1 and non-SDN#2. Non-SDN#1 has high development overhead but good performance due to dense coupling, while non-SDN#2 costs less overhead, but it may have relatively low performance.

Figure 3 indicates non-SDN based network management model requiring quite less development and verification overheads compared to the first model in that both controller and DvNOC only need to accommodate standard east/westbound interface and APIs of SDN scheme. The other standard interfaces such as north/southbound interface/APIs in Fig. 4 are also supposed to be equipped with controller in general SDN architecture. Therefore, only does DM need to be additionally implemented for controller and DvNOC. The inadequacy of the second model is that it doesn't conduct as much supervisory management as the first

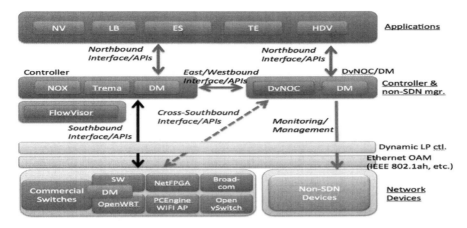

Fig. 3 The second model: non-SDN-oriented SDN management integration

model does, because DvNOC communicates parallel with controller in the same layer using east/westbound interfaces.

In the meanwhile, for the bright side, the DM implemented in DvNOC may interact with SDN hardware when it comes to fault and failure management using cross-southbound interface and APIs. In this way, DvNOC is acting like a second-controller regarding breakdown management, which enhances overall integrated network reliability by providing an alternative to solve single point of failure problem of one centralized controller. In addition, both SDN and non-SDN environment are closely coupled with Ethernet OAM layer functionalities (IEEE 802.1ag, IEEE802.ah, OpenFlow based CCM exchanges) and dynamic circuit control systems. This coupling makes it possible for users to achieve the implementation and deployment of end-to-end virtual networks, high-end application networks, dedicated QoS networks, etc., from SDN to non-SDN and vice versa.

5 Conclusion and Future Works

This paper proposed two network management approaches for integrated SDN and non-SDN environment, since there are user, application, and service demands to incorporate two different network infrastructure, mostly for the purpose of advanced researches and experiments. The proposed approaches mainly describe how SDN controllers and DvNOC, a specialized non-SDN network management framework, can interact with each other in order to maintain reliable end-to-end user oriented Future Internet environment. Our future works will evaluate implementation aspects of prototypes based on proposed network management models, considering each model's tradeoff in terms of performance, development overheads, and reliability.

References

1. Clark D (1988) The design philosophy of the DARPA internet protocols. ACM SIGCOMM Comput Commun Rev 18(4):106–114
2. McKeown N, Anderson T, Balakrishnan H, Parulkar G, Peterson L, Rexford J, Shenker S, Turner S (2008) OpenFlow: enabling innovation in campus networks. ACM SIGCOMM Comput Commun Rev 38(2):69–74
3. Levin D, Wundsam A, Heller B, Handigol N, Feldmann A (2012) Logically centralized? State distribution trade-offs in software defined networks. In: 1st workshop on hot topics in software defined networks, pp 1–6
4. Kim D (2008) User oriented virtual network management based on dvNOC environment. Int J Comput Sci Netw Secur 8(10):59–65
5. SATW Initiative. http://groups.geni.net/geni/ticket/913
6. K-GENI Initiative. http://groups.geni.net/geni/wiki/K-GENI
7. IEEE 802.1ag Standard. http://standards.ieee.org/getieee802/download/802.1ag-2007.pdf
8. Pol R (2012) Ethernet OAM integration in OpenFlow. In: 27th NORDUnet conference
9. Song S, Hong S, Guan X, Choi B-Y, Choi C (2013) NEOD: network embedded on-line disaster management framework for software defined networking. In: IFIP/IEEE international symposium on integrated network management. Accepted for publication
10. Ham J, Dijkstra F, Grosso P, Pol R, Toonk A, Laat C (2008) A distributed topology information system for optical networks based on the semantic Web. Opt Switch Networking 5(2–3):85–93
11. OpenFlow Reference. http://www.openflow.org/wp/tag/reference-implementation/
12. NOX Controller. http://www.noxrepo.org/
13. Beacon Controller. https://openflow.stanford.edu/display/Beacon/Home
14. MAESTRO Platform. http://code.google.com/p/maestro-platform/
15. Trema Controller. http://trema.github.com/trema/
16. RouteFlow Controller. https://sites.google.com/site/routeflow/

Analysis and Design of a Half Hypercube Interconnection Network

Jong-Seok Kim, Mi-Hye Kim and Hyeong-Ok Lee

Abstract This paper proposes a new half hypercube interconnection network that has the same number of nodes as a hypercube but reduces the degree by approximately half. To evaluate the effectiveness of the proposed half hypercube, its connectivity, routing, and diameter properties were analyzed. The analysis results demonstrate that the proposed half hypercube is an appropriate interconnection network for implementation in large-scale systems.

Keywords Half hypercube · Hypercube variation · Interconnection network

1 Introduction

The need for high-performance parallel processing is increasing because modern engineering and science application problems require many computations with real-time processing. A parallel processing system can connect thousands of processors with their own memory, or even more via an interconnection network enabling inter-processor communication by passing messages among processors through the network. An interconnection network can be depicted with an

---

J.-S. Kim
Department of Computer Science, University of Rochester,
Rochester 14627, USA
e-mail: Rockhee7@gmail.com

M.-H. Kim
Department of Computer Science Education, Catholic University of
Daegu, Daegu, South Korea
e-mail: mihyekim@cu.ac.kr

H.-O. Lee (✉)
Department of Computer Education, Suncheon National University, Suncheon, South Korea
e-mail: oklee@sunchon.ac.kr

J. J. (Jong Hyuk) Park et al. (eds.), *Multimedia and Ubiquitous Engineering*,
Lecture Notes in Electrical Engineering 240, DOI: 10.1007/978-94-007-6738-6_65,
© Springer Science+Business Media Dordrecht(Outside the USA) 2013

undirected graph. The most common parameters for evaluating the performance of interconnection networks are degree, connectivity, diameter, network cost, and broadcasting [1, 2].

In an interconnection network, degree (relevant to hardware cost) and diameter (relevant to message transmission time) are correlated. In general, the throughput of an interconnection network is improved with a higher degree because the diameter of the network is increased when its degree is increased. However, a parallel computer design increased the hardware costs of an interconnection network, because of the increased number of processor pins. An interconnection network with a lower degree can reduce hardware costs, but its latency and throughput are degraded because the message transmission time is increased. Due to such characteristics of interconnection networks, the network cost (= degree × diameter) is a typical parameter used to evaluate interconnection network performance [3].

The hypercube is a typical interconnection network topology and is widely used in both research and commercial fields due to its advantages that can easily provide a communication network structure as required in various application areas. The hypercube is node- and edge-symmetric with a simple routing algorithm, maximum fault tolerance, and simple recursive structure. Additionally, it can be easily embedded in various types of existing interconnection networks [4, 5]. However, it has the drawback of increasing network costs associated with the increased degree when the number of nodes increases. To improve this shortcoming, a number of variations of the hypercube have been proposed, such as multiple reduced hypercube [3], twisted cube [6], folded hypercube [7], connected hypercube network [8], and extended hypercube [9]. This paper proposes a new variation of the hypercube that reduces the hypercube degree by approximately half with the same number of nodes: the Half Hypercube (HH). We denote an n-dimensional half hypercube as HH_n. To validate the effectiveness of the proposed HH, performance measurement parameters were analyzed, such as connectivity, routing, and diameter.

This paper is organized as follows. Section 2 presents the definition of the proposed HH and discusses its properties, including connectivity. Section 3 proposes and analyzes a simple routing algorithm and the diameter of the HH. Section 4 summarizes and concludes the paper.

2 Definition and Properties of the Proposed Half Hypercube

The hypercube Q_n ($n \geq 2$) is defined as an n-dimensional binary cube where the nodes of Q_n are all binary n-tuples. Two nodes of Q_n are adjacent to each other if and only if their corresponding n-tuples differ in one bit at exactly one position [10]. Q_n is an n-regular graph with 2^n nodes and its diameter is n. In this paper, $\bar{S}$ indicates the complement of the binary string $S(= s_n s_{n-1} \ldots s_1)$; that is, it is obtained by inverting all the bits in the binary number (inverting 1's for 0's and vice versa).

We denote an n-dimensional half hypercube as HH_n and represent its node with n binary bits. Let the address of node S in HH_n be $s_n s_{n-1} s_{n-2} \ldots s_i \ldots s_3 s_2 s_1 (n \geq 3)$. There are two types of edge in HH_n: the h-edge, which connects node $S(= s_n s_{n-1} s_{n-2} \ldots s_i \ldots s_3 s_2 s_1)$ to a node that has the complement in exactly one h position of the bit string of node $S(1 \leq h \leq \lceil n/2 \rceil)$, and the sw-edge (i.e., swap-edge), which connects node $S(= s_n s_{n-1} s_{n-2} \ldots s_h s_{h-1} \ldots s_3 s_2 s_1)$ to a node in which the $\lfloor n/2 \rfloor$ leftmost bits of the bit string of S and the $\lfloor n/2 \rfloor$ bits on the right-side of S starting from the $\lfloor n/2 \rfloor$ bit are swapped ($h = \lfloor n/2 \rfloor$). For example, when n is even, node $S(= s_n s_{n-1} s_{n-2} \ldots s_{\lfloor n/2 \rfloor + 1} s_{\lfloor n/2 \rfloor} s_{\lfloor n/2 \rfloor - 1} s_{\lfloor n/2 \rfloor - 2} \ldots s_3 s_2 s_1)$ is connected to node $(s_{\lfloor n/2 \rfloor} s_{\lfloor n/2 \rfloor - 1} s_{\lfloor n/2 \rfloor - 2} \ldots s_3 s_2 s_1 s_n s_{n-1} s_{n-2} \ldots s_{\lfloor n/2 \rfloor + 1})$ in which $(s_n s_{n-1} s_{n-2} \ldots s_{\lfloor n/2 \rfloor + 1})$ and $(s_{\lfloor n/2 \rfloor} s_{\lfloor n/2 \rfloor - 1} s_{\lfloor n/2 \rfloor - 2} \ldots s_3 s_2 s_1)$ of S are exchanged. When n is odd, node S is connected to node $(s_{\lfloor n/2 \rfloor} s_{\lfloor n/2 \rfloor - 1} s_{\lfloor n/2 \rfloor - 2} \ldots s_3 s_2 s_n s_{n-1} s_{n-2} \ldots s_{\lfloor n/2 \rfloor + 1} s_1)$ in which $(s_n s_{n-1} s_{n-2} \ldots s_{\lfloor n/2 \rfloor + 1})$ and $(s_{\lfloor n/2 \rfloor} s_{\lfloor n/2 \rfloor - 1} s_{\lfloor n/2 \rfloor - 2} \ldots s_3 s_2)$ of S are swapped. However, if two parts of $\lfloor n/2 \rfloor$ bits to exchange in the bit string of node S are the same, the node S connects to node $s_n s_{n-1} s_{n-2} \ldots s_{\lfloor n/2 \rfloor + 1} s_{\lfloor n/2 \rfloor} s_{\lfloor n/2 \rfloor - 1} s_{\lfloor n/2 \rfloor - 2} \ldots s_3 s_2 s_1$, which is the one's complement of the binary number of node S. Figure 1 shows an example of a 5-dimensional half hypercube (HH_5). The degree of HH_n is $\lceil n/2 \rceil + 1$, which adds the number of h-edges ($1 \leq h \leq \lceil n/2 \rceil$) and one of the sw-edges. Table 1 presents the degree of HH according to the dimension of HH graphs.

Lemma 1 *An HH_n graph is expanded with recursive structures.*

Proof An HH_n graph is constructed with the nodes of two $(n-1)$-dimensional HH_{n-1} graphs by adding one sw-edge or h-edge. The address of each node in an HH_j graph is represented as j bit strings with binary numbers $\{0, 1\}$ (i.e., with a

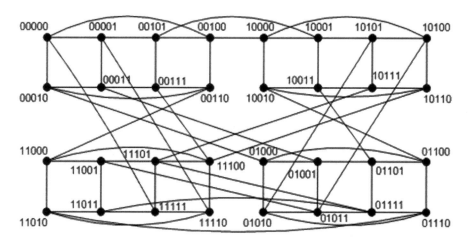

Fig. 1 Example of a 5-dimensional half hypercube (HH_5)

Table 1 Degree of HHn according to its dimension

Dimension		3	4	5	6	7	8	...	13	14	15	16	...	n
Edge Type	h-edge	2	2	3	3	4	4	...	7	7	8	8	...	$\lceil n/2 \rceil$
	sw-edge	1	1	1	1	1	1	...	1	1	1	1	...	1
Degree		3	3	4	4	5	5	...	8	8	9	9	...	$\lceil n/2 \rceil + 1$

binary number of length j). We denote HH_j^0 when the bit at the $j + 1$ position of the address of a node (i.e., at the $j + 1$ position of the bit string of a node) is binary 0, and denote HH_j^1 when it is binary 1. Let us examine the expansion of HH$_n$ by dividing it into two cases: when the dimension n is even and n is odd.

Case 1 When expanded from an odd-dimension (j) to an even-dimension (k), $k = j + 1$.

Let us construct an HH$_k$ graph by connecting a node of HH_j^0 and a node of HH_j^1 in HH$_j$ ($k = j + 1$). When expanded from an HH$_j$ graph to an HH$_k$ graph, the sw-edges in HH$_j$ will be replaced with new sw-edges in the expanded HH$_k$ graph. A node of HH_j^0 the address bit of which is binary 1 at the $n/2$ position, will be connected to a node of HH_j^1 through an sw-edge in HH$_k$. When the bit is binary 0, the node will be connected to a node of HH_j^0.

Case 2 When expanded from an even-dimension (k) to an add-dimension (l), $l = k + 1$.

We construct an HH$_l$ graph by connecting a node of HH_k^0 and a node of HH_k^1 in HH$_k$. When expanded from an HH$_k$ graph to HH$_l$, the sw-edges in HH$_k$ will be replaced with new sw-edges in the expanded HH$_l$ graph. A node of HH_k^0, the address bit of which is binary 1 at the $\lceil n/2 \rceil$ position, will be connected to a node of HH_k^1 via an sw-edge in HH$_l$. When the bit is binary 0, the node will be connected to a node of HH_k^0.

Lemma 2 *There exist $2^{\lfloor n/2 \rfloor} \lceil n/2 \rceil$-dimensional hypercube structures in an HH$_n$ graph.*

Proof An h-edge of HH$_n$ connects node $S(= s_n s_{n-1} s_{n-2} \ldots s_h \ldots s_3 s_2 s_1)$ to a node, the address bit of which is the complement of the bit string of node S at exactly one h position ($1 \leq h \leq \lceil n/2 \rceil$). The h-edge of HH$_n$ is equal to the edge of a hypercube. Therefore, the structure of a partial graph constructed from HH$_n$ via the h-edge is the same as the structure of a $\lceil n/2 \rceil$-dimensional hypercube. Let us assume that a partial graph that has the same structure as an $\lceil n/2 \rceil$-dimensional hypercube is $HH_n^{\lceil n/2 \rceil}$ in HH$_n$ and refer to it as a cluster. Here, the address of each node in a cluster is $s_h \ldots s_3 s_2 s_1$. The number of clusters consisting of the HH$_n$ graph is $2^{\lfloor n/2 \rfloor}$ because the number of bit strings that can be configured by the bit string $s_n s_{n-1} s_{n-2} \ldots s_h$ is $2^{\lfloor n/2 \rfloor}$ Node (or edge) connectivity is the minimum number of nodes (or edges) that must be removed to disconnect an interconnection network to

Analysis and Design of a Half Hypercube Interconnection Network

two or more parts without duplicate nodes. If a given interconnection network remains connected with the removal of any arbitrary k-1 or fewer nodes, but the interconnection network becomes disconnected with the removal of any arbitrary k nodes, then the connectivity of the interconnection network is k. When the degree and node connectivity of a given interconnection network are the same, we say that the interconnection network has maximum fault-tolerance [3]. It has been proven that $k(G) \leq \lambda(G) \leq d(G)$ where the node connectivity, edge connectivity, and degree of interconnection network G are denoted as $k(G)$, $\lambda(G)$, and $d(G)$, respectively [6, 11]. Through the proving process of $k(HH_n) = \lambda(HH_n) = d(HH_n)$ in Theorem 1, we will demonstrate that the proposed HH_n has maximum fault-tolerance.

Theorem 1 *The connectivity of* $HH_{n,k}(HH_n) = \lceil n/2 \rceil + 1 (n \geq 3)$.

Proof Let us prove that HH_n remains connected even when n nodes are deleted from HH_n. Through Lemma 2, we know that an HH_n graph is composed of clusters, and all clusters in HH_n are hypercubes and two arbitrary nodes are connected via an sw-edge. Assuming that X is a partial graph of HH_n where $|X| = n$, it will be proven that $k(HH_n) \geq \lceil n/2 \rceil + 1$ by demonstrating that HH_n remains connected even after the removal of X. This will be done by dividing two cases in accordance with the location of X. We denote the HH_n in which X is deleted as HH_n-X and a node of HH_n as S.

Case 1 When X is located in one cluster of HH_n:

The degree of each node in a cluster is $\lceil n/2 \rceil$. If $\lceil n/2 \rceil$ nodes adjacent to an arbitrary node S of the cluster are the same as the nodes to be deleted from X, HH_n is divided into two components: an interconnection network HH_n-X and a node S. However, all nodes in a cluster are linked to other clusters in HH_n via sw-edges, and $2^{\lfloor n/2 \rfloor} - 1$ clusters in which X is not located are also connected to other clusters via sw-edges. Therefore, HH_n-X is always connected when X is included in only a cluster of HH_n.

Case 2 When X is located across two or more clusters of HH_n:

As the nodes of X to be deleted are included across two or more clusters of HH_n, the number of nodes to be deleted from a cluster is at most $\lceil n/2 \rceil - 1$. However, even if $\lceil n/2 \rceil - 1$ nodes adjacent to an arbitrary node S of a cluster are deleted, the nodes of the cluster in which node S is included remain connected because the degree of a node in a cluster is $\lceil n/2 \rceil$. Although the other node to be removed is located in a different cluster, and not in the cluster that includes node S, it is still clear that HH_n remains connected. Therefore, $k(HH_n) \geq \lceil n/2 \rceil + 1$ because HH_n always remains connected after removing X from any clusters in HH_n and $k(HH_n) \leq \lceil n/2 \rceil + 1$ because the degree of HH_n is $\lceil n/2 \rceil + 1$. Consequently, the connectivity of $HH_{n,k}(HH_n) = \lceil n/2 \rceil + 1$.

3 Routing Algorithm and Diameter of HH_n

This section analyzes a simple routing algorithm and diameter of HH_n. We assume that an initial node S is $s_n s_{n-1} s_{n-2} \ldots s_{n/2} \ldots s_3 s_2 s_1$ and a destination node T is $t_n t_{n-1} t_{n-2} \ldots t_{n/2} \ldots t_3 t_2 t_1$. A simple routing algorithm can be considered in two cases depending on whether n is an even or an odd number.

Case 1 When n is an even number
If an initial node $S(= s_n s_{n-1} s_{n-2} \ldots s_{n/2} s_{n/2-1} \ldots s_3 s_2 s_1)$ is presented with two $\lceil n/2 \rceil$ bit strings $A(= s_n s_{n-1} s_{n-2} \ldots s_{n/2})$ and $B(= s_{n/2-1} \ldots s_3 s_2 s_1)$, node S can be denoted as AB. In the same way, a destination node $T(= t_n t_{n-1} t_{n-2} \ldots t_{n/2} t_{n/2-1} \ldots t_3 t_2 t_1)$ can be denoted as CD where $C(= t_n t_{n-1} t_{n-2} \ldots t_{n/2})$ and $D(= t_{n/2-1} \ldots t_3 t_2 t_1)$ (i.e., C is the $\lceil n/2 \rceil$ leftmost bits and D is the $\lceil n/2 \rceil$ rightmost bits in the bit string of node T).

Simple routing algorithm-even:

(1) Convert the bit string of B in node $S(= AB)$ with the bit string C in node $T(= CD)$ using h-edge $(1 \leq h \leq \lceil n/2 \rceil)$.
(2) Exchange the bit string of A with that of C in node $S(= AC)$ using sw-edge.
(3) Convert the bit string of A in node $S(= CA)$ with D of node $T(= CD)$ using h-edge.

In Phases (1) and (3), the bit string B is converted with C and A is converted with D using the hypercube routing algorithm.

Corollary 1 When n is even, the length of the shortest path is $2 \times \lceil n/2 \rceil + 1 = n + 1$ by the above *simple routing algorithm-even*.

Case 2 When n is an odd number
Let the initial node be $S(= s_n s_{n-1} s_{n-2} \ldots s_{\lfloor n/2 \rfloor+1} s_{\lfloor n/2 \rfloor} s_{\lfloor n/2 \rfloor-1} \ldots s_3 s_2 s_1)$, the $\lfloor n/2 \rfloor$ leftmost bits of the bit string of S be $A(= s_n s_{n-1} s_{n-2} \ldots s_{\lfloor n/2 \rfloor+1})$ and the $\lfloor n/2 \rfloor$ bits on the right-side of S starting from the $\lfloor n/2 \rfloor$ bit be $B(= s_{\lfloor n/2 \rfloor} s_{\lfloor n/2 \rfloor-1} \ldots s_3 s_2)$. Then, node S can be denoted as (ABs_1). In the same way, a destination node $T(= t_n t_{n-1} t_{n-2} \ldots t_{\lfloor n/2 \rfloor+1} t_{\lfloor n/2 \rfloor} t_{\lfloor n/2 \rfloor-1} \ldots t_3 t_2 t_1)$ can be denoted as (CDt_1), where $C(= t_n t_{n-1} t_{n-2} \ldots t_{\lfloor n/2 \rfloor+1})$ and $D(= t_{\lfloor n/2 \rfloor} t_{\lfloor n/2 \rfloor-1} \ldots t_3 t_2)$.

Simple routing algorithm-odd:

(1) Convert the bit string of B in node $S(= ABs_1)$ with the bit string C in node $T(= CDt_1)$ using h-edge $(1 \leq h \leq \lfloor n/2 \rfloor)$.
(2) Exchange the bit string of A with that of C in node $S(= ACs_1)$ using sw-edge.
(3) Convert the bit string of As_1 in node $S(= CAs_1)$ with Dt_1 of node $T(= CDt_1)$ using h-edge $(1 \leq h \leq \lfloor n/2 \rfloor)$.

In Phases (1) and (3), the bit string B is swapped with C and As_1 is swapped with Dt_1 using the hypercube routing algorithm.

Analysis and Design of a Half Hypercube Interconnection Network 543

Corollary 2 When n is odd, the length of the shortest path is $2 \times \lfloor n/2 \rfloor + 2$ by the above *simple routing algorithm-odd*.

Through the proposed simple routing algorithms, we can see an upper bound for the diameter of HH_n, thus proving Theorem 2.

Theorem 2 *The upper bound on the diameter of HH_n is $n + 1$ when n is even and $2 \times \lfloor n/2 \rfloor + 2$ when n is odd.*

4 Conclusion

This paper proposed a half hypercube interconnection network HH_n (a new variation of the hypercube) that reduced the degree by approximately half, $n/2$, even though it has the same number of nodes as a hypercube. To evaluate the effectiveness of the proposed half hypercube, we analyzed its connectivity and diameter properties. We also analyzed a simple routing algorithm of HH_n and presented an upper bound for the diameter of HH_n. These results demonstrate that the proposed half hypercube is an appropriate interconnection network for implementation in a large-scale system.

Acknowledgment This research was supported by Basic Science research program through the National research Foundation of KOREA (NRF) funded by the Ministry of Education, Science and Technology (2012R1A1A4A01014439).

References

1. Leightoo FT (1992) Introduction to parallel algorithms and architectures: arrays, hypercubes. Morgan Kaufmann Publishers, San Francisco
2. Mendia VE, Sarkar D (1992) Optimal broadcasting on the star graph. IEEE Trans Parallel Distrib Syst 3(4):389–396
3. Sim H, Oh JC, Lee HO (2010) Multiple reduced hypercube (MRH): a new interconnection network reducing both diameter and edge of hypercube. Int J Grid Distrib Comput 3(1):19–30
4. Saad Y, Schultz MH (1988) Topological properties of hypercubes. IEEE Trans Comput 37(7):867–872
5. Seitz CL (1985) The cosmic cube. Commun ACM 26:22–33
6. Abraham S (1991) The twisted cube topology for multiprocessor: a study in network asymmetry. J Parallel Distrib Comput 13:104–110
7. EI-Amawy A, Latifi S (1991) Properties and performance of folded hypercubes. IEEE Trans Parallel Distrib Syst 2(1):31–42
8. Ghose K, Desai KR (1995) Hierarchical cubic network. IEEE Trans Parallel Distrib Syst 6(4):427–435
9. Kumar JM, Patnaik M (1992) Extended hypercube: a hierarchical interconnection network of hypercubes. IEEE Trans Parallel Distrib Syst 3(1):45–57
10. Livingston M, Stout QF (1988) Embedding in hypercubes. Math Comput Model 11:222–227
11. Akers SB, Harel D, Krishnamurthy B (1987) The star graph: an attractive alternative to the N-Cube. In: Proceedings of the international conference on parallel processing, pp 393–400

Aperiodic Event Communication Process for Wearable P2P Computing

Tae-Gyu Lee and Gi-Soo Chung

Abstract Wearable computing has been proposed as an alternative to the best computing interfaces and devices for the ubiquitous computing. A digital wear can be a main element of wearable computers. This study shall apply digital yarn as a material of data communications for the purpose to take advantage a digital garment. Wearable P2P application communications are consisted of periodic or aperiodic methods. This paper proposes an aperiodic event process for wearable P2P computing. It shows the transmission process that collects from a digital garment at random time. Specially, the process supports the recovery transfer process when the aperiodic event messages are failed.

Keywords Aperiodic communication · Wearable computing · P2P communication · Digital garment

1 Introduction

Nowadays, as we enter the era of ubiquitous computing, the computing appliances are more gradually closer to human and the using time of information devices has exponentially increased. A wearable computing has been proposed as an alternative to the best mobile computing devices for these ubiquitous computing [1].

A digital garment accounts a principle element of wearable computing. This study shall apply digital yarn as a material of data communications for the purpose to take advantage a digital garment [3]. This conductive micro-wire digital yarn can be applied in a general garment knitting or weaving process as a lightweight weaving textile unlike the existing communication lines. The digital fiber was

---

T.-G. Lee (✉) · G.-S. Chung
Korea Institute of Industrial Technology (KITECH), Ansan 426-791, Korea
e-mail: tigerlee88@empal.com

J. J. (Jong Hyuk) Park et al. (eds.), *Multimedia and Ubiquitous Engineering*, Lecture Notes in Electrical Engineering 240, DOI: 10.1007/978-94-007-6738-6_66, © Springer Science+Business Media Dordrecht(Outside the USA) 2013

already developed, but still is in incomplete status to be used as a communication standard configuration and transport platforms [2, 3].

Wearable P2P application communications are consisted of periodic or aperiodic methods. Periodical communications support the regular collection of information, and aperiodic communications support the transfer of information when a particular event has occurred. This paper proposes an aperiodic event process for wearable P2P computing. It shows the transmission process that collects from a digital garment at random time. Specially, the process supports the recovery transfer process when the aperiodic event messages are failed [4–6].

An *event message channel* is a temporary storage location for data while the data is being transferred. The event channel is often used for supporting a recovery of loss data frames. There have the problems that if the channel size were too large, it makes the efficient use of channel resources worse and otherwise, if it has the smaller channel capacity, it makes the available channel bandwidth inefficient due to a waste of the network failure of digital yarn. Therefore, it is necessary to set up the optimal size of the event channel capacity.

Typically, an event process consists of First In First Out (FIFO) communication structure. The event message model presented in this paper is based on the sender-receiver communication channel of Ethernet, which is widely used in communication systems.

This paper describes about the overview of aperiodic message communication system in Sect. 2. Section 3 shows an aperiodic P2P event processes. Section 4 shows the application scenarios for wearable P2P commutations. Finally, Sect. 5 concludes this paper.

2 Aperiodic Message Communication

Wearable aperiodic P2P communication system supports the infrastructure of information exchange to build the wearable embedded computing services of a mobile wearable user. In order to build these systems, two or more nodes (MSS: Mobile Service Station) of sharing information have been configured and the wired and wireless communication channels should be organized for the information exchange. More particularly, this study focuses on the wear-embedded wired communication system based on digital yarn.

For aperiodic messages, their arrival durations usually are irregular, however it is assumed that there is a minimum inter-arrival time in order to guarantee its temporal constraint. Aperiodic messages can be classified into *urgent* and *normal* states.

The wearable aperiodic P2P communication is a transferring process which sends the information initiated by the sender-peer to the receiver-peer. The aperiodic message transmission system considers the link propagation delay and multiplexing method as a factor which affects to the transmission performance and the channel efficiency.

Figure 1 shows the aperiodic P2P computing system on P2P communication channels. Any one peer terminals become an initiator and the other peer terminal become a correspondent.

The P2P system configuration for applications supports a distributed P2P MSS and a dual P2P MSS. First, the distributed MSS applications are corporately performed for providing different computing service on two more terminals. As a distributed P2P system, the cross-work configuration of left-peer and right-peer can be applied to the distributed applications crossover terminals, respectively. Second, the dual P2P MSS supports the same computing services on two more terminals, and one MSS peer performs the active computing as the foreground services and the other MSS peer performs stand-by backup computing. As a dual P2P communication system, the front-peer and rear-peer configuration can be employed to each of application terminal and backup terminal, respectively. The rear-peer can be optionally used as standby terminal.

The wearable aperiodic P2P computing system has the needs to meet the following requirements. First, the P2P computing on two more peers can provide the interactive event message at any time. Second, it can support the mission-critical projects more than single terminal with the same event process image. Third, it can support the load balancing service for transferring the event messages by load-distributing method. There are military wear, police wear, and firefight wear as the mission-critical applications based on the wearable P2P system. These applications are important to consider computing performance as well as safety issues.

As the differentiation with other P2P systems, the existing P2P systems provide the efficient resource sharing and load balancing on wide-area Internet, but the proposed P2P communication system realizes the distributed asynchronous P2P transmission system among the limited specific terminals on wearable local network. Thus, the wearable P2P communication system has the low complexity of transmission links. And it has the high frequency of P2P transfer between the specific terminals. The system should consider the performance aspects and resource-efficient aspects together.

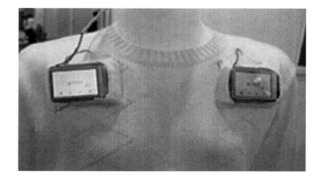

Fig. 1 Aperiodic P2P communication system in wearable computing

This paper only considers dual P2P communication system for the aperiodic communication performance and system safety. When the dual P2P system organization is considered for experiment and analysis model, the main factors that influence communication performance are asynchronous P2P communication link. The P2P link supports bidirectional communication between front-peer and rear-peer.

To improve the communication performance, we consider the propagation delay of the asynchronous dual P2P link. Also, we will consider the multiple levels of the P2P links to enhance the communication safety.

3 Aperiodic P2P Event Process

In the normal event process, the aperiodic P2P event can be defined of a set as event tuple $E < DN_j, E < SN_i, M_k \gg$. E means the event transfer function, DN_j is a destination peer node, SN_i is a start peer node, and M_k indicates an event message of SN_i. The following process of Fig. 2 shows these normal event processes. Initialization method of checkpoint cycle uses a time period, or employs a message count period.

In the failure event process, the P2P fault-recovery events can be processed as the following process of Fig. 3. A destination node DN_j discards the duplicate event messages M_k from the same start peer node SN_i. It determines that the message was lost when the k index value of the incoming event messages M_k is not increased sequentially as skipping. In a sender SN_i, if there is no response message to the outgoing message from a receiver DN_j, we may decide the transmitted message was lost.

The checkpoint is a snapshot of process control block as resource information for process recovery or synchronization.

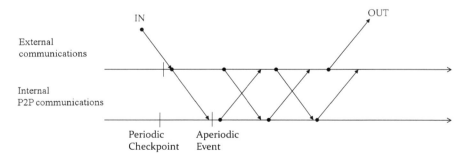

Fig. 2 Normal process in aperiodic P2P communication system

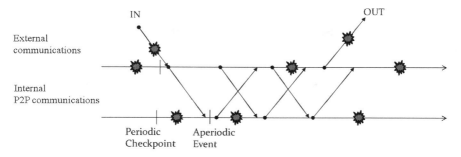

Fig. 3 Fault-recovery process in aperiodic P2P communication system

The process events are classified into real-time and non-real-time data. First, the real-time data means the continuous information as streaming sensing signals. It can be received from the correspondent and saved to storage devices periodically for easy data management. Second, the non-real-time data means the discrete information as general intermediate messages. It can be received from the correspondent and saved to storage devices a-periodically for easy data management.

The following process sequences describe the sender and receiver communication processes, and present the fault-recovery processes.

Sender process in an starter peer node, SN_i:

Start and initialize with its SN_i and the correspondent's DN_j;
Check a new event and save the process aperiodic checkpoints on any random time;
Select data resource in collected data;
Send the event message including data resource to the correspondent, DN_j;
Receive the confirm message from Receiver;
If (the confirm message $= NAK$) then retry;
Otherwise, wait for next event;

Receiver process as a correspondent peer, DN_j:

Start and initialize with its DN_j and the correspondent, SN_i;
Receive the event message from Sender, SN_i;
Check a new event and save the process checkpoints on aperiodic time;
Reply the confirm message with event sequence information to Sender, SN_i;

Event fault-recovery:

Each Sender, SN_i, and Receiver, DN_j, transfer its process sequences respectively;
Save the event message into the backup memory of the Sender, SN_i;
Remove the messages of mismatched sequences;
Remove orphan or duplicate messages;

Fault detection();

Detect the crash fault using beacon signals by correspondent;
Detect the temporal fault using the inform message by faulty peer;

Standby state();

Wait for the alive message of faulty peer;
Skip or block the P2P messages;

Alive state();

Rollback to last checkpoint;
Send alive message including checkpoint position;
Receive the confirm message with synchronization;
Perform the recovery computing and communication using the message backup memory of the sender SN_i;

P2P event synchronization();

Send event sync message with the event sequence numbers to the correspondents;

4 Wearable Aperiodic P2P Application

In this section, as a wearable P2P application, the mission-critical operations are performed such as police and military missions as shown in Fig. 4. The wearable system collects aperiodic event of non-real-time messages. For example, the military applications with digital garment have the interactive messages between a MSS peer and the other MSS peer as the following scenario sequences.

These are aperiodic P2P application scenarios as the mission-critical services for military digital garment.

(1) Start top-down commands of orders and mission commitment;
(2) Determine the mission and communicate internal P2P communication links;
(3) Read circumstances using embedded sensors and start reporting;

Aperiodic Event Communication Process for Wearable P2P Computing 551

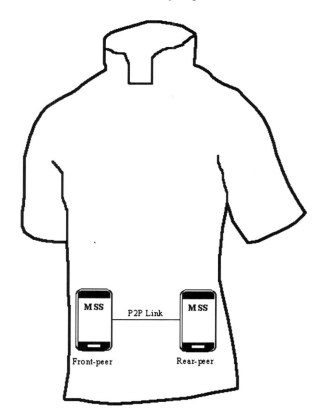

Fig. 4 P2P applications for wearable communication system

(4) Execute mission monitoring using internal computing and P2P communication link;
(5) Real-time message reporting of special situations as aperiodic event;
(6) Mission can be changed and aborted optionally;
(7) Mission is reconfirmed and a self-assessment results is performed;
(8) Perform the expanded mission;
(9) Mission off commands are delivered between a peer and the correspondent peer by the commander;
(10) Sleep for waiting the new missions after a current mission is off;

The components of wearable embedded mobile smart station (MSS) are consisted as follows. First, as a dual P2P communication system, the front-peer and rear-peer configuration is employed to each of application terminal and backup terminal, respectively. The rear-peer can be selectively used for standby dual terminal.

Second, as a distributed P2P system, the cross-work configuration of left-peer and right-peer is applied to the distributed applications crossover terminals, respectively.

When the dual P2P system organization is considered for experiment and analysis model, the key factors that influence communication performance are P2P communication link. The P2P link supports P2P bidirectional communication between front-peer and rear-peer.

5 Conclusions

This work has presented the aperiodic P2P communication logic in the wearable computing fields using digital yarn. Then it has proposed the multiple links and the fault-recovery process as the methods for enhancing the performance and feasibility of the P2P communication system. Finally, the aperiodic communication application scenarios have been shown the needs of multiple links and the effectiveness of fault-recovery process simultaneously.

References

1. Lee T (2012) Information life cycle design and considerations for wearable computing. EMC Technol Serv LNEE 181:501–508
2. Vassiliadis S, Provatidis C, Prekas K, Rangussi M (2005) Novel fabrics with conductive fibers. In: Intelligent textile structures—application, production & testing international workshop, GREECE, May 2005
3. ChungGS (2009) Digital garment for data communication using digital yarn. In: 2009 Korean-German smart textile symposium, pp 57–67, Sept 2009
4. Lee T, Chung G (2012) Wearable P2P communication system organization on digital yarn. Ubiquitous information technologies and applications. Lect Notes Electr Eng 214:601–609
5. Wang Z, Shen X, Chen J, Song Y, Wang T, Sun Y (2005) Real-time performance evaluation of urgent aperiodic messages in FF communication and its improvement. Elsevier Comput Stand Interfaces 27:105–115
6. Kato S, Fujita XY, Yamasaki N (2009) Periodic and aperiodic communication techniques for responsive link. In: 15th IEEE international conference on embedded and real-time computing systems and applications, pp 135–142

Broadcasting and Embedding Algorithms for a Half Hypercube Interconnection Network

Mi-Hye Kim, Jong-Seok Kim and Hyeong-Ok Lee

Abstract The half hypercube interconnection network, has been proposed as a new variation of the hypercube, reducing its degree by approximately half with the same number of nodes as an n-dimensional hypercube, Q_n. This paper proposes an algorithm for one-to-many broadcasting in an n-dimensional half hypercube, HH_n, and examines the embedding between hypercube and half hypercube graphs. The results show that the one-to-many broadcasting time of the HH_n can be accomplished in $n + 1$ when n is an even number and in $2 \times \lceil n/2 \rceil$ when n is an odd number. The embedding of HH_n into Q_n can be simulated in constant time $O(n)$ and the embedding of Q_n into HH_n in constant time $O(1)$.

Keywords Half hypercube · One-to-many broadcasting · Embedding

1 Introduction

There is increasing interest in parallel processing as a technique for achieving high-performance owing to the need for many computations with real-time data processing in modern applications [1, 2]. A parallel processing system can connect hundreds of thousands of processors with their own memory via an interconnection

M.-H. Kim
Department of Computer Science Education, Catholic University of Daegu,
Daegu, South Korea
e-mail: mihyekim@cu.ac.kr

J.-S. Kim
Department of Computer Science, University of Rochester, Rochester 14627, USA
e-mail: Rockhee7@gmail.com

H.-O. Lee (✉)
Department of Computer Education, Sunchon National University, Sunchon, South Korea
e-mail: oklee@sunchon.ac.kr

J. J. (Jong Hyuk) Park et al. (eds.), *Multimedia and Ubiquitous Engineering*,
Lecture Notes in Electrical Engineering 240, DOI: 10.1007/978-94-007-6738-6_67,
© Springer Science+Business Media Dordrecht(Outside the USA) 2013

network. The overall performance of the system is dependant on the performance of each processor and the architecture of the interconnection network used [1–3]. Many interconnection network topologies have been described in the literature, such as star, mesh, bubble-sort, and pancake graphs.

The hypercube, Q_n, is a typical topology and is an n-regular and node- and edge-symmetric graph with 2^n nodes and diameter n ($n \geq 2$). The hypercube Q_n has simple routing algorithms and recursive structures with maximum fault-tolerance. In addition, it has the advantage that its network structure can easily be embedded in various types of commonly used interconnection networks. With such advantages, it is widely used in various application areas [1, 4, 5]. However, its network cost is increased considerably in relation to the increased degree when the number of nodes increases. To resolve this drawback, several hypercube variations have been introduced. We have proposed the half-hypercube interconnection network, reducing its degree by approximately half, even though it has the same number of nodes as a hypercube Q_n. In this paper, we propose an algorithm for one-to-many broadcasting in an n-dimensional half hypercube, HH_n, and analyze the embedding method of a half hypercube graph into a hypercube graph and vice versa.

The most common properties for measuring the performance of interconnection networks include degree, diameter, connectivity, fault tolerance, broadcasting, and embedding [6, 7]. In [1], we analyzed the degree, diameter, connectivity, and fault-tolerance parameters of the half hypercube. Here, we examine the broadcasting and embedding properties of a HH_n to strengthen its effectiveness. Broadcasting is one of the major primitives for communication of parallel processing involving message disseminating from an origin node to all the other nodes (one-to-many broadcast) or among the nodes (many-to-many broadcast) in an interconnection network. Embedding is to evaluate the relative performance of two arbitrary interconnection networks. This is of interest, because the properties and algorithms developed in a certain topology can easily be adapted to anther network at less cost [8].

The organization of this paper is as follows. Section 2 presents the definition of the half hypercube HH_n. Section 3 proposes and analyzes a broadcasting algorithm for an n-dimensional half hypercube, HH_n. Section 4 examines the embedding algorithms between hypercube and half hypercube graphs. Section 5 concludes the paper.

2 Definition of the Half Hypercube

The half hypercube HH_n ($n \geq 3$) is defined as an n-dimensional binary cube where the nodes of HH_n are all binary n-tuples in the same way as the hypercube Q_n ($n \geq 2$). That is, an n-dimensional half hypercube is denoted as HH_n and each node is represented with n binary bits. The degree and node connectivity of HH_n are $\lceil n/2 \rceil + 1$ and $\lceil n/2 \rceil + 1 (n \geq 3)$, respectively. An HH_n graph is expanded with recursive structures and has $2^{\lfloor n/2 \rfloor} \lceil n/2 \rceil$-dimensional hypercube structures with maximum fault-tolerance [1].

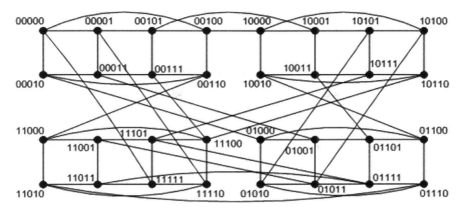

Fig. 1 Example of a 5-dimensional half hypercube (HH_5) [1]

In Q_n, an edge exists between two arbitrary nodes S and S' if and only if their corresponding n-tuples differ in exactly one k position of the bit strings of S and S' ($1 \leq k \leq n$) [9]. On the other hand, in HH_n, two types of edge exist: the h-edge, which connects node S to a node that has the complement in one bit at exactly one h position ($1 \leq h \leq \lceil n/2 \rceil$), and the *swap*-edge (shortly, *sw*-edge), which connects node S to a node where the $\lfloor n/2 \rfloor$ leftmost bits of the bit string of S and the $\lfloor n/2 \rfloor$ bits on the right-side of S starting from the $\lfloor n/2 \rfloor$ position are exchanged ($h = \lfloor n/2 \rfloor$). However, if two parts of $\lfloor n/2 \rfloor$ bits to swap in the bit string of node S are the same, the node S connects to node $\bar{S}$, which is the one's complement of the binary number of node S [1]. Note that $\bar{S}$ indicates the one's complement of the binary number of node S in this paper.

At node $S(=s_n s_{n-1} \ldots s_{\lfloor n/2 \rfloor+1} s_{\lfloor n/2 \rfloor} s_{\lfloor n/2 \rfloor-1} \ldots s_2 s_1)$ of HH_n, the address of node S' adjacent to node S via an *sw*-edge is considered using two cases depending on whether n is even or odd. For instance, when n is even, node $S(=s_n s_{n-1} \ldots s_{\lfloor n/2 \rfloor+1} s_{\lfloor n/2 \rfloor} s_{\lfloor n/2 \rfloor-1} \ldots s_2 s_1)$ is adjacent to node $S'(=s_{\lfloor n/2 \rfloor} s_{\lfloor n/2 \rfloor-1} \ldots s_2 s_1 s_n s_{n-1} \ldots s_{\lfloor n/2 \rfloor+1})$ where $(s_n s_{n-1} \ldots s_{\lfloor n/2 \rfloor+1})$ and $(s_{\lfloor n/2 \rfloor} s_{\lfloor n/2 \rfloor-1} \ldots s_2 s_1)$ of S are swapped. When n is odd, node S is adjacent to node $S'(=s_{\lfloor n/2 \rfloor} s_{\lfloor n/2 \rfloor-1} \ldots s_2 s_n s_{n-1} \ldots s_{\lfloor n/2 \rfloor+1} s_1)$ where $(s_n s_{n-1} \ldots s_{\lfloor n/2 \rfloor+1})$ and $(s_{\lfloor n/2 \rfloor} s_{\lfloor n/2 \rfloor-1} \ldots s_2)$ of S are swapped. Figure 1 presents a 5-dimensional half hypercube (HH_5) [1].

3 Broadcasting Algorithm for HH_n

Broadcasting is a basic data communication technique for interconnection networks involving message transmission between nodes and is used by parallel algorithms [7, 10]. There are two types of broadcasting communication: one-to-many transmission, which transmits messages from a node to all the other nodes, and many-to-many transmission, which transmits messages among nodes. Here, we will demonstrate that the one-to-many broadcasting time of HH_n is $n + 1$ when n is even and the broadcasting time is $2 \times \lceil n/2 \rceil$ when n is odd.

Theorem 1 *When n is even, the one-to-many broadcasting time of HH_n is $n + 1$ and when n is odd, the one-to-many broadcasting time of HH_n is $2 \times \lceil n/2 \rceil$.*

Proof Each cluster of HH_n represents a hypercube and the one-to-many broadcasting time of a hypercube Q_m is m. A cluster is connected to all the other clusters in HH_n by an external sw-edge. The broadcasting process is divided into three phases as follows:

(1) Phase 1: Node S transmits messages to all the other nodes within its cluster
(2) Phase 2: All nodes within the cluster to which node S belongs, including node S transmit messages to an arbitrary node in all the other clusters of HH_n using an external sw-edge.
(3) Phase 3: Repeat the process of Phase 1 in each cluster of HH_n.

When n is even, the one-to-many broadcasting time is as follows. As the broadcasting time of an internal cluster of HH_n is the same as the one-to-many broadcasting time of a hypercube, the broadcasting time of Phase 1 is $n/2$. As broadcasting is performed only once in Phase 2, its broadcasting time is 1. As Phase 3 repeats the process of Phase 1, the broadcasting time is $n/2$. Therefore, the one-to-many broadcasting time of HH_n is $n/2 + 1 + n/2 = n + 1$ when n is even.

When n is odd, the one-to-many broadcasting time is as follows. As the broadcasting time of an internal cluster of HH_n is the same as the one-to-many broadcasting time of a hypercube, the broadcasting time of Phase 1 is $n/2$. As broadcasting is performed only once in Phase 2, its broadcasting time is 1. If n is odd, the number of the sw-edges connecting clusters is 2 or 4. Thus, the number of nodes that initiate a message transmission is 2 or 4 in Phase 3. If the number of start nodes is 2, the one-to-many broadcasting time of a hypercube is reduced by 1. Therefore, the broadcasting time of Phase 3 is $\lceil n/2 \rceil - 1$. Consequently, the one-to-many broadcasting time of HH_n is $\lceil n/2 \rceil + 1 + \lceil n/2 \rceil - 1 = 2 \times \lceil n/2 \rceil$ when n is odd.

4 Embedding Between Hypercube and Half Hypercube Graphs

Numerous parallel processing algorithms are being designed to solve many problems in a variety of interconnection network structures. Whether such algorithms designed for a specific interconnection network structure can be run on different interconnection network structures is an important issue in parallel processing. One of the most widely used measuring methods for this issue is embedding [10, 11], which involves mapping the processors and communication links of an interconnection network into those of another interconnection network.

Broadcasting and Embedding Algorithms

We can represent an interconnection network as a graph $G(V, E)$, where $V(G)$ and $E(G)$ are the set of nodes and edges of graph G, respectively, and the set of paths of graph G is $P(G)$. The embedding of an interconnection network $G(V, E)$ into another interconnection network $G'(V', E')$ is defined as a function (Φ, ρ), where Φ maps the set of vertices $V(G)$ one-to-one into the set of vertices $V'(G')$ and ρ maps the set of edges $E(G)$ into the set of paths $P'(G')$; that is, $\Phi: V \rightarrow V'$ and $\rho: E \rightarrow P(G')$. The representative measurement parameters for embedding costs are dilation and congestion. Dilation is the length of the shortest path from node S' to node T' in G' when the nodes S and T of an edge (S, T) in G are mapped to nodes S' and T' of G'; i.e., the number of edges comprising the shortest path from node S' to node T' in G'. Congestion is the number of edges in G that pass an edge e in G' when G is mapped to G' [3, 4]. In this section, we analyze the embedding between a hypercube Q_n and a half hypercube HH_n using dilation.

Theorem 2 *An n-dimensional hypercube Q_n can be embedded into an n-dimensional half hypercube HH_n with dilation 3.*

Proof We can analyze the dilation of this mapping through the number of edges of HH_n required to map the k-dimensional edge $(1 \leq k \leq n)$, which represents the adjacent relationships of the nodes in Q_n, into edges in HH_n. Theorem 4 is proven by dividing the k-dimensional edge of Q_n into two cases depending on the dimension of k.

Case 1 k-dimensional edge, $1 \leq k \leq \lceil n/2 \rceil$

It can be easily observed that the k-dimensional edge of hypercube Q_n $(1 \leq k \leq \lceil n/2 \rceil)$ are the same as the h-dimensional edge of half hypercube HH_n $(1 \leq h \leq n/2\rceil)$. Therefore, it is clear that the embedding of an n-dimensional hypercube Q_n into an n-dimensional half hypercube HH_n is possible with dilation 1 when the two adjacent nodes via a k-dimensional edge in Q_n are mapped to two adjacent nodes through an h-dimensional edge in HH_n.

Case 2 k-dimensional edge, $\lceil n/2 \rceil + 1 \leq k \leq n$

The address of node S' adjacent to an arbitrary node $S(= s_n s_{n-1} s_{n-2} \ldots s_k \ldots s_{n/2} \ldots s_3 s_2 s_1)$ of Q_n via a k-dimensional edge has the complement at exactly one k position of the bit string of node S (i.e., bit s_k). An edge of HH_n that has the same role as the k-dimensional edge of hypercube Q_n can be presented by sequentially applying the following edge sequence: $<sw$-edge, $(k - \lceil n/2 \rceil)$-edge, sw-edge$>$ $(\lceil n/2 \rceil + 1 \leq k \leq n)$. That is, it reaches a node with the address $s_n s_{n-1} s_{n-2} \ldots \bar{s}_k \ldots s_{n/2} \ldots s_3 s_2 s_1$ when the edge sequence $<sw$-edge, $(k - \lceil n/2 \rceil)$-edge, sw-edge$>$ is applied sequentially to node $S(= s_n s_{n-1} s_{n-2} \ldots s_k \ldots s_{n/2} \ldots s_3 s_2 s_1)$ of HH_n. Let a node S' adjacent to node $S(= s_n s_{n-1} s_{n-2} \ldots s_k \ldots s_{n/2} \ldots s_3 s_2 s_1)$ of hypercube Q_n via a k-dimensional edge $(\lceil n/2 \rceil + 1 \leq k \leq n)$ be $S'(= s_n s_{n-1} s_{n-2} \ldots \bar{s}_k \ldots s_{n/2} \ldots s_3 s_2 s_1)$.

The address of node $sw(S)$ adjacent to node $S(= s_n s_{n-1} s_{n-2} \ldots s_k \ldots s_{\lceil n/2 \rceil + 1} s_{\lceil n/2 \rceil} \ldots s_3 s_2 s_1)$ of HH_n via an sw-edge is $s_{\lceil n/2 \rceil} \ldots s_3 s_2 s_1 s_n s_{n-1} s_{n-2} \ldots s_k \ldots s_{\lceil n/2 \rceil + 1}$. To invert

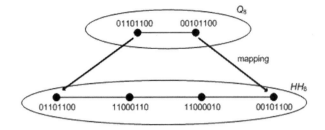

Fig. 2 Embedding example between Q_8 and HH_8 with dilation 3

the bit s_k in the bit string of node $sw(S)$ to the complement, we take a node $S''(=s_{\lceil n/2 \rceil}\ldots s_3 s_2 s_1 s_n s_{n-1} s_{n-2}\ldots \bar{s}_k \ldots s_{\lceil n/2 \rceil+1})$ adjacent to node $sw(S)$ via a $(k - \lceil n/2 \rceil)$-dimensional edge. The address adjacent to node S'' through an sw-edge is $s_n s_{n-1} s_{n-2} \ldots \bar{s}_k \ldots s_{\lceil n/2 \rceil+1} s_{\lceil n/2 \rceil} \ldots s_3 s_2 s_1$. Therefore, the embedding of a k-dimensional edge of Q_n into a half hypercube HH_n is possible with dilation 3. Figure 2 presents an example of embedding between Q_8 and HH_8 with dilation 3.

Theorem 3 *A half hypercube HH_n can be embedded into a hypercube Q_n with dilation n.*

Proof There exist two types of edges in half hypercube HH_n. Thus, we proved Theorem 5 by dividing it into two cases: h-edge and sw-edge.

Case 1 h-edge, $1 \leq h \leq \lceil n/2 \rceil$

The address of node S' adjacent to node $S(=s_n s_{n-1} s_{n-2}\ldots s_{n/2}\ldots s_3 s_2 s_1)$ of half hypercube HH_n via an h-edge is $s_n s_{n-1} s_{n-2}\ldots s_{n/2}\ldots \bar{s}_h \ldots s_3 s_2 s_1$ $(1 \leq h \leq \lceil n/2 \rceil)$. The address of node S' adjacent to node $S(=s_n s_{n-1} s_{n-2}\ldots s_k \ldots s_{n/2}\ldots s_3 s_2 s_1)$ of hypercube Q_n through a k-dimensional edge $(1 \leq k \leq n)$ is $s_n s_{n-1} s_{n-2}\ldots \bar{s}_k \ldots s_{n/2}\ldots s_3 s_2 s_1$. Accordingly, the dilation of this embedding is 1 because the h-edge in half hypercube HH_n and the k-dimensional edge in Q_n are equivalent $(1 \leq h, k \leq \lceil n/2 \rceil)$.

Case 2 sw-edge

The sw-edge of half hypercube HH_n can be divided into two cases depending on the address of node $S(=s_n s_{n-1} s_{n-2}\ldots s_{n/2}\ldots s_3 s_2 s_1)$. Here, we prove the case with dilation n. If the n-address bits of node S are all binary 0, the n-address bits of node S' adjacent to node S through an sw-edge are all binary 1. The shortest path from node $S(=s_n s_{n-1} s_{n-2}\ldots s_{n/2}\ldots s_3 s_2 s_1)$ to node $S'(=\overline{s_n s_{n-1} s_{n-2}\ldots s_{n/2}\ldots s_3 s_2 s_1})$ in hypercube Q_n is the same as the path to which the k-dimensional edges are all applied. Therefore, a half hypercube HH_n can be embedded into a hypercube Q_n with dilation n $(1 \leq k \leq n)$.

5 Conclusion

This paper proposes a one-to-many broadcasting algorithm for the half hypercube interconnection network, HH_n that we proposed in [1], and proved that the one-to-many broadcasting time is $n + 1$ when n is even and $2 \times \lceil n/2 \rceil$ when n is odd in HH_n. We also showed that it is possible to embed an n-dimensional hypercube Q_n into an n-dimensional half hypercube HH_n with dilation 3, and that it is possible to embed HH_n into Q_n with dilation n. These results suggest that our half hypercube interconnection network HH_n has potential for implementation in large-scale systems for parallel processing.

References

1. Kim JS, Kim M, Lee HO (2013) Analysis and design of a half hypercube interconnection network. ATACS 2013, LNEE. Springer, Heidelberg (will be appeared)
2. Kim M, Kim DW, Lee HO (2010) Embedding algorithms for star, bubble-sort, rotator-faber-moore, and pancake graphs. HPCTA 2010, LNCS, vol 6082. Springer, Heidelberg, pp 348–357
3. Lee HO, Sim H, Seo JH, Kim M (2010) Embedding algorithms for bubble-sort, macro-star, and transposition graphs. NPC 2010, LNCS, vol 6289. Springer, Heidelberg, pp 134–143
4. Saad Y, Schultz MH (1988) Topological properties of hypercubes. IEEE Trans Comput 37(7):867–872
5. Seitz CL (1985) The cosmic cube. Commun ACM 26:22–33
6. Leightoo FT (1992) Introduction to parallel algorithms and architectures: arrays, hypercubes. Morgan Kaufmann Publishers, San Francisco
7. Mendia VE, Sarkar D (1992) Optimal broadcasting on the star graph. IEEE Trans Parallel Distrib Syst 3(4):389–396
8. Feng T (1981) A survey of interconnection networks. IEEE computer 14:12–27
9. Bettayel S, Cong B, Girou M, Sudborough IH (1996) Embedding star networks into hypercubes. IEEE Trans Comput 45(2):186–194
10. Hedetniemi SM, Hedetniemi T, Liestman AL (1988) A survey of gossiping and broadcasting in communication networks. Networks 18:319–349
11. Hamdi M, Song SW (1997) Embedding hierarchical hypercube networks into the hypercube. IEEE Trans Parallel Distrib Syst 8(9):897–902

Obstacle Searching Method Using a Simultaneous Ultrasound Emission for Autonomous Wheelchairs

Byung-Seop Song and Chang-Geol Kim

Abstract A method to locate an obstacle and calculate the distance to it is proposed. The proposed method utilizes multiple ultrasound emitters that generate signals of identical frequencies and intensities. Corresponding sensors detect the reflected ultrasound signals, and the position of the obstacle is calculated based on the time of flight (TOF) of the ultrasound wave. This method is suitable for autonomous wheelchairs as it facilitates detection of the nearest obstacle, and yields more accurate estimation of the position.

1 Introduction

Assistive technology (AT), which aids the disabled and elderly people, is garnering worldwide attention, and latest IT and robotics technologies are being integrated with AT, resulting in novel devices [1–3]. One such life-changing device for the severely disabled people who cannot transport themselves is the autonomous wheelchair, the commercialization of which is still impending.

The overall purpose of an autonomous wheelchair is to transport the user to a destination safely and precisely. To operate the device without external aids, autonomous wheelchairs employ voice recognition, automatic control, radar, navigation, and robotics technologies. These technologies were used in several blind guide systems [3–9]. An essential capability that every autonomous wheelchair must have for user safety is obstacle detection and automatic avoidance. The obstacle detection and automatic avoidance technology, which is also used in autonomous mobile robots, emits ultrasound waves, locate obstacles based on the detected reflection, and adjust the path to avoid them.

B.-S. Song (✉) · C.-G. Kim
Department of Rehabilitation Science and Technology,
Daegu University, 201 Daegudae-ro Jillyang, Gyeongsan, Gyeongbuk 712-714, Korea
e-mail: bssong@daegu.ac.kr

J. J. (Jong Hyuk) Park et al. (eds.), *Multimedia and Ubiquitous Engineering*,
Lecture Notes in Electrical Engineering 240, DOI: 10.1007/978-94-007-6738-6_68,
© Springer Science+Business Media Dordrecht(Outside the USA) 2013

An autonomous wheelchair uses multiple ultrasound sensors. Each sensor sequentially emits and detects ultrasound waves. In the traditional method, an obstacle in the direction of each sensor is detected based on the time of flight (TOF) of the wave, and a map of surrounding obstacles is generated [10, 11]. However, this method has a drawback in calculating accurate distances when the signal is emitted from a moving wheelchair because the error in obstacle detection grows as the source moves, and each emitter sends signals at different positions and times. One of many efforts to resolve this problem is by the error eliminating rapid ultrasonic firing (EERUF) technique designed at the University of Michigan [12]. EERUF controls the sequence of ultrasound emission to effectively reduce the error. This method shows significant reduction in the error; however, EERUF is not the ultimate solution as the sources of error, i.e., the delay time, still exists.

This paper proposes a simultaneous ultrasound emission technique that utilizes "crosstalk" as additional information to counter this problem. Crosstalk is an interference signal at a sensor in a multi-sensor configuration, which occurs due to simultaneous signal waves being emitted by many emitters. The proposed method emits ultrasound signals of identical frequency and intensity from all emitters at the same time, to reduce the error due to time delay that a sequential emission technique inherently generates. Each sensor detects the very first reflected signal from an obstacle, and in this way, the distance to the closest obstacle can be calculated with significantly improved accuracy levels.

2 Method

The proposed method emits simultaneous ultrasound signals of the same frequency and intensity from multiple emitters, and in the process detects the nearest obstacle. Each ultrasound sensor has an effective emission angle of 60°, and the searching system collects information from multiple sensors for the front of the wheelchair. For instance, the 3-sensor configuration shown in Fig. 1a, which considers the wheelchair motion and the overlapping area, can cover an angle of about 120°. The distance from the sensor to the detected object is calculated using the TOF method. The duration between the emission of the signal and the detection of the reflection is used in the distance calculation. Assuming that all the sensors emit signals at the same time, the reflected signal will be detected first at the closest sensor as shown in Fig. 1b. For an object at a longer distance, each sensor may capture reflected waves from other sensors that have longer flying time.

The traditional method regards any reflected signal from other emitting sources detected at a sensor as crosstalk noise. However, the proposed method considers crosstalk as additional information.

In the proposed method, the distance between each sensor and the obstacle is calculated as follows:

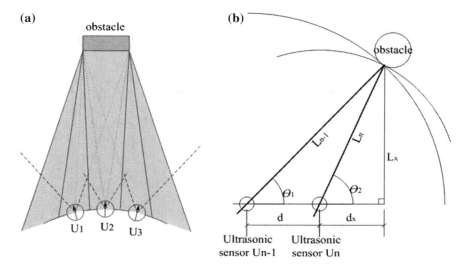

Fig. 1 Example of obstacle detection with three ultrasound sensors: **a** Simultaneous emission and reflection pattern, **b** Distance calculation model

$$L_n = c\frac{T_{n,n}}{2} \quad (1)$$

$$L_{n-1} = c\frac{T_{n,n-1} + (T_{n,n-1} - T_{n,n})}{2} \quad (2)$$

where, c is the speed of ultrasound wave, L_n is the distance between sensor 'n' and the obstacle, L_{n-1} is the distance between sensor 'n−1' and the obstacle, $T_{n,n}$ is the TOF for a wave emitted from sensor 'n' to return to sensor 'n' and $T_{n,n-1}$ is the TOF for a wave emitted from the sensor n to return to the sensor 'n−1'.

3 Experiments

The proposed method was validated and compared with the traditional method using sensors installed on a wheelchair. The scan rate of the sensors used in the experiment was 30 ms, and the center frequency of the emitted ultrasound signal was 40 kHz. Five sensors were placed in the front portion of the wheelchair, 15 cm apart from each other, and a 56 × 62 cm wooden panel was used as the obstacle.

The first measurements were taken using both the methods for a stationary wheelchair with the obstacle placed 2 and 3 m ahead. Next, the obstacle was placed 4 m ahead, and the distance was measured from the moving wheelchair as it approached 2 and 3 m distances from the obstacle at 90 and 120 cm/s. The test speed of the wheelchair was determined based on the speed during actual operation, which is around 1 m/s. Each test segment was repeated three times, and the average distance was considered. For the scan method, ultrasound signals were sequentially emitted from the U1 sensor on the left to U2–U5 at an interval of 30 ms, and the distance was calculated using the measured TOF at each corresponding sensor. For the simultaneous emission method, distances from each sensor to the obstacle were calculated using the measured TOF of simultaneously emitted signals from each sensor. An 8-bit, ATmega 128 microprocessor was used to control the sensors, and to calculate the distances using the measured times. The microprocessor calculated the distances in real time and saved the data on a computer. Figure 2 shows the percentage error of the calculated distances to the measured values.

4 Discussion and Conclusion

As shown in Fig. 2, both methods yielded errors within 1–2 % with no significant difference for the stationary case, except for the 5th sensor that resulted in a larger error for the scan method. The moving wheelchair cases yielded larger errors with no noticeable correlation with the speed.

However, the errors still stayed within 2 % for the simultaneous emission method, while the scan method yielded generally larger errors, with some sensors showing much larger error jumps. The error jumps were significant on U1 and U5, and Fig. 2b shows the jump on U4 as well. It is inferred that the cause for the larger errors is the inherent time delay in the scan method. In general, the proposed method shows better results in terms of the error in distance calculation than the traditional scan method.

If the distances from three or more sensors to the obstacle are known, the obstacle can even be located in 3-D space. Assuming that each sensor is at the center of a sphere of which the radius is the distance between the obstacle and sensor, the intersection of the spheres that the sensors generate denotes the location of the obstacle. Therefore, with three sensors solving three sphere equations yields the location of the obstacle, which is an efficient method for an autonomous wheelchair. Since accurate estimation of distances is crucial in 3-D calculations, the proposed method is suitable for application to 3-D mapping of obstacles.

However, the proposed method is not superior to the scan method in every aspect. The proposed method detects only the closest object, while the scan method can detect multiple objects at the same time. Searching for multiple obstacles is beneficial because a wheelchair encounters many in a practical environment. On the other hand, from a user's perspective, information about the closest object is most important. Therefore, it is critical for the safe and convenient operation of autonomous wheelchairs to detect precisely and avoid the closest object.

Fig. 2 Percentage error of calculated distances to measured values for various wheelchair speeds:
a Stationary, **b** 60 cm/s, **c** 90 cm/s, **d** 120 cm/s

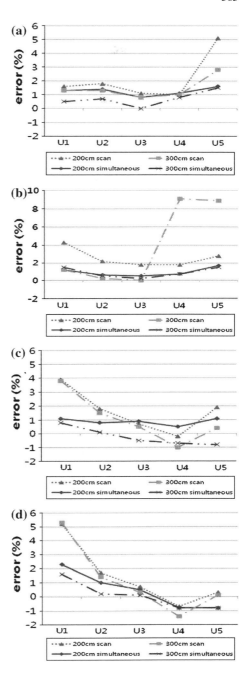

In summary, the proposed obstacle detection method based on a simultaneous emission strategy exhibits excellent performances on autonomous wheelchairs.

Acknowledgments This work was supported by the National Research Foundation of Korea Grant funded by the Korean Government (NRF-2010-013-D00091).

References

1. Brabyn J, Gerrey W, Fowle T, Aiden A, Williams J (1989) Some practical vocational aids for the blind. In: Proceedings of 11th annual international conference of IEEE engineering in medicine & biology society, pp 1502–1503
2. de Acevedo RLM (1999) Electronic device for the blind. In: IEEE AES systems magazine, pp 4–7
3. Tachi S, Komorya K, Tanie K, Ohno T, Abe M (1981) Guide dog robot-Feasibility experiments with Meldog Mark III. In: Proceedings of 11th international symposium on industrial robots, pp 95–102
4. Tachi S, Tanie K, Komoriya K, Abe M (1985) Electrocutaneous communication in a guide dog robot (MELDOG). IEEE Trans Biomed Eng 32(7):461–469
5. Kim CG, Lee HG, Kang JH, Song BS (2007) Research of wearable walking assistive device for the blind. Korean J Vis Impair 23(1):147–164
6. Borenstein J, Ulrich I (1997) The GuideCane-a computerized travel aid for the active guidance of blind pedestrians. In: Proceedings of the IEEE conference on robotics & automation, New Mexico, pp 20–25
7. Shoval S, Ulrich I, Borenstein J (1998) The Navbelt-A computerized travel aid for the blind based on mobile robotics technology. IEEE Trans Biomed Eng 45(11):1376–1386
8. Kang JH, Kim CG, Lee SH, Song BS (2007) Development of walking assistance robot for the blind. J Korean Sens Soc 16(4):286–293
9. Rentschler AJ, Cooper RA, Blasch B, Boninger ML (2003) Intelligent walkers for the elderly: Performance and safety testing of VA-PAMAID robotic walker. J Rehabil Res Dev 40(5):423–432
10. Shoval S, Ulrich I, Borenstein J (2003) Robotics-based obstacle-avoidance systems for the blind and visually impaired, NavBelt and the GuideCane. IEEE Robotics Autom Mag 10:9–20
11. Moon CS, Do YT (2005) Design of range measurement systems using a sonar and a camera. J Korean Sens Soc 14(2):116–124
12. Borenstein J, Koren Y (1995) Error eliminating rapid ultrasonic firing for mobile robot obstacle avoidance. IEEE Trans Robotics Autom 11(1):132–138

Part X
Future Technology and its Application

A Study on Smart Traffic Analysis and Smart Device Speed Measurement Platform

Haejong Joo, Bonghwa Hong and Sangsoo Kim

Abstract In recent years, with a fast spread of smart phones, the number of users is rapidly increasing, causing the saturation of various kinds of traffic in mobile networks and access networks. This implies that although not only conventional Internet traffic but also many smart phone applications create much traffic, the development of network analysis technologies has yet to be furthered. As a method to analyze and monitor the traffic types of smart devices, the method of using signature and classifying traffic is used, and the speed measurement agent is used to measure the Internet speeds in smart devices. This paper aims to classify the smart device traffic and Internet traffic, analyze the traffic use amount, measure Internet speeds in smart phones amid traffic being created, and thus propose a platform designed to measure user service quality.

Keywords Smart device traffic analysis · Smart device speed measurement · Traffic and speed monitoring

---

H. Joo
Department of LINC, Dongguk University, #710, 82-1 Pil-dong 2-ga,
Jung-gu, Seoul 100-272, Korea
e-mail: hjjoo@dongguk.edu

B. Hong (✉)
Department of Information and Telecommunication, Kyung Hee Cyber University,
1 Hoegi-Dong, dongdaemun-Gu, Seoul 130-701, Korea
e-mail: bhhong@khcu.ac.kr

S. Kim
Contents Vision Corp, #613, 82-1 Pil-dong 2-ga, Jung-gu, Seoul 100-272, Korea
e-mail: cqsky@paran.com

J. J. (Jong Hyuk) Park et al. (eds.), *Multimedia and Ubiquitous Engineering*,
Lecture Notes in Electrical Engineering 240, DOI: 10.1007/978-94-007-6738-6_69,
© Springer Science+Business Media Dordrecht(Outside the USA) 2013

1 Introduction

In recent years, with the wireless Internet rapidly growing, there is a growing access to broadband services through new devices such as smartphones, netbooks, and mobile Internet devices (MIDs). Online video service traffic is increasingly spreading to mobile services through these devices, significantly burdening the network [1].

Studies on mobile traffic reveal that mobile traffic is characterized by a high ratio of HTTP traffic and many applications with the client–server structure [2]. To classify these mobile applications, information on each mobile application, like with conventional traffic classification, needs to be gathered and analyzed. However, mobile devices, unlike general PCs, are very slow in handling speeds, and have limited memory. Thus, this paper proposes a configuration designed to classify and analyze mobile traffic in the general PC environment, as well as a platform designed to measure the traffic handling speed in smart devices through speed measurement agents.

This study proposes a platform by which a monitoring device is installed in the access network to monitor the traffic of both smart devices and Internet so as to classify traffic and to measure the speed of the WiFi Internet services in smart devices.

2 The Proposed Structure of Traffic Analysis and Classification

This Chapter explains the proposed traffic analysis structure. Section 2.1 describes the network traffic gathering structure, and Sect. 2.2 explains the traffic analysis structure which uses the open source snort.

2.1 Gathering of Traffic Data

The method of extracting flow data according to the environment of measurement points is outlined as follows [3, 4].

- Agent method: the method designed to set a flow monitoring function in the data gathering device to gather flow data.
- Probe method: the method designed to use the tapping or monitoring method and extract the flow information instead of extracting flow information directly from the gathering device.

In order to gather raw data flowing in the network and analyze them, this study uses the probe method and gathers traffic data (Fig. 1).

A Study on Smart Traffic Analysis and Smart Device Speed Measurement Platform 571

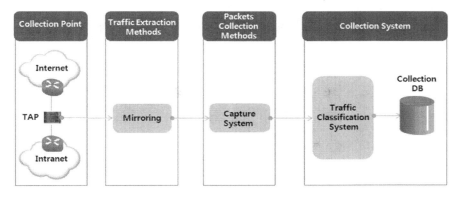

Fig. 1 Traffic data collection

2.2 Classification and Analysis of Smart Traffic

This study uses packet sniffer and packet logger among snort functions and classifies raw data gathered from the gathering system [5, 6].

The following explains the general functions of snort.

- Packet sniffer: the function designed to read and show packets of the network
- Packet logger: the function designed to store monitored packets and leave them in the logger
- Network IDS: the function to analyze network traffic and explore attacks

This study uses the packet sniffer function, transforms network traffic (TCP, UDP, ICMP IP, etc.,) into easily analyzable patterns, and uses them in analyzing network properties (Fig. 2).

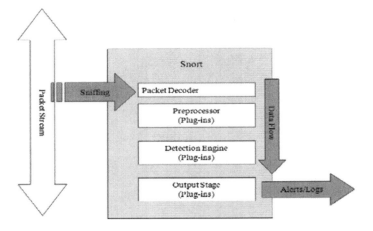

Fig. 2 Structure of SNOT

Table 1 Classification of traffic using the snort signature

Rule header							Rule option
Action	Protocol	IP address	Port	->	IP address	Port	(Option)
1	2	3	4	5	6	7	8

2.3 Classification of Traffic Using the Snort Signature

Data, processed through snort packet decoder, are classified through the preprocessor, and this study uses the snort signature's headers and rule options and classifies data [3]. The gathered data can be classified through two methods, namely, the method of classifying packets on the payload, and the statistical method. In order to analyze the content of data, this study uses the method of classifying packets on the payload (Table 1).

3 Structure of Measuring Speeds in Smart Devices

This study explains the speed measurement structure in smart devices. However, since 3G and 4G networks are separated from wire data networks, only Wi-Fi service speeds are measured (Fig. 3).

The Wi-Fi quality indices are measured using the indices provided by the National Information Society Agency.

In order to measure the speeds in I-Phone, this study uses the CFNetwork and DSC socket standard SDK (Fig. 4).

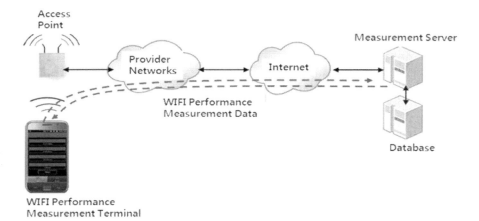

Fig. 3 The Wi-Fi speed measurement environment in smart devices

Fig. 4 API structure

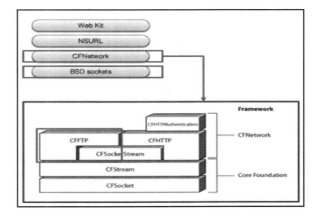

4 Structure of Traffic Analysis and Speed Measurement

Figure 5 shows the structure of the entire system for the traffic analysis and speed measurements in smart devices.

- Quality measurement server: Use the server, and measure downloading, uploading and delays in order to measure the quality linked to smart devices.
- TAP: Device designed to gather data flowing between the school network and the Internet.
- Monitoring system: The system designed to analyze data flowing in through TAP and to monitor the performance of network.

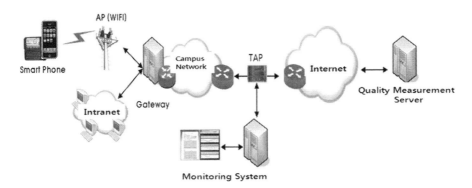

Fig. 5 The total system structure

5 Conclusion

With many smart device applications currently being served, diverse traffic patterns are created. Thus, it is increasingly important to analyze mobile traffic in the Internet environment.

To monitor the mobile and wire traffic, this study proposes a method designed to use the snort signature, classify and monitor data, as well as an environment in which to measure the performance of smart devices in the monitoring environment. When huge data flow into the network from smart devices, the method can analyze how the performance in smartphones will change. The proposed method is expected to apply to mobile networks (3G, 4G).

Acknowledgments This research was supported by the Kyung Hee Cyber University a sabbatical Year in 2013.

References

1. National IT Industry Promotion Agency (2011) Weekly Technical Trends, July 15
2. Maier G, Schneider F, Feldmann A (2010) A first look at mobile hand-held device traffic. Passive and active measurement. Zurich, Switzerland, pp 161–170, Apr 7–9
3. WAN and Application Optimization Guide, chapter 5 'Traffic classification'
4. National IT Industry Promotion Agency (2012) Weekly Technical Trends, April 11
5. ITU-T Recommendation I.350 (1993) General aspects of quality of service and network performance in digital networks, including ISDNs, Mar 1993
6. ITU-T Recommendation Y.1541 (2003) Network performance objectives for IP-based services, Feb 2003
7. ITU-T Recommendation Y.1543 (2007) Measurements in IP networks for inter-domain performance assessment, Nov 2007

Analysis and Study on RFID Tag Failure Phenomenon

Seongsoo Cho, Son Kwang Chul, Jong-Hyun Park and Bonghwa Hong

Abstract RFID tag failure analysis test involves analysis of general devices and interpretation of analysis results, which suggests failure mechanism. Hence, it can only be performed by failure analysis experts with rich expertise and experience. Interpretation of causal relationship between a failure phenomenon and explainable causes of the failure has to rely on failure analyzers' knowledge and experience. Analyzers have the capability to figure out the precise failure mechanism because they know which type of a failure is associated with which failure mechanism. Failure mechanism by major failure type frequently observed in RFID tag and analysis method should be fully recognized by analyzers.

KeyWords RFID · Failure site · ESC · Solvent crack

1 Introduction

RFID (Radio Frequency Identification) is a non-contact identification system in which a small chip is attached on an object to transmit and process the object's

---

S. Cho (✉) · S. K. Chul · J.-H. Park
Department of Electronic Engineering, Kwangwoon University,
20 Kwangwoon–ro, Nowon-gu, Seoul 139-701, Korea
e-mail: css@kw.ac.kr

S. K. Chul
e-mail: kcson@kw.ac.kr

J.-H. Park
e-mail: world78u@hanmail.net

B. Hong
Department of Information Communication, Kyunghee Cyber University,
Dongdaemun-gu, Seoul 130-701, Korea
e-mail: bhhong@khcu.ac.kr

J. J. (Jong Hyuk) Park et al. (eds.), *Multimedia and Ubiquitous Engineering*,
Lecture Notes in Electrical Engineering 240, DOI: 10.1007/978-94-007-6738-6_70,
© Springer Science+Business Media Dordrecht(Outside the USA) 2013

information wirelessly. Recognized as the most dramatic data identification technology, RFID is rapidly advancing and taking a firm position in the market driven by advances in semiconductor and wireless communication as well as global standardization. The minute semiconductor chip embedded in RFID helps transmit information about an object and the surrounding environment via radio-frequency. It consists of a tag, reader and antenna, etc. RFID tag can be automatically identified and tracked anywhere, anytime and the embedded memory enables information to be updated and revised. Furthermore, unlike a bar code limited by environmental factors like rain, snow, fog and pollution and a smartcard only identified from a short distance, RFID tag can be identified even in the most constrained environment. It can even identify moving objects [1–4].

RFID tag's failure analysis, which is a specialized area with long history and evolved with advances in science and technology, has apparently made a big contribution to delivering mutually beneficial output in both physics-of-failure and reliability tests. As one of the key factors for evaluating a product's improvement process and reliability, it focuses on finding the root cause of issues regarding quality or reliability of products or parts. The ability to secure stable performance of new parts and materials is critical since it is directly related to reliability of parts and materials. But, loss cost can be kept to minimum just by forecasting the most vulnerable areas and type of failures for further improvement [5–9].

Failure analysis based on physics-of-failure adopted a comprehensive approach to figure out root causes of a failure. Failure mode and failure site can be identified through data analysis gained from performance evaluation, nondestructive analysis and destructive physical analysis technology, based on which failure mechanism, which is the mechanical, electrical and physical process causing a failure, can be revealed. The guidelines in this study proposed failure analysis methods for electric and electronic parts based on physics-of-failure.

2 Failure Analysis

2.1 Basic Technology of Failure Analysis

Failure analysis technology is still far from complete. Rather, development is still ongoing in all areas and electronic parts, in particular. In advanced technology, ultramicro analyzer, atom-level analyzer and device for property of a matter and failure analysis (computational analysis) is necessary. Worse, there are many areas that do not render analysis and it even tends to fall behind the speed of advance in electronic parts. Basic elements of failure analysis are observation, opening and analysis. Observation may be taken for granted but it is actually the start of a failure analysis since it takes a special perspective. Opening is to observe the interior structure of electric and electronic parts. To date, there is no technical publication on opening as opposed to heaps of literature and studies that can be

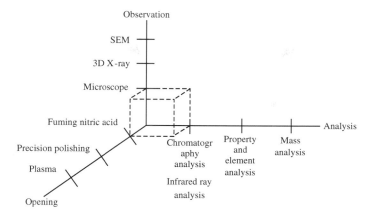

Fig. 1 Three elements of failure analysis

used as a reference for product manufacturing. Being such, the whole process of disassembling a part one by one without destructing for observation is still quite a challenge. Furthermore, even the most advanced technology to date cannot open all parts. Analysis is a technology required to check chemical change but there is no way to analyze a minute area and its chemical composition with a simple device. For now, an expensive device analyzer is the only available means for analysis. The basic elements are shown in Fig. 1 and the area surrounded by the dotted line is the only scope in which failure analysis can be carried out with a simple tool.

Failure analysis starts off with exterior observation of a specimen, property measurement, and observation of the interior structure on a sequential basis and finally discovers a failed site or the cause of a failure. Failure analysis technology refers to observation; opening (disassembly) to observe the structure interior; and finally analysis to investigate the cause of a failure.

2.2 Cause of a Failure

In general, failures don't occur by accident. Rather, major factors of a failure include stress, material, structure and geometry. In other words, a failure can occur by many causes, which can be tracked. A failure can be classified into initial failure, accidental failure and abrasion failure depending on the period failure occurs during product operation. Initial failure is mainly caused by a manufacturer, which can be prevented with quality control and screening. Accidental failure is mainly caused by user's misuse or poor design. Abrasion failure is mainly caused by poor reliability or durability that was not properly considered by the designer.

2.3 Data Analysis

Data analysis is a process to analyze data collected using an integrated technology and put them into a system. Cause and Effect Diagram, Ishikawa Diagram or Fishbone Analysis specifically provides conditions to find the root cause of a failure or all the possible causes related to the failure. Failure mode and effect analysis is a qualitative reliability forecast and failure analysis method surveying impact of failure mode of parts implementing a system on other parts, system and users with a bottom-up approach. It is a tool used in the design and process stage. Failure tree analysis uses logic symbols (AND or OR) on causal relationship between a system's failure phenomenon and its cause to draw a failure tree shaped in tree branches. It is a quantitative failure analysis and reliability evaluation method adopted to improve system's reliability by calculating system's failure probability. Pareto diagram indicates issues that potentially require the most improvement by showing relative frequency or critical [10, 11].

2.4 Procedure of Failure Analysis

Order of failure analysis is extremely critical. Improper order of failure analysis could ruin analysis just as a product manufacturing requires a certain manufacturing order. Unlike products, which can be remanufactured, failure analysis cannot be conducted again, in most cases, if the analysis is not done in the right order because the work designed to identify failure mechanism includes the nonreciprocal destructive element. Order of failure analysis and its details differ from one electric and electronic part to the other but the general procedure for failure analysis is shown in Fig. 2.

3 Analysis of Environmental Stress Cracking

Environmental stress cracking (ESC) refers to when a stress (environmental stress) works on a solvent crack phenomenon at the same time to cause uniform destruction in a short period of time. Stress is divided into mechanical stress, which is external, and residual stress that exists inside the high molecules during

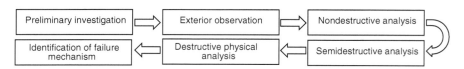

Fig. 2 Procedure of failure analysis

forming processing, which is internal. Residual stress that occurs in the interior during formation exists in high-molecular materials and these high molecules in which residual stress exists can face destruction from even the slightest external stress when they contact chemical substances like solvent, release agent, surfactant and machine oil.

The general difference of crack shape of solvent crack and ESC lies in the shape of failure surface and high-molecular materials even though distinguishing one from the other is not straightforward. Shape of a failure surface is a solvent crack (a smooth failure surface on a mirror) and ESC (brittle fracture or mixture of brittle fracture and failure surface on a mirror). High-molecular material is solvent crack (on non-crystalline high molecules) and ESC (on non-crystalline high molecules with relatively higher crystallizability than solvent crack).

3.1 Failure Phenomenon

ESC incurs a crack failure mixed with solvent crack as environmental stress simultaneously acts on the solvent crack.

Solvent crack refers to a crack caused by contact with plastics and chemicals (oil, detergents and blooming of other high-molecular additives) under stress situation (residual stress during forming processing, etc.) and it is largely observed in non-crystalline plastic (PC, PMMA, ABS, etc.).

Craze phenomenon occurs in high molecules when a certain level of critical stress acts and molecules locally take an orientation (molecules are locally elongated and arranged towards the same direction). A micro destruction occurs in the crazed area, which later cracks. High-molecular products usually have residual stress resulting from forming processing even when there is no external stress. The mechanism working on solvent crack as shown in Fig. 3 occurs when chemical agents infiltrate into high molecules caused by weakening of van der Waals force between molecules of high-molecular substances due to environmental stress. The solvent crack shows a mirror-face shape (smooth gloss).

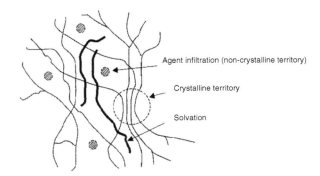

Fig. 3 Type illustration of molecular structure after solvent crack

Fig. 4 Plane shape of brittleness fracture

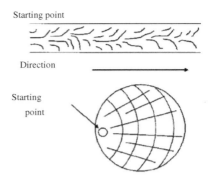

When solvent contacts with high molecule surroundings, it creates a strong bond with high-molecular substances, which separates mutual cohesion of high-molecular substances and causes solvent to infiltrate in between high-molecular substances and swell. In extreme cases, it could dissolve the substances. Naturally, the degree of solvent crack differs depending on whether chemical substances just infiltrated into high-molecular substances or actually caused dissolution. ESC behavior of high molecules varies by exposure time, exposure temperature, and density of stress destructive materials and stress load degree of high molecules.

3.2 Analysis Method

Failure analysis caused by ESC took a relatively simple process that involved analysis of the quality of material, observation of fracture plane, analysis of fracture plane's surface component and reproducibility test. Fractured plane is observed with SEM or stereoscopic microscope and interpretation of fracture plane shape can help analyze the direct cause of fracture. Brittle fracture is caused by shock fracture as shown in Fig. 4. In Chevron Pattern shaped like a chevron, the starting point is in chevron peak and crack occurs in the opposite direction to the peak as indicated in $< \rightarrow < \rightarrow < \rightarrow$.

Striation shaped like a shell has a unique stripe otherwise called beach mark. It shows a failure surface. Striation, which is produced by repeated stress, occurs when crack develops and recedes on a regular basis. Small solvent crack produced from a contact surface or a surface infiltrated with a solvent usually occurs vertically to the direction where a mechanical load works. The crack surface is mostly smooth like a mirror shown in Fig. 5 and is thus dubbed as mirror face.

Analysis and Study on RFID Tag Failure Phenomenon

Fig. 5 Solvent crack and ESC section

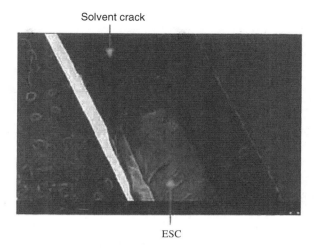

Fig. 6 Exterior observation of clothing tag **a** front side of clothing tag **b** back side of clothing tag

(**a**) Front side of clothing tag (**b**) Back side of clothing tag

4 Case of RFID Tag Failure

4.1 Exterior Observation

When a product code is printed on RFID tag chip, fault occurs due to chip's volume and foreign substances appear on the back of where chip is attached (Fig. 6).

4.2 Microscopic Observation

Figures 7 and 8, a detailed observation of chip condition via optic microscope.

As presented in Fig. 7, it was hard to observe chip due to adhesive used to apply ACP and attach a tag. Crack in (a) poor specimen in Fig. 8 could be observed with optic microscope but crack could not be observed with optic microscope after removing ACP and adhesive in (b) normal specimen in Fig. 8.

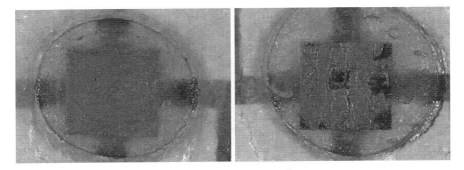

Fig. 7 RFID chip observation via optic microscope

Fig. 8 Chip surface ACP and removal of adhesive **a** poor specimen **b** normal specimen

5 Conclusion

The cause of RFID failure and development of the failure that prompted the cause of the failure to turn into failure phenomenon can be explained as failure mechanism. Analysis tests of failure analysis can be performed by anyone who can analyze general equipment and has good understanding of analytical chemistry. However, interpretation of analysis results to suggest mechanism of a failure is a job that can only be carried out by failure analysis experts built with expertise and experience. Interpretation on the causal relationship between a failure phenomenon and explainable failure causes has to rely on knowledge and experience of failure analyzers and thus a rich understanding on numerous failure mechanisms by each type of failure is required to come up with the precise failure mechanism. In this sense, failure mechanism by major type of failure frequently observed in RFID tag and ways of analysis should be properly understood and ways to combine and summarize results gained from procedures of failure analysis should hopefully put to good use in industries.

Series of test results associated with failure analysis are limited in suggesting improvement measures capable of suppressing or preventing failures. Furthermore, there is potential of proposing measures that have serious error since the proposed improvement measures are based on inferences not demonstrated with tests. Even so, failure analysis is meaningful since it deals with improvement measures aimed at suppressing or preventing a failure to identify causes of a failure and their development into a failure phenomenon. Hence, it plays an effective role in developing alternatives to suppressing or preventing failures.

References

1. Subramanian V, Chang PC, Huang D, Lee JB, Molesa SE, Redinger DR, Volkman SK (2006) All-printed RFID tags: materials, devices, and circuit implications. VLSI Design
2. Leachman RC, Hodges DA (1996) Berkeley semiconductor manufacturing. IEEE Trans Semicond Manuf
3. Bloomsburg J (2002) RFID tag manufacturing. MIT UROP
4. RFID Guardian (2005) A battery-powered mobile device for RFID privacy management. In: Proceedings of the Australasian conference information security and privacy. http://www.cs.vu.nl/~melanie/rfid_guardian/papers/acisp.05.pdf
5. Wagner U, Franz J, Schweiker M (2001) Mechanical reliability of MEMS-structures under shock load. Microelectron Reliab 41(9–10):1657–1662
6. McCluskey P (2002) Design for reliability of microelectro-mechanical systems. In: Proceedings of the electronic components and technology conference, pp 760–762
7. Müller FR, Wagner U, Bernhard W (2002) Reliability of MEMS- a methodical approach. Microelectron Reliab 42(9–11):1771–1776
8. Anderson TL (1995) Fracture mechanics: fundamentals and applications 2nd edn. CRC Press, Boca Raton
9. Finn J (1995) Electronic component reliability. Wiley, USA, pp 21–153
10. Yiping W (2004) RFID tag manufacturing. Glob SMT Packag 4(6):4–7
11. An B, Cai XH, Chu HB (2007) Flex reliability of RFID inlays assembled by anisotropic conductive adhesive. In: Proceedings of the 9th International IEEE CPMT Symposium on high density design, packaging and microsystem integration, pp 60–63

Administration Management System Design for Smart Phone Applications in use of QR Code

So-Min Won, Mi-Hye Kim and Jin-Mook Kim

Abstract With the development of information and communication technology, the social infrastructure is secured in order to pass and use anytime, anywhere the information we want. Smart phone application occupies an important place in Ubiquitous Environments. More efficient and sensible way are needed across the medical field taking and getting the prescribed medications according to the disease. This paper present the smart phone application based medication management system using the QR code. We can prevent the duplication prescription and reduce the side effects such as exasperating the disease from cases of overdose or under-taking due to the failure of recognizing the times or time of taking the medication or augmenting immunity to the drug schedule through this model. Also, More efficient and sensible research are needed such as connection problems between the most basic home diagnostic equipment and smart phone application, the algorithm about avoiding drug duplication, time management algorithm of taking the medication and medication delivery algorithm base on this model.

Keywords Medication management system · Mobile · Smartphone application · QR code

---

S.-M. Won (✉) · M.-H. Kim
Department of Computer Science, Chungbuk National University, Cheongju-si, South Korea
e-mail: wsm012@nate.com

M.-H. Kim
e-mail: mhkim@cbnu.ac.kr

J.-M. Kim
Division of Information Technology Education, Sunmoon University,
Cheonan, South Korea
e-mail: calf0425@sunmoon.ac.kr

J. J. (Jong Hyuk) Park et al. (eds.), *Multimedia and Ubiquitous Engineering*,
Lecture Notes in Electrical Engineering 240, DOI: 10.1007/978-94-007-6738-6_71,
© Springer Science+Business Media Dordrecht(Outside the USA) 2013

1 Introduction

Together with the development of information and communications, the social infrastructure ensuring provision, receive and confirmation of information without time and locational limitations has been secured. Under such circumstances, smart phone applications have taken significantly important positions in the modern society. Among the medical business, a more efficient and rational measure is required in the overall process of drug prescription and administration in line with diseases.

This study proposes a model for drug administration management system for smart phone applications. This model is expected to prevent duplications in the drug prescription process and to reduces cases of worsening diseases due to overdose or under-dose from recognition failure for drug dose number of times or time or enhancing immunity to certain drugs.

In addition, a study is required based on this model regarding the relation issue between most basic domestic diagnosis apparatus and smart phone applications, algorithm for drug redundancy prevention, time management algorithm for drug dose and concrete and efficient algorithm for delivery system after taking prescription.

Together with the development of information and communications, the social infrastructure ensuring provision, receive and confirmation of information without time and locational limitations has been secured. Under such circumstances, smart phone applications have taken significantly important positions in the modern society. Among the medical areas, a more efficient and rational measure is required in the overall process of drug prescription and administration for diseases.

In a case where a patent gets medical services from different private clinics for several diseases, he or she might be prescribed for an identical drug by each clinic and then, redundant dose might be caused for the patient due to drug prescription redundancy. In the process of prescription drug dose, it is practically difficult to make drug dose in a punctual manner and there are many cases it is difficult to make correct judgment on whether dose times are well abided by. Such cases are unanimous regardless age and gender. Therefore various side-effects might be caused in disease treatment attributable to drug overdose from redundant dose caused by incorrect recognition drug dose time and number of times; drug dose deficiency from drug dose omission cause by recognition failure of drug dose and excessive drug dose caused by repetitive dose.

Therefore this study proposes a model for smart phone applications based on the drug administration management system using QR code which is a new method for prevention of disease worsening while providing proactive contributions to treatment of patients through effective administration management. With application of ubiquitous concept, this model enables patents to get administration information and treatment whenever and wherever; to receive individual prescription through their smart phone by using QR code; and execute administration management based on transferred data.

2 Relevant Researches

Largely attributable to recent development of information and communication technologies, Korea has witnessed active researches for technology convergence in each professional area with digital devices based on quality high-speed communication infrastructure better than those in other countries. The medical industry is yearning for Ubiquitous Healthcare (hereinafter referred to as U-Health Care) for monitoring individual's health status 'wherever and whenever' and providing customized health management service by utilizing development of digital device and wireless communication technologies and various kinds of bio signal measuring sensors which are small and portable. In addition, various solutions have been provided for automatic medical services. The representative example is a method using RFID with a purpose to prevent relapse of disease and following re-hospitalization with management of defined drug therapy. Researches are conducted to prevent side effects from redundant drug doses by patients attributable for redundant prescription in the process of administration prescription.

U-Health Care enables individuals not to recognize medical services since U-Health Care Service by itself makes real-time monitoring on individual's health status and automatically takes actions at a time requiring treatment or management. Therefore individuals are able to sustain best health conditions and get convenient and precise medical service without time and location restriction when it is required. Moreover for well-being which is considered most interested item of people; and for preparation against upcoming 'aged society', conventional level of medical service is not appropriate and a more sophisticated medical service focusing on management and prevention is in a desperate need.

U-Health Care refers to health management and medical service provided by collection, process, delivery, and management of health related information without time and location limits which is ensured by application of applies information and communication technologies (ubiquitous computing) to the public health and medical industry. Ubiquitous computing refers to a condition realizing computing whenever and wherever with various computers coexisting with human beings, physical objects and environment and linked each other.

In other words, it might be considered as a comprehensive medical service ranging from remote management of patient's disease to daily health management as it is ensured by application of information and communication technologies such as wired and wireless networking to the medical industry for convenience and efficiency of medical service usage [1].

U-Health Service is a business providing services such as self-diagnosis, remote monitoring and management of medical professionals based on real time transfer of health index of patients such as blood pressure, blood sugar, pulse, and body fat to medical institutions in use of wired or wireless health measurement devices linked by network of cell phones, PDA, and the Internet [2].

Studies conducted to prevent potential redundant description which might be occurred in the process of drug prescription in hospitals are as follows;

In a case where a patient gets drug prescription in more than two hospitals, redundant prescription gets highly likely with increased dose opportunities of various drugs so that the measures to secure medical safety against side effects of drugs or overdose. Medical institutions have recently tried measures to enhance quality of drug treatment with systematic supports to prevent and pre-check potential errors which might be caused in the process of drug prescription of doctors such as drug overdose, drug allergy, drug prohibited during pregnancy and breast feeding, drug side effects, drug interactions, prohibited drugs by age group and dose control for kidney patients in a way of application of Clinical Decision Support System (CDSS) to Computerized Physician Order Entry (CPOE) [3].

The representative example of provision of automatic medical services is an administration management in use of RFID. The introduction of automatic solution using RFID embedded cell phones, tag and web based servers. The target of this solution is to reduce the number of heart patients who are re-hospitalized from failure to follow defined drug therapy. To discharged patients, eMedonline cell phone attached with RFID reader is provided. The cell phone has eMedonline software application and is linked with web based server. In addition, the patient gets RFID label with names of prescribed drugs. At a time of drug dose, the patient gets information on which drug should be taken vis eMedonline cell phone. It is followed by questions on dose time, change prescription and his or her health status and those questions can be responded through the touch screen.

When a patient approaches prescribed drug bottle with a tag to RFID reader of a cell-phone in couple of inches, the reader reads ID No. of the tag and transfer to the web-based sever. EMedonline software checks whether this ID No. matches with the drug in which the patient should take and displays the photo of relevant pills on the phone to help the patient identify the correct drug while it also saves information on the level of drug therapy obeisance of the patient, time, health status and behaviors. When a patient fails to take the prescribed drug, relevant message is transferred to a hospital and the hospital is able to notify such failure to a guardian of the patient.

3 Proposed Model

This study proposes an administration management system which is able to provide more effective medical services through convergence of various communication devices and technologies for medical services based on high-speed communication network infrastructure.

The system in Fig. 1 calls data of patients saved in DB pool of a hospital server and transfers it to a pharmacist to help drug prescription and supports patients to get messages on alarms and information in line with the cycle of prescribed drugs via smart phone. In addition, patients are able to take drugs of their prescription at home rather than taking it in pharmacy. By doing so, inconvenience and time for sick patients to get to pharmacy to take drugs can be minimized. A patient who

Administration Management System Design for Smart Phone Applications 589

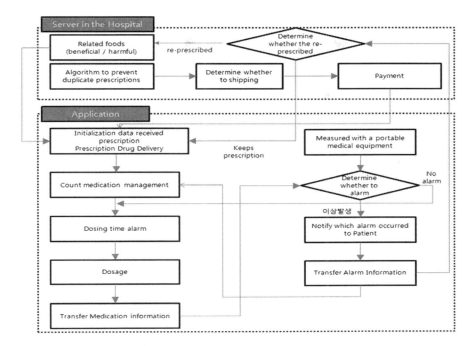

Fig. 1 Flow of administration management system

received prescription, food-related information, and cautions related to prescribe drugs also are entitled to measuring device information, prescription information, administration alarm and relevant food function services provided by smart phone applications.

Each function provided by smart phone applications is structured as data collection stage and analysis stage as in Fig. 2.

In the data collection stage, check can be made for prescription information of patients transferred from hospital server, relevant food information which should be referred to depending on disease severity of each patient, drug administration timing and contents in accordance with prescription, basic measuring information on portable medical devices like blood sugar device and link status of medical devices.

The data analysis stage can identify dose time, remaining times, omission of administration and abnormality occurrence with administration record as an information generated and provided based on information collected in the data collection stage.

The system is categorized into hospital server and smart phone applications and functions of each process in the hospital server are as follows;

Data Collection and Calibration carries out a function collecting administration timing and relevant information of patients from smart phone applications of patients. As it is the case for Data Collection, Alarm Collection and Notification

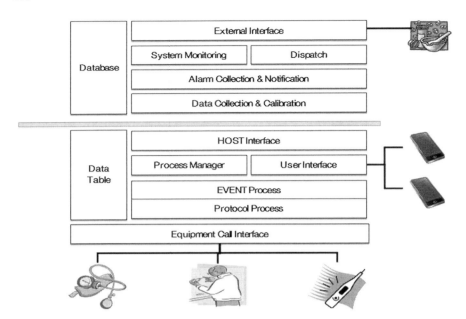

Fig. 2 System architecture

collect data from smart phone applications but for abnormality occurrence related to administration and alarm information. System Monitoring executes functions of activating whole process, operation supervising and monitoring. Dispatch check redundant prescription for prescriptions issued by doctors, and generates data on which delivery information can be transferred to pharmacists collected via smart phone applications of patients. External Interface executes a function of transferring data generated by Dispatch to pharmacists.

Next, the process diagram for smart phone is as follows;

Equipment Call Interface carries out a role of data transfer and receive between portable medical devices like blood sugar device used by patients and smart phones. Protocol Process as a sub-element of Equipment Call Interface plays a role of Protocol converter which is able to deal with individual transfer methods of each portable medical device. Event Process check abnormality based on patients data collected by Equipment Call Interface and transfer the information to hospital server though HOST Interface which is for data transfer and receive with hospital server. Process Manager activates whole processes of smart phone and executes monitoring on operation supervise. User Interface functions data illustration through which patients are able to search and check relevant data and alarm on screen.

Figure 3 is about functions in smart phone applications. It is able to save data in smart phone applications which is record by linking measuring devices for blood pressure, body temperature and blood sugar with smart phone; to analyze such data, notify hospital server when abnormality is occurred and to request re-

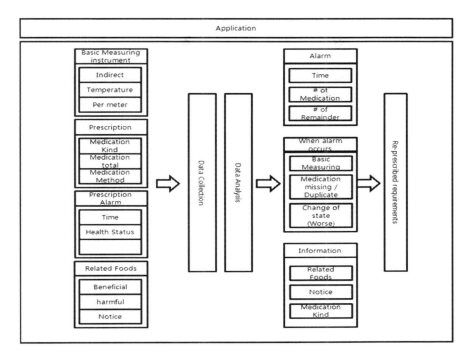

Fig. 3 Functions in smart phone application

prescriptions. Regarding to prescription, the information of administration types can be transformed, times and method of administration might be delivered and it is possible for patients to check information in a convenient manner via smart phones whenever he or she desires to do. The administration alarm function notifies patients on administration schedule and times and gives warning depending on critical mass for certain timing and times upon failure to abide by administration schedule and times while the critical mass is broken, it notifies the abnormality occurrence to hospital server and gets information related to re-prescription. In addition, It gets information on relevant foods, cautions during drug dose and foods avoided and based on such information, gives more effective supports for the recovery of the patient. Moreover, it is possible for a patient to request re-prescription. In such case, the patient feels worsening status rather than improving.

The comprehensive functions of smart phone applications are to collect and analyze information on basic measuring device, prescription, administration alarm function and relevant foods; to request re-prescription to hospital server when abnormality occurs; and to gets the prescription.

4 Conclusion

As smart phone health care systems have recently used in hospitals and public health clinics, various smart phone health care systems are under development; but in reality, health care system supporting patients in need of regular management is in short.

Therefore, this study proposes a model for drug administration management system for smart phone applications to prevent duplications in the drug prescription process and to reduces cases of worsening diseases due to overdose or under-dose from recognition failure for drug dose number of times or time or enhancing immunity to certain drugs. In addition, the delivery system is proposed to get rid of inconvenience of patients who should get prescribed drugs by themselves, which enables collection, management and application of individualized information by user and further optimized administration management by person through learning.

As for future research project, a study will be conducted for linking issue between most basic domestic diagnosis devices and smart phone applications, algorithm for drug redundancy prevention and concrete and efficient algorithm on delivery system after getting prescription.

References

1. Features and applications, and the need of [U-healthcare] Ubiquitous Healthcare (u-health care), Features and meaning of U-healthcare, Development practices and the types of services. http://www.sysbase.co.kr/s_faq/rs232.htm
2. Direction and development of an effective response of U-health
3. Development and evaluation of drug duplication alerting (2007) 녀 Ewha Woman University
4. u-Health New Business Model (2010)

Use of Genetic Algorithm for Robot-Posture

Dong W. Kim, Sung-Wook Park and Jong-Wook Park

Abstract Robot-posture with genetic algorithm is presented in this paper. As a robot platform walking biped robot is used. To cope with the difficulties and explain unknown empirical laws in the robot, practical robot walking on a descending sloped floor is modeled by genetic architecture. These results from the modeling strategy is analyzed and compared.

Keywords Genetic algorithm · Robot posture · Comparison analysis

1 Introduction

To achieve walk realization, the foot should be controlled well but generally it cannot be controlled directly but in an indirect way, by ensuring the appropriate dynamics of the mechanism above the foot. Thus the overall indicator of the mechanism behavior is the point where the influence of all forces acting on the mechanism can be replaced by one single force. This point was termed the zero moment point (ZMP). Recognition of the significance and role of the ZMP in the biped artificial walk was a turning point in gait planning and control.

In this paper, practical robot walking on a descending sloped floor is employed as robot platform, neuro-fuzzy system, and evolutionary architecture are also used as intelligent modeling strategies. In addition, these results from two methods are shown, analyzed and finally compared.

D. W. Kim (✉)
Department of Digital Electronics, Inha Technical College,
Incheon, South Korea
e-mail: dwnkim@inhatc.ac.kr

S.-W. Park · J.-W. Park
Deptartment of Electronics, University of Incheon, Incheon, South Korea
e-mail: jngw@incheon.ac.kr

J. J. (Jong Hyuk) Park et al. (eds.), *Multimedia and Ubiquitous Engineering*,
Lecture Notes in Electrical Engineering 240, DOI: 10.1007/978-94-007-6738-6_72,
© Springer Science+Business Media Dordrecht(Outside the USA) 2013

Fig. 1 Joint angle representation of the robot

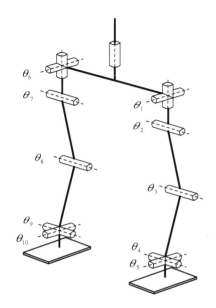

2 Robot Platform

The identical robot platform employed in [7, 8] is also used. The robot has 19 joints (three DOFs are assigned to each arm, three and two DOFs are assigned to the hip and the ankles, respectively, and one to each of the two knees). But only 10 dominant joints are used for input candidates. The locations of the joints are shown in Fig. 1. The height and the total weight are about 445 mm and 3 kg, respectively. Each joint is driven by the RC servomotor that consists of a DC motor, gear, and simple controller.

3 Usage of Genetic Algorithm

The genetic algorithms (GA) is an efficient tool for searching solutions in a vast search space. By the GA, proper type of MF, number of MF, type of consequent polynomial, set of input variables, and dominant inputs among input candidates are likely to be found in the case where the 10 input candidates have complex correlation.

These parameters stated above in designing a fuzzy system are determined in advance by the trial and error method. But in this paper, the key factors for optimal fuzzy system are specified by GA automatically. When designing a fuzzy system using the GA, the first important consideration is the representation strategy, which is how to encode the fuzzy system into the chromosome. We employ binary coding for the available design specifications. The chromosomes encoded information for fuzzy system are made of five sub-chromosomes. The first one has one bit and presents type of membership function. Two types of MF, Triangular and Gaussian MF, are used as the MF candidates. Each is represented by a bit 0 and 1. If the gene in the first sub-chromosome contains 0, the corresponding type of MF is Triangular type. If it contains 1, the MF is Gaussian type. The second sub-chromosome has two bits for number of MF. If many number of MF is selected for certain input variables then fuzzy rules and computational complex can be increased. So we constrain the number of MF to vary only between 2 and 4 for each input variable. The 3rd sub- chromosome has two bits and represents types of polynomial. A total of four types of polynomial are used as candidates and each candidate is represented by two bits. Selection of dominant input variables which is greatly contributed to the output and number of these variables is very important. Research of appropriated method for the input selection is still under investigation. In this paper, we handle these problems using the fourth and fifth sub-chromosomes.

For the best x-coordinate, produced string information is [0, 4, 3, 4: 2, 4, 8, 9]. This string means triangular MF type, 4 MFs, Type 3, and 4 inputs, second, fourth, eighth, ninth, are selected to get good performance. 3.9877 of MSE value is obtained from this string information. For the best y-coordinate, [1, 4, 2, 4: 2, 7, 8, 9] of string information is obtained. So Gaussian MF type, 4 MFs, Type 2, and 4 inputs, second, seventh, eighth, ninth, are selected by evolutionary algorithm and 5.5385 of MSE value is obtained. The corresponding walking trajectory for a descent floor based on the model output is shown in Fig. 2.

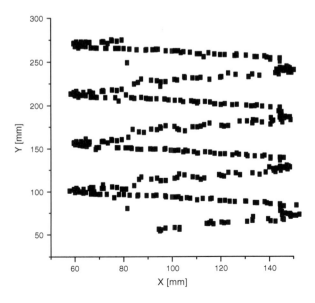

Fig. 2 Walking trajectory on a descent slope

4 Conclusion

Robot posture strategy with genetic algorithm is presented and it results are analyzed in this paper. As a robot platform walking biped robot which has 19 joints is used and trajectory of the zero moment point (ZMP) is employed for the important criterion of the balance. To cope with the difficulties and explain unknown empirical laws in the robot, practical robot walking on a descending sloped floor is modeled by genetic architecture. In this paper, Fig. 2 is finally from the genetic algorithm based model.

References

1. Vukobratovic M, Brovac B (2004) Zero-moment point-thirty five years of its life. Int J Humanoid Rob 1:157–173
2. Vukobratovic M, Andric D, Borovac B (2005) Humanoid robot motion in unstructured environment-generation of various gait patterns from a single nominal. In: Kordic V, Lazinica A, Merdan M (eds) Cutting edge robotics. In Tech
3. Hirai K, Hirose M, Haikawa Y, Takenaka T (1998) The development of Honda humanoid robot. In: Proceedings of the IEEE international conference robotics and automation, pp 1321–1326
4. Vukobratovic M, Brovac B, Surla D, Stokic S (1990) Biped locomotion. Springer, New York
5. Kim D, Seo SJ, Park GT (2005) Zero-moment point trajectory modeling of a biped walking robot using an adaptive neuro-fuzzy systems. IET Control Theory Appl 152:411–426
6. Kim D, Park GT (2007) Advanced humanoid robot based on the evolutionary inductive self-organizing network. Humanoid robots-new developments, pp 449–466
7. Kim D, Park GT (2010) Intelligent walking modeling of humanoid robot using learning based neuro-fuzzy system. J Inst Control Rob Syst 16(10):963–968
8. Kim DW, Silva CW, Park GT (2010) Evolutionary design of Sugeno-type fuzzy systems for modeling humanoid robots. Int J Syst Sci 41(7):875–888
9. Chun BT, Cho MY, Jeong YS (2011) A study on environment construction for performance evaluation of face recognition for intelligent robot. J Korean Inst Inf Tech 9(11):81–87
10. Shin JH, Park JG (2012) Implementation of an articulated robot control system using an On/Off-line robot simulator with TCP/IP multiple networks. J Korean Inst Inf Tech 10(01):37–45

Use of Flexible Network Framework for Various Service Components of Network Based Robot

Dong W. Kim, Ho-Dong Lee, Sung-Wook Park and Jong-Wook Park

Abstract These days wide variety of platforms and frameworks were researched for network based robot (NBR) but it is difficult to apply those platforms and frameworks to the NBR, because the NBR and its service components keep developed. In other works, the interfaces and execution environments of the NBR and the service components are changed and upgraded very often, and at times, the service components are totally reconstructed. Consequently, a new flexible and reliable network framework that adapts to various service components is necessary. To handle these problems, a solution is suggested in this paper.

Keywords Network based robot (NBR) · Platform and frameworks for service component

1 Introduction

TCP/IP (Transmission Control Protocol/Internet Protocol) is a well known and the most popular communications protocol for Internet connection. TCP/IP is a traditional method of integrating components via a network. It is reliable and stable, and it is very easy to develop its components. Various architectures and integration schemes have been researched. In this paper, communications interfaces based on TCP/IP are designed for an NBR.

D. W. Kim (✉)
Department of Digital Electronics, Inha Technical College, Incheon, South Korea
e-mail: dwnkim@inhatc.ac.kr

H.-D. Lee
Korea Institute of Science and Technology, Seoul, South Korea

S.-W. Park · J.-W. Park
Department of Electronics, University of Incheon, Incheon, South Korea
e-mail: jngw@incheon.ac.kr

J. J. (Jong Hyuk) Park et al. (eds.), *Multimedia and Ubiquitous Engineering*,
Lecture Notes in Electrical Engineering 240, DOI: 10.1007/978-94-007-6738-6_73,
© Springer Science+Business Media Dordrecht(Outside the USA) 2013

Fig. 1 Internet protocol stack

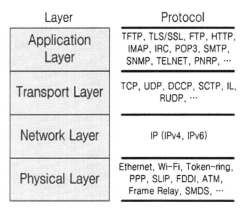

2 Communications Interface Design

TCP/IP consists of an IP (Internet protocol), which is an Internet protocol that uses the packet communication method, and a TCP (transmission control protocol).

Figure 1 shows the TCP/IP protocol's stack and related protocols. The hierarchical structure is also shown.

Environment data such as video streams and audio streams from an NBR are sent to each resource via the main server. Thus, if the number of service components increases, then the data throughput of the main server is increased. This situation is shown in Fig. 2.

To develop the network framework described previously, a network core that can be configured into the network with a tree structure was designed. Also, the network core serves as a server and a client simultaneously. In Fig. 3, the network core that was designed and implemented is shown.

Figure 4 shows a block diagram of the network core. As shown in Fig. 4, the sender transmits data packets to other network cores. The receiver receives

Fig. 2 Case of clients receiving information from the main server

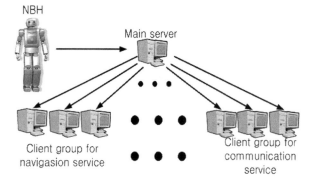

Fig. 3 Basic concept of a network core structure

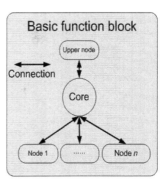

Fig. 4 Block diagram of network core

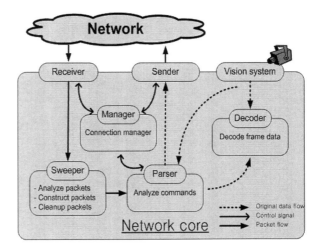

the data packet from other network cores. The sweeper assembles the packets from the receiver to the data. The parser is responsible for processing the data. A network core has these four basic functions and management roles.

3 Performance Analysis and Discussion

The transfer rate is measured between cores through a wired LAN, a wireless LAN, and a local machine, respectively. It shows also quick network core data transfer. Note that during the data transfer, the network core partitions and assembles the data.

References

1. Schuehler DV, Moscola J, Lockwood J (2003) Architecture for a hardware based, TCP/IP content scanning system. In: Proceedings of the high performance interconnects 2003, pp 89–94
2. Tang W, Cherkasova L, Russell L, Mutka MW (2001) Customized library of modules for STREAMS-based TCP/IP implementation to support content-aware request processing for Web applications. In: Advanced issues of E-Commerce and web-based information systems, WECWIS 2001, pp 202–211
3. Hansen JS, Riech T, Andersen B, Jul E (1998) Dynamic adaptation of network connections in mobile environments. IEEE Internet Comput 2(1):39–48
4. Lee KB, Schneeman RD (1999) Internet-based distributed measurement and control applications. IEEE Instrum Measur Mag 2(2):23–27
5. Liao R-K, Ji Y-F, Li H (2006) Optimized design and implementation of TCP/IP software architecture based on embedded system. In: Proceedings of the machine learning and cybernetics 2006, pp 590–594

China's Shift in Culture Policy and Cultural Awareness

KyooSeob Lim

Abstract Since the modern age, China's major historical events have always been related to culture and it has been centered to discussions on the national development. Since adopting the reform and openness policy, China has been emerging as a new power and enlarging influence in international politics, economics, and military. However, China recognizes that its cultural influence is still light. Thus, the Chinese government exercises cultural policy which enables China to expand Chinese cultural clout and soft power externally and, at the same time, integrate society by establishing a clear national identity internally. China's cultural heritage, which once was target to be overthrown, is now inherited and developed. China is changing its cultural awareness.

Keywords Culture policy · Cultural system reform · Cultural heritage · Soft power · China

1 Introduction

We can say that Chinese history is in line with Chinese cultural history. Since the modern times, China's major historical events have always been related to its culture [1], and culture has been centered to the discussions. Broadly, culture exercises influence in 'people' who are main agents of politics and economics. For such reasons, culture consists of one system providing phenomenon related to

K. Lim (✉)
Department of Foreign Language and Culture, Institute of International Education,
Kyung Hee University, 26 Kyunghee-daero, Dongdaemun gu,
Seoul 130-701, South Korea
e-mail: lks@khu.ac.kr

J. J. (Jong Hyuk) Park et al. (eds.), *Multimedia and Ubiquitous Engineering*,
Lecture Notes in Electrical Engineering 240, DOI: 10.1007/978-94-007-6738-6_74,
© Springer Science+Business Media Dordrecht(Outside the USA) 2013

politics and economics [2]. Even now, in contemporary China, culture interacts with politics and economics and develops its rules and functions.

Since the new cultural movement in China, China has defined culture as comprehensive concept including not only literature and arts but also knowledge, ideology, ethics, and values. China also stressed culture saying 'cultural transformation' should be proceeded in order to bring a political, economic, and social change. The Chinese government estimated that to transform politics, economics, and society, overall changes in people's value should be preceded. For that reasons, in the contemporary Chinese history, in every major transitions for political, economic, and, social change, there have been discussions on the culture [3]. There is a saying that 'cultural change should be put first' and the tradition is still influential. As a result, after the reform and openness policy, the Chinese government emphasizes 'cultural reformation' to overcome internal and external conflicts and crises and to transform its politics and society.

In the past, the Chinese government and Chinese leaders regard culture as means for propaganda and thought indoctrination. However, escaping from this recognition, the Chinese government started to rethink the value and meaning of culture in a different way. Specially, reform and openness makes progress in conjunction with cultural 'marketization' and 'popularization' and consequently the Chinese government's cultural awareness has fundamentally changed. The central question that I intended to examine in this article is 'why the Chinese government is responsive to culture issues?' In the following, I will try to examine backgrounds to Chinese cultural policy and its contents and analyze China's awareness shift on culture.

2 Backgrounds to China's Culture Policy

2.1 Building Legitimacy of Communist Party and Social Integration

China's cultural heritage which is combined to Chinese nationalism has been major means of stabilizing the Chinese Communist regime. China's cultural strategy has been used to remove threats on the Chinese Communist regime rather than means to overcoming national crisis. In other words, as Chinese economy rapidly expands through reform and openness, Chinese people recover national confidence while China's national ideology was weakened in the globalization and the post-Cold War era. So, the Chinese government tried to find the Chinese national identity through succession of national culture and unite and integrate the nation through cultural heritage [4]. The Chinese government still maintains Marxism-Leninism and Maoism socialist system perfunctorily, however, it stresses Confucius idea to expand and develop the Chinese culture. The Chinese government which uses Confucian ideas tries to maintain its regime by constantly announcing that it is a legitimate successor of the Chinese nation [5]. As ideological function of socialism which once played as a key role in social integration weakened, China intended to form a new national ideology from Confucianism in order to maintain economic development and steadily integrate society.

2.2 Attempting Shift in Economic System

Since adopting Western capitalism, the essential crisis of the cultural awareness has been not only about its values but also about economy or capital. In other words, Western culture industry with the gigantic capital strength and global competitiveness seemed to encroach on China's culture market [6]. Owing to the growing economic importance of cultural industry, the Chinese government gave shape to culture as one sector of major policies and strategies [2]. Beijing put stress on economic value of culture for economic development and brought fundamental change of 'means to the economic development in the future.'

2.3 Resisting Western Culture Invasion

Against expanding Western culture invasions such as culture standardization, China is experiencing 'cultural crisis', so it carries out aggressive culture system reformation mainly towards to Western advanced capitalism countries [6]. In the current world system under the Western political ideology and values and capitalistic market economy, China should observe western values-dominant international order. China thinks the Western powers attempt to gradually change the Chinese social system by using these values and orders. Thus, China appreciates unless it deters the Western infiltration, Chinese identity and loyalty to Chinese culture would be deteriorated and the Chinese culture would come to a crisis by being eroded assimilated with the Western culture. Beijing makes ceaseless effort to get over Western centrism. China aims to maintain identity and independence of Chinese civilization against the Western culture and its first priority aim is executing culture policy.

2.4 Forming Peaceful National Image

The western countries and neighboring nations express concerns on stronger China. These concerns greatly impede Chinese national development and its leverage [7]. For that reasons, China started to claim new ideology calling 'harmonious society' and 'harmonious world' and make better national image internationally using soft power. To eliminate negative awareness such as 'Chinese throat' and make better national image, the Chinese government highly pushes ahead with the cultural policy.

2.5 Developing Toward a Culture Power

In line with economic, military, and international political clouts from success of reform and openness policy and cultural outcomes from rich history, China hopes to develop to a culture power nation. The Chinese government wants to resolve unbalanced development between economy and culture, and also hopes to enlarge soft power in the world. In other words, China started to regard both hard power development plans and soft power national development plans as important. The Chinese government emphasizes cultural revival to become a culture power and a powerful nation at the same time.

3 China's Culture Policy

3.1 Culture Policy

In the era of globalization, the Chinese government promotes 'globalizing Chinese culture' and seeks to new national identity through traditional cultural elements. Especially, since 2000, the Chinese government highlights the importance of culture in various fields and regards culture as crucial factor for the national development. As a matter of fact, the Chinese government accentuates culture-related policies such as culture security, soft power, culture power, national image, cultural diplomacy, Confucius revival, and Confucius Institute.

The Chinese government advocates it should strengthen soft power claimed by Joseph Nye in diplomatic policy. Currently, Beijing proceeds with its soft power policy by 'actively engaging with national competency.' The Beijing Olympics in 2008 and Expo 2010 Shanghai China are good examples. Particularly, the Chinese government established the first Confucius Institute on November 21, 2004, so as to publicize Chinese culture to the world and expand international leverage of the Chinese culture. Since then, until July 2012, China establishes and operates some 387 Confucius Institutes in 108 countries.[1]

3.2 Government Official Papers

Since the reform and openness, we could the witness awareness change in culture at the Chinese government released official papers. Since <the 12 Chinese Communist party congress>, China has emphasized the importance of 'the culture'. In <the 12th report> of the 1982 Chinese Communist Party congress, China

---

[1] http://www.hanban.edu.cn/

separated cultural construction and ideology construction from the socialist civilization construction and put an emphasis on culture saying it is a part of the socialist civilization construction. In <the 14th report> of the 1992 Chinese Communist Party congress, China released the cultural construction line. In essence, China intended to 'actively promote cultural system reform and perfectly implement economic policy relating to culture for the sake of flourishing socialist culture. At the time, the Chinese government firstly insisted that it should promote the cultural system reform on the basis of economic and political system reformation.

In <the 15th report> of the 1997 Chinese Communist Party congress, the Chinese government suggested that it should build culture construction with Chinese socialism trait, and since then, it has been the basis of the Chinese Communist party's cultural line [8]. In <the 16th report> of the 2007 Chinese Communist Party congress in October, the Chinese government explained cultural construction and cultural system reform claiming that 'according to the socialist civilization construction, it should implement cultural system reformation along with the socialistic market economy development.' China stressed that it should perfectly prepare cultural market system and its management mechanism [9]. In <the 17th Chinese Communist Party congress> in 2007, Hu Jin-Tao raised that 'we should lead greater development and prosperity of the socialistic culture.[2]' At the event, Hu Jin-Tao firstly included 'Chinese culture' to the Chinese Communist Party's official paper. In <the 6th general meeting of the 17th Chinese Communist Party congress> in October, 2011, Chinese government selected the cultural reformation for an agenda, and passed 'the resolution on crucial issues to intensifying cultural system reformation of central Chinese Communist Party and promote greater development and prosperity of the socialistic culture.[3]' At the meeting, the Chinese government decided to improve Chinese identity and confidence on the Chinese culture by keeping cultural security and strengthening its soft power in succession to the China's economic development. The report said that China is faced to the problem of protecting cultural security and improving soft power and international clouts of the native culture.

4 Changing Awareness in Culture

4.1 Cultural Status in the Traditional China

In the traditional China, culture was the basis for the Chinese order. In the warring states period (Xian Qin period), Chinese 'Mun (文, knowledge)' tradition which opposed to Mu (武, martial arts) laid the basis of the Chinese Hwaism (華夷論).

---

[2] http://news.163.com/07/1020/07/3R7TGC240001124J.html

[3] http://www.gov.cn/jrzg/2011-10/25/content_1978202.htm

Chinese people thought that they are the center of the world and they enjoy the most advanced culture in the world. Chinese people developed Hwaism, another kind of Sinocentrism, which ignores and neglects other cultures [10]. In the traditional China, the concept of culture included 'discriminatory structure', and Chinese culture is top of other barbarous cultures. They believed Chinese culture is related to their nature spirit, and there is no comparison.

4.2 Changing Status of Culture Since the Modern Age

Since the modern times, by shocking from the Western civilization, Chinese myth that Chinese culture is unique has been collapsed. Also, all the national crises of China brought awareness change in 'culture' [1]. Especially, there had constantly been arousing discussions on 'Chinese cultural heritages' since the May Fourth movement (the new cultural movement) and it was the key issues on the national development. In fact, in the May Fourth movement and the Cultural Movement period, the Chinese cultural heritage was denial and traditional ideas and values was target to be overthrown. In the May Fourth movement period, Confucius was accepted as an ultimate cause of national crisis, and in the Cultural Movement period, the traditional heritage was considered as an obstacle for establishing socialist regime by accepting proletarian culture [3].

At the discussion of 'the culture fever' in 1980s, discussion on the traditional China and the contemporary China, Chinese fundamental cognition on the Chinese culture was unchanged. Chinese regarded feudal culture, before the May Fourth movement culture, as old tradition and ultraleftism culture in Mao Zedong times, as neo tradition, and considered them to be overthrown [11]. By the time, China thought tradition as an opposite to the moderns, and there was prevailing opinion that tradition should be defeated to build a modern nation.

4.3 Culture Awareness During Reform and Openness Times

After the reform and open policy, China began to overcome economy fell, but the policy caused numerous side effects i.e. mass inflow of the Western culture, deepening economic bipolarization, lacking political democratization, deficit of the new sense of values, and cultural identification [12]. Therefore, China sought for an alternative ways that could lead social stabilization and integration and constant economic development simultaneously, and started to regard highly traditional culture such as Confucius. In 1990s, from the negative standpoint on the traditional cultures, China started to recognize cultural heritage as a target to be succeed and developed. In particular, the identity crisis, evoked from the 1989 Tiananmen Massacre, and anti-West and nationalism emotion arose, and that prepared the ground for 'the sinology fever' [13].

The Chinese government concentrated on the Confucianism, representative Chinese traditional culture, and deemed it positive to the Chinese modernization and economic development. Especially, China considered that culture would resolve a great many negative factors caused by the introduction of western capitalistic market economy system [12]. After 2000, China heightens confidence on economic power and pays much attention to culture in both governmental and private sector.

4.4 Relationship Between Culture and Economy

The report 'several view on enforcing the cultural market' from the Department of Culture on February 4, 1994, said 'development of the cultural market promotes the interaction between culture and economy and it establishes developing cultural industry.[4]' In this report, the Chinese government emphasized the relations between culture and economy on an equal footing. The culture policy, especially the culture industry, includes not only cultural but also economic traits. That is to say, it consists of the two fields, 'upper structure' and 'economic foundation.' The Chinese government made strategic choice to integrate economic and cultural value at the same time through culture industry [6]. In the past, in the emphasis on politics with class strife, culture was a tool for spreading national policy and socialist ideology under the slogan, 'serving workers · farmers · soldiers and national politics. On the other hands, in the present market economic system, culture is a tool for creating 'profit' and consuming 'culture' itself at the same time. By forming a new version of cultural market, the Chinese government's cultural awareness, unlike to the past, has fundamentally changed.

4.5 Relationship Between Culture and Politics

Since the New China, Chinese leaders have taken culture as a secondary way for political strife. The Chinese government cognizes culture not comprehensive but limited concepts that is helpful to attain 'specific purpose'. During the Cultural Revolution period, culture worked as an ideological tool for class strife and political strife. The Chinese government considered and developed the relation between politics and culture as a master-servant relationship. However after the reform and openness in 1979, China changed its national development strategy focusing on economic development and Chinese awareness in culture gradually changed. Namely, though a master-servant relationship between politics and

---

[4] Department of Culture report: http://www.34law.com/lawfg/law/6/1187/law_251625388934.shtml.

culture still exists, the relationship slowly escaped from 'the one-sided relationship [14].' The Chinese government used culture's public ripple effect to maintain and enforce national ideology. It produced 'culture' and expressed cultural political acts and then increased clouts in the culture market. The Chinese government changed cultural identity to political or national identity so as to integrate China using the national emotion. The political intervention changed contexts and rules of culture, and the transformed culture once again provided the basis of legitimacy to the political power.

5 Conclusion

Although China attained success in reform and openness policy, there have been lots of side effects all over the politics, economy, and society. To settle contradictions in the Chinese society and consolidate the Chinese Communist party regime, the Chinese government actively promoted national development strategy in the culture area. The Chinese government carried out many different culture policies such as culture security, soft power, national image, cultural diplomacy, Confucius revival, and Confucius Institute. As a result, Chinese awareness in culture gradually changed. In other words, China started to recognize cultural heritage, which was once seen in a negative light, as a target to be succeeded and developed. Also, China's viewpoint between culture and economy and culture and politics has been changed. Currently, it is likely that the Chinese government and the Chinese society simultaneously attempt to make cultural transformation. The 'top-down' cultural transformation led by China is still occurring.

References

1. Seo K (2008) Reviewing Chinese civilization and its language in the early 20th century. Chin Lang Overv 27:51–74
2. Kim K (2012) China's cultural policy. East Asia Briefing 7.2(24)
3. Lee Y (2006) Discussions on the Chinese culture and cultural development strategy in the age of globalization. Contemp Chin Lit 37
4. Lee K (2009) Chinese cultural nationalism and the practical strategy. Korean Northeast J 52
5. Zhao C-S, Lin Y-J (2012) Chinese Marxism and Chinese Communist Party's interpretation on culture. Prospect Investig (Taibei) 5.5
6. Kwon K (2012) China's nation vision in the 21st century and its strategy for developing cultural industry. Res Contem China 14.1
7. Pan B, Nam C, Chang Y (2011) The process of the Chinese culture diplomacy and problems. Unification Res 15.2
8. Zhang S-L (2012) Historical contexts on cultural reformation and development of the Chinese Communist Party. Chinese Communist Party YunNan (Yunnan), vol 13.1
9. Long X-M (2012) China's culture system reformation and development. BaiHuaChao (Beijing), vol 8

10. Choi S (2011) Chinese culture, how can we understand and which approach could we take? Sinol J 4
11. Tang Y-J (1998) Three problems of traditional culture research. XinHuaWenZhai (Beijing) 1988, vol 3
12. Yeon J (2012) Implications and prospects of the restoration movement in the Tang Dynasty. Korean Philos J 30
13. Zhang X-C (2002) Discussion on practicing multiculturalism. 21st Century (Hongkong), June 2002 vol 71
14. Lee J (2009) Research on the Chinese socialist market economy and cultural policy change. Chin Lit 60

China's Cultural Industry Policy

WonBong Lee and KyooSeob Lim

Abstract In the globalized world, the cultural industry has emerged as a new promising field. Countries are accelerating their competition to become a culture power nation. In the past, China used culture in the purpose of electing ideology. Since China's entry into the World Trade Organization (WTO), China started to foster the cultural industry. Since 2001, China started to carry out the cultural industry policy. In 2009, after 'Cultural Industry Promoting Plan', China propelled cultural industry to promote policy in earnest. In 2011, the Chinese government claimed it would nurture the cultural industry as a national strategic industry. In China, cultural industry has been stressed as a part of its soft power strategy. The culture has been emerged as a mean for spreading ideology and a new growth engines for industry.

Keywords Chinese culture · Cultural Industry · Cultural Strategy · Cultural system reform

1 Introduction

In the globalized world, 'culture' is combined to 'industry.' At the same time, countries have been forming 'the cultural industry' by systemizing and industrializing culture for supply. Since 1999, the Chinese government has given an

---

W. Lee (✉)
Department of Chinese Studies, Kyung Hee Cyber University, 1 Hoegi-Dong, Dongdaemun Gu, Seoul 130-701, South Korea
e-mail: wblee@khcu.ac.kr

K. Lim
Department of Foreign Language and Culture, Institute of International Education, Kyung Hee University, 26 Kyunghee-daero, Dongdaemun gu, Seoul 130-701, South Korea
e-mail: lks@khu.ac.kr

J. J. (Jong Hyuk) Park et al. (eds.), *Multimedia and Ubiquitous Engineering*, Lecture Notes in Electrical Engineering 240, DOI: 10.1007/978-94-007-6738-6_75, © Springer Science+Business Media Dordrecht(Outside the USA) 2013

important on the status of culture, and in the 2000s, the government accelerated importance of the culture. Since China's reform and openness policy, China's economy has been growing and it brought changes in consuming culture. Therefore, demands on the culture products are skyrocketed and Beijing realized lack of cultural products which should be consumed in various mass media [1]. Also, as China emerged as an economic power, the Chinese government accentuated fostering cultural industry in order to enhance China's status to the higher level in the world [2].

Although China's cultural industry largely developed in line with the economic growth, cultural industry is given little weight than other industries. In 1996, the added value of the cultural industry was 21.184 billion yuan (RMB), accounting 0.3 % of GDP. Moreover, in 2009, the added value of the cultural industry was 103.77 billion yuan (RMB), accounting 0.3 % of GDP. Since 2000, China's cultural industry has comparably maintained stable growth [3]. China sets up the cultural industry as an emerging industry and considers it as a driving force for the national development in the tide of globalization. China aims to enhance social and cultural level through the cultural industry policy and wants to emerge as a culture power. Also, the Chinese government tries to establish Chinese own cultural industry system. The focal point of our discussion will examine the forming factors of the Chinese cultural industry and its development strategies with current state of the cultural industry on the regional basis. This paper will also analyze characteristics of China's cultural industry.

2 Cultural Industry Policy

2.1 Culture and Cultural Industry

Culture refers to the people's way of living or the way of thinking at a certain place. Culture is used extensively including food, clothing, shelter, language, religion, knowledge, arts, and institutions [4]. Culture riches people's lives by changing natural state of human being artificially with skills and labors. In other words, culture endows new value to the nature. Culture is a creation of both spiritual value and material value [5]. Recently, countries put an importance on culture as a key factor when it comes to evaluating comprehensive nation's power. Every nation tries to heighten its cultural soft power.

Generally, the United Nations Educational, Scientific, and Cultural Organization (UNESCO) defines the cultural industry as 'one business's production, reproduction, storing, and distribution of cultural products or services and its steering those cultural products or services in commercial way. In other words, the cultural industry means using culture for economic interests rather than cultural development itself [4].' The cultural industry connotes one nation's value comprehensively. Therefore, culture plays important role in building one country's national identity. The spiritual culture provides foundation for the cultural industry and the cultural industry exercises a great effect on people's mindset.

The cultural industry is a new phenomenon which is evoked in the process of integrating culture, economy, and technology. The cultural industry creates synergy effect through integrating environment-friendly industries, higher value-added businesses, and other industries. Also, the cultural industry develops new markets and jobs with small money. As the cultural industry does not harm the environment, the cultural industry draw attention to many nations as a new emerging industry. At the same time, the cultural industry is a promising industry with higher practical ability with relations to other industries and higher synergy effect. The cultural industry became a key industry for social development and economic growth. The competition among nations to become a culture power has become cutthroat. The cultural industry market in the world over 1 trillion dollar, and it will be one of industry upgrade in the future.

2.2 Decision Factor of the Chinese Cultural Industry Policy

Before the reform and openness, China recognizes culture as a mean for spreading ideology rather than industry. Also Beijing did not put in any efforts to consider culture as an industry. Culture was managed by the government budget and used only for the ideological purpose. Namely, culture was accentuated only for the key propaganda tool for the Chinese politics. As a result, China's culture could not sharpen its competitiveness.

According to the China's rapid economic growth, national power also enhanced. China's international standing also largely surged. However, the competitiveness of the Chinese culture was not commensurate with China's international standing. Therefore, we could say that the Chinese leadership awareness change on culture started from outside factors. China's entry into the World Trade Organization (WTO) is a good example. Since China's entry into the World Trade Organization (WTO), China was necessary to open its cultural market. According to the change, China needed to protect its cultural sovereignty. Unless China grew its competitiveness of the cultural industry, it destined to face consumer market from outside influences. Since the entry into the World Trade Organization (WTO), the Chinese government has consistently issued policies to nurture cultural industry.

Therefore, along with the economic growth, China experienced changes of consumption patterns. With the changes, the Chinese government keeps enlarging demands on pop culture. Based on the economic ability, the Chinese government started to promote cultural industry aggressively. Beijing has strong will to combining its 5000 years rich history, which was an incomparable culture power in Asia, with culture and nurturing cultural industry.

3 Development Process and Development Strategy of China's Cultural Industry Policy

3.1 Categorizing Cultural Industry Policy

Cultural industry includes some 10 types of industry i.e. publishing, radio and television, newspapers and magazines, commercial display, entertainment, exhibition, and network. Those industries share commons as well as different characters. The Bureau of Statistics of China categorizes cultural industry according to their industrial fields for statistics process. The key field of cultural industry is four parts: news paper, publishing and copyright, movie and drama, and cultural arts. Departments in charge are the Bureau of Culture, the Bureau of Newspapers and publishing, and the Bureau of Photoelectric.

In 2012, the Bureau of Statistics of China reissued 'Category for Culture and relating industries' and separates the cultural industry with two parts depending on the industrial relevance.[1] The first sector is 'production of cultural goods' including newspapers, publishing, broadcasting, TV, movie, cultural arts, culture and information release, cultural creation and establishment, leisure and entertainment service, and crafts arts. The second sector is 'production of culture-related goods' including production of auxiliary materials for cultural goods (publication right, print, and copy etc.) and production of cultural goods (office stationary, music instrument, and plaything etc.).

3.2 Development Process of China's Cultural Industry Policy

Before 1999, China used the term 'cultural business' rather than 'cultural industry.' The term 'cultural industry' firstly emerged in the Chinese society after 2000. China's cultural industry policy corresponds with development stages of its cultural industry. China's cultural industry had its earliest beginning at 1978. We can call the period the simple supporting stage (1978–1992). In the 1990s, as the Chinese economic development accelerated, development of the cultural industry also became faster. This period is the promotion stage. Since 2000, the Chinese government has implemented policies so as to supporting cultural industry strategically.

The simple supporting stage was the first step to the cultural industry. In the period, culture manufacturing businesses and cultural service industries emerged with some advertisement companies. In the promotion stage, the Chinese government intentionally encouraged the development of the cultural industry. In this

---

[1] http://www.stats.gov.cn/tjbz/t20120731_402823100.htm

period, policies focusing on reforming cultural system established. Also, all sorts of regulations were enacted. At the 5th Plenary Session of the 15th Central Committee of the Communist Party of China, the Chinese official papers firstly used the definition of 'the cultural industry' and 'the cultural industry policy.'

Since 2001, we call this period the strategically supporting period. After the twenty first century, cultural competition among nations has been deepening. Since China's entry into the World Trade Organization (WTO) in 2001, China started to firmly consider the cultural industry's statistic standing. Since then, the Chinese government released announcements to lead cultural system reformation and the development of the cultural industry.

In 'the 10th Five-Year Plan (2001), the Chinese government put an emphasis on electing the cultural industry policy and promoting development of the cultural industry. In 'the 11th Five-Year Plan (2006), the Chinese government established ordinances and regulations to develop cultural industry. In 'the 12th Five-Year Plan (2011), the Chinese government decided on to nourish cultural industry. In 'the 11th Five-Year Plan and cultural development planning', the Chinese government mentioned it would promote movie, publishing, printing, advertisement, entertainment, exhibition, digital contents and character, and animation industries. The statement intended to specify cultural industry relating policies. 'The cultural industry promoting plan' was issued in September 2009, and it provided a basis for the cultural industry promoting policy. The 12th Five-Year Plan contained plan for nourishing cultural industry as a major industry. To achieve the goal, the Chinese government implemented several supportive policies such as inviting major enterprises and investors [6] (Table 1).

3.3 Regional Distribution of the Cultural Industry

China's cultural industry is largely different from one region to another. On added value basis in 2009, Shanghai, Zhejiang province, Guangdogn province, Jiangsu province, Sichuan province, Shandong province are of great importance when it comes to culture. We could say added value of the cultural industry in East coast region is relatively greater than middle or western region [3].

Table 1 Development process of China's cultural industry policy

	Period	Major issues
Initial stage	10th Five-Year Plan (2001–2005)	Establishing cultural industry policy Categorizing cultural-related industry
Embodiment stage	11th Five-Year Plan (2006–2010)	Development of cultural industry Strengthening governmental control
Deepening stage	12th Five-Year Plan (2011–2015)	Fostering culture as a major industry Electing policies and strategies to be a 'culture power'

In order to develop regionally and ethnically characteristic cultural industry complex, China try to its competitiveness of the culture industry by enlarging its industrial volume. In the 'Table 2' we could find out 8 regions with high added value of the cultural industry during the period of 'the 11th Five-Year Plan' (2006–2010). Through the Chinese government's management systems and policies which are suited for each region, most regions develop the cultural industry in different fields.

4 Characteristics of the Chinese Cultural Industry

4.1 Future Industry Combining Value of Economy and Culture

At the 6th Plenary Session of the 17th Central Committee of the Communist Party of China, the Chinese government set up a goal to nourishing cultural industry as a key industry in economy by 2020 in order to realizing the nation's vision. In China, cultural industry enjoys its standing not only as a key industry but also so called 'future industry' or 'next-generation industry'. The strategic value of the cultural industry lies not just in creating economic value, but in reforming

Table 2 During the 11th Five-Year Plan, Characteristics of cultural industry major cities

Region	Main characteristics	Major industry
Beijing	Securing strength on resources with rich history and culture Providing abundant cultural assets on the development of cultural culture creating industry	Publishing, news paper, movie, TV, entertainment industry
Shanghai	Developing culture focused on advertisement and service	Record and video, TV, play, mobile game
Guangdong	Taking the No. 1 ranking in China in printed material, broadcasting, digital publishing, printed publishing	Media cultural industry, amusement, TV
Hubei	Development of animation, online game, online optional service, new media	Book, publishing, record and video, exhibition
Zhejiang	Having the largest private cultural enterprises, with the largest movie studio established by the private capital China's largest animation industrial complex	Education, animation, movie
Shan dong	Geographically taking the advantageous position to spread culture in another region	Broadcasting, publishing, record and video
Sichuan	Western cultural industrial region with the center market	Internet game, media
Shanxi	Promoting cultural policy aiming to cultural industry and tourism	Play, Internet, amusement

industrial structure. That is the fundamental transformation of 'the way of the economic development [7]. Moreover, China intends to promote the cultural industry combining 'the economic development' and 'the proud 5000 years cultural power [8].' China considers cultural industry is intensely related to the China's national strategy in the future as well as China enhance strategic value of the cultural industry.

4.2 Changing Attitude from Control and Management to Revival

In 1998, the Chinese government reorganized the Bureau of Culture and established 'the Department of cultural Industry.' It meant that the Chinese government's awareness on 'culture' had been changed from 'the market' to 'the industry.' Using the word 'industry' in 'culture' means the Chinese government's attention to the culture pays more attention on the production of cultural goods and service and its circulation and distributions to the market. In the end, the process is centered on the sales to the customers and the economic market. The related policies also stressed on the 'industrial aspects' of culture rather than 'its cultural aspects.' The changing viewpoint represents the cultural policy is more 'promoting prosperous' rather 'imposing control.'

4.3 The Chinese Communist Party and the Chinese Government's Leadership

In China, the government encourages the cultural market [6]. Especially, the Chinese cultural industry is leaded by 'national enterprises', and Beijing is enlarging the number of private and public joint cultural corporations. Depending on the government's progressive reformation of the cultural system, most national enterprises claim free competition [8]. In the future, the major actor of the Chinese cultural industry will gradually move on from the government to the private sector such as individuals and private businesses.

4.4 Enlarging Publicness and Public Interest

China aims to reform overall cultural system in the end. Beijing wants to transform national cultural company into business management system. However, as China is a communist nation and is high lightening universal welfare system, it wants to enlarge cultural industry for publicness and public interest.

4.5 Supporting Policy

China's cultural industry policy is more of supportive rather than control. The Chinese government considers the cultural industry as a promising industry and basic industry in the future. The government sees cultural industry as a national strategic industry and implements various supportive policies to promote it.

4.6 Gradual Opening Strategy

The spiritual culture provides foundation for the cultural industry and the cultural industry exercises a great effect on people's mindset. Therefore, the Chinese government is very cautious when it comes to the external opening of the cultural industry. The Chinese government attentively accepted other cultures and cultural industries and made deliberate choice on induction of foreign capital and inviting foreign capital. In order to protect its own cultural industry, China gradually carries into openness policy [8].

4.7 Promoting Different Industrial Policy in Accordance with Industries

In China, there is none 'unified cultural industry policy' which can lead the development of the whole cultural industry [9]. Most industrial policies aim to specific fields, and it has certain regulation so as to developing the fields. It is closely related to the separated development of the Chinese cultural industry for a long time.

4.8 Implementing the Cultural Industry Policy with Regional Traits

Through management systems and policies that are suited for each region, most regions is developing cultural industry differently according to their traits [6]. Through developing cultural industry with regional and ethnic characteristics, China aims to enlarge industrial scale and enhance competitiveness. China's industrial policy is different from regions to regions.

5 Conclusion

In the current international society, each nation tries to enhance national image through the cultural industry. At the same time, countries make an effort to earn economic benefits from cultural industries. In the globalized world, competitions among nations to become a culture power are deepening. Cultural industry serves as the important foundation for forming national identity. The spiritual culture provides foundation for the cultural industry and the cultural industry exercises a great effect on people's mindset.

Since China's entry into the World Trade Organization (WTO), China consistently put an emphasis on the development of cultural industry. The Chinese government accentuated culture as a pivotal factor for the national competitiveness as cultural ability represents for the ethnic vital force and cohesiveness. The Chinese government considers cultural industry as a significant foundation in establishing ethnic identity. Also, China deems cultural industry as a major factor in forming people's mindset. In 2009, after 'Cultural Industry Promoting plan', Beijing's cultural industry promoting policy was implemented in earnest.

For China, cultural industry is not only creating new values, but also resolving weaknesses of existing industries. Cultural industry in China is emerging as a part of soft power strategy. In other words, culture is not only the means for spreading ideology, but also driving forces for the national development. The Chinese leadership not only elevates strategic value of culture industry, but also considers it is closely related to the China's future national strategy.

References

1. Park J (2006) China's cultural industry and Korean wave. Asia Pac Trend 30
2. Kim B, Lee J (2012) China's cultural industrial policy and human resources science the reform and openness. Int Labor Brief 15
3. Kim S (2011) Comparing national productivity of China's cultural industry. China Stud 57
4. NAVER http://terms.naver.com, 2013.1.2-2013.1.30
5. Cho C World cultural history. Parkyoungsa 2079
6. Oh H (2012) Research on the policy development and major characteristics of China's cultural industry science the reform and openness. China Center in Busan University, China Research, 13
7. Kwon K (2012) China's nation vision in the 21 century and its strategy for developing cultural industry. Res Contemp China 14(1)
8. Seon J (2011) Characteristics and strategical goal of China's cultural industry. Korean Research Center of Korean University. Korean Studies 37
9. Yang J (2007) On defects in cultural industry policy. Bus Adm (Beining) 3(3)
10. Seon J (2012) Current state and prospect of China's cultural. Sinol J 37
11. Yang G (2011) Research on the Chinese cultural industry and knowledge production. Chin Lit 57
12. Jeong W (2009) Research on market participant in China's cultural: focusing on 3C (company, customer, and competitor). Curr China Res 11:1

Development of Mobile Games for Rehabilitation Training for the Hearing Impaired

Seongsoo Cho, Son Kwang Chul, Chung Hyeok Kim and Yunho Lee

Abstract This research is to suggest a mobile game program for rehabilitation of the hearing impaired. The suggested program is based on the characteristics of hearing-impaired children and the classification of hearing loss. The voice recognition technology is the most intimate way of delivery of information. The suggested program does not require any additional learning or training course, but enables the hearing impaired to do vocal exercises through games and helps them to enjoy rehabilitation training.

Keywords Auditory training · Voice recognition · Rehabilitation training · Hearing-impaired children · Mobile game

1 Introduction

All the games have background stories, helping developing objectives (goals), rules, adaptability, problem-solving skills and interaction [1–3].

S. Cho (✉) · S. K. Chul · C. H. Kim
Department of Electronic Engineering, Kwangwoon University,
20 Kwangwoon-ro, Nowon-gu, Seoul 139-701, Korea
e-mail: css@kw.ac.kr

S. K. Chul
e-mail: kcson@kw.ac.kr

C. H. Kim
e-mail: hyeokkim@kw.ac.kr

Y. Lee
Department of Social Welfare, Kyung Hee Cyber University, 1 Hoegi-Dong,
Dongdaemun-Gu, Seoul 130-701, Korea
e-mail: anne6@khcu.ac.kr

J. J. (Jong Hyuk) Park et al. (eds.), *Multimedia and Ubiquitous Engineering*,
Lecture Notes in Electrical Engineering 240, DOI: 10.1007/978-94-007-6738-6_76,
© Springer Science+Business Media Dordrecht(Outside the USA) 2013

Among the hearing-impaired children, those who are completely deaf and must use sign language for communication are only 10 %. Children with low hearing impairment can have a command of a language if they receive auditory training and language reinforcing training during the critical period for language development (36–40 months old). Hearing-impaired children receive various kinds of treatment, including play psychotherapy, art therapy, music therapy and cognitive therapy, as well as auditory training and language reinforcing training, at rehabilitation centers for the disabled, clinics and hospitals, and rehabilitation clinics, but it is found that their satisfaction levels with those therapies have not been high [4].

This research describes characteristics of hearing-impaired children, classifies them based on level of hearing loss, provides plan and design for mobile games through analysis of needs, and finally suggests the direction of future research.

2 Related Studies and Technologies

2.1 Characteristics of Hearing-Impaired Children

Children who lost their hearing severely before the language acquisition period have very limited accommodation of language through speech sound, so internalization of a language can be delayed in comparison with normal children. Schum (1991) found that 18-month-old hearing-impaired children use 0–9 words only in average, while 22-month-old hearing-impaired children receiving an oral training program can command approximately ten spoken words. Griswiod and Cmmings (1974) found that hearing-impaired children at the age of 4 command approximately 158 spoken words, while normal children at the same age use approximately 2,000 words. In a research on the gap of language functioning between hearing-impaired children and normal children, it was found that only 75 % of 5-year-old hearing-impaired children know the words which are known by all 4-year-old normal children. Hearing-impaired children at the age of 3 and 4 use gestures and hand signs. In a pointing test, it was revealed that only a part of 5-year-old hearing-impaired children give answers for questions which are answered by all 4-year-old normal children, and it was not possible for hearing-impaired children at their age of 3 and 4 to give answers for the same questions. The results show that hearing-impaired children have a limited vocabulary, and 5-year-old hearing-impaired children have no sufficient understanding of abstract words [5, 6].

2.2 Grades of Hearing Impairment

Hearing impairment is divided into grades based on the degree of loss of hearing. Loss of hearing is measured with internationally-approved audiometers. Table 1

Development of Mobile Games for Rehabilitation Training

Table 1 Classification of hearing-impaired children based on the degree of loss of hearing

Grade	Average loss of hearing (dB)	Characteristics
Low hearing impairment	27–40	Difficult to speak clearly and understand words spoken in a low voice
Medium hearing impairment	41–55	Understand a face-to-face conversation with a person at a distance of 1 m, but do not understand about 5 % of a group discussion
Medium–high hearing impairment	56–70	Understand a conversation in a loud voice only, but difficult to participate in a group discussion
High hearing impairment	71–90	Understand words spoken loudly in ears Distinguish some vowels but no consonants at all
Deafness (top hearing impairment)	91 or higher	Sense a loud voice through vibration rather than tone, and have a language defect Depend on vision rather than sound for communication

shows the classification of hearing-impaired children based on the standard established by International Organization for Standardization (ISO).

3 Voice Recognition Technology and its Development

The voice recognition technology is a kind of pattern-recognition process, through which input voice is analyzed by a computer, and is converted into a similar command through the voice model database. Because each person has his/her own tone, pronunciation and intonation, the standard pattern is created based on common characteristics extracted from voice data from people as many as possible.

Research on voice recognition technology was first started when AT&T Bell Laboratories developed the single-voice number recognition system called 'Audrey'. In 1963, IBM released the world's first voice recognition device called Shoebox, which recognized 16 English words and supported simple calculation. Since then, government laboratories of the United States, the United Kingdom, Japan and the Soviet Union have developed exclusive hardware devices that can recognize human verbal output, extending the voice recognition technology to four vowels and nine consonants. From 1971 to 1976, Defense Advanced Research Projects Agency (DARPA) under US Department of Defense implemented the largest voice-recognition research project ever in history, called Speech Understanding Research. As voice portal services became popular in 2000s, Voice VML (VXML) 1.0 was established as the standard language for voice-based Internet use [7]. The voice recognition technology has been widely adopted in various fields, including home appliances, computer and information device.

4 Implementation of Mobile Game Design

The suggested game is developed based on Android as OS, JDK 1.7, Eclipse 3.72 and Android SDK as development tools, and Android OS SmartPhone as the execution environment.

The suggested functional game is designed for speaking training and phonetic correction. It is designed for the hearing-impaired to enjoy the rehabilitation training through smart phone games. This Android-based game gradually increases its level of difficulty, enabling users to exercise, while competing with each other.

Figure 1 shows the principles of the voice recognition technology and Fig. 2 shows the sources for the voice recognition. The rehabilitation training game convinces users of its treatment effect, and helps users to overcome a sense of uneasiness and fear.

Players start the game at the basic level, and as the level grows, acquire cognitive skills while competing with each other.

At the starting level of the game, the number of bugs which obstruct the play increases as the play time gets longer. At the second and the third level, longer and harder-to-pronounce words are given as the play time gets longer. The game layout consists of the main screen, game-selection screen, and game-play screen. On the

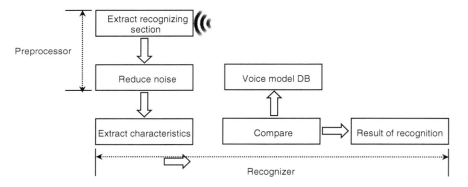

Fig. 1 Principle of voice recognition technology

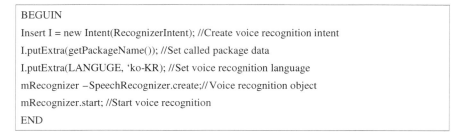

Fig. 2 Sources for voice recognition

main screen, users select the game title and the athletic location as the background. Touch the screen to display the game-selection screen. The game-selection screen enable users to select a game with a flip gesture, and lists the best score of the game. The first game screen displays the graph on the right to measure the decibel and the score at the top of the screen. The second and the third game screen show Life and Score at the top left of the screen.

5 Conclusion

The mobile functional game program proposed in this research is designed for rehabilitation training for the hearing-impaired. The functional game is designed for rehabilitation treatment through phonetic correction and speaking training, and is implemented to run on smart phones so that users can play anytime and anywhere. The functional games are expected to be very effective for rehabilitation. To further enhance the effect of rehabilitation treatment, it is required to develop a hearing training program, and to have the hearing impaired to finish the hearing training course before phonetic correction and speaking training course. It is required to develop phonetic correction games for language training, in addition to the functional games, through research in the hearing training field. This research is expected to promote utilization of the positive aspects of mobile games.

References

1. Federation of American Scientists (2006) R&D challenges in games for learning. Report of the learning federation. http://www.fas.org
2. Gee JP (2003) What video games have to teach us about learning and literacy, computers in entertainment (CIE)—theoretical and practical computer applications in entertainment archive, vol 1(1). Palgrave/Macmillan, New York
3. McFarlane A, Sparrowhawk A, Heald Y (2002) Report on the educational use of games. TEEM (Teachers Evaluating Educational Multimedia). http://www.teem.org.uk
4. Nickes L, Howard D (2004) Dissociating effects of number of phonemes, number of syllables, and syllabic complexity on word production in aphasia: it's the number of phonemes that counts. Cogn Neuropsychol 21(1):57–78
5. Metz DE, Samar VJ, Schiavetti N, Sitler RW, Whitehead RL (1985) Acoustic dimensions of hearing-impaired speakers intelligibility. J Speech Hear Res 28:345–355
6. Beate P, Carol S-G (2005) Timing errors in two children with suspected childhood apraxia of speech (sCAS) during speech and music-related tasks. Clin Linguist Phonetics 19(2):67–87
7. Juang B-H, Furui S (2000) Special issue on spoken language processing. Proc IEEE 88(8):1139–1141

A Study to Prediction Modeling of the Number of Traffic Accidents

Young-Suk Chung, Jin-Mook Kim, Dong-Hyun Kim and Koo-Rock Park

Abstract Traffic accidents are one of the big problems of modern society. The social damage caused by the traffic accidents are increasing. So, there have been a variety of research analyses to predict the traffic accidents. But there are few studies to predict the frequency of traffic accidents. In this paper, the modeling proposes applying the Markov chain modeling to predict the traffic accidents. In this paper, it is expected that the proposed traffic accident prediction modeling to predict the number of traffic accidents.

Keywords Simulation · Crime statics · Predictive model · Traffic accident

1 Introduction

Despite the efforts to reduce traffic accidents, the Korea still has many traffic accidents occur.

Looking at the current status of the OECD in case of an accident in 2009, the number of traffic accident deaths per 100,000 is 12.0 people. Greece following up

---

Y.-S. Chung · K.-R. Park (✉)
Division of Computer Science and Engineering, Kongju National University, Cheonan, Korea
e-mail: ecgrpark@kongju.ac.kr

Y.-S. Chung
e-mail: merope@kongju.ac.kr

J.-M. Kim
Division of Information Technology Education, Sunmoon University, Asan, Korea
e-mail: calf0425@sunmoon.ac.kr

D.-H. Kim
Department of IT Management, Woosongn University, Daejeon, Korea
e-mail: dhkim@wsu.ac.kr

J. J. (Jong Hyuk) Park et al. (eds.), *Multimedia and Ubiquitous Engineering*, Lecture Notes in Electrical Engineering 240, DOI: 10.1007/978-94-007-6738-6_77, © Springer Science+Business Media Dordrecht(Outside the USA) 2013

of the OECD countries is high [1]. The study was conducted in order to reduce traffic accidents. After confirming the relationship of a traffic accident occurs, the type and severity of traffic accidents, the risk of type is presented, and the characteristics of the driver and the studies were to investigate the relationship between traffic accidents [2]. There is a traffic accident forecasting model based on the curve radius, gradients, such as road traffic accident that occurs at the highway turnoff for linear elements to Study [3]. Prediction and detection system design for the proposed studies have the bridge section of the road freezing [4]. However, so far, the progressed traffic accidents studies analyzed the type of traffic accidents. And the study was conducted according to the forms of the road traffic accidents relating to the prediction. There are a few studies to predict the incidence of traffic accidents.

This paper proposes modeling to predict the number of traffic accidents. The implementation is being used to studies the various predictions by applying a Markov chain modeling to predict the number of traffic accidents. This paper is organized as follows. Section 2, related to the studies of Markov chains is discussed. Section 3, there will be discussed proposed of a traffic accidents prediction modeling, in this paper. Finally, conclusions and future research are discussed.

2 Markov Chain

Markov processes in discrete stochastic process representing the Markov chain are called [5]. In any case, the previous state to the current state of the Markov chain will affect and from the state's past does not affect the probability of the process.

A random time $t_1 < t_2 < \cdots < t_k < t_{k+1}$,

About $X(t)$ when discrete value, the Markov chain.

The following formula (1), expressed as

$$\begin{aligned} P[a < X(t_{k+1}) &= x_{k+1} | X(t_k) = x_k, \ldots, X(t_1) = x_1] \\ &= P[X(t_{k+1}) = x_{k+1} | X(t_k) = x_k] \end{aligned} \tag{1}$$

Equation (1) then t_k: the current point in time, t_{k+1}: Future point in time, $t_1, \ldots, t_{k+1}$: Past the point.

Markov chain is a set of states (group of the state), the initial probability (the initial probability vector), each transition matrix (transition between each state) the configuration [6]. Markov chain to predict the future has been used in various fields. To predict the movement of the occupants within the House of Commons in order to maintain comfortable indoor air, has been applied to studies [7]. And Rubber tired AGT system of vehicle operating condition has been applied to studies modeling to predict the reliability and availability [8].

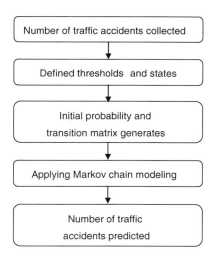

Fig. 1 The number of traffic accidents occur predictive modeling

3 The Number of Traffic Accidents Occurs Predictive Modeling

The number of traffic accidents occurs predictive modeling is shown in Fig. 1.
Each step is executed as follows.

First, collect the number of traffic accidents.
Second, Thresholds are set to the number of traffic accidents by analyzing the collected.
And define the status of each of the thresholds.
Third, the number of traffic accidents and the state defined mapping.
Fourth, Mapped, using state of the initial and transition probability matrix is generates.
Fifth, Initial probability and transition matrix is applied to the Markov chain modeling.

And, calculate the probability of traffic accidents occur in the future.

Finally, the predicted value of the probability of occurrence traffic accidents occurred recently.

Applied to the average of the number of traffic accidents occur. And predict the number of Traffic accidents.

4 Conclusion

In this paper using a traffic accident statistics, the number of traffic accidents prediction modeling is proposed. Prediction utilized in the foreseeable future study applied the Markov chain. If you predict the number of traffic accidents is expected to be helpful for policy formulation to reduce the damage caused by a traffic accidents.

References

1. Traffic accident statistics. http://taas.koroad.or.kr/index.jsp
2. Shim K-B (2009) The determination of risk group and severity by traffic accidents types—focusing on Seoul City. J Korea Soc Road Eng 11(2):195–203
3. Choi Y-H, Oh YT, Choi K, Lee CK, Yun I (2012) Traffic crash prediction models for expressway ramps. J Korea Soc Road Eng 14(5):133–143
4. Sin G-H, Song Y-J, You Y-G (2011) Ridge road surface frost prediction and monitoring system. J Korea Contents Assoc 11(11):42–48
5. Grinstead CM (1997) Introduction to probability, 2nd revised edn. American Mathematical Society, Providence, pp 405–406 (in press)
6. Kim Y-G, Baek Y, Peter In H, Baik D-K (2006) A probabilistic model of damage propagation based on the Markov process. J KIISE 33(8):524–535
7. Kim Y-J, Park C-S (2008) Prediction of occupant's presence in residential apartment buildings using Markov chain. Korea Institute of Architectural Sustainable Environment and building System. 2008 autumn conference, pp 116–121
8. Ha C-S, Han S-Y (2004) Reliability evaluation of AGT vehicle system using Markov chains. 2003 autumn conference and annual meeting of the Korean society for railway, pp 91–96

Part XI
Pervasive Services, Systems and Intelligence

A Wiki-Based Assessment System Towards Social-Empowered Collaborative Learning Environment

Bruce C. Kao and Yung Hui Chen

Abstract The social network has been a very popular research area in the recent years. Lot of people at least have one or more social network account and use it keep in touch with other people on the internet and build own small social network. Thus, the effect and the strength of social network is a very deep and worth to figure out the information delivery path and apply to digital learning area. In this age of web 2.0, sharing knowledge is the main stream of the internet activity, everyone on the internet share and exchanges the information and knowledge every day, and starts to collaborate with other users to build specific knowledge domain in the knowledge database website like Wikipedia. This learning behavior also called co-writing or collaborative learning. This learning strategy brings the new way of the future distance learning. But it is hard to evaluate the performance in the co-writing learning activity, researchers still continue to find out more accurate method which can measure and normalize the learner's performance, provide the result to the teacher, assess the student learning performance in social dimension. As our Lab's previous research, there are several technologies proposed in distance learning area. Based on these background generation, we build a wiki-based website, provide past exam question to examinees, help them to collect all of the target college or license exam resource, moreover, examinees can deploy the question on the own social network, discuss with friends, co-resolve the questions and this system will collect the path of these discussions and analyze the information, improve the collaborative learning assessment efficiency research in social learning field.

B. C. Kao (✉)
Department of Computer Science and Information Engineering, Tamkang University, New Taipei, Taiwan, People's Republic of China
e-mail: acebruce@gmail.com

Y. H. Chen
Department of Computer Science and Networking Engineering, Lunghwa University of Science and Technology, Taoyuan, Taiwan, People's Republic of China
e-mail: cyh@mail.lhu.edu.tw

J. J. (Jong Hyuk) Park et al. (eds.), *Multimedia and Ubiquitous Engineering*, Lecture Notes in Electrical Engineering 240, DOI: 10.1007/978-94-007-6738-6_78, © Springer Science+Business Media Dordrecht(Outside the USA) 2013

Keywords Social learning · Social network · Wiki · Past exam · Co-writing · Collaborative learning · Assessment

1 Introduction

'Wiki' is the Hawaiian word for 'quick'. Broadly, wiki is a open and convenient editing tool for participants to visit, edit, organize and update website. The most well-known public wiki is Wikipedia [1], which is an online encyclopedia. The most distinguished feather of wiki is that anyone can create knowledge on wiki at any time anyplace, and the intelligence of more creators is much greater that individual creators, it offers multidirectional Communications among creators [2].

After Web 2.0 technology has been proposed, wikis have been widely used in the realm of education, and serve as a medium for collaborative learning. In a scenario of wiki-based collaboration, students are divided into groups and assigned tasks. Liu et al. [3], and, everyone who use internet service in the world usually have one or more accounts of social network site in recent years, social network behavior in cyber world is become more and more important part in peoples life, if we can combine these two different but have same basic ideas web service, apply on blended learning to improve the learning efficiency.

In Taiwan, the higher education's examinations past question just provide the questions but without answers, so the examinees must calculate the answer by itself. Or spend lots of money to join the tutorial, even search the answer on Wikipedia, also can not understand the problem solving process and learning with friends. Thus this MINE wiki service provide a community and customized edit tool to help examinees discuss and discover, make sure the answer and record the

The main method we apply in this research is based on the Prof. Trentin's research [4], this paper is focus on the evaluation of collaborative learning project, he design a formula to calculate the contribution of each student in the learning group, also use the Wiki-like system to do the experiment, but most of this evaluation procedure needs the manual assessment by teacher. So there are many procedures can be improved, like system assist data mining, recording and calculating, reduce the assessment time for teacher, and help the student to learn the co-writing skill in social dimension then assess the contribution correctly.

In this study, we describe a Wiki Based Web system which provide online past exams about admission of master's degree or PhD degree. And apply the co-writing assessment theory provide by [4] in this system to help other user understand which answer is the best one of the questions, is not learning in the group, but to use the social network, Students may search and have discussion with other register users in this system, if users doesn't know how to resolve some questions, he or she can deliver to own social network to other friends who maybe know the answers or forward this message to their own social network until the question is resolved.

2 Related Work

In this section, we will discuss the past research about social learning or the system use wiki-based system to enhance distance learning.

In Web 2.0, one of the emerging visions is the "collective intelligence" where the folks are motivated to contribute their knowledge to solve common problems or to achieve common goal [5].

A Wiki is a type of social software that allows users to write, share and edit content real-time, with only rudimentary skills in web page creation. Anyone can edit and manage the content of Wiki, coordinate and create knowledge in collaboration with other members. The essence of Wiki consists of opening up, cooperation, equality, creating, and sharing. As a collaborative authoring platform or an open editing system, the most fundamental characteristic of Wiki is the teamwork and open editing [6].

In Marija's research, He described and evaluated two consecutive trials of the use of wiki technology as a support tool for curriculum delivery and assessment, as well as for students' learning [7]. The common characteristic of all trials was that they were based on weekly wiki (MediaWiki) updates by students that were triggered by tutor-set questions and assessed. The details of the assessment strategy for wiki contributions have been discussed in [8].

Wiki is considered the latest web innovation on content management and sharing [4]. Using any web browser, a user can visit a Wiki site—a web site running Wiki software, and by using simple Markup text, the user can create new pages, edit existing pages, or restructure page hierarchy and links. The simplicity and flexibility of Wiki make it an appealing tool for content sharing and online collaboration [9].

In Chang's research [1], a case study is based on an optional curriculum called social technology and tools in Beijing Normal University. The participants are fresh students or sophomore students with various majors. There are totally 75 students participating in this activity. This study is based on the open software 'Mediawiki'. They also choose a definite subject for this activity which is 'google products', and providing several different products for the students to choose which one they intend to edit [2].

Research on using computers to assess and support the development of social and emotional skills has focused on a range of populations including children with particular needs [10], and a major purpose of fully collaborative writing is to ease the dysfunctional anxiety of the individual solitary student when confronted with a blank piece of paper [11]. Traditional learning hopes student can learning spontaneously, but ignored the social dimension, but how to evaluate the individual contribution of each student is the first question we must surmount, but the assessment method of this area still not have enough related research can fully support the teacher to evaluate the students contribution automatically or more conveniently.

In view of these related researches, we can comprehend the collaborative learning and Wiki-based learning process or framework can help leaner improve their learning efficiency through collaborative learning; social network also can integrate with wiki service to promote the scale of social learning.

3 System Designs

In this section, we will present the system designs of this Wiki-based past exam system.

$$P_{total} = \Sigma P_{norm} = P_{forum,\ norm} + P_{peer-review,\ norm} + P_{links,\ norm} + P_{content,\ norm} \quad (1)$$

The main assessment formula we applied in this system is shown above, it is from the Trentin's research [4], and the system will accord this method and the assessment data table but improve the scoring process from manual to the system automatic, try to exclude the human factor from the teacher or other users, the main scope of this study is assess the separate contribution of each question through this method and display it to other users.

The MINE Wiki site is built by Mediawiki: the open source wiki-site software, as shown in Fig. 1, is the portal page of MINE wiki, when the user access the web site, the main pager will shows the broadcast message in middle of the windows, and the navigation area on the left side, can allow users access into the main page to start the learning sequence.

Fig. 1 The portal page of MINE Wiki

A Wiki-Based Assessment System

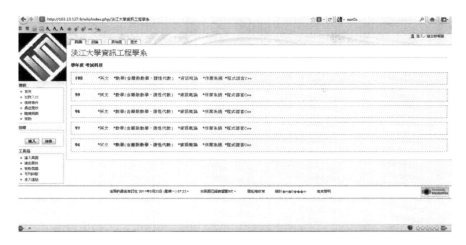

Fig. 2 Subject select page

If the user is a unregistered, the user only can allow to search and watch the past exam questions, can not edit or deliver any answer on the page and wiki, even user was invited by register user, still need create an account in this wiki.

When user login the community of MINE Wiki, the first page entered was the list of the four regions in Taiwan, in this research, MINE Wiki's default content we provide is the every universities in Taiwan, in other words, this site have the all past examination question file about Master degree and PhD. Degree's admission, users can search all examination's file in this wiki, in Taiwan, the higher education's examination's past question just provide the questions but without answers, so the examinees must calculate the answer by itself.

After user choose the region, it will enter the university select list page, when user choose an university, the system will navigate to the collage list page shown in Fig. 4, this page shows the whole colleges in this university.

When user selects the college, as shown in Fig. 2, system will list the all subject in this collage at least past five years, user may choose any subject which he needs here.

The most important part in this wiki based site, is the past examination papers, as shown in Fig. 3, the PDF files will shows in the middle of the page, users can not watch the questions in this page, and tag or mark the specific question in this file, also will save in user's learning profiles, allow users to share or take down. The bottom of this file, is the message board, any users who ever access this files and leave the question, answers or messages all will save in this place to modeling the learning process of this past examination paper.

The last part of this site, shows in Figs. 4 and 5, the ELGG social site, a open source social network website, just like the Facebook, in our Lab., we use this open source software to build a social network learning website, test and collect the social learning data to analyze the learner's learning efficiency.

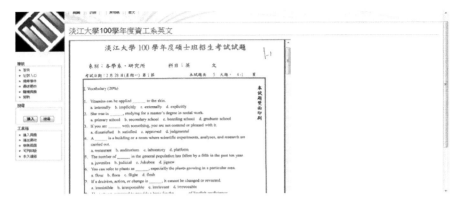

Fig. 3 Past examination paper

Fig. 4 ELGG social site

The MINE Wiki also merge this social network website, when users can't figure out the answers in the Wikis, he may publish the message to the wall in ELGG, the friends of the users will see the issue in their wall, if anyone know the answers or meet someone who may have a direct thinking way to the right answers, just need to link or forward to other friend's walls, the system will record this path of spread, build a social learning model in this Wiki.

Figure 5 shows the assessment result of each past exam paper, the system will record the contribution and the interact with other users, this result will save in the account profile, every user can access the contribution made by the user in each past exam subject, teacher also can according this result to evaluate the student performance in this social co-writing learning process.

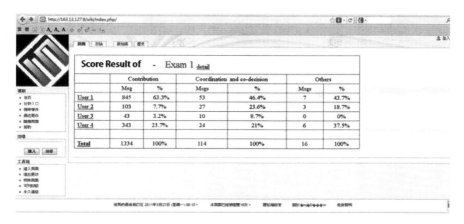

Fig. 5 Assessment result

4 Conclusion

In this study, we proposed a Wiki-based past exam System which can provide examiners search and discuss with other people in the website, this Wiki system provide lots of past exam questions of all Taiwan University's Master Degree admission, if user can not find the past examination questions which he need, user can upload any question PDF files by himself, and build the wiki page about these questions, then deploy to his own social network website like Facebook, twitter, pluk, Elgg...etc. to ask friends who can resolve the questions or join the discussion, even they are all can not figure out the answer, they still can deploy to their own social network to ask another friends, use the strength of the social network, extend the co-writing learning field to the internet.

The system will record this process, and build a learning process of this past examination questions file, analyze the efficiency of learning. Construct the analyze result to the teacher or other users, allow teacher to use these resource to understand what kind of skills and gain student learned during this co-writing learning activity, it also can let other users can understand their knowledge contribution turn into normalization numbers, compare with others, comprehend their knowledge level they reach.

Our future work includes: (a) find out more appropriate user model or education theory to support this wiki based web service even blended social learning; (b) after build a robust user model of this learning process, we may design a assessment mechanism with strong education background support to evaluate users learning efficiency more accurate.

References

1. Chang B, Zhuang X (2008) Wiki-based collaborative learning activity design: a case study. In: International conference on computer science and software engineering
2. Tseng S-S, Weng J-F (2009) Wiki-based design of scientific inquiry assessment by game-based sratch programming. In: IEEE international conference on advanced learning technologies (ICALT)
3. Liu B, Chen H, He W (2008) Wiki-based collaborative learning: incorporating self-assessment tasks. In: Proceedings of the 4th international symposium on Wikis (WikiSym), ACM
4. Trentin G (2009) Using a wiki to evaluate individual contribution to a collaborative learning project. J Comput Assist Learn 25(1):43–55
5. Lu Q, Chen D, Hu H (2010) Wiki-based digital libraries information services in China and abroad. In: Wireless communications networking and mobile computing (WiCOM)
6. Leuf B, Cunningham W (2001) The wiki way: quick collaboration on the web. Addison-Wesley, Boston
7. Cubric M (2007) Wiki-based process framework for blended learning. In: Proceedings of the 4th international symposium on Wikis (WikiSym), ACM
8. Cubric M (2007) Using wikis for summative and formative assessment. In: International online conference on Re-engineering assessment practices (REAP), May 2007
9. Xu L (2007) Project the wiki way: using wiki for computer science course project management. J Comput Sci Coll 22(6)
10. Jones A, Issroff K (2005) Learning technologies: affective and social issues in computer-supported collaborative learning. J Comput Educ 44(4):395–408
11. Sutherland JA, Topping KJ (1999) Collaborative creative writing in eight-year-olds: comparing cross-ability fixed role and same-ability reciprocal role pairing. J Res Reading 22(2):154–179
12. http://www.en.wikipedia.org

Universal User Pattern Discovery for Social Games: An Instance on Facebook

Martin M. Weng and Bruce C. Kao

Abstract With the population of social platform, such as Facebook, Twitter and Plurk, there are lots of users interest in playing game on social platform. It's not only the game style they want to play, but also high interaction with other users on social platform. Hence, the development of social network makes social games as a teaching tool becomes an emerging field, but in order to use the social network games for teaching, it needs to understand the game flow as an appropriate reference to be judged. In this paper, we use simulation games and social utility games of Facebook as examples. Using the flowcharts and the triangulation methods theory to analyze the characteristics of the flows by finding verbs and goals, and obtaining the differences by comparing social behavior, accumulate of experience, items collection system and tasks.

Keywords Social game · Flowchart · Simulation game · Social utility game

1 Introduction

Web 2.0 application, such as Facebook and Twitter, have dramatically increased within the last 5 years for social purposes. The provision of common channel for increasing social interactions has revealed the major attractive issue, The survey [quote the related papers] also presents that the entertainment part (e.g. add-on games) has the potential for raising the interactions.

M. M. Weng · B. C. Kao (✉)
Department of Computer Science and Information Engineering, Tamkang University,
New Taipei, People's Republic of China
e-mail: acebruce@gmail.com

M. M. Weng
e-mail: wm25@hotmail.com

J. J. (Jong Hyuk) Park et al. (eds.), *Multimedia and Ubiquitous Engineering*,
Lecture Notes in Electrical Engineering 240, DOI: 10.1007/978-94-007-6738-6_79,
© Springer Science+Business Media Dordrecht(Outside the USA) 2013

We found many attractive features in social games. But how to use the advantages of online social network for education? And how to use social games as a new game-based learning method? It is still an untapped research issues. However, there are many types of social games in social community. If we want to find out the developing prototype in various of social games which consistent with the educational purposes, we need to comprehend the process in different type of social games first.

In this paper, we use simulation games and social utility games from Facebook as examples, and generalize the game frameworks of the two types we mentioned before. Finally, we make analysis and comparison. Facebook is one of the emerging social websites in recent years. A social network is a set of clustered nodes (group of people) or single nodes interconnected to transmit the information from one cluster or node to another. If we compare Facebook with other social websites, we can find that Facebook offers many services and applications. And one of the popular application and service is social game. Because of the information transformant and sharing functions in social networks are cooperative, social games are designed to use the way of cooperative games [1]. In this contention, author consider that social games have the social property of interaction and cooperation.

2 Related Works

2.1 Triangulation Method

Aki Järvine proposeda Verbs–Goals–Network play model of triangulation, which helps to define 'game mechanics' [2]. The verbs as mechanics are linked with the goals of the game, which are the means to reach the ends. In social network games, the system of the social network is as a whole, consisting of the service, individual players, and the community. The dynamic within these elements can be conceptualized as a triangle with the three elements, around the user experience that start to emerge as play (Fig. 1).

This paper refers to the model as the method which analyzes the features of the game process. On social platform, we only focus on Facebook which is the most popular social platform in social community. Then looking for the commonality among player's actions or behaviors in flowcharts, sort out the corresponding verbs to reach the goals. Verbs and goals are the basis in this research which use to judging whether game processes are similar or not.

Universal User Pattern Discovery for Social Games 643

Fig. 1 Verbs–goals–network play model

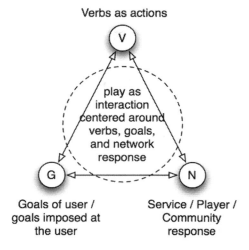

3 Features of the Games

Since the current game types are becoming more and more composite. The basis for subjects' selection in this study is selecting the games which do not mix with the elements in other game types significantly. This paper selects three simulation games: Restaurant city, My Fishbowl, Happy Harvest, and three social utility games: Hugged, Give Hearts, The King of Kidnappers, to be the subjects of this study We use those selecting games to sort out their own game flowcharts, extracts common features and compares them.

3.1 Simulation Games

The following contents are the three simulation games which we select in this paper:

- Restaurant city

  Restaurant city is a restaurant simulation game, players open their own restaurant, and find some ways to collect ingredients, increase their menu and keep the operation of the restaurant. When the level up, it increase the scale of the restaurant gradually.

- My Fishbowl

  My Fishbowl is a fish raising game. Players can see their own fishbowl when login to the game, and obtain treasures in the process of raising fish or access friends' fishbowls to steal treasure. Treasures in the game can be sold into money for buying more kinds of fish or fish feeds, which continue to operate their own fishbowl.

- Happy Harvest

Happy Harvest is a farming game, players' main work is planting crops and selling them into money. They can also go to friends' house to steal crops or hinder crops grow. The money can be use to buy items needed for farm operation.

3.2 Social Utility Games

The following contents are three social utility games which select in this paper:

- Hugged

Hugged is a simple game to communicate emotions with friends. Players select their friends and send them a type of hug. And the receiver will receive the hug and the associated picture by message.

- Give Hearts

Give Hearts is a game of heart gifting, players select hearts and gift to their friends. In the game, we can see the types of hearts he collect and the ranking of hearts number friends' get.

- The King of Kidnappers

The King of Kidnappers is a game of kidnapping and rescuing friends. Players use all kinds of funny ways and tools to kidnap their friends. Players may become captives by their friends, and use the way which guess the hiding place to escape or rescue friends.

3.3 Features of the Games

- The use of currency

The reason why virtual currency is developed have been mentioned by Hui Peng and Yanli Sun. Then main reason is mentioned as below: 'The demands of virtual economy development', 'Technical progress and internet development', 'Demands from users', and 'The profit maximization of issuers' [3]. Those games are developed on Facebook platform. It will certainly be developed and use virtual currency due to the demands from users and issuers, with sufficient support of network technology.

- Social behavior

The nature of social networks is built during play [4]. Because the games are developed on the social networking platform, spontaneously be inseparable with community interaction. But they stressed that the players interact with the user must have any relationship with other users (such as friends). This means that social games have some standard to the scope of relationships between interaction players and the user. Only players who have a certain relationship with the user can participate. This can avoid unacquainted players allot the incomes which get from interactions in the games (steal resources). Nabeel Hyatt has pointed out how social network games can 'rely heavily on social context (namely school, department, and residence loyalties) to provide a framework for alliances, game playing and motivation' [5]. Social context is the interpersonal relationships in the player's real life, and social networking platform extends the relationships into the games. This also becomes the main scheme of the interactions in simulation games.

According to Aki Järvinen's argument, game design techniques are needed to integrate competition, challenge, and tension into those acts of socializing, and mostly of the integration would improve the playfulness of the games [2]. The competition in social games are established in the interactions with friends, but maintain the idea of non-zero-sum games basically, with certain risk control. Friends can compete with each other, but don't come to win or lose directly; friends can steal each other's resources, but there are certain quantitative restrictions, this protects the players' incomes.

- Accumulate experience points to upgrade and get reward

Kirman, Lawson, and Linehan have mentioned that in Fighters' Club and Familiars, player score (Street Credit in FC) is a function of the social behavior of the person within the game [4]. This paper considers that in most social games which use experience points, one of the sources of experience points is also a function of the social behavior of the person within the game. In addition, the experience points are also the incomes of a player's labor in the game. As the same with score, the more a player works, the more experience points he will gets.

Valentina Rao lists fast rewards for player actions, abundance of positive feedback, no negative consequences for exploration [6]. Experience point is an income in fast rewards, which won't be reduced by player actions. This encourages players to act in the game positively. The value of the experience points that player gains will enhance his status and reputation in social network games. As the triangulation method mentioned, this cyclic process of development plays the core mechanics of the games [2]. The incomes of the players' operations will reach the goals. Social behavior such as visiting friends, is one of the source of experience points, and in this process, exchange and compare with friends will be spontaneous.

- Item collection system

Kim has identified certain core game mechanics, i.e. player actions, such as collect and exchange [7]. Those actions promote the games to run. Both of the items collected and accumulated are experience points that are incomes from players' actions in the game, the difference is that items collected in some games may burden risk because the players cannot always keep in the game, for example, goods are stolen by friends, but experience points cannot be stolen by using any channel; This is the difference between item collection and experience points accumulation.

Item collection also led the need to add features to the game. Hence, it would evolve with players' needs, achievements in the form of different badges, new types of cars, etc. [2].

- Task

Task has a feature that is various situations. No matter how many changes are in the form of the tasks, generally around the verbs, goals theories of triangulation method, simply changing the way of the tasks to add more fulfillment.

4 Conclusion

In recent year, social network issues have become very popular and the application of social network also play an important role on related research issues.

In this paper, we used the flowcharts and triangulation method theory to analyze and compare the characteristics of simulation games and social utility games in Facebook. As a contribution, the discovered pattern can be applied onto the design of social learning games by integrating customized factors. The result will be provided as suggestions when developing the related social games with educational purposes. In the future, we will focus on more social game on Facebook or other social platform, and analyze the flowchart of game and show the compare results of different games in similar categories.

References

1. Reddy YB (2009) Role of game models in social networks. In: Conference on computational science and engineering (CSE), pp 1131–1136
2. Järvinen A (2009) Game design for social networks: interaction design for playful dispositions. In: Proceedings of the ACM SIGGRAPH symposium on video games, pp 95–102
3. Peng H, Sun Y (2009) Network virtual money evolution mode: moneyness, dynamics and trend. In: Conference on information and automation (ICIA), pp 550–555

4. Kirman B, Lawson S, Linehan C (2009) Gaming on and off the social graph: the social structure of facebook games. In: Conference on computational science and engineering (CSE), pp 627–632
5. Hyatt N (2008) What's wrong with facebook games? http://nabeel.typepad.com/brinking/2008/01/whats-wrongwit.html
6. Rao V (2008) Playful mood: the construction of facebook as a third place. In: Proceedings of the 12th international conference on entertainment and media, Mindtrek 2008, pp 8–12. http://portal.acm.org/citation.cfm?id=1457199.1457202&coll=Portal&dl=GUIDE&CFID=24746181&CFTOKEN=85617762
7. Kim AJ (2008) Putting the fun in functional, applying game mechanics to functional software. http://www.slideshare.net/amyjokim/putting-the-fun-infunctiona?type=powerpoint
8. Sharabi A (2007) Facebook applications trends report, 19 Nov 2007. http://no-mansblog.com/2007/11/19/facebookapplications-trends-report-1/
9. Järvelin K, Kekäläinen J (2004) IR Evaluation Methods for Retrieving Highly Relevant Documents. In: ACM international conference on information retrieval
10. Jensen D, Neville J (2002) Data mining in social networks. In: Proceedings of national academy of sciences symposium on dynamic social network analysis
11. Thomas LS (1990) Decision making for leaders-the analytic hierarchy process for decisions in a complex world. RWS Publications, Pittsburgh
12. Andersson N, Broberg A, Bränberg A, Janlert L-E, Jonsson E, Holmlund K, Pettersson J (2002) Emergent interaction—a pre-study. UCIT, Department of Computing Studies, Umea University, Umea, Sweden
13. Caillois R (1962) Man, play and games. Thames and Hudson, London, p 12
14. Carroll JM, Aaronson AP (1988) Learning by doing with simulated intelligent help. Commun ACM 31(9):1064–1079
15. Chalmers M, Galani A (2004) Seamful interweaving: heterogeneity in the design and theory of interactive systems. In: Proceedings of the ACM designing interactive systems (DIS2004)
16. Dourish P (2004) What we talk about when we talk about context. Pers Ubiquit Comput 8(1):19–30
17. Prensky M (2001) Digital game based learning. McGraw-Hill, New York
18. Sierra JL, Fernández-Valmayor A, Fernández-Manjón B (2006) A document-oriented paradigm for the construction of content-intensive applications. Comput J 49(5):562–584

Ubiquitous Geography Learning Smartphone System for 1st Year Junior High Students in Taiwan

Wen-Chih Chang, Hsuan-Che Yang, Ming-Ren Jheng and Shih-Wei Wu

Abstract In recent years, ubiquitous learning becomes more and more popular. Geography learning can be adapted into smart phone platform to be very useful learning system. Junior high school students can assess the smartphone to study geography in Taiwan. With simple test items, the system will generate individual learning profile and test analysis report for students.

Keywords Ubiquitous learning · Geography learning · Smartphone

1 Introduction

Over the years, the progress of the E-Learning shows the significant development in content digitalization and learning technologies obviously. During this progress, the kernel value of E-Learning is created by constructing new learning style with new technologies such as network or multimedia. Meanwhile, some traditional pedagogic theories shifted to underpin the development of learning style.

Many studies tried to apply auxiliary elements to conventional e-learning for improving the learning efficiency and enriching the learning motivation. In addition, the mobile technology is now widely used and the related mobile facilities are

---

W.-C. Chang · M.-R. Jheng · S.-W. Wu
707, Sec.2, WuFu Rd, Hsinchu, 30012 Taiwan, People's Republic of China
e-mail: yilan.earnest@gmail.com

M.-R. Jheng
e-mail: jhengmingren@gmail.com

S.-W. Wu
e-mail: sware1786@gmail.com

H.-C. Yang (✉)
152, Sec. 3, Beishen Rd, Shenkeng dist, New Taipei City, 222 Taiwan People's Republic of China
e-mail: hsuanche.yang@mail.tnu.edu.tw

J. J. (Jong Hyuk) Park et al. (eds.), *Multimedia and Ubiquitous Engineering*, Lecture Notes in Electrical Engineering 240, DOI: 10.1007/978-94-007-6738-6_80, © Springer Science+Business Media Dordrecht(Outside the USA) 2013

also affordable to the public. As a result, more and more research put emphasis on the integration of e-learning and mobile technology, and the terms of "m-learning" (mobile learning) and "u-learning" (ubiquitous learning) then become the magic words while talking about the technology enhanced learning. The mobile learning allows the learning activities to be performed no longer limited to specific location, and accordingly provides more learning opportunities for learners with the mobile technology. Churchill and Churchill [1] pointed out the mobile devices play the roles of multimedia-access, connectivity, capture, representational, and analytical tools for mobile learning activities. Eschenbrenner and Nah [2] also revealed that the learning performance and efficiency can be much more improved with the benefits brought by mobile technologies. They also found that the mobile learning activity can be applied to encouraging the ability of problem solving, and to providing the opportunity of self-regulation learning. Furthermore, with the mobile technology supported, it's more practical to achieve the collaborative learning. The mobile learning is not only changing the way of traditional learning style and behaviors, but also impacting the way in future learning. Most of the m-learning applications and researches mainly focus on the "mobility", the first half of "m-learning", including the accessibility, the transferability, and the content delivering through the mobile learning devices, which makes learning activities no longer limited to specific time and space. However, there is the other half in m-learning, and that is the "learning" which is usually disregarded in the learning context. As a result, it's hard for people to find out the interrelationship between mobility and learning activity. An approach to this issue is typically called the location-aware learning that integrates both location-based and contextual learning activities.

Geography learning needs map and geographic background knowledge for beginner. Geography forms the basis for understanding our political and physical realities. A great of geography education can be effectively taught through geography interactive learning materials and maps. Combining the mobile devices, the geography can be more fun and attractive.

The organization of this paper is as follows. Section 2 introduces related technologies and background knowledge in this work. In Sect. 3, we discuss the design of the smart phone learning applications for geography learning topic, and we also take account some significant pedagogical methods to support our proposed ubiquitous learning environment. In Sect. 4 comes the conclusion.

2 Related Works

2.1 Geography Learning

Geography is a Obligatory course from elementary education. Geography learning makes people learn human, physical and geography environment in the real life. In traditional geography learning, teacher used the map and figures assist learners in the classroom.

Hakan et al. [3] designed and developed a three-dimensional educational computer game for learning about geography by primary school students. Twenty four students in fourth and fifth grades in Ankara, they learnt about world continents and countries through this game for 3 weeks. The effects of the game environment on students' achievement and motivation made significant learning gains by participating in the game-based learning environment. These positive effects on learning and motivation, and the positive attitudes of students and teachers suggest that computer games can be used as an ICT tool in formal learning environments to support students in effective geography learning (Fig. 1).

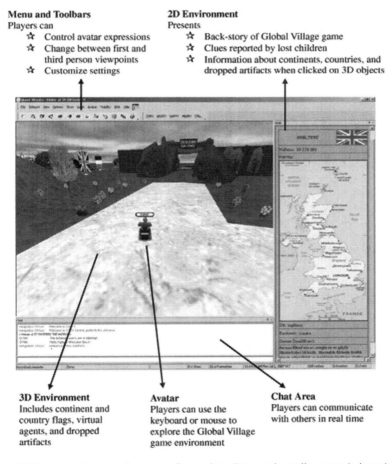

Fig. 1 MAP game and geography game. *Source* http://www.sciencedirect.com/science/article/pii/S0360131508000985

2.2 Smartphone Platforms

Generally speaking, the mobile devices include notebook, Tablet PC, Ultra Mobile PC (UMPC), Smartphone, Personal Digital Assistant (PDA) and other portable devices with computing capability. With the phenomenal growth of GPS technology, the mobile devices equipped with the GPS functionality are now wildly used to take a huge leap toward location-aware computing. It also facilitates the integral services of complex computing and personal information seamlessly. In order to realize our proposed ubiquitous learning environment, we use the Smartphone device as the location-aware learning platform due to its flexibility and expansibility.

Some Smartphone provide open architecture as desktop computer with standard Application Program Interfaces (APIs) to allow the varied developments from the third parties. Therefore the Smartphone is so-called an open operating system and is also a mobile phone with the capability of running applications. Typically the Smartphone has the network capabilities and is often equipped with a build-in or slide-out QWERTY keyboard (Fig. 2).

2.3 Mobile Learning

In environment of traditional classroom, students are restricted in closed space. The way students to absorb knowledge is from what Teacher teaches, And context of books are boring to students, which lessens the efficiency of learning. The above-mentioned learning way is typical passive learning [3]. Learning is no more restricted to traditional classroom and the knowledge resources are no more to

Worldwide smart phone market
Shipments by platform, Q4 2011

Platform	Q4 2011 shipments (millions)	Share (%)	Growth Q4'11/Q4'10
Total	158.5	100.0%	56.6%
Android	81.9	51.6%	148.7%
iOS	37.0	23.4%	128.1%
Symbian	18.3	11.6%	-40.9%
BlackBerry	13.2	8.3%	-9.7%
bada	3.8	2.4%	39.1%
Windows Phone	2.5	1.6%	-14.0%
Others	1.8	1.1%	117.9%

Source: Canalys estimates © Canalys 2012

Worldwide smart phone market
Shipments by platform, full year 2011

Platform	Full year 2011 shipments	Share (%)	Growth Q4'11/Q4'10
Total	487.7	100.0%	62.7%
Android	237.8	48.8%	244.1%
iOS	93.1	19.1%	96.0%
Symbian	80.1	16.4%	-29.1%
BlackBerry	51.4	10.5%	5.0%
bada	13.2	2.7%	183.1%
Windows Phone	6.8	1.4%	-43.3%
Others	5.4	1.1%	14.4%

Source: Canalys estimates © Canalys 2012

Fig. 2 Analysis on market shares of smart phone by operating system in full year 2011. *Source* http://paidcontent.org/2012/02/04/419-canalys-worldwide-smartphone-shipments-overtake-pctablet-market/

textbooks. Nowadays, people can learn ubiquitously via internet, which makes learning more flexible and enjoyable [4]. Technological hardware is a must for school learning environment; meanwhile, we cannot ignore the importance of learning in natural environment because technological learning cannot be a substitute for interaction with nature while learning [5].

2.4 Ubiquitous Learning

With the development of embedded system and sensor technology, people obtain useful information through sensor network around our environment. According to individual need, the systems provide the adaptive needs automatically which called ubiquitous computing [6, 7]. Taiwan government promotes "U Taiwan" which uses RFID and wireless services integrating on digital family and internet. Under the infrastructure and hardware support, ubiquitous learning becomes a new learning trend. The ubiquitous computing makes ubiquitous learning more easily. Based on ubiquitous learning system support, students can learn more around the living world. U-learning (ubiquitous learning) connects the back end database; analyze knowledge cognition distribution and supports immediate learning feedback after students respond.

3 The Ubiquitous Geography Learning System

We proposed the geography ubiquitous learning system which is composed of population, industry (I), industry (II) and traffic learning content for junior high school students. The following shows the learning content.

(1) Population

It can not only provide adequate human resources but also be good development of a country or region by appropriate population, healthy population structure and excellent quality of the population. Due to the high population density, living environment is getting more and more pressure and potential impact. In recent years, the changes of population structure is caused by "the low birth rate", "aging" and "international migration". The main content is composed of Taiwan's population size and distribution, Taiwan's population growth, Taiwan population migration, Taiwan's demographic composition and Taiwan's population problems.

(2) Industrial (I)

Industrial is in response to a variety of human needs, of which agriculture is the basis of human beings for living. It provides good growth environment for crop by variety of terrain, and warm and humid climate in Taiwan. With the rapid economic development, Taiwan's agriculture has been towards the refinement and

Fig. 3 Population choice test item

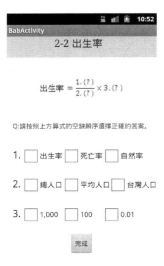

Fig. 4 Population choice test item

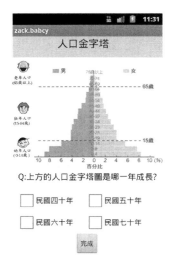

development of leisure and tourism. A wide variety range of agricultural products enhance the added value of agriculture and create a new style for Taiwan's agriculture. The main content is composed of Taiwan's industrial structure, Taiwan's primary industrial and Globalization of Taiwan's agricultural.

(3) Industrial (II)

In the past, Taiwan had the name "banana kingdom" and then it is loud of this name "Leather shoes kingdom". Nowadays we strengthen industrial restructuring actively and effort to create "Boutique of Taiwan". It is the direction of Taiwan's industrial development to build brand, pursuit of high value-added and further

Fig. 5 Industry (I) choice test item

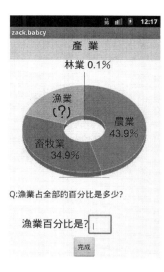

Fig. 6 Chapter score summary

more promote economic growth. The main content is composed of Industry, Characteristics of Taiwan's industrial development, Service sector and Globalization of Taiwan's service sector.

(4) Traffic

With technological advances, transport is an indispensable part of modern life. Modern transport brings us comfort and convenience, as well as the pace of life is more compact. Taiwan's transportation develops rapidly and its efficacy is rising in many forms. It is an important foundation for Taiwan's internal and external contact. The main content is composed of traffic types, Taiwan's transport and Taiwan's communications (Figs. 3, 4, 5, 6).

4 Conclusion

The Geography ubiquitous learning system provides smart phone platform for junior high school students learning. We completed four chapters for first year junior high schools in Taiwan. It is composed of population, industry (I), industry (II) and traffic. The system adapted multiple choice problem, fill-in blank problem, match problem types. Students can assess this system offline. It promotes students learning motivation and learning confidence.

Acknowledgments We would like to thank the NSC for funding this research under grants NSC-97-2511-S-032-006.

References

1. Churchill D, Churchill N (2008) Educational affordances of PDAs: a study of a teacher's exploration of this technology. Comput Educ 50(4):1439–1450
2. Eschenbrenner B, Nah FF-H (2007) Understanding highly competent information system users. In: SIGHCI 2007 proceedings
3. Tüzün H, Yilmaz-Soylu M, Karakuş T, Inal Y, Kizilkaya G (2009) The effects of computer games on primary school students' achievement and motivation in geography learning. Comput Educ 52(1):68–77
4. Lin T, Chen T (2007) The study of instructional design and learning performance in context-aware mobile learning environments. In: 2007 Information education and technological applications conference (IETAC)
5. Chen C-H, Chen Y-M (2006) The study of integrating global positioning system into mobile learning. National Pingtung University E-Learning 2006, Pingtung
6. Chen T-S, Chiu P (2007) A study of learner's behavioral intentions in a context-aware mobile learning environment. In: 2007 National computer symposium, pp. 20–21, Dec 2007
7. McDonald DS (2004) The influence of multimedia training on users' attitudes: lessons learned. Comput Educ 42(2):195–214
8. Mark W (1999) Turning pervasive computing into mediated space. IBM Syst J 38:677–692

Housing Learning Game Using Web-Based Map Service

Te-Hua Wang

Abstract Nowadays E-learning is widely used in teaching and training purposes, and it also facilitates conventional classroom-based learning activities with advanced information technologies. On the other hand, such advanced information technologies also make digital learning content easily applied to some acknowledged learning theories, such as Behaviorism, Cognitivism, Constructivism, and make e-learning being more practical and diverse in many pedagogical purposes.

Keywords Game-based learning · Web map service · Self-regulated learning · Learning motivation

1 Introduction

Many acknowledged learning theories, such as Behaviorism, Cognitivism, and Constructivism, can be realized with the improving cutting-edge e-learning technologies. The aim of E-learning has gradually turned into the process-oriented learning approach from the traditional content-based learning materials. In this paper, we proposed an online game-based learning platform utilizing the web-based map service technology. The game element lies in the completion and challenges of specific housing game missions. In addition, given the diversity of online maps mash-up services and the rise of online virtual community gaming platforms, we merge the learning content with the game-based housing missions into the proposed digital map. Learners are able to make an arrangement for acquiring the assigned game-based learning object for specific game-based learning activities according to their own preferences. With the accomplishment of

---

T.-H. Wang (✉)
Department of Information Management, Chihlee Institute of Technology,
New Taipei City, Taiwan
e-mail: tehua@mail.chihlee.edu.tw

J. J. (Jong Hyuk) Park et al. (eds.), *Multimedia and Ubiquitous Engineering*,
Lecture Notes in Electrical Engineering 240, DOI: 10.1007/978-94-007-6738-6_81,
© Springer Science+Business Media Dordrecht(Outside the USA) 2013

the learning activities, learners then can get visualized learning feedback on the game-based e-map. By introducing essential game elements to web-based learning activities, we aim at providing an attractive learning platform to motivate learners to get more involved and engaged during the online learning process. Furthermore with the storyline of the housing missions, we expect learners to develop attitudes toward active and self-regulated learning, and to realize the importance of accumulating knowledge is similar to accumulating the housing budgets, just as the old saying goes, "many a little makes a mickle".

2 Related Works

One of the most significant issues for e-learning lies in the way of enriching the learning motivation. Accordingly many outstanding e-learning researches proposed various mechanisms to motivate learners via multimodal learning activities and found that the essential elements of gaming, including challenge, curiosity, fun, instant feedback, and achievement, provide the best solution.

Gaming itself attracts people with interesting storyline and fantastic interactions, which make the players to be willing to dedicate the attention to specific gaming scenario. Game-based learning can be applied to enhance the interactivity and richness of digital learning content. By combining the game situation and the aims of education, the learning motivation can be stimulated and the attractiveness of the learning process can be enhanced as well [8]. Another interesting study also pointed out that the different game types will have different effects on the pedagogical objectives [1]. All the first and second year college students in biology department participated in the experiment; four independent groups of subjects each were applied to different types of game-based learning, including simulation games, strategy games, narrative-driven adventure games and first person shooter games. The study concluded that the adventure games and strategy games are much more acceptable for learners to get better learning performance, since these two types of games require more logical thinking, memory, imagination, and problem-solving ability. And these factors also play the essential roles of game-based learning. The research in [3] defines and conceptualizes the key factors of successful game-based learning, such as Identity, Interaction, Risk Taking, Customization, Situated Meanings, System Thinking and etc. Another work proposed a comparative analysis between entertainment and learning and found that the most challengeable and attractive game for learning is puzzle game. Puzzle game is good for understanding the appearance of the object, and the process of the puzzle develops learners' organizing ability, as well as the ability for pattern analysis [4]. Squire found that the incompletion of game might be due to the insufficient of prerequisite knowledge and is helpful for motivating learning to conquer the game barriers [10]. A significant example of successful game-based learning can be found in [7]. The authors pointed out light game for learning should contain some essential characteristics as serious game, including awarding, challenging, curiosity, fantasy, objectives, competition, cooperation and achievement, to improve intrinsic motivations for learning. Another point to note is the possible mash-up services covered in web-based technologies. Such mash-up applications can be easily found in Web-based Map Service (WMS), such as e-commerce, traffic, broadcasting, parking, online society, and etc. Map provides information by instinct, and has the attribute of leading direction. People can learn from environment about the location-aware information, including landmark knowledge, route knowledge and survey knowledge [6, 9]. To avoid learning astray, a map could be used to represent the overall interrelation among specific subjects.

3 Educational Game Design

The current trend of e-learning is towards edutainment, providing an interactive and attractive learning environment. In this work, we target at the intuition of the e-map service to enrich both the learning motivation and the learning accomplishment. With respect to the game design, the aggressiveness of being rich or having lots of houses would become a symbol of great achievement. Accordingly, we put the aggressiveness of getting rich into the game-based learning activities on the e-map. Learners are able to learn various topics on the e-map, and the acquisition of knowledge and the feedback of learning competence can be considered to accumulate the property as the housing funds for different types of buildings. Eventually, we look forward to realizing the old sayings "reading brings us everything," by using our game-based e-map learning platform. Furthermore, the learning sequence in the e-map can be self-regulated according to the learning interests. Learners are able to select the appropriate sequence of learning activities and develop the ability to seek knowledge. So that learners can intuitively understand the learning targets by using e-map and plan appropriate sequence to complete the learning tasks. And not only the objectives of learning content are obtained, but the logical thinking skills and self-learning ability are also enhanced. The task-oriented learning strategy is a constructivist teaching theory suitable for developing self-learning ability, as well as the problem-solving ability. Bae et al. revealed that learners are able to achieve predefined learning objectives via specific tasks, missions or barriers in game scenario [2]. As a result, activity-based and task-oriented learning strategy can be easily applied to edutainment. Another research defined activity theory as a philosophy process and interdisciplinary approach describing human development [5]. In other words, activity theory utilizes various exercises and practices to acquire knowledge, and affiliates learning objects with learning experience.

The abovementioned game-based learning research and practical teaching strategies highlight the process of the game and provide training activities allowing learners to find a better way to achieve the objectives according to gaming rules and storylines. A well-designed educational game provides various ways to achieve the learning objectives, and thus, different gaming process will lead to corresponding learning competency. In this study, we set up a game-based e-map to facilitate web-based learning activities by using housing mission. Eventually, by providing such learning platform, learners are able to develop the ability with self-regulated learning strategies through the task-solving process.

4 System Architecture and Implementation

The proposed learning platform mainly delivers learning content, and in addition, during the task solving process, learners are able to develop problem-solving ability, and gradually to accumulate the knowledge of the learning content.

4.1 System Architecture

The system architecture is discussed in three components. The first component is design and management of the backend database, which contains three main data tables to maintain the learning content, learning activity, and the learning portfolio. The second one includes the functionalities and services of the game-based e-map, such as the interactive e-map module, learning resource management module and learning activity management module. The third component aims at the analysis of learning competency and performance. It provides instructors and learners a review module respectively to examine the degree of the accumulation of obtained knowledge. The system architecture and functionalities are illustrated in Fig. 1, and each component can be discussed in detail as follows.

Design and Management of the Backend Database

- Learning Content: To record the information of learning content, including title, description, author, routing information, additional learning resource, difficulty, hierarchy, and knowledge domain.
- Learning activity: To connect learning content and the specific house type. Each learning subject can be considered as an activity unit, and instructors can assign the relationship between learners and learning activities.
- Learning portfolio: To track the corresponding feedback in the proposed game-based e-map learning platform, including learner id, obtained house, amount of house, the longitude and latitude information, time, duration, and the unsolved activities.

Functionalities and Services of the Game-Based E-Map

- Interactive E-Map Module: to provide the interactive event and to manage the game barrier using the Google Maps APIs. Learners are able to receive learning

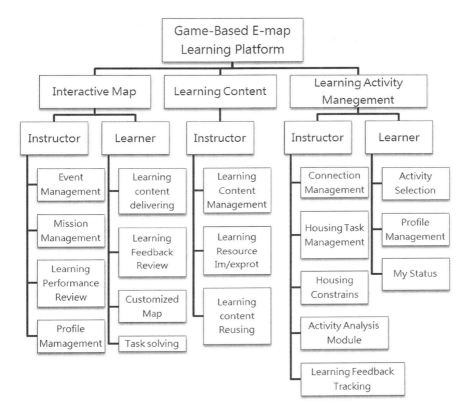

Fig. 1 Game-based e-map system architecture

content, to review the learning outcomes, and eventually to customize the look-and-feel of e-map.
- Learning Resource Management Module: to manage learning resource and to allow instructor to upload, to update, to delete the specific learning resource. Most of the learning resource is based on multimedia, such as audios, videos, and web pages. As a result, the learning resource can be sharable and reusable in this platform.
- Learning Activity Management Module: to relate the learning content and the learning activity. In addition, this module confines the housing task to self-regulated learning sequence. And learners can choose particular learning activity according to their needs.

Analysis of Learning Competency and Performance

- Learning Feedback Tracking Module: To examine the learning performance of each involved leaner. And the learning process will be sent to the backend server and kept in the learning portfolio module. Instructors are able to analyze and summarize the learning performance.

- Learning Activity Analysis Service: to ensure the quality of service in the game-based e-map learning platform. Also this service can be applied to analyze the performance of learning content delivering.

4.2 Game Design of Housing Learning Missions

The educational game design in the proposed system is discussed in two directions. The first one is about the learning content design, and the other is about leading gaming factor into the learning activities. We chose "Data Structure" as the experimental subject in the proposed learning platform, and the learning content on the e-map is delivered to learners with the methods supported in Google Maps APIs. To realize the evaluation of learning performance, a corresponding assessment will be triggered when a learning activity ends up, and the result will then be sent to the backend server for updating the learning portfolio.

Another point to note is the way leading gaming factors into the learning activities to improve the learning motivation. To serve this goal, we allow instructors to arrange corresponding house type according to the difficulty and importance of the learning content. And the arrangement represents the relationship between learning content and the requirement of purchasing house on the e-map. Learners have to start the learning activity from the elementary housing level, and accumulate adequate learning property (i.e. houses on the e-map) in successive learning activities. For instance, learners should get a brick house before having three thatched cottages gained from corresponding learning content. Learners can make their own arrangements of the learning activities according to their needs. If learners try to enter the restricted activity without having adequate learning property, the learning management system will alert learners with a popup dialog box. Eventually learners can review the status of housing mission and the amount of different type of obtained houses. In addition, learners are able to view the learning status of other learner with permission.

4.3 Interaction and Learning Feedback

The interaction and feedback can be considered as the key factors of a successful game. Learners in a game-based learning system might have different skills and sequences to complete the assigned mission. Therefore, in the proposed game-based e-map platform, we allow learners to determine the order for acquiring the same type or the same level of houses according to their preference or arrangement (Fig. 2).

In the proposed e-map, the instant learning feedback can be displayed on the e-map in the form of various house icons (as shown in Fig. 3). These icons on the e-map represent the results after passing corresponding learning activity. As long

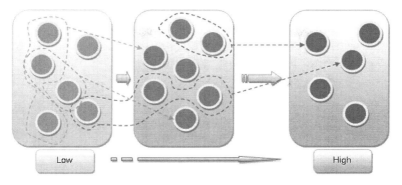

Fig. 2 Self-regulated routing strategy

Fig. 3 Game-based e-map in satellite view

as a learner enters a new learning activity, the learning content not only improves the property in knowledge but also the property in housing.

5 Conclusion

A game-based e-map learning platform is proposed in this study. Learners are able to acquire learning content after having the predefined learning activities on the e-map with housing missions. The gained knowledge content can be transformed into the property for corresponding buildings or houses. The participating learners can view personal learning status and compare the learning results as a contest with others.

Game-based learning to current educational methods is proven to improve learning motivation, situated cognition and problem-solving skills by providing a seamless virtual world affiliated with learning experience. In real life, housing for most of the young students is an unattainable activity, and they need to accumulate sufficient assets to achieve the purpose of housing. In the proposed learning platform with the WMS technology, learners can gradually accumulate their knowledge assets. And through the game scenario, learners can be easier to realize the philosophy of a penny saved is a penny earned.

References

1. Amory A, Naicker K, Vincent J, Adams C (1999) The use of computer games as an educational tool: identification of appropriate game types and elements. Br J Educ Technol 30(4):311–321
2. Bae Y, Lim J, Lee T (2005) Work in progress—a study on educational computer, games for e-learning based on activity theory. In: Frontiers in education, 2005. FIE '05. Proceedings 35th annual conference, pp F1C-18
3. Gee JP (2003) What video games have to teach us about learning and literacy. Palgrave Macmillan, New York
4. Koster R (2005) A theory of fun for game design. Paraglyph Press, Scottsdale
5. Kuutti K (1995) Activity theory as a potential framework for human-computer interaction research. In: Nardi B (ed) Context and consciousness: activity theory and human-computer interaction. MIT Press, Cambridge, pp 17–44
6. Lynch K (1960) The image of the city. MIT Press, Cambridge
7. Malone TW, Lepper MR (1987) Making learning fun: a taxonomy of intrinsic motivations for learning. Aptitude Learn Instr 3:223–253
8. Sandford R, Williamson B (2005) Games and learning: a handbook. NESTA Futurelab, Bristol
9. Siegel AW, White SH (1975) The development of spatial representations of large-scale environments. Adv Child Dev Behav 10:9–55
10. Squire K (2005) Changing the game: what happens when video games enter the classroom? J Online Educ

Digital Publication Converter: From SCORM to EPUB

Hsuan-pu Chang

Abstract The resources and applications about ebook have been changing people's way of reading due to the popularization of ebook readers. It also has significantly changed the traditional rules and concepts of publication. Therefore, an important issue we have to face consequentially is how to create qualified digital publications for readers. In fact, there are a lot of excellent e-learning contents have been produced and stored in repositories or management systems. But unlike the convenience of enjoying ebooks with ebook readers, these excellent learning contents have much more complicated design issues and have to be put on specific learning management systems (LMS) due to the conformance of learning strategies and management requirements. Moreover, many excellent learning contents and courses are produced by the teachers who spent a lot of time and energy. They also have the expectation of converting these learning contents to their own private publications. As the result, we propose a file converter which is able to convert the SCORM compliant courses into EPUB publications. The system consist of four modules; Presentation Transforming Module, Metadata Transforming Module, Sequencing and Navigation Transforming Module and Packaging Transforming Module. We look forward to seeing these excellent SCORM learning contents can be wildly distributed and enjoyed with the EPUB format and publications.

Keywords Ebook · Digital publication · SCORM · EPUB · Converter

---

H. Chang (✉)
Department of Information and Library Science, Tamkang University, Taipei, Taiwan
e-mail: musicbubu@gmail.com

J. J. (Jong Hyuk) Park et al. (eds.), *Multimedia and Ubiquitous Engineering*,
Lecture Notes in Electrical Engineering 240, DOI: 10.1007/978-94-007-6738-6_82,
© Springer Science+Business Media Dordrecht(Outside the USA) 2013

1 Introduction

Every university department has its particularly professional knowledge and skills applied for employment. Accordingly, department professional competences are set as learning targets that students are expected to possess after graduating from school. Many scholars have attempted to define competence and numerous studies have developed, examined or applied assessment and validation techniques for evaluating performance by analyzing competences [1].

Some scholars [2, 3] described the core intent of competency as "an underlying personal characteristic which results in effective and/or superior performance in a job" or "an underlying characteristic of an individual that is causally related to criterion-referenced effective and/or superior performance in a job or situation." Competency is defined as the "underlying characteristics" that can be used to predict the job performance of specific professionals. Competences are long-lasting characteristics that comprise motives, traits, self-concept, knowledge and skills, and can be demonstrated in thinking, behaviors or onsite responses. Furthermore, researchers defining competences agreed that they can be trained and improved [3–6]. The trainability of competencies has major implications for training departments and educational institutions.

2 The Components of the Evaluation System

The major purpose of this system is assessing students' department professional competences after a serial of courses designed and arranged by department course developing committee. Generally, in order to represent a student's learning performance a score is used to express whether the student is qualified for this course's learning objectives. But in fact a course may include plural professional competences that a student need to learn and pursue. In order to inspect whether a student is mastering professional competences in a course more or less, the relation among course, assessment and professional competence as the Fig. 1 shows.

The following sections detail how we connect the three parts.

Fig. 1 Three parts for constructing the evaluation system

2.1 Build the Relation Between Course and Professional Competence

In this section, a Course-Competence table will be introduced to describe the relation between course and professional competence that a department course committee needs to conduct a discussion for accomplishing it.

2.1.1 Construct Course-Competence Table

The first step of carrying out the competence evaluation system is constructing the Course-Competence table that describes the relation between courses and competences. Because a course may contain different professional knowledge and skill for learning, which means more than one professional competences may exist in a course only the matter of ratio. The course committee needs an entire picture of department development for constructing the Course-Competence table. For instance, Table 1 is the department of Digital Information and Library Science (DILS) in Tamkang University (TKU) in Taiwan, the first column lists the a few example course titles of the department and the rest of columns are professional competence index and their ratio contained in these courses. The details of the eight competences A–H are described in the Table 2.

2.1.2 Add Competence Information in Course Syllabus

As Fig. 2 shows, while teacher is designing a course syllabus on line, system automatically retrieves the competence information from database and adds it to teacher's syllabus. It is a significant step that reminds teachers or instructors what competences should be learned after taking this course. The competence information gives teachers a direction to prepare teaching content and design learning activities. Meanwhile it's also an opportunity to reconsider whether the relation set between the competences and the course is properly matched.

Table 1 Course-competence

Course	Competence ratio (%)							
	A	B	C	D	E	F	G	H
Introduction to librarianship & information science	20	20	15	20	20	0	5	0
Statistics for library science	0	30	35	0	35	0	0	0
Introduction to innovative publishing industry	0	0	0	20	0	0	40	40
Archive management and development	0	25	15	0	20	40	0	0

Table 2 Professional competences of DILS, TKU

Competence	Description
A	Have the competences to know the library and information science principles and trends
B	Have the competences to develop, organize, archive and integrate various information resources.
C	Have the competences to realize information theories and apply information systems
D	Have the competences to communicate and coordinate information services
E	Have the competences to manage information services in various libraries and institutions
F	Have the competences to manage digital documents and file archives.
G	Have the competences to integrate traditional publication affairs and library works
H	Have the competences to integrate library works and digital content industry

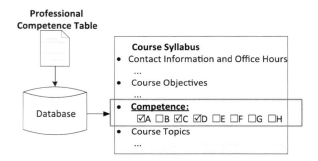

Fig. 2 System retrieves competence information from database then adds it to a course syllabus

2.2 Build the Relation Between Competence and Assessment

In the previous section, the Course-Competence table has been built to describe the relation between course and competence. The competences are not only the specific learning objectives but also taken into account when we are evaluating student's learning. In this section we focus on how to infuse the competence factors into traditional exam process.

2.2.1 Providing Exam Authoring Tool

The connection between exam and competence includes two parts; first is connecting a competence-oriented item bank, the other is an exam authoring interface which primarily helps teachers picking up questions from item bank. The authoring tool architecture is illustrated in Fig. 3.

Digital Publication Converter

Fig. 3 Exam authoring tool architecture

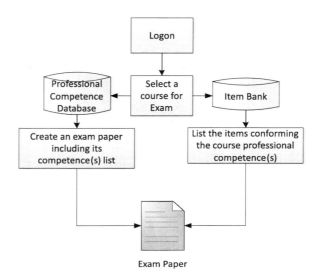

2.2.2 Provide Competence Ratio Information for Teacher Editing Exam Paper

As Fig. 4 shows, that's similar to the competence information added in a syllabus. While teacher is starting from a blank exam paper, the professional competence ratio information of this course is automatically retrieved from database then added on an exam paper that reminds a teacher the competences should be evaluated in this course.

But the notable point here is that the competence ratio information is not used to instruct teachers to create an exam completely conforming the ratio but remind teacher the ratio could be a reference kept in mind while editing the exam. The teacher still has the flexibility to edit the exam paper content according to their teaching progress and professional knowledge.

Fig. 4 A course professional competence ratio added on exam paper

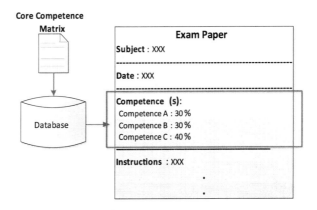

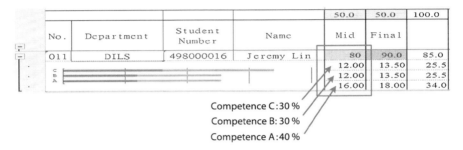

Fig. 5 Grade sheet with competence calculation

2.3 Build the Relation Between Competence and Grade Sheet

A teacher generally proposes a few testing activities to evaluate student's learning, such as paper works, presentations, exams etc. In order to express how students have learned the competences, a grade is separated to its corresponding competence grades.

Take Fig. 5 for example, a student who gets 80 in midterm which can be extended to display his/her each competence score. When a teacher fills in the score which will be automatically separated to the three competences grade according to ratio defined in the Course-Competence table. The grade of competence A in midterm is 16 because the midterm contains 50 % of the entire course grade and the ratio of the competence A is 40 %, so the result is multiplying 80 by 50 % and 40 %.

The grade sheet also allows a teacher to manually fill in or adjust the grade for each competence; the default competence score is calculated according to the ratio of competences of this course.

3 Conclusion and Future Work

We provide a prototype system of evaluating college student's professional competences that they have learned in their department. Students may take a serial of training courses for acquiring the professional knowledge and skills in a particular department, but the results can't be completely represented or measured by a grade. Through constructing Course-Competence table, the relation between a course and its corresponding competence(s) can be assured as well as the designing of the course syllabus. Through constructing the exam authoring tool, teachers can easily pick up required questions corresponding to competences set in this course and have the information of proper competence ratio distribution while creating the exam paper. Through constructing the relation between competence and grade sheet that can not only present students learning grade but also a description of how they have learned the competences in this course. So the proposed evaluation system can used to understand a student professional competences learning situation. Moreover the evaluation feedbacks can not only help a teacher inspects the relation between their teaching and these professional competences but also provide valuable information for reviewing the entire department course structure. Our future works will include the visualizing the evaluation results and resolving the calculation problems caused by elective courses and data normalization.

Acknowledgment The authors would like to thank the anonymous reviewers for their insightful comments on an earlier version of this paper. The work described in the paper has also been supported by the National Science Council of Taiwan under Grant NSC 101-2221-E-032-064.

References

1. Parry SB (1998) Just what is a competency? (And why should you care?). Training 35(6):58–64
2. Boyatzis RE (1982) The competent manager: a mode for effective performance. Wiley, New York
3. Spencer LM, Spencer SM (1993) Competence at work: models for superior performance. Wiley, New York
4. Clarke N (2010) The impact of a training programme designed to target the emotional intelligence abilities of project managers. Int J Project Manag 28(5):461–468
5. Parry SB (1996) The quest for competencies. Training 33(7):48–56
6. Yeomans WN (1989) Building competitiveness through HRD renewal. Train Dev J 43(10):77–82

An Intelligent Recommender System for Real-Time Information Navigation

Victoria Hsu

Abstract People like to attend exhibition activities, but hard to enter into the information effectively. We build new system with wireless internet and mobile device to guide visitor into the core information initiatively and effectively. The mobile guide system could classify visitor base on exhibition information and personal information that provide more suitable for users. Our system combined with semantic web technology to connect items data which users' markup the type or property information in our system to created human portfolio. Our system is in compliance with human portfolio and metadata method to provide user information automatically and appropriately.

Keywords Mobile guide · Mobile device · Wireless internet · Semantic web · Human portfolio

1 Introduction

Many people visit the exhibitions or museums for their leisure time. Most of the museums and the exhibitions will provide the corresponding information to people for their visiting. Now, the technologies of mobile devices and wireless network could provide visitors their own style visiting via mobile devices which devices may be provided from the organizers or their own.

This research aims to propose a scheme of the guide system which first proposes a data storage format such that the exhibition organizers can store all the exhibition data simply. Second, the visitors can simply describe some personal information, which will be evaluated by the best appropriate recommendation

---

V. Hsu (✉)
Department of Computer Science and Information Engineering,
Tamkang University, Taipei, Taiwan
e-mail: saintvoice.1981@gmail.com

method (BAR) we proposed in this paper, and this guide system will provide the contents or information that is fit for the visitor by wireless network technique. And this information will be shown on the mobile device to the visitor.

Mobile devices are small computational equipment [1]. Users can get information from internet or telecommunication networks and execute some program by these devices, e.g. cell phone, PDA, notebook, iPad and so on. Metadata of the information is very important to mobile devices. Metadata is first defined in the conference Metadata workshop [2] and applied to data storage, data retrieving and so on. Dublin Core is a simple [3], efficient and popular metadata standard. It can fast organize the network resources, improve the precise of data search and retrieving, provide a metadata format to describe the network resources by many experts from different areas, and the network resources will be divided to 15 categories.

Categories for the description of works of art, CDWA, are a popular metadata definition to art exhibitions and museums categories [4]. It is proposed by Art Information Task Force, AITF, of J. Paul Getty Trust. CDWA provides a scheme to describe the content of works of art such that we can establish a database of the works of art by these describes. There are 27 main categories and 233 subcategories in CDWA.

After establishing the metadata, the ontology and the semantic web will be the critical techniques to develop our BAR method. Ontology was used to some specified and existed type or the well-described statements in philosophy [5, 6]. In computer science, ontology represents knowledge as a set of concepts within a domain, and the relationships between those concepts. Common components of ontologies include individuals, classes, attributes, relations, function terms, restrictions, rules, axioms and events. Semantic web is the concept proposed by Berners-Lee in W3C [7]. The main idea of semantic web is to let computers can "understand" the text files on the internet, that is, to know the semantics of the text files. By using the techniques of semantic web, the search engine can use a unique and precisely vocabulary and mark to the text files they searched without confusing.

2 Intelligent Recommender System

2.1 System Procedure

For an exhibition or a museum, we first establish the database of the works of art by Dublin Core and CDWA. A visitor has to describe some of his/her personal data to the recommendation system before using this recommendation system. The recommendation system evaluates these personal data by the BAR method and finds some works of art will be recommended to the visitor. Then, the visitor will get the information about the recommended works of art via the mobile device he/she takes.

An Intelligent Recommender System 675

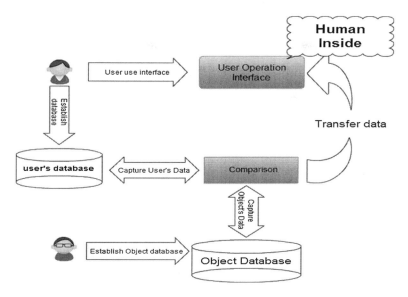

Fig. 1 Procedure of the recommendation system

With this recommendation system in the mobile device, the visitor can mark the works of art he/she likes during the visiting. The recommendation system will record the works of art to the database of the visitor's profile. The more marks the visitor made, the more precisely our recommendation system will be, by the BAR evaluation. Figure 1 is the procedure of the recommendation system.

2.2 Database Establishment

The database of the recommendation system includes the following tables: art_detail, art_relation, type, human_relation, human.

By CDWA, the table art_detail stores the information about the works of art including: title, author, date, format, material, and description. The table human stores the visitors' personal information including: name, sex, birthday, telephone number, e-mail address, address, and education degree. The table type stores the information of all kinds of types in this system. The table art_relation stores the information about the types of the works of art in the exhibition. The table human_relation stores the relationship of the visitor and the work of art. The visitor likes a work of art and mark it in the system that will be store in this table. Figure 2 represents the database structure of the recommendation system.

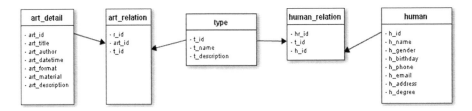

Fig. 2 Database structure of the recommendation system

2.3 Best Appropriate Recommendation Method

The BAR method includes two phases: (1) when the visitor establishes his/her personal data initially, BAR evaluates to determine the initial weights for recommendation; (2) when the visitor mark a work of art, BAR evaluates the attributes of the work of art and the personal data to update the weights to make more precisely commendation.

In phase 1, we first evaluate the initial weights by the following variables of personal data:

UA   user's age
UD   user's education degree
UI   user's interest
WUA   weight of user's age
WUD   weight of user's education degree
WUI   weight of user's interest
$P = \{H_i | 1 \leq i \leq n\}$ a set of human_portfilio database where Hi means the ith human data from human_portfilio database.

The user's age is divided into 3 intervals: less than 19, from 20 to 40, greater than 40, and we assign weight 1, 2, 3, to each interval respectively. The user's education degree is classified by primary school, junior high school, senior high school, collage, educated school, and we assign weight from 1 to 5 for each degree respectively. The user's interest options include art, music, sport etc. and are divided into art related and not art related, and assign weight 2 and 1 respectively.

```
Input user's personal data values in database
Output user's weight values WP
Set new user, WP = 0, WUA = 0, WUD = 0,
WUI = 0
//evaluate weight of user's age
If 0 < UA ≤ 19
   WUA = WUA + 1
else If 20 ≤ UA ≤ 39
   WUA = WUA + 2
```

An Intelligent Recommender System

```
else
   WUA = WUA + 3
//evaluate weight of user's education degree
If UD = ''primary school''
   WUD = WUD + 1
else If UD = ''junior high school''
   WUD = WUD + 2
else If UD = ''senior high school''
   WUD = WUD + 3
else If UD = ''collage''
   WUD = WUD + 4
else
   WUD = WUD + 5
//evaluate weight of user's interest
If UI = art related options
WUI = WUI + 2
else
WUI = WUI + 1
//add all weights
WP = WUA + WUD +WUI
Store WP to user's profile
```

The initial weight is evaluated in phase 1 and the recommendation system recommends the works of art to the visitor depending on it before the visitor marks some works of art he/she likes. Then, when the visitor starts to visit the exhibition and marks some works of art he/she likes, the recommendation system receives the marks and BAR begins evaluating to update the weight of the visitor. That is the phase 2 of BAR method. Some variable definitions used in phase 2 BAR is shown as follows.

$A = \{A_i | 1 \leq i \leq n\}$ set of art database where Ai means the ith author data from author database.
$S = \{S_i | 1 \leq i \leq n\}$ a set of art database where Si means the ith style data from style database.
$MAI = \{VA_i | VA_i \in A, 1 \leq i \leq k\}$author's identification of the work of art which user marked
$MSI = \{VS_i | VS_i \in S, 1 \leq i \leq k\}$style identification of the work of art which user marked
$RAI = \{RA_i | 1 \leq i \leq n\}$Record the frequency of author's identification of the work of art which user marked
$RSI = \{RS_i | 1 \leq i \leq n\}$Record the frequency of style identification of the work of art which user marked

In phase 2 of BAR method, when the visitor marks a work of art, the recommendation system retrieves the attributes author and style from database. After the visitor finishing this visiting, the recommendation system evaluates that which author and which style the visitor marked most, then store this author and this style information in the visitor database. Therefore, when the visitor visits another exhibition next, the recommendation system can make good recommendation by these data.

2.4 Linking Semantic Web

The mobile device which is taken by the visitor during the visiting can receive the marks he/she made and provides the information and recommends of the works of art to the visitor. Moreover, by using the techniques of semantic web and wireless network, the visitor can get more information by the mobile device by connecting to other websites.

```
Input user marked author MAI,
user marked style MSI
Output the highest author and style value
For i = 1 to i = k do
Set RAi = 0, RSi = 0
//count the frequency of authors and styles that the visitor
marked
For i = 1 to i = k do
For j = 1 to j = n do
IF VAi = Aj do RAi += 1
IF VSi = Sj do RSi += 1
//find which author and which style that the visitor likes
most
```

$$RAI_{\max} = \arg\max\{RA_i|\ 1 \le i \le n\}$$
$$RSI_{\max} = \arg\max\{RS_i|\ 1 \le i \le n\}$$

```
Store RAImax, RSImax
```

3 Concluding Remarks

Using navigation system by mobile devices is very popular for exhibitions and museums recently. The wireless network technique, the semantic network technique and personal mobile devices are also well-developed. Visitors can get more information by different mobile devices than before. The appropriate recommendation system we designed can provide suitable information to visitors fast and convenient by their own mobile devices.

The appropriate recommendation system can be improved by considering the visitors' own experiences in the BAR method evaluation such that the system can recommend works of art to visitors more precisely. The appropriate recommendation system can be extended to be a community system. The visitors can share and exchange their experiences and make more commands to the exhibitions or museums on the system. These are all the future researches.

References

1. Roschell J (2003) Unlocking the learning value of wireless mobile devices. J Comput Assist Learn 19:260–272
2. http://zh.wikipedia.org/wiki/Metadata
3. Hillman D (2005) Using dublin core.http://dublincore.org/documents/usageguide/#whatis
4. Agbabian MS, Masri SF, Nigbor RL, Ginell WS (1988) Seismic damage mitigation concepts for art objects in museums. In: Proceeding of ninth world conference on earthquake engineering
5. Gruber TR (1993) A translation approach to portable ontology specifications. Knowledge Systems Laboratory, Palo Alto, pp 199–220
6. Arvidsson F, Flycht-Eriksson A (2008) Ontologies I. http://www.ida.liu.se/~janma/SemWeb/Slides/ontologies1.pdf. Accessed 26 Nov 2008
7. Berners-lee T, Connolly D, Kagal L, Scharf Y, Hendler J (2008) N3logic: a logical framework for the world wide web. theory and practice of logic programming, vol 8. Cambridge University Press, New York, pp 249–269

Part XII
Advanced Mechanical and Industrial Engineering, and Control I

Modal Characteristics Analysis on Rotating Flexible Beam Considering the Effect from Rotation

Haibin Yin, Wei Xu, Jinli Xu and Fengyun Huang

Abstract This paper deals with modal frequencies of rotating flexible beam. To investigate effects on the modal frequencies from rotation, the mathematical models are derived by using three descriptions on deformation: the conventional approach, the quadratic approach, and a synthetical approach. The theoretical solutions of modal frequencies based on the three methods are used to compare and draw some summaries.

Keywords Flexible beam · Dynamic modeling · Modal characteristics · Rotation

1 Introduction

Because of light weight, small inertia, high operating speed, and low energy consumption, flexible beam has many promising applications such as helicopter propellers, flexible robot, etc. However, flexible beam also has its shortages one of which is vibration. There are a lot of studies on modeling and vibration control of flexible manipulators [1].

So far, there are three classifications on description of elastic deformation for flexible beam during modeling. The common and most widely used method is the conventional linear deformation method [2]. In the past decades, some papers discussed the quadratic deformation approach, such as Abe investigation on trajectory planning based on dynamic model, which adopted the quadratic method to describe the elastic deformation of flexible beam [3]. In 2005, a synthetical method had been proposed to derive the dynamic model of flexible robots by Lee [4]. In 2011, the synthetical method was extended to two-link flexible manipulator by Yin

---

H. Yin (✉) · W. Xu · J. Xu · F. Huang
School of Mechanical and Electronic Engineering, Wuhan University of Technology,
122 Luoshi Road, Wuhan, Hubei, People's Republic of China
e-mail: chinaliuyin@whut.edu.cn

J. J. (Jong Hyuk) Park et al. (eds.), *Multimedia and Ubiquitous Engineering*,
Lecture Notes in Electrical Engineering 240, DOI: 10.1007/978-94-007-6738-6_84,
© Springer Science+Business Media Dordrecht(Outside the USA) 2013

et al. where the synthetical method was deemed to be a better approach for flexible manipulator at high speed than conventional method [5].

Above three approaches are used to derived dynamic model of a rotating flexible beam and solve the modal frequencies in consideration of the effect from rotation in this report. In recent years, some researchers have investigated the effects on modal characteristic from rotation of flexible beam. Mei proposed differential transform method (DTM) to analyze the modal shape functions of a centrifugally stiffened Timoshenko beam, where author concluded as the modal shape functions were affected by rotation [6]. Gunda addressed a rational interpolation functions to analyze rotating beam and concluded that the shape functions were not only functions of positions but also functions of rotational speed [7]. Additionally, Kaya et al. studied the modal frequencies of rotating cantilever Bernoulli–Euler beam by using DTM, where the modal frequencies increase with angular velocity [8]. However, these published modal analyses on rotating flexible beam would focus on digital method rather than base on theoretical solution. This paper proposed the theoretical solutions of modal frequencies based on three models in consideration of the rotating effects.

2 Mathematical Modeling

Figure 1 shows the schematic diagram of deformation for the flexible beam. In the conventional deformation, the displacement v of an arbitrary point at the flexible beam is vertical to initially undeformed beam shown as Fig. 1a. In the quadratic deformation, the length from the arbitrary point at the flexible beam to original point remains unchanged before and after deformation, shown as Fig. 1b. The synthetical deformation describes that the displacement v of an arbitrary point at flexible beam is vertical to the vector r direction, which firstly proposed by Lee shown as Fig. 1c.

The coordinate O-XY represents the inertial reference frame with the original point O at the center of the hub, while the o_1-x_1y_1 is the local coordinate system fixed to the root of the flexible beam rotating around the hub. The x-axis is oriented along the beam in undeformed configuration. The radius of the hub is assumed to be negligible and the hub rotates around the global Z-axis, with angle θ. The three models have uniform beam with cross-section A and density ρ, the length l and flexural rigidity EI. The vector r denotes the position vector of arbitrary point at flexible beam after deformation. The flexible beams are modeled as Bernoulli–Euler beam.

Conventional Deformation. In the Fig. 1a, the vector r in the global frame is expressed as:

$$r = \begin{bmatrix} x \cos \theta - v \sin \theta \\ x \sin \theta + v \cos \theta \end{bmatrix}. \tag{1}$$

Modal Characteristics Analysis on Rotating Flexible Beam

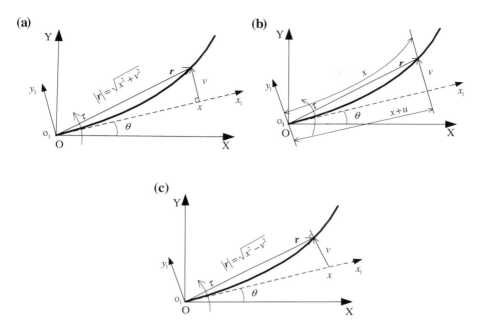

Fig. 1 Schematic diagram on description of deformation. **a** Conventional deformation. **b** Quadratic deformation. **c** Synthetical deformation

The terms related to the kinetic energy can be expressed as:

$$\dot{r}^T\dot{r} = (x^2 + v^2)\dot{\theta}^2 + \dot{v}^2 + 2x\dot{v}\dot{\theta}, \tag{2}$$

where · denotes derivative with respect to time t.

The dynamic equation associated with flexible deformation is derived through the Hamilton's principle as follows:

$$\int_0^l [\rho A\ddot{v}(x,t) + EIv''''(x,t) - \rho Av(x,t)\dot{\theta}^2 + \rho Ax\ddot{\theta}]dx = 0, \tag{3}$$

where $'$ denotes derivative with respect to position x.

Quadratic deformation. In the Fig. 1b, the parameter u denotes the axial shortening. The vector r in the global frame is expressed as:

$$r = \begin{bmatrix} (x+u)\cos\theta - v\sin\theta \\ (x+u)\sin\theta + v\cos\theta \end{bmatrix}. \tag{4}$$

The terms related to the kinetic energy can be expressed as:

$$\dot{r}^T\dot{r} = [(x+u)^2 + v^2]\dot{\theta}^2 + \dot{v}^2 + 2(x+u)\dot{v}\dot{\theta} + \dot{u}^2 - 2v\dot{u}\dot{\theta}. \tag{5}$$

The flexible dynamic equation is derived through the Hamilton's principle as follows:

$$\int_0^l \left\{ \rho A \ddot{v}(x,t) + EI[1+(v')^2]v''''(x,t) - \rho A v(x,t)\dot{\theta}^2 + \rho A(x+u)\ddot{\theta} + \rho A \dot{u}\dot{\theta} \right. $$
$$\left. + EI[4v'v''v''' + (v'')^3] \right\} dx = 0. \tag{6}$$

Synthetical deformation. In the Fig. 1c, the position r is expressed as:

$$r = \begin{bmatrix} (x - \frac{v^2}{x}) \cos\theta - \frac{\sqrt{x^2-v^2}}{x}\sin\theta \\ (x - \frac{v^2}{x}) \sin\theta + \frac{\sqrt{x^2-v^2}}{x}\cos\theta \end{bmatrix}. \tag{7}$$

The terms related to the kinetic energy can be expressed as:

$$\dot{r}^T r = (x^2 - v^2)\dot{\theta}^2 + \frac{x^2}{x^2-v^2}\dot{v}^2 + 2\sqrt{x^2-v^2}\dot{v}\dot{\theta}. \tag{8}$$

Considering the displacement v is very small. So Eq. (8) can be simplified as:

$$\dot{r}^T r = (x^2 - v^2)\dot{\theta}^2 + \dot{v}^2 + 2x\dot{v}\dot{\theta}. \tag{9}$$

The flexible dynamics is derived through the Hamilton's principle described as:

$$\int_0^l [\rho A \ddot{v}(x,t) + EIv''''(x,t) + \rho A v(x,t)\dot{\theta}^2 + \rho A x \ddot{\theta}]dx = 0. \tag{10}$$

3 Theoretical Solution on Modal Frequencies

According to Eqs. (3) and (10), the governing differential equations of vibration including rotational effect and non-conservative force are respectively described as follows:

$$\rho A \ddot{v}(x,t) + EIv''''(x,t) - \rho A v(x,t)\dot{\theta}^2 = f_1, \tag{11a}$$

$$\rho A \ddot{v}(x,t) + EIv''''(x,t) + \rho A v(x,t)\dot{\theta}^2 = f_1, \tag{11b}$$

where non-conservative force is represented by:

$$f_1 = -\rho A x \ddot{\theta}. \tag{12}$$

Neglecting the high order infinitesimal such as $(v')^2$, the governing differential equation of vibration including rotational effect and non-conservative force. Based on Eq. (6) is expressed as:

$$\rho A \ddot{v}(x,t) + EIv''''(x,t) - \rho A v(x,t)\dot{\theta}^2 = f_2, \tag{13}$$

Modal Characteristics Analysis on Rotating Flexible Beam

where non-conservative force is defined as:

$$f_2 = -\rho A[(x+u)\ddot{\theta} + \dot{u}\dot{\theta}]. \tag{14}$$

To obtain the homogeneous solutions of above three differential equations of vibration, the f_1 and f_2 are set equal to zero in Eqs. (11a) and (13) as following unified equation:

$$\rho A\ddot{v}(x,t) + EIv''''(x,t) \pm \rho Av(x,t)\dot{\theta}^2 = 0, \tag{15}$$

where the operator "+" of the third term in left side is based on the synthetical deformation, the operator "−" of the third term in left side is based on the conventional and quadratic deformations.

Assuming the harmonic vibration with centrifugally affected angular frequency ω in flexible dynamics, the unified differential equations are written as:

$$EIv'''' - \rho A(\omega^2 \mp \dot{\theta}^2)v = 0. \tag{16}$$

Defining an equivalent angular frequency W, which is angular frequency of flexible beam in static structural dynamics and the equivalent differential equation of vibration is described as:

$$EIv'''' - \rho AW^2v = 0, \tag{17}$$

where W is dependent on boundary conditions and mechanical parameters of flexible beams, the centrifugally affected angular frequency is represented by:

$$\omega_i = \sqrt{W_i^2 \pm \dot{\theta}^2}, \tag{18}$$

where i is the modal order number; the operator "+" is based on the synthetical method, the operator "−" is based on the conventional and quadratic methods.

4 Calculated Sample

This section gives calculated sample based on following parameters of a flexible arm: the mass density of the beam $\rho = 7.70 \times 10^3$ kg/m^3 and the sectional area of the beam $A = 1.85 \times 10^{-5}$ m^2. The elasticity modulus is $E = 200$ Gpa, and the moment of inertia $I = 5.40 \times 10^{-13}$ m^4. The beam length is $l = 0.40$ m. In static structural dynamics, the first three frequencies $(i = 3)$ of cantilever beam are $W_1 = 19.22$ rad/s, $W_2 = 120.28$ rad/s and $W_3 = 333.11$ rad/s, respectively. Considering the rotational effect from angular velocity, the centrifugally affected natural frequencies based on three models are shown in Fig. 2. Figure 2a, b denote the first two and third modal frequencies, respectively. The dash lines show the negative variation with angular velocity based on the conventional and quadratic

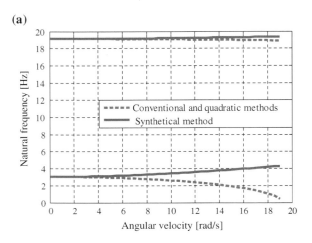
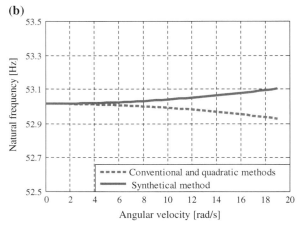

Fig. 2 Centrifugally affected natural frequencies from the angular velocity. **a** The first and second mode. **b** The third mode

methods, the solid lines depict the positive relationship with angular velocity based on the synthetical method. The first frequencies are most significantly affected by rotational velocity.

5 Summary

This research has derived the differential equations of flexible vibration based three models, and the theoretical solution of centrifugally affected frequencies was proposed. The modal frequencies based on conventional and quadratic methods decrease with rotational velocity, but the modal frequencies based on synthetical method increase with rotational velocity. The theoretical solution was based on conditional simplification; consequently, further study is required in future work.

Acknowledgments The authors acknowledge the support of the National Natural Science Foundation of China (Grant No. 11202153).

References

1. Dwivedy SK, Eberhard P (2006) Dynamic analysis of flexible manipulators, a literature review. Mech Mach Theory 41:749–777
2. Edelstein E, Roson A (1998) Nonlinear dynamics of a flexible multirod system. J Dyn Syst Meas Control 120:224–231
3. Abe A (2009) Trajectory planning for residual vibration suppression of a two-link rigid-flexible manipulator considering larger deformation. Mech Mach Theory 44:1627–1639
4. Lee HH (2005) New dynamic modeling of flexible-link robots. J Dyn Syst Meas Control 127:307–309
5. Yin H, Kobayashi Y, Hoshino Y (2011) Modeling and vibration analysis of flexible robotic arm under fast motion in consideration of nonlinearity. J Syst Des Dyn 5:219–230
6. Mei C (2006) Differential transformation approach for free vibration analysis of a centrifugally stiffened Timoshenko beam. J Vib Acoust 128:170–175
7. Gunda JB, Ganguli R (2008) New rational interpolation functions for finite element analysis of rotating beams. Int J Mech Sci 50:578–588
8. ÖZdemir Ö, Kaya MO (2006) Flapwise bending vibration analysis of a rotating tapered cantilever Bernoulli-Euler beam by differential transform method. J Sound Vibr 289: 413–420

The Simulation Study on Harvested Power in Synchronized Switch Harvesting on Inductor

Jang Woo Park, Honggeun Kim, Chang-Sun Shin, Kyungryong Cho, Yong-Yun Cho and Kisuk Kim

Abstract Different piezoelectric harvester interface circuits are demonstrated and compared through SPICE simulation. The simulations of the effect of switch triggering offset and switch on time duration on SSHI's power are performed. The inductor's quality factors in synchronized switch harvesting on inductor interface have important effect on the harvested power. Parallel SSHI shows the optimal output voltage to harvest the maximum power varies according to the Q severely. It is concluded that switch triggering offset has more impact on the s-SSHI than p-SSHI and the switch on-time duration is more important in case of the p-SSHI. p-SSHI shows when the on-time duration becomes more than 1.3 times or less than 0.7 times of exact duration time, the harvested power gets negligible. s-SSHI reveals the characteristics that when less than 1.5 times exact on-time duration, the harvested power varies significantly with the on-time duration.

Keywords Piezoelectric · Energy harvesting · SSHI

J. W. Park (✉) · H. Kim · C.-S. Shin · K. Cho · Y.-Y. Cho
Department Information and Communication Engineering,
Sunchon National University, Suncheon 540-950, Republic of Korea
e-mail: jwpark@sunchon.ac.kr

H. Kim
e-mail: khg_david@sunchon.ac.kr

C.-S. Shin
e-mail: csshin@sunchon.ac.kr

K. Cho
e-mail: jkl@sunchon.ac.kr

Y.-Y. Cho
e-mail: yycho@sunchon.ac.kr

K. Kim
Power Engineering Co., Ltd, Gwangyang 540-010, Republic of Korea
e-mail: marohyun@hanmail.net

J. J. (Jong Hyuk) Park et al. (eds.), *Multimedia and Ubiquitous Engineering*,
Lecture Notes in Electrical Engineering 240, DOI: 10.1007/978-94-007-6738-6_85,
© Crown Copyright 2013

1 Introduction

The recent development of ultra-low power applications in ubiquitous sensing and computing demands low cost, long lifetime, small volume and light weight and especially eliminating the battery. Some ubiquitous applications can reduce the average power consumption to the level of tens to hundreds of microwatts, which results in energy harvested from environments to be used as an alternative power [1, 2]. Sustainable power generation can result from converting ambient energy into electrical energy. Mechanical energy conversion is one of the common sources for energy harvesting applications and exists almost everywhere. It is estimated that mechanical vibrations inherent in the environment can provide a power density of tens to hundreds of microwatt per cm^3, which is sufficient to sustain operations of a sensor node [3]. In Mechanical energy conversion, while electromagnetic and electrostatic generators have been developed [4, 5], piezoelectric generators [6, 7] are of major interest due to solid-state integration abilities.

While conventional power supplies and batteries typically have very low internal impedance, internal impedance of the piezoelectric generators is relatively high, which restricts the amount of output current driven by the piezoelectric source to the micro-amp range. The relatively low output voltage of the piezoelectric device is another challenge of this power source. This low output voltage poses a difficult on developing efficient rectifier circuits. The piezoelectric element subjected to a vibration generates the alternating voltage. However most of the electronic sensor nodes and circuits need the DC voltage. So called the standard interface has been widely used, where the interface consists of full-bridge rectifier and storage element. Some techniques to increase significantly the amount of energy by piezoelectric harvesters have been proposed, which are derived from called "synchronized switching damping (SSD) [8]". The SSD technique is based on a non-linear processing on the voltage delivered by the piezoelectric element. This process increases the electrically converted energy resulting from the piezoelectric mechanical loading cycle. From SSD, parallel [8, 9] and serial [8] synchronized switching harvesting on inductor (SSHI) have been proposed. The techniques have increased the harvested power several times more than the standard technique. The nonlinear processing of SSHI consists in inductor and a switch in series and then needs the strict switching action of the switch.

In this paper, different harvesting interface circuits including standard interface, standard interface with a switch, parallel SSHI, and serial SSHI are simulated and compared with LT-SPICE® [10]. The voltage and current waveforms helps the comprehension of the interfaces. Then, effect of the switching time of SSHI on the harvested power is examined through simulation. Especially, it is also studied how the switch on time offset and on-time duration deviation have an effect on the power harvested from SSHIs.

2 Standard Interface Circuits for Piezoelectric Harvesters

An input vibration applied on to a piezoelectric material causes mechanical strain to develop in the device which is converted to electrical charge. The piezoelectric laminate is mechanically forced to vibrate and thus works as a generator to transform the mechanical energy into electrical energy for micro-power generation. At or close to resonance, the piezoelectric element can be modeled in electrical domain. When excited by sinusoidal vibrations, the piezoelectric element can be modeled as a sinusoidal current source in parallel with a blocking capacitance C_0 which represents the plate capacitance of the piezoelectric material. The amplitude I_0 of current source depends on a displacement and frequency of the vibration.

$$i_S = I_0 \sin \omega_0 t \quad (1)$$

where $\omega_0 = 2\pi f_0$ and f_0 is the frequency with which the piezoelectric harvester is excited.

Because the power output by the piezoelectric harvester is not in a form which is directly usable by load circuits, the voltage and current output by the harvester needs to be conditioned and converted to a form usable by the load circuits. The power conditioning and converting circuits should also be able to extract the maximum power available out of the piezoelectric energy harvester.

Figure 1 shows the standard interface using full-bridge rectifier and the simulated voltage and current of the piezoelectric harvester. For the sake of this analysis, assume that the value of C_L is so large compared to C_0 that the voltage at the output of the rectifier (V_{DC}) is essentially constant. During the interval from t_0 to t_{off}, the piezoelectric current source is charging its capacitor C_0 to the V_{DC} and all diodes in the bridge rectifier are reverse-biased. And then, in the interval between t_{off} and $t_{T/2}$, the bridge rectifier will be on, the piezoelectric source provides the current to the load. We can know the reduction in the duration for charging the C_0 can allows the power delivered to the load to be maximized.

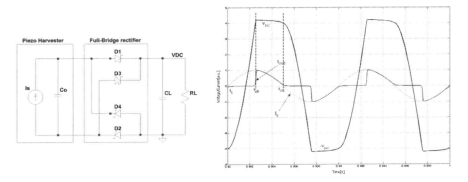

Fig. 1 Standard interface to extract power: circuit schematic and waveforms

To increase the power from the piezoelectric harvester, several interfaces have been proposed. Synchronized switching harvesting on inductor (SSHI) consists of a non-linear processing circuit. There are two types of SSHI, one is parallel-SSHI (p-SSHI) where the non-linear processing circuit is connected across the piezoelectric harvester and a full-bridge rectifier, and the other is series-SSHI (s-SSHI) where the non-linear processing circuit is connected between the piezoelectric harvester and a full-bridge rectifier in series. The non-linear processing circuit is composed of an inductor and a switch in series. This interface utilizes the synchronous charge extraction principle which consists in removing periodically the electric charge accumulated on the blocking capacitor C_0 of the piezoelectric element, and to transfer the corresponding amount of electrical energy to the load or to the energy storage element. Figures 2 and 3 shows the two interface circuits and the voltage and current waveforms in them.

The electronic switch is briefly turned on when the current source of the piezoelectric elements crosses zero. This moment is when the mechanical displacement reaches maxima. At these triggering times, an oscillating electrical circuit $L - C_0$ is established, where the electrical oscillation period is chosen much smaller than the mechanical vibration period T. The switch is turned off after a half

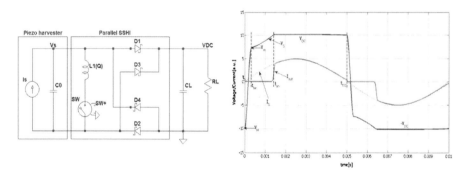

Fig. 2 Parallel synchronized switch harvesting on inductor interface

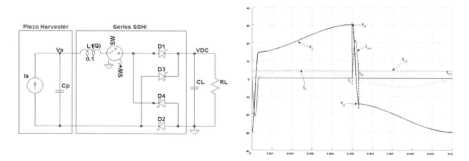

Fig. 3 Parallel synchronized switch harvesting on inductor interface

electrical oscillating period, resulting in a quasi-instantaneous inversion of the voltage V. The time interval, t_{sw} during the switch is on is expressed as:

$$t_{sw} = \pi\sqrt{LC_0} \qquad (2)$$

The voltage relation between before the switch is on and after the switch is off depends on the quality factor Q of inductor.

$$V_m = -V_M e^{-\pi/2Q} \text{ for p} - \text{SSHI} \qquad (3)$$

$$(V_m + V_{DC}) = -(V_M - V_{DC})e^{-\pi/2Q} \text{ for s} - \text{SSHI} \qquad (4)$$

where V_m is the voltage after the switch is off, V_M is the voltage right before the switch is on and Q is the quality factor of inductor.

3 The Effect of Switching Time on Harvested Power of SSHI

As expected, in SSHI interfaces, the operation of the switch is very important. The switch has to turn on exactly when the displacement reaches maxima and then has to stay on only very short duration, a half period of $L - C_0$ oscillation period. The switch triggering offset which is the switch on time deviation from the ideal on time and the on-time duration deviation have an important effect on the harvested power.

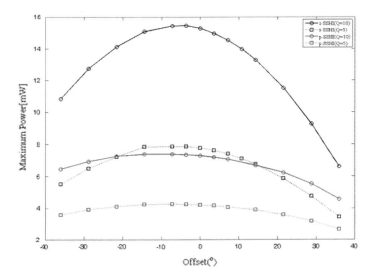

Fig. 4 The harvested power as a function of the switch triggering offset

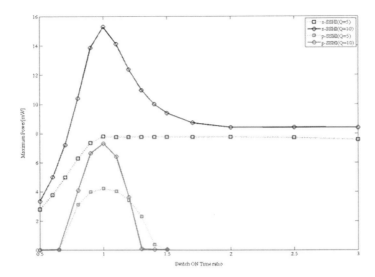

Fig. 5 Harvested power as a function of the switch on time duration deviation

In Fig 4, the effect of the switch triggering offset on harvested power is shown. This figure shows s-SSHI is more dependent on the triggering time than p-SSHI. Figure 5 shows the harvested power depending on the switching on-time duration. In this figure, switching on time duration ratio is the ratio of the real on-time duration and exact on-time duration as Eq. 2. The power harvested in p-SSHI interface is very dependent on the on-time duration deviation, where when the on-time duration becomes over 1.3 times of exact on-time duration expressed by Eq. 2, the harvested power goes zero. However, s-SSHI shows when the deviation gets larger, the harvested power does not depend on the on-time duration. In s-SSHI, on-time duration gets larger, piezoelectric voltage gets smaller, however when on-time duration becomes over two times of exact duration, the voltage waveform does not change.

4 Conclusion

In this paper, we demonstrate and compare different piezoelectric harvester interface circuits using SPICE simulation. We consider the standard interface, the standard interface with a switch, parallel and serial synchronized switching harvesting on inductor. Switch triggering offset and switch on-time duration are very interest in calculating the power in SSHI. It is conformed that switch triggering offset has more impact on the s-SSHI than p-SSHI. It is recommended that the switch triggering offset is kept smaller than 10 % of piezoelectric element vibration period. The switch on-time duration is more important in case of the

The Simulation Study on Harvested Power

p-SSHI. p-SSHI shows when the on-time duration becomes more than 1.3 times or less than 0.7 times of exact duration time, the harvested power gets zero. Because on-time duration is very small, careful consideration on the duration will be needed. s-SSHI reveals the characteristics that when less than 1.5 times exact on-time duration, the harvested power varies significantly with the on-time duration, however larger than 1.5 times exact on-time duration has scarcely influenced on the harvested power. Inductor's Q also has contributed on these characteristics.

Acknowledgments This work was supported by the Industrial Strategic technology development program, 10041766, Development of energy management technologies with small capacity based on marine resources funded by the Ministry of Knowledge Economy (MKE, Korea) and this work (Grants No. R00045044) was supported by Business for Cooperative R&D between Industry, Academy, and Research Institute funded Korea Small and Medium Business Administration.

References

1. Mateu L, Moll F (2005) Review of energy harvesting techniques and applications for microelectronics. In: The proceedings of the SPIE microtechnologies for the new millenium, pp 359–373
2. Roundy S, Wright PK, Rabaey J (2003) A study of low level vibrations as a power source for wireless sensor nodes. Comput Commun 26(11): 1131–1144
3. Arms SW, Townsend CP, Churchill DL, Galbreath JH, Mundell SW (2005) Power management for energy harvesting wireless sensors. In: SPIE international symposium on smart structures and smart materials. San Diego, CA
4. Glynne-Jones P, Tudor MJ, Beeby SP, White NM (2004) An electromagnetic vibration-powered generator for intelligent sensor systems. Sens Actuators A 110:344–349
5. Mitcheson PD, Green TC, Yeatman EM, Holmes AS (2004) Architectures for vibration driven micropower generators. IEEE J Microelectromech Syst 13:429–440
6. Roundy S, Wright PK (2004) A piezoelectric vibration based generator for wireless electronics. Proc Smart Mater Struct 13:1131–1142
7. Ottman G, Hofmann H, Bhatt A, Lesieutre G (2002) Adaptive piezoelectric energy harvesting circuit for wireless remote power supply. IEEE Trans Power Electron 17(5):669–676
8. Lefeuvre E, Badel A, Richard C, Petit L, Guyomar D (2006) A comparison between several vibration-powered piezoelectric generators for standalone systems. Sens Actuators A 126:405–416
9. Guyomar D, Badel A, Lefeuvre E, Richard C (2005) Toward energy harvesting using active materials and conversion improvement by nonlinear processing. IEEE Trans Ultrason Ferroelectr Freq Control 52:584–595
10. LTspice IV, http://www.linear.com/designtools/software/

An Approach for a Self-Growing Agricultural Knowledge Cloud in Smart Agriculture

TaeHyung Kim, Nam-Jin Bae, Chang-Sun Shin, Jang Woo Park, DongGook Park and Yong-Yun Cho

Abstract Typically, most of the agricultural works have to consider not only fixed data related with a cultivated crop, but also various environmental factors which are dynamically changed. Therefore, a farmer has to consider readjust the fixed data according to the environmental conditions in order to cultivate a crop in optimized growth environments. However, because the readjustment is delicate and complicated, it is difficult for user to by hand on a case by case. To solve the limitations, this paper introduces an approach for self-growing agricultural knowledge cloud in smart agriculture. The self-growing agricultural knowledge cloud can offer a user or a smart agricultural service system the optimized growth information customized for a specific crop with not only the knowledge and the experience of skillful agricultural experts, but also useful analysis data, and accumulated statistics. Therefore, by using the self-growing agricultural knowledge cloud, a user can easily cultivate any crop without a lot of the crop growth information and expert knowledge.

T. Kim (✉) · N.-J. Bae · C.-S. Shin · J. W. Park · D. Park · Y.-Y. Cho
Department of Information and Communication Engineering, Sunchon National University,
413 Jungangno, Suncheon, Jeonnam 540-472, Korea
e-mail: taehyung@sunchon.ac.kr

N.-J. Bae
e-mail: bakkepo@sunchon.ac.kr

C.-S. Shin
e-mail: csshin@sunchon.ac.kr

J. W. Park
e-mail: jwpark@sunchon.ac.kr

D. Park
e-mail: dgpark6@sunchon.ac.kr

Y.-Y. Cho
e-mail: yycho@sunchon.ac.kr

J. J. (Jong Hyuk) Park et al. (eds.), *Multimedia and Ubiquitous Engineering*,
Lecture Notes in Electrical Engineering 240, DOI: 10.1007/978-94-007-6738-6_86,
© Crown Copyright 2013

Keywords Ubiquitous agriculture · Agricultural cloud · Smart service · Knowledge-based

1 Introduction

Recently, to improve labor-intensive working environments, to secure economic feasibility, and to enhance productivity and quality in the fields of agriculture, many researchers are concentrating on the convergence of information technologies into the agricultural environments. Now, the agricultural environment is preparing new advancement. In agricultural environment, many studies about the IT-agriculture convergence have included optimum growth monitoring and growth environmental controlling system. The works are generally based on situation conditions from various sensors on ubiquitous sensor networks, which are deployed around the cultivation facilities or grounds. Existing studies for the optical crop growth information are based on a few fixed environmental data, which are temperature, humidity, illuminations, etc. However, crops are living organisms. Because environmental conditions are affected with each other in a very detailed and complex relationship, the crop growth status may be different in the same environmental conditions. Therefore, these studies about the optical crop growth information are underway constantly. That is, for smart agricultural environments, we need a method which can control dynamically and efficiently the environmental information about the crops.

A smart service in agricultural environments has to be able to use various environmental conditions automatically and organically as decision conditions on the service execution without human's interference. However, because it is so difficult to orchestrate the useful information from a lot of data related with specific crops and so hard to apply to other crops, getting stable and meaningful information to cultivate a specific crop is not simple. Recently, there have been a few of interesting researches to apply situation information into agricultural environments. However, most of the current agricultural systems using the situation information cannot make the best use of the great store of agricultural knowledge.

In this paper, we introduce an approach for self-growing agricultural knowledge cloud in smart agriculture. Knowledge DBs in the self-growing agricultural knowledge cloud have a lot of defined situation conditions, which are called contexts [1]. In smart agricultural environments, a context is one of very important elements to make an agricultural service autonomous and smart. The proposed self-growth agricultural knowledge supports a knowledge cloud architecture based on the various kinds of the knowledge DBs, which may be very far apart from each other and can offer a user or a smart agricultural service system the optimized growth information customized for a specific crop. The information contains not only the knowledge and the experience from users, researchers, and experienced farmers, but also useful analysis data and statistics accumulated into the knowledge DBs. Especially, the knowledge cloud can be plentiful more and more after the lapse of time, and can be growing by itself. Therefore, users can access to the knowledge cloud system through the Internet, and obtain various and useful information for agricultural works.

2 Related Work

Smart service in agriculture environments. Generally, an agricultural environment that provides smart services is called ubiquitous agricultural environment. And there have been many studies about the convergence of state of the art information technology to build a new agricultural environment in many countries well aware about importance of agriculture. At its most basic, the monitoring services based on the wireless sensor network are designed. One of them [2] uses only the data from sensors in agricultural environment and another service [3] uses weather monitoring network and on-farm frost monitoring network and another one [4] uses Geographic Information Systems (GIS). And there are services for precision agriculture.

Smart services are well suited to apply the greenhouse. And these smart services in agricultural environment, designed by context-aware service models for the systematical definition and extensibility [1].

Many studies are underway to provide smart services in agricultural environment, but all have limitations in the domain that can be used. Farmers want to receive services that given the best choice regardless of the type of crops and the environmental characteristics. For this purpose, we need a service that can offer the knowledge and the experience of skillful agricultural experts rather than relying on system algorithms.

Knowledge-based services. A knowledge-based is a special kind of database for knowledge management. A knowledge base provides a means for information to be collected, organized, shared, searched and utilized. Therefore, the knowledge-based services help machines to have a decision-making like a human's decision-making. Recently, many knowledge-based services are provided in various fields. And it makes it possible to provide an appropriate service without human's interference. To do that, the knowledge-based services are designed by the context models based-on the ontology language [5]. For the smart home service, commonsense knowledge base is defined and designed by context modeling [6]. The field of e-learning investigates the integration of e-Learning systems and knowledge management technology to improve the capture, organization and delivery of both traditional training courses and large amounts of corporate knowledge [7].

To provide services without human intervention, it should be based on the knowledge-base. Therefore, the smart services in agricultural environment also offer the knowledge-based services for the ubiquitous agricultural environment.

A cloud service in smart environments. Cloud computing provides a new way to build applications on on-demand infrastructures instead of building applications on fixed and rigid infrastructures. Cloud computing solutions only requires access to the Internet and a Web browser, and the heavy lifting of the software and hardware of the individual computer workstation is removed [8]. Therefore, cloud service is provided to users through the Internet anytime and anywhere. Cloud service is the ideal way to build ubiquitous environment. So Many cloud services are provided in various ubiquitous environments. Recently, A few researches have begun to cloud services in agricultural environment. A service model based-on an agricultural expert cloud is introduced [9]. This service is based an expert system, in which the knowledge and the experience of the various fields related in agriculture is accumulated. Because cloud service in agricultural environment is in the beginning step, more research is needed.

A self-growing agricultural knowledge cloud is designed as a unified architecture between knowledge-based service and cloud service. Through this, user can successfully and stably cultivate any crop by adopting the automatically suggested service to their agricultural systems anytime and anywhere.

3 A Self-Growing Agricultural Knowledge Cloud

Figure 1 shows a conceptual view for a self-growing agricultural knowledge cloud in smart agriculture. As shown in Fig. 1, the various groups can access the self-growing agricultural knowledge cloud just through the Internet anytime and anywhere. Therefore, a user in these groups can take the valuable agricultural information from users in other groups. Then, this information is applied to the user's agricultural environment. In addition, users can upload their own information of experience and knowledge to Knowledge DBs in the self-growing

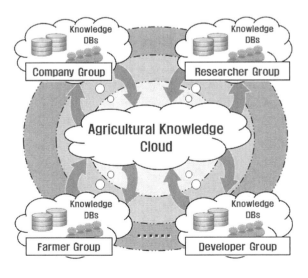

Fig. 1 A conceptual view for self-growing agricultural knowledge cloud

An Approach for a Self-Growing Agricultural Knowledge Cloud

agricultural knowledge cloud. As time passes, the agricultural knowledge cloud will grow more valuable itself because of the accumulating the various information of experience and knowledge from the various groups continually. So, we named it the self-growing agricultural knowledge cloud.

Figure 2 shows a brief conceptual architecture of the proposed approach for a self-growing agricultural knowledge cloud in Smart Agriculture. The architecture consists three layers, which are the Application Layer, the Middle Layer, and Physical Layer. The Physical Layer generates low-level data, which are crop conditions, real sensed data, status of control devices and user profile, etc. The Middle Layer consists of the ServiceProvider, the ContextProvider and the DBManager. The ContextProvider defines low-level data from the Physical Layer into Context data and uploads it to a self-growing agricultural knowledge cloud DBs. And it can also take agricultural information which is experience and knowledge from various other groups in knowledge DBs. A self-growing agricultural knowledge cloud can be accessed through the Internet anytime and anywhere. The ServiceProvider provides the best service through the ContextInterpreter based-on context data from user's agricultural environment and agricultural knowledge cloud. The Application Layer supports that a developer composes a smart agricultural service application easily and quickly. To do this, the layer offers a GUI-based development toolkit and an agricultural ontology, which is used to make a smart service scenario by a developer and a context by a ContextProvider appeared in Fig. 2. The service scenario may focus on the growth environment of crops, a growth rate of crops and a consumption of energy, etc.

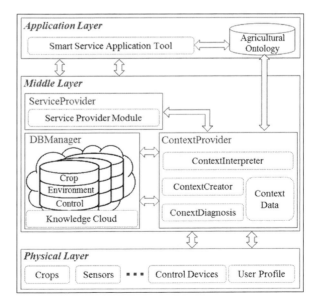

Fig. 2 A brief conceptual architecture of the proposed approach

4 A Sample Smart Service Scenario

In this section, we show a sample smart service scenario for the self-growing agricultural knowledge cloud in smart agriculture. The following example scenario in the Table 1 is Mark's situation in the common agricultural environment.

To meet the hope of Mark, a smart service system using the self-growing agricultural knowledge cloud may need various contexts as situation conditions. Of cause, the contexts have to include not only simple sensed contexts, but also high-level contexts composed from knowledge and experience which he has. Now, let's suppose that Mark's greenhouse needs a temperature control system in order to maintain an optimal growth environment of lettuce. Because Mark has no experience of lettuce cultivation, he can take Contexts of lettuce cultivation from the agricultural knowledge cloud. This is shown in Fig. 3a. Then, the temperature control service will be executed when the current temperature is sensed by a temperature sensor and recognized as a context by the smart service system. Again, let's suppose when the temperature is low enough to start the heating service but is a little bit high to do that in the seasonal and geographical aspect. Again, let's suppose when the temperature is low enough to start the heating service but not affects lettuce in the seasonal and geographical aspect. In this case, for the cultivation goal, Mark accesses to the agricultural knowledge cloud using the 'economical way' as a keyword. In this case, for the cultivation goal, Mark accesses to the agricultural knowledge cloud using the 'economical way' as a keyword. So, he can take a lot of information of environment, growth and control, etc. from experienced farmers who cultivate lettuce in the similar situation. Finally, he cannot have to operate the heating service using the best economical control context. This is shown in Fig. 3b.

Table 1 The example scenario

Mark is an experienced farmer who cultivates various crops, except the lettuce in his greenhouse. Now he wants to cultivate lettuce
He should refer to the method of cultivation or the advice of an experienced farmer
He frequently comes into his greenhouse and checks the environmental information and the growth condition of lettuce
Then, he has to take proper action according to the conditions
If the temperature is below the normal values, he has to turn on the temperature control systems. In this case, the operation on heating system affects the humidity in greenhouse. So, the windows open and shut system and the irrigation system should also be considered
He should invest a lot of money in order to maintain an optimal growth environment of lettuce. Therefore, He wants a smart growth service system to do the tiresome works automatically without his intervention and conserve cost using the economical way

An Approach for a Self-Growing Agricultural Knowledge Cloud

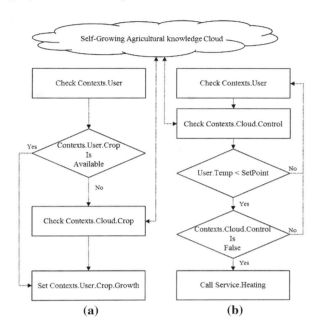

Fig. 3 A process path for the crop searching and heating service

5 Conclusion and Future Works

Typically, most of the works in the agricultural environments are affected by various conditions, which tend to be unsystematically and dynamically changed. For the smart agricultural service in ubiquitous agricultural environment, the knowledge and experience about crops growth and agricultural skills have to be offered to users immediately through the Internet and networks. In this paper, we propose an approach for a self-growing agricultural knowledge cloud in smart agriculture. To support various and valuable agricultural information anytime and anywhere to service users, the introduced approach uses an agricultural knowledge cloud, which is various knowledge DBs stored by other expert group users through the Internet or networks. Therefore, the proposed approach can be used for users to provide appropriate various smart services for higher productivity and better quality in ubiquitous agricultural environments without service user's interference.

As a future work in this paper, we will focus on the studies about real implementation of an efficient smart service framework or system based on the suggested architecture in the field of ubiquitous agricultural environments. And, in order to sufficiently testify to efficiency of the implemented smart service system using the suggested architecture, we will try to compose useful smart service applications and to adopt them into real agricultural environments with various sensors and computing devices.

Acknowledgments This work was supported by the Industrial Strategic technology development program, 10040125, Development of the Integrated Environment Control S/W Platform for

Constructing an Urbanized Vertical Farm funded by the Ministry of Knowledge Economy (MKE, Korea). And this research was supported by Basic Science Research Program through the National Research Foundation of Korea (NRF) founded by the Ministry of Education. Science and Technology (2011-0014742).

References

1. Cho Y, Moon J, Kim I, Choi J, Yoe H (2011) Towards a smart service based on a context-aware workflow model in u-agriculture. IJWGS 7:117–133
2. Zhou Y, Yang X, Guo X, Zhou M, Wang L (2007) A design of greenhouse monitoring and control system based on ZigBee wireless sensor network. In: international conference on wireless communications, networking and mobile computing, WiCom 2007, pp 2563–2567, Sept 2007
3. Pierce FJ, Elliott TV (2008) Regional and on-farm wireless sensor networks for agricultural systems in Eastern Washington. Comput Electron Agric 61:32–43
4. Ayday C, Safak S (2009) Application of wireless sensor networks with GIS on the soil moisture distribution mapping. In: Symposium GIS Ostrava 2009—seamless geoinformation technologies, Ostrava, Czech Republic
5. Kumar H, Park P (2010) Know-ont: a knowledge ontology for an enterprise in an industrial domain. IJDTA 3(1):23–32
6. Kawasar F, Shaikh M, Park J, Mitsuru I, Nakajima T (2008) Augmenting user interaction in a smart home applying commonsense knowledge. IJSH 2(4):17–31
7. Qwaider W (2011) Integrated of knowledge management and e-learning system. IJHIT 4(4):59–70
8. Caytiles R, Lee S, Park B (2012) Cloud computing: the next computing paradigm. IJMUE 7(2):297–302
9. Cho Y, Cho K, Shin C, Park J, Lee E (2012) An agricultural expert cloud for a smart farm. FutureTech 164:657–662

Determination of Water-Miscible Fluids Properties

Zajac Jozef, Cuma Matus and Hatala Michal

Abstract The paper presents the main specifications for a monitoring of metal-working fluids and cleaners for mass industries. The implementation of fluid outsourcing in these industries. There is presented method o flaboratory determination and diagnostic in industry of water-miscible fluids.

Keywords Fluid management · Waste management · Fluid contamination · Metalworking cutting fluids

1 Introduction

Concentration posses a quality of leading parameter amongst other parameters of water-miscible fluids in their applying in manufacturing processes. For operational detection of product concentration two time-modest methods are used:

- *Refractometric method*
- *Titrimetric method*

Refractometric method is a simple way of operational concentration detection with manual or desk-top refractometer. The method is based on light beam refraction in optical refracting prism. Refractometer can be easily calibrated with a

---

Z. Jozef (✉) · C. Matus · H. Michal
Department of Manufacturing Technologies, Faculty of Manufacturing Technologies,
Technical University in Košice, Bayerova 1, 080 01 Presov, Slovakia
e-mail: jozef.zajac@tuke.sk

C. Matus
e-mail: matus.cuma@tuke.sk

H. Michal
e-mail: michal.hatala@tuke.sk

J. J. (Jong Hyuk) Park et al. (eds.), *Multimedia and Ubiquitous Engineering*,
Lecture Notes in Electrical Engineering 240, DOI: 10.1007/978-94-007-6738-6_87,
© Springer Science+Business Media Dordrecht(Outside the USA) 2013

given water sample and adjuster screw. For water-miscible fluids it is sufficient to use refractometers scaled from 0 to 15 %. Concentrations of these fluids exceeding 15 % are very rare. If the fluid is contaminated by higher percentage of tramp oil, it is very difficult to obtain relevant information about real value of product concentration in used mix. In such a case the titrimetric method is applied.

The titrimetric method is based on oil-contaminated water titration. For measuring of total alkalinity titrimetric set is often used, e.g. TA-kit. The set contains two solutions labeled L (HCL) and K and two syringes for measuring out of the product and for measuring out of alkaline concentrate. Resultant value must be modified.

Monitoring and maintaining fluid quality are crucial elements of a successful fluid management program. A fluid must be monitored to anticipate problems. Important aspects of fluid monitoring include system inspections and periodic measurements of fluid parameters such as concentration, biological growth, and pH. Changes from optimal fluid quality must be corrected with appropriate adjustments (such as fluid concentration adjustments, biocide addition, tramp oil and metal cuttings removal, and pH adjustment). It is important to know what changes may take place in your system and why they occur. This allows fluid management personnel to take the appropriate steps needed to bring fluid quality back on-line and prevent fluid quality problems from recurring.

The pH value determines acidity (0–7) or alkalinity (7–14) number. For metalworking water-miscible fluids pH values range from 8.8 to 9.5 in order to ensure corrosion and bacteriological protection. The pH value is a fast indicator of applied mix condition.

If pH value is under 8.8 it can be assumed that the fluid contains bacteria and mix becomes unstable, corrosion protection reduces, and an odor may occur… To increase pH value it is vital to apply additive (e.g. Additive 63). If value is above 9.5 the mix is contaminated by alkaline compounds (e.g. washing and cleaning media).

In Germany it is required before implementing water-miscible fluids to carry out test TRGS611 every week because nitrites react with secondary amines into nitroamines which are included in the list of carcinogenic agents in #2 category. Progressive manufacturers do not use nitrites (e.g. soda). The only contaminations are nitrites already present in applied water. Maximum value of nitrites quantity in processing fluids is 20 ppm. In the case the value is higher, it is necessary to change water source or modify the water.

The new metalworking cutting fluids generation brings also higher requirements for the system of fluid performance monitoring and control.

The simplest way for bacteria quantity identification is applying of the set for complex determination of mould, bacteria and fungi content with straps that are immersed for 5 s in measured medium and successively for 72 h are these so called "dipslides" placed in incubation apparatus with constant temperature. When fungi, mould or bacteria are present in greater amount, cultivation on dipslides would grow, consequently the cultures can be identified and quantified. Based on the result a proper additive can be applied (e.g. Kathon, Additive 63, etc.).

Oil used for machine parts lubrication and hydraulic oils are common contaminants in water-miscible fluids and they can drastically change performance of metalworking fluids especially their washing up and cleaning abilities.

If the operating system contains more tramp oils than 1.8 %, following problems from operating contaminated medium can be expected:

- *Cooling capability decrease*
- *Degradation of filtrating*
- *Reduced stability of the mix*
- *Smoke creation in cutting zone*
- *Growth of bacteria, mould and fungi*
- *Problems with determination of refractometric concentration*
- *Deterioration of workplace environment and increase of fluid skin-aggression*
- *Tramp oils that leak into the systems with water-miscible fluids can be divided to:*
- *Free oils*
- *Emulsifiable oils*

The main task of water-miscible processing fluids is to dissipate heat from the cutting zone. Effective removal of the heat increases tools life-cycle and product dimensions stability. Water has a better capability to take away heat than oil, however water in cutting zone causes corrosion of machined parts as well as the machine components which are in contact with the water during manufacturing process. Corrosion may occur whole year round but the higher probability is when there is a great temperature difference between day and night, or when both temperature and humidity are high. When temperature is rising, so is chemical activity including oxidation processes. According to long-time experience with applying water-miscible fluids the critical season is late April—early May and September ("Indian summer"). Prevention is accomplished by concentration increase of approximately one third. If concentration increase is not possible (e.g. foam creation, worsened wash up or workmen skin problems), it is necessary then to use additives for better anti-corrosion protection.

Corrosion protection tests are lengthy and could be realised only in laboratory conditions. Corrosion problems can be avoided by observing pH values, quality of input water, bacteria content and mix concentration.

Metalworking fluids having pH values of 9.0 and higher should be satisfactory for ensuring short-term anti-corrosion protection (three weeks) for iron-based alloys as well as for non-ferrous metals alloys (aluminium, copper, tin, etc.).

Water containing more than 80 ppm of chlorides and more than 250 ppm of sulphides is considered to be aggressive. These compounds rapidly decrease anti-corrosion abilities of fluids. It is necessary to check out the compounds content every week and implement proper measures when their concentration increases (adding de-mineralised or distilled water).

For observing manufacturing process mix quality it is necessary to measure hardness of the mix in the system. After certain operating time the fluid hardness increase (e.g. when central system volume is 30 m^3—for bearing rings grinding—the hardness increase of cooling semi-synthetic fluid with 31 % mineral oil is from 15 to 20°GH in a month). Water hardness above 20°GH causes decrease of anti-corrosion protection ability of water-miscible fluids.

If the mix contains higher amount of bacteria that "feed" on its components and constantly create organic acids, pH value decreases and anti-corrosion properties of fluids reduce. When pH values in individual tanks drop below value of 8.8, it is possible to apply alkaline system cleaner with concentration 0.1 %, when pH in central system drops, it is inevitable to apply biocides and makrobiocides for bacteria growth regulation.

2 Proposal for Method of Processing Fluids Monitoring

Based on theoretical analyses, customer requirements and operational state of processing medium survey, it is vital to observe on regular basis:

- *Appearance of processing medium*
- *pH value*
- *Total alkalinity of mix in %*
- *Anionic mix concentration*
- *Refractometric mix concentration*
- *Overall oil content in the mix*
- *Free oil content in the mix*
- *Bacteria, fungi, mould content*
- *Bacteriocide content*
- *Corrosion test applied at final user*
- *Corrosion test according DIN 51 360/2*
- *Nitrites content*
- *Nitrates content*
- *Overall hardness*
- *Borates*
- *Chlorides*
- *Contaminants content*

For monitoring of fluid condition it is required to determine limit content values of particular characteristics and contaminants in processing medium (see Table 1). When limits are overrun, it is essential to carry out measures concerning fluid condition, filtration effectiveness and according to overall oil content in the fluid it is possible to determine leak of tramp oil from machinery or other sources. For the observation of fluids in central systems it is appropriate to use tabular graphic fluid conditions outputs.

Table 1 xxx

Parameters	Given values Min.	Opt.	Max.	5	6	7	8	9	12	13	14	15	16	19	20	21	22	292	Interventions since
Numb. of days from filling	275	365	545															292	July 28, 2012 additives for improving conservation and bacteriocidal properties
Appearance	–	–	–	o.k.	o.k.	o.k.	o.k.	o.k.	o.k.	o.k.	o.k.	o.k.	o.k.	o.k.	o.k.	o.k.	o.k.	o.k.	
pH	8.9	9.1	9.5	9.2	9.2	9.2	9.2	9.2	9.2	9.1	9.2	9.2	9.2	9.2	9.2	9.2	9.2	9.2	
Total alkalinity in % of mix	4.0	5.0	6.0	4.6						5.1						5.6			
Total alkalinity in % NaOH	0.196	0.245	0.300	0.224						0.252						0.277			
Anionic conc. in % of mix	2.0	3.0	4.0	4.0						4.0						3.5			
Refr. conc. in % of mix	3.0	4.0	5.0	4.0	4.0	4.1	4.2	4.3	4.3	4.3	4.4	4.5	4.5	4.8	5.0	5.0	5.2	3.5	
Total oil in %	0.7	1.3	2.0	0.8						1.0						1.1			
Free oil in %	0.0	0.5	1.3	0.0						0.1						0.3			
Bacteria, fungi, mould	0-0-0	–	5-2-2	0-0-0						0-0-0						0-0-0			
Bakteriocide in mg/l	120	150.0	250	200						200						200			
Corrosion test ZVL	0		20	0	1	0	0	0	0	0	0	0	0	0	0	0	0	0	
Corros. test DIN 51 360/2	k = 0	k = 0	k = 2	0						0						0			
Nitrites in ppm	0	–	40	0															
Nitrates in ppm	0	–	100	10															
Overall hardness in ° Ger	18	30.0	45	16.0															
Borates ppm	–	–	150																
Chlorides in ppm	20	–	150	6						14									
Contam. cont. in mg/l	0	10	30						9.7	11.3	10.2	11.7	17.2	17	17.7	12.3	11.5	12	

3 Summary

Verification of fluid management in bearings factories shows that proposed method of processing media care decreases overall costs by one third. In next year the expected gain comparing to last for these factories from fluid management project is 15 % in servicing two central systems with water-miscible products and one central system with non-water-miscible product.

References

1. Gots I, Zajac J, Vojtko I (1995) Equipment for measuring the degree of wear to cutting tools. Tech Mess 1:8–11
2. Zajac J (2003) Accession at answer of synergy grinding and fluids in grinding. Manuf Eng 2–3:14–16
3. Zajac J (2007) Develop trends in research of processing fluids. Manuf Eng 2:36–38
4. Chao Wu et al (2009) Study on green design and biodegradability of B-containing water-based cutting fluid. Key Eng Mater 407–408:309–312
5. Harnicarova M, Zajac J, Stoic A (2010) Comparison of different material cutting technologies in terms of their impact on the cutting quality of structural steel. Tehnicki Vjesnik 3:371–376
6. Dima IC, Gabrara J, Modrák V, Piotr P, Popescu C (2010) Using the expert systems in the operational management of production. In: MCBE '10, p 307–312
7. Mohamed WANW et al (2011) Thermal and coolant flow computational analysis of cooling channels for an air-cooled PEM fuel cell. Appl Mech Mater 110–116:2746
8. Dima IC, Modrák V, Duică A, Goldbah IR (2011) The method of optimisation of the service of several tools, using the "mechanisation coefficient". In: IC-SSSE-DC '11, pp 152–160
9. Wang Fei et al (2012) Management of drilling waste in an environment and economic acceptable manner. Adv Mater Res 518–523:3396–3402
10. Čuma M, Zajac J (2012) The impact analysis of cutting fluids aerosols on working environment and contamination of reservoirs. Tehnicki Vjesnik 2:443–446
11. Novak-Marcincin J, Novakova-Marcincinova L, Janak M (2012) Simulation of flexible manufacturing systems for logistics optimization. In: LINDI-2012, vol 631950, pp 37–40

Influence of Technological Factors of Die Casting on Mechanical Properties of Castings from Silumin

Stefan Gaspar and Jan Pasko

Abstract Die casting represents the highest technological level of metal mould casting. This technology enables production of almost all final products without necessity of further processing. The important aspect of efficiency and production is a proper casting parameters setting. In the submitted paper following die casting parameters are analyzed: casting machine plunger speed and increase pressure. The studied parameters most significantly affect a qualitative castings dimension and they influence the most a gained porosity level f as well as basic mechanical properties represented by tensile strength R_m and ductility A_5.

Keywords Die casting · Technological factors · Casting · Mechanical properties

1 Introduction

The die casting is a method of precise casting and it meets requirements for transformation from basic material into a ready product. Liquid metal is pressed at high speed (10–100 m.s^{-1}) into a cavity of a divided recursive mould (Fig. 1). The liquid metal high speed can be obtained by making intake groove more narrow at high pressure [13]. When liquid metal thoroughly fills the mould cavity, during a short time before it completely hardens, a increased pressure—increase pressure affects on it. It has to substitute gravitational setting of melting into empty mould cavities and thus suppress agglutinating and expansion of gas bubbles during cast

---

S. Gaspar (✉) · J. Pasko
Faculty of Manufacturing Technologies, Technical University Kosice,
Slovakia, Bayerova 1, 080 01 Presov, Slovakia
e-mail: stefan.gaspar@tuke.sk

J. Pasko
e-mail: jan.pasko@tuke.sk

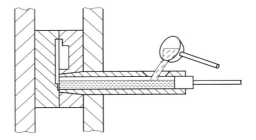

Fig. 1 Die casting machine with cold horizontal chamber

crystallization process (the necessary condition is a sufficient hydraulic connection of the mould with an intake system). The die casting products are of very precise dimensions, smooth surface, thin walls and very good mechanical properties [1, 2].

The die casting products quality is influenced by basic technological factors, i.e. pressing speed during casting cycle, melting specific pressure—increase pressure, filling chamber temperature and mould temperature [1–5].

Increased attention is paid also to casting internal quality which is characterized by a kind and extend of foundry faults. The most important casting faults are exogenous cavities being generated by gases and air capturing which pass through the melt in turbulent stream. Pressing plunger speed in filling chamber of die casting machine is the most important factor of die casting. Metal stream speed in intake groove, which determines mould filling mode and thus affects both internal and external casts quality, depends on the plunger's velocity. If the main technological parameters are considered, then the die casting products quality affects also pressure acting on the cast being hardened in the casting cycle last phase—increase pressure [1, 6–8].

The aim of the paper is to analyze basic technological factors (plunger pressing speed and increase pressure of die casting and to set relations of their influence on selected mechanical properties: porosity f, tensile strength R_m and ductility A_5. Knowing these relations can be used in a new product design phase as well as in production process which can help increase manufacturing productivity and quality.

2 Material for Experiments, Methodology of Experiments and Used Devices

Influence of filling mode based on plunger pressing speed and increase pressure changes on selected mechanical properties study was performed on experimental samples (Fig. 2) appointed for static tensile test. For this experiment, a melting process had been carried. Its chemical composition is given in Table 1 and is in accordance with Standard EN 1706.

During the test, a wide extend of plunger speed in filling chamber within five levels was set: ($v_1 = 1.9$ m.s^{-1}; $v_2 = 2.3$ m.s^{-1}; $v_3 = 2.6$ m.s^{-1}; $v_4 = 2.9$ m.s^{-1},

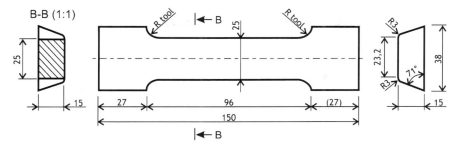

Fig. 2 Scheme of experimental sample

Table 1 Chemical composition of the experimental cast of the applied alloy

Chemical composition of the experimental cast of the applied alloy % of elements content											
Al	Si	Fe	Cu	Mn	Mg	Cr	Ni	Zn	Pb	Sn	Ti
85.27	12.02	0.71	1.19	0.21	0.13	0.02	0.02	0.35	0.02	0.03	0.03
According to EN 1706											
The rest	10.5–13.5	Max. 1.5	0.7– 1.2	Max. 0.55	Max. 0.35	Max. 0.1	Max. 0.3	Max. 0.55	Max. 0.2	Max. 0.1	Max. 0.2

$v_5 = 3.2$ m.s^{-1}). The increase pressure values extend was set also within five levels: ($p_1 = 13$ MPa; $p_2 = 15$ MPa; $p_3 = 18$ MPa; $p_4 = 22$ MPa; $p_5 = 25$ MPa). Beside factors which effect was monitored experimentally, there were factors whose effect had been reduced by keeping on the constant level. They were: melting temperature 708 °C and mould temperature 199 °C.

The static tensile test was performed on device ZDM 30/10 with jaw shift speed 10 mm.min^{-1}. The ductility test was then performed on these rods. Near fracture areas generated after static tensile test (Fig. 3), samples for a microscopic analysis were taken. The analysis was performed with a microscope Olympus GX51, magnifying 100 x. The samples were processed by PC with ImageJ program, which evaluated a proportional part of porosity from the studied cut (Fig. 4).

3 Reached Results and Their Analysis

Table 2 shows recorded reached values of porosity f, tensile strength R_m and ductility A_5 in relation of studied casting parameters, i.e. plunger pressing speed and increase pressure. Figures 5, 6, 7, 8, 9 and 10 represent individual shapes of relations between selected mechanical properties and studied casting parameters.

Figures 11, 12, 13, 14, 15 and 16 represent metallographical cuts of selected macroscopic samples where dark spots show pores in the cast.

Measured porosity values f, breaking tensile strength R_m and ductility A_5 presented in Table 2 show considerable variance in relation to plunger pressing speed and to increase pressure.

The result figures confirmed the fact that the lowest rate of porosity of A-category samples was found in samples with the lowest plunger speed in the filling chamber of pressing cast machine. Simultaneously, these samples represent the highest values of breaking tensile strength and ductility. Increasing of plunger speed means also increasing of porosity and decreasing of breaking tensile strength as well as ductility. Thus it can be said that the melt speed in the mould depends on plunger speed in the filling chamber. The wide range of filling plunger speed was related to test different filling modes. On the base of achieved results, studied macrostructures of the samples from point of view of porosity size and placement, the mould filling mode can be defined. At plunger speed in the filling chamber 1.9 m.s^{-1} the filling mode was laminar (Fig. 11), i.e. the front of flowing melt was continuous, homogenous and without whirlpools. When speed increases to 2.3–2.6 m.s^{-1} the filling mode changes from laminar to turbulent (Fig. 12). Turbulence of melt pulling air from the filling chamber out and its locking in cast walls. Further plunger speed increasing (2.9–3.2 m.s^{-1}) creates dispersive mixture of air and liquid metal, so called dispersive filling mode (Fig. 13).

When studying affect of increase pressure to porosity, breaking tensile strength and ductility, its positive influence on analyzed parameters can be unambiguously confirmed. At the lowest increase pressure values, the highest rate of porosity (Fig. 16) and the lowest values of breaking tensile strength and ductility were reported. So, increasing of increase pressure means reducing porosity (Figs. 14 and 15) and increasing breaking tensile strength and ductility. Increase pressure is recently one of the most discussed factors of press casting. On the one side, its high value reduces lifetime of the moulds and increases downtime periods, but on the other side it improves casting, reduces air volume locked in solid casts (porosity) and thus increases their quality (strength, tightness etc.).

4 Conclusion

The measured values of porosity f, breaking tensile strength R_m a ductility A_5 of analyzed pressure cast samples produced on the pressure cast machine with cold horizontal chamber confirmed affect of plunger pressing speed and increase pressure on cast quality. It can be said that cavity filling speed defining mould filling mode is depended on plunger speed in the filling chamber. The casts made under higher pressure during hardening present better mechanical properties with the lowest porosity proportional rate. Performed studies of internal porosity dimensional evaluation and distribution enables to compare measured values with these parameters. Firstly, the studies showed that breaking tensile strength and ductility correlate with pores size which make the cast cross-section more weak.

Influence of Technological Factors of Die Casting 717

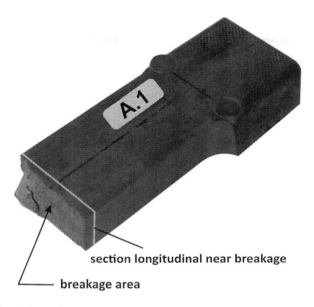

Fig. 3 Experimental sample

Fig. 4 Evaluation of the porosity of the sample no. A.3 s sample porosity evaluation by the Image J software porosity 5.85 %

Table 2 Reached results of studied mechanical properties

Studied technological casting parameters			Studied mechanical properties measured values		
			Porosity "f" [%]	Breaking tensile strength "R_m" [MPa]	Ductility "A_5" [%]
Plunger pressing speed [m.s^{-1}]	No. A.1	1.9	0.65	169	2.8
	No. A.2	2.3	2.13	143	2.7
	No. A.3	2.6	5.85	124	2.5
	No. A.4	2.9	8.2	115	2.3
	No. A.5	3.2	14.8	103	2.2
Increase pressure [MPa]	No. B.1	13	6.18	121	2.3
	No. B.2	15	5.04	127	2.4
	No. B.3	18	4.32	133	2.5
	No. B.4	22	2.42	140	2.6
	No. B.5	25	0.89	153	2.7

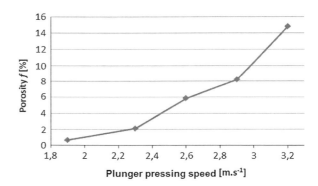

Fig. 5 Relation between plunger pressing speed and porosity "f"

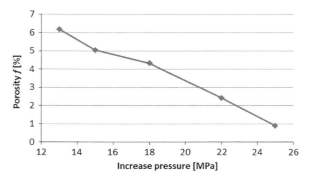

Fig. 6 Relation between increase pressure and porosity "f"

Fulfilling of pressing plunger defined speed and of increase pressure values is decisive for quality casts production.

This article has been prepared within the project VEGA No. 1/0593/12: Research of Technological Parameters Influence of Die Castings and Design

Influence of Technological Factors of Die Casting

Fig. 7 Relation between plunger pressing speed and breaking tensile strength "R_m"

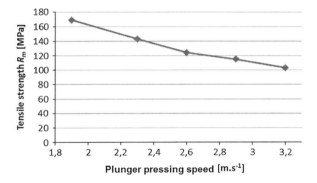

Fig. 8 Relation between increase pressure and breaking tensile strength "R_m"

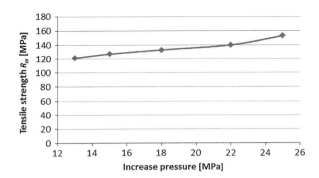

Fig. 9 Relation between plunger pressing speed and ductility "A_5"

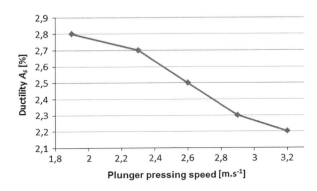

Modification of the Die System of the Casting Machine on Mechanical Properties of Die Castings of Lower Mass Category Made of Silumin.

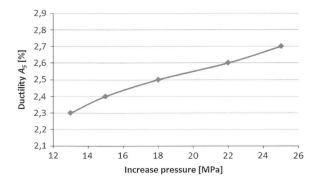

Fig. 10 Relation between increase pressure and ductility "A_5"

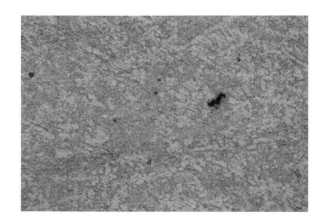

Fig. 11 Sample no. A1 porosity 0.65 %

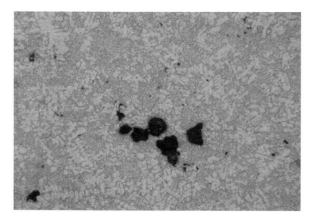

Fig. 12 Sample no. A2 porosity 2.13 %

Influence of Technological Factors of Die Casting 721

Fig. 13 Sample no. A.5 porosity 14.08 %

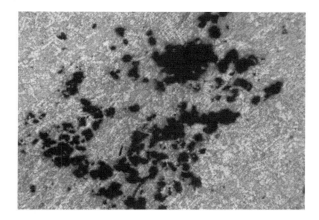

Fig. 14 Sample no. B5 porosity 0.89 %

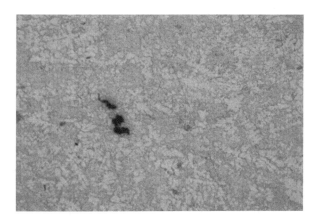

Fig. 15 Sample no. B4 porosity 2.42 %

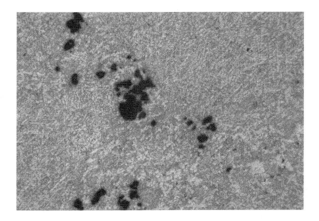

Fig. 16 Sample no. B1 porosity 6.18 %

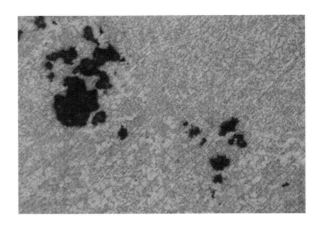

References

1. Gaspar S, Mascenik J, Pasko J (2012) The effect of degassing pressure casting molds on quality of pressure casting. Adv Mater Res 428:43–46
2. Semanco M, Fedak M, Rimar M, Ragan E (2012) Equation model to evaluate fluidity of alluminium alloys under pressure die-casting condition. Adv Mater Res 505:190–194
3. Stailcek L, Bytysev AI, Caplovic L, Batysov KA (2007) Metallographic verification of the model of the flow enforced during solidification under high external pressure. Die Cast Eng 51:56–60
4. Puskar M, Bigos P (2012) Method for accurate measurements of detonation in motorbike speed racing engine. J Int Meas Confed 45:529–534
5. Ragan E (2007) Liatie kovov pod tlakom, Presov
6. Vaskova I, Malik J, Futas P (2009) Tests of moulding mixture by using various clay binder granularity. Arch Foundry Eng 9:29–32
7. Kocisko M, Novak-Marcincin J, Baron P, Dobransky J (2012) Utilization of progressive simulation software for optimization of production systems in the area of small and medium companies. Tech Vjesn 19:983–986
8. Choi JC, Kwon TH, Park JH, Kim JH, Kim CH (2012) A study on development of a die design for die casting. Int J Adv Manuf Technol 20:1–8

Active Ranging Sensors Based on Structured Light Image for Mobile Robot

Jin Shin and Soo-Yeong Yi

Abstract In this paper, we propose a ring array of active structured light image-based ranging sensors for a mobile robot. Since the ring array of ranging sensors can obtain omnidirectional distances to surrounding objects, it is useful for building a local distance map. By matching the local omnidirectional distance map with a given global object map, it is also possible to obtain the position and heading angle of a mobile robot in global coordinates. Experiments for omnidirectional distance measurement, matching, and localization were performed to verify the usefulness of the proposed ring array of active ranging sensors.

1 Introduction

Localization is the estimation of current position and heading angle, i.e., the posture of the mobile robot. Ranging sensors for the measurement of distances to surrounding objects are required for localization. There exist many kinds of ranging sensors, such as ultrasonic sensors, infrared laser sensors, laser scanners, stereo cameras, and active structured light image-based sensors [1]. Among these sensors, the structured light image-based sensor can effectively acquire distance

---

This research was supported by Basic Science Research Program through the National Research Foundation of Korea (NRF) funded by the Ministry of Education, Science and Technology (2011-0009113).

J. Shin · S.-Y. Yi (✉)
Seoul National University of Science and Technology, Seoul, Republic of Korea
e-mail: suylee@seoultech.ac.kr

J. Shin
e-mail: gomlands@naver.com

J. J. (Jong Hyuk) Park et al. (eds.), *Multimedia and Ubiquitous Engineering*, Lecture Notes in Electrical Engineering 240, DOI: 10.1007/978-94-007-6738-6_89, © Springer Science+Business Media Dordrecht(Outside the USA) 2013

information [2]. Bulky laser equipment and long image processing time have discouraged the use of the structured light image-based method in the past, but recent advancements in semiconductor laser equipment and faster processors have made this system more viable and economical.

In this paper, a new ring array of ranging sensors is presented for the localization of a mobile robot. The ring array sensor has four active structured light image-based ranging sensors attached to the mobile robot. It is clear that omnidirectional distance acquisition is much more useful for a mobile robot than unidirectional distance acquisition. A ring array structure of ranging sensors that covers all directions had been used with ultrasonic ranging sensors [1]. In case of the ultrasonic ring array of sensors, however, it is impossible to activate multiple ultrasonic sensors at the same time because of signal crosstalk, which slows the distance measurement rate. In contrast, the structured light image-based sensors in a ring array can measure omnidirectional distances in one shot, without any mutual interference. To alleviate the computational burden in the main controller of a mobile robot, we developed structured light image-based ranging sensor modules by embedded image processor and arranged them in a circular pattern on the mobile robot. Each ranging sensor module transmits distance data to the main controller of the mobile robot after structured light image processing, and the main controller estimates the posture of the robot by matching the omnidirectional distance data with a given global object map.

2 Structured Light Image-Based Distance Measurement with a Ring Array

As shown in Fig. 1, a structured light image-based ranging sensor consists of a camera and a structured light source. In order to obtain horizontal object distances under the assumption of robot motion on two-dimensional ground, a horizontal sheet of structured laser light is used in this study. By using the time difference of two images with modulated structured laser light, it is possible to extract the structured light pixel image, as shown in Fig. 2 [3].

Figure 2 shows the image illuminated by structured light and the extracted structured light pixel image obtained through image processing. From the center line of the image in Fig. 2b, structured light pixel distance p is detected in the vertical direction. From the pixel distance p, measurement angle ρ is given as follows:

Fig. 1 Distance measurement based on structured light image

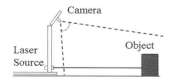

Active Ranging Sensors Based on Structured Light Image for Mobile Robot

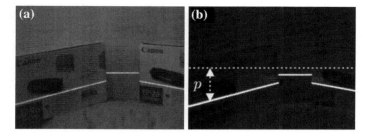

Fig. 2 Extraction of structured light pixel image: **a** structured light image. **b** Extracted structured light pixel image (*solid line*)

$$\rho = \tan^{-1}\left(\frac{p}{\lambda}\right) \quad (1)$$

where λ represents the focal length of the camera. From the distance measurement model in Fig. 3a, the distance l to an object can be obtained as follows:

$$l = b \cdot \cot\left\{\theta - \tan^{-1}\left(\frac{p}{\lambda}\right)\right\} \quad (2)$$

In Fig. 3a, λ is the camera focal length, θ represents the camera view angle, and b denotes the baseline.

The well-known CMUcam3 [4] and a 660 nm wavelength infrared semiconductor laser are adopted to develop the structured light image-based ranging sensor module. The embedded processor in CMUcam3 performs all of the image processing and only transmits the distance data to the main controller of the robot. Figure 3b shows the ring array of the ranging sensor modules attached to the mobile robot, which can measure omnidirectional object distances.

3 Posture Estimation from Omnidirectional Distance

From the structured light pixel images of the ranging sensor array shown in Fig. 4 and Eq. (2), it is possible to acquire a local omnidirectional distance map, as

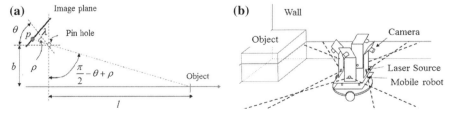

Fig. 3 Distance measurement model and omnidirectional ranging through ring array. **a** Distance measurement model. **b** Ring array of ranging sensor module

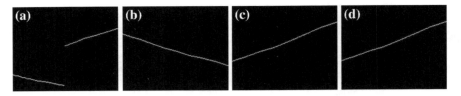

Fig. 4 Structured light pixel images from ranging sensor array. **a** From camera 1. **b** From camera 2. **c** From camera 3. **d** From camera 4

Fig. 5 Omnidirectional distance data. **a** Mobile robot environment. **b** Measured local distance map

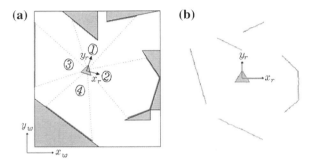

shown in Fig. 6. In Fig. 5a, the circled numbers denote each camera in the ring array corresponding to each image in Fig. 4. Figure 5b shows the local distance map in the moving coordinates of the mobile robot. The local distance map consists of a set of measured points (x_m, y_m) in the moving coordinates. When the estimated posture of the robot is $(\hat{x}_r, \hat{y}_r, \hat{\theta}_r)$ in world coordinates, the measured local distance data can be transformed into world coordinates as follows:

$$\begin{bmatrix} x_w \\ y_w \end{bmatrix} = R(\hat{\theta}_r) \begin{bmatrix} x_m \\ y_m \end{bmatrix} + T(\hat{x}_r, \hat{y}_r) \qquad (3)$$

where $R(\theta)$ and $T(x, y)$ represent the rotation and translation, respectively, as follows:

$$R(\theta) = \begin{bmatrix} \cos(\theta) & -\sin(\theta) \\ \sin(\theta) & \cos(\theta) \end{bmatrix}, T(x, y) = \begin{bmatrix} x \\ y \end{bmatrix} \qquad (4)$$

Posture estimation should be updated by matching the real-time omnidirectional distance map with a given global object map. There have been many studies of the matching problem. In [5] and [6], a least-squared-error-based matching algorithm was suggested to associate real-time distance data with a given global map. Since the matching algorithm considers every measured points individually, it requires a significant number of computations. In order to improve computational efficiency, a matching algorithm is developed in this paper by modifying the algorithms in [5] and [6]: line segments are obtained from the measured local

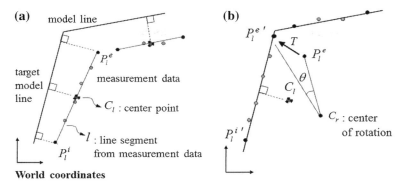

Fig. 6 Matching algorithm. a Before matching. b After matching

distance map first and only two end points of a line segment are matched with the global map, rather than all of the measured points.

The matching algorithm is described in Fig. 6 where P_l^i and P_l^e denote the two end points of a line segment l from the measured local distance map and $P_l^c = (P_l^i + P_l^e)/2$ is the center point of the segment. Here, we assumed that those points are described in world coordinates by the transformation (3). Among the all model line segments from the given global map, the target model line segments nearest to the center point of a segment can be found. The target line segment satisfies the following (5).

$$P \cdot \mathbf{u}_l = r_l. \tag{5}$$

where P is a point on the target line segment, $\mathbf{u}_l$ is the unit normal vector, and r_l is a real number.

Rotation by $\Delta\theta$ about the present estimated position, $C_r = \begin{bmatrix} \hat{x}_r & \hat{y}_r \end{bmatrix}^t$, of the robot and translation by $(\Delta x, \Delta y)$ makes two end points of segment l as follows:

$$P'_l = R(\Delta\theta)(P_l - C_r) + C_r + T(\Delta x, \Delta y). \tag{6}$$

where P_l and P'_l represent two end points before and after the transformation in world coordinates. Then, the total matching error is defined by the sum of the squared distance between the transformed points P'_l of the all line segments l and the target line (5) as follows:

$$S = \sum_l \left(P_l^i \cdot \mathbf{u}_l - r_l\right)^2 + \left(P_l^e \cdot \mathbf{u}_l - r_l\right)^2 \tag{7}$$

By the well-known gradient method to minimize the total matching error (7), it is possible to get the transformation parameters, $(\Delta x, \Delta y, \Delta\theta)$, which is used to update the estimation of robot's posture as follows:

$$(\hat{x}_r, \hat{y}_r, \hat{\theta}_r) \leftarrow (\hat{x}_r + \Delta x, \hat{y}_r + \Delta y, \hat{\theta}_r + \Delta\theta) \tag{8}$$

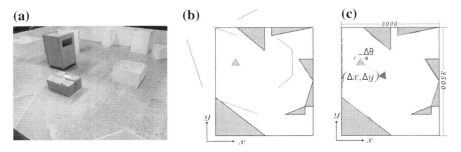

Fig. 7 Data matching experiment. **a** Experimental environment. **b** Before matching. **c** After matching

4 Experimental Results

We performed experiments to verify the effectiveness of the proposed ring array ranging sensor and the matching algorithm. As shown in Fig. 7a, some polygonal objects are placed in the mobile robot's environment. The omnidirectional distance data measured at an unknown robot posture are depicted in Fig. 7b, c and shows the resultant robot posture after matching and updating by the algorithm described in (5) through (8). In Fig. 7b, the transformation parameters to update the robot's posture are $(\Delta x, \Delta y, \Delta \theta) = (630, 320, 16.54°)$, as obtained from the matching algorithm.

5 Conclusion

The ring array of the structured light image-based ranging sensors proposed in this paper is able to obtain omnidirectional distances in one shot for fast localization of a mobile robot. Compact cameras with embedded processors used for the ring array of the ranging sensor in this paper send only final distance data to the main controller of the robot, thereby lowering its computational burden. Matching between the omnidirectional distance data from the proposed ranging sensors and the given global object map is required for localization of the mobile robot. A least-squared error-based algorithm was developed in this paper to associate line segments extracted from the omnidirectional distance data with the polygonal model of the global object map. Since the matching algorithm in this paper uses only two end points of a measured line segment to associate with the reference line segment of the polygonal world model, efficiency of computation is greatly improved. The proposed ring array of active structured light image-based ranging sensors and the matching algorithm in this paper were verified through experiments on local omnidirectional distance data acquisition, localization.

References

1. Cameron S, Probert P (1994) Advanced guided vehicles-aspects of the Oxford AGV Project, World Scientific, London
2. Noh D, Kim G, Lee B (2005) A study on the relative localization algorithm for mobile robots using a structured light technique. J Inst Control, Robot Syst 11(8):678–687
3. Jain R, Kasturi R, Schunck BG (1995) Machine vision. McGraw-Hill, New York
4. http://www.cmucam.org
5. Cox I (1991) Blanche-an experiment in guidance and navigation of an autonomous robot vehicle. IEEE Trans Robot Autom 7(2):193–204
6. Cox I, Kruskal J (1988) On the congruence of noisy images to line segment models. In: Proceedings of international conference on computer vision, pp 252–258

Improved Composite Order Bilinear Pairing on Graphics Hardware

Hao Xiong, Xiaoqi Yu, Yi-Jun He and Siu Ming Yiu

Abstract Composite-order bilinear pairing has been applied in many cryptographic constructions, such as identity based encryption, attribute based encryption, and leakage resilient cryptography. However, the computation of such pairing is relatively slow since the composite order should be at least 1024 bits. Thus the elliptic curve group order n and the base field are large. The efficiency of these pairings becomes the bottleneck of the schemes. Existing solutions, such as converting composite-order pairings to prime-order ones or computing many pairings in parallel cannot solve the problem. The former is only valid for certain constructions and the latter is only helpful when many pairings are needed. In this paper, we make use of the huge number of threads available on Graphics Processing Units (GPUs) to speed up composite-order computation, both between pairings and within a single pairing. The experimental result shows that our method can speed up paring computation. Our method can also be ported to other platforms such as cloud systems.

Keywords Composite-order bilinear pairing · GPU · Cryptography construction

1 Introduction

A bilinear pairing is in the form of $e : G \times G \to G_T$, a bilinear pairing is said to be over a composite-order group if the order of G (and G_T) is composite. Composite-order bilinear pairing has been used in many cryptographic constructions, such as identity based encryption, attribute based encryption, functional encryption and leakage resilient cryptography. However, computing a pairing over a composite-

---

H. Xiong · X. Yu · S. M. Yiu
Department of Computer Science, The University of Hong Kong, Hong Kong, Hong Kong

Y.-J. He (✉)
R&D Centre for Logistics and Supply Chain Management, Hong Kong, Hong Kong

J. J. (Jong Hyuk) Park et al. (eds.), *Multimedia and Ubiquitous Engineering*,
Lecture Notes in Electrical Engineering 240, DOI: 10.1007/978-94-007-6738-6_90,
© Springer Science+Business Media Dordrecht(Outside the USA) 2013

order group is much more expensive compared to its prime-order counterpart and other building blocks. For example, according to [4], to achieve the same 80 bits AES security level, the composite order should be at least 1024 bits, while only 160 bits are needed in its prime-order counterpart. Since more bits are needed in composite order pairing, the underlying finite field, elliptic curve operations and the pairing evaluating algorithm itself become much slower. Freeman [4] shows that the composite-order pairing is 50X slower than prime-order pairing. Due to the above reasons, composite-order pairing computation easily becomes the bottleneck of a cryptographic scheme, especially where large amounts of such pairings need to be computed (e.g., [6]). Several solutions have been proposed to solve this problem recently. Freeman [4] proposed a method to convert a scheme based on composite-order pairing to a prime-order pairing based scheme. The shortcoming of this method is that it is only valid for certain cryptographic constructions. In fact, [8] points out that some schemes inherently require composite-order groups. Another solution is to compute composite-order pairing in parallel, either in the way where many pairings are computed in parallel or compute one single in parallel. In the early era, the hardware needed in parallel computing is expensive to most organizations and make this solution unreasonable. However, due to the rapid development of technology recently, we can apply GPU (Graphics Processing Unit) which is a relatively cheaper hardware in parallel computing. The huge number of threads available on GPU can be leveraged to speed up the composite-order bilinear pairing computation. Zhang et al. [11] proposed the first composite-order bilinear pairing computation algorithm based on GPU. It shows that it can achieve a 20-fold speedup compared to the state-of-the art CPU implementation. However, the method in [11] only computes many pairings in parallel and little work is done on the parallel computation of one single pairing. In this paper, we consider the parallelism both within and between pairings. We compute each pairing on a block of threads and compute each part of the pairing in parallel on these threads within this block, while we concurrently run many blocks to compute many pairings in parallel. We implemented 32 bit modular addition, subtraction and multiplication on each thread. The corresponding operations (addition, subtraction and multiplication) on F_q are conducted on a block of threads via Residue Number System (RNS). The corresponding multiplication and square operations on extension field F_{q^2}, addition and double operations on an elliptic curve are implemented upon F_q operations, which are based on a block of threads. Besides, we compute the part of $g_{u,v}(\varphi(Q))$, which is the most time consuming part in the main loop of one single pairing in parallel. Combining all these techniques, we are able to speed up the computation. Our method is a general method to compute the composite-order pairing in parallel and can serve for all cryptographic schemes constructed in composite-order pairing. As far as we know, our work is the first one which can truly achieve the parallel computation both within and between composite-order pairings on GPU. It is non-trivial to convert

Improved Composite Order Bilinear 733

existing CPU-version into a GPU-version due to the different levels of parallelism provided by GPU and CPU. Besides, it is difficult to divide one single pairing operation into several parts which can be computed in parallel.

2 Mathematics of Composite Order Bilinear Pairing

This section focuses on the basics of bilinear pairing and the group on which a bilinear pairing is defined. We describe the relationship between group size and security level and explain the reason why composite order in a bilinear pairing should be much larger than a prime order.

Let G_1 and G_2 be two cyclic additive groups and G_T a cyclic multiplicative group. A bilinear map (of order $1 \in N$) is defined as follows with three properties:

$$e_l : G_1 \times G_2 \rightarrow G_T \tag{1}$$

(a) bilinearity: for all $P \in G_1$, and $Q \in G_2$, $e_t(aP, bQ) = e_t(P, Q)^{ab}$; (b) non-degeneration: $e_t(P, Q) \neq 1$ for some P and Q, where 1 is the identity element of G_T; and (c) computable: there is an efficient algorithm to compute $e_t(P, Q)$ for any $P \in G_1$ and $Q \in G_2$. If there exists a distortion map: $G_1 \rightarrow G_2$, we can define a symmetric bilinear pairing $e_l : G_1 \times G_2 \rightarrow G_T$ so that $e_l(P_1, P_2) = e_l(P_1, \varphi(P_2))$ for any $P_1, P_2 \in G_1$.

Let E be an elliptic curve which is defined over a finite F_q where $q = p^m$, $p, m \in N$ and p is the characteristic of F_q. Let O be the point at infinity for E. For a nonzero integer 1, the set of points P in $E(F_q)$ such that $lP = O$ is denoted as $E(F_q)[l]$. The group $E(F_q)[l]$ is said to have security multiplier or embedding degree k for some $k > 0$ if $l | q^k - 1$ and $l \nmid q^s - 1$ for any $0 < s < k$. The Tate pairing of order 1 is a map

$$e_l : E(F_q)[l] \times E(F_{q^k})[l] \rightarrow F_{q^k} \tag{2}$$

The pairing-friendly elliptic curve used in a composite order bilinear pairing is a super singular elliptic curve in the following form defined over a prime field.

$$E : y^2 = x^3 + (1 - b)x + b, \ b \in \{0, 1\} \tag{3}$$

The group order 1 is composite and the embedding degree k is 2. There exists a distortion $\varphi : E(F^q) \rightarrow E(F_{q^k})$ which allows us to define a symmetric bilinear map as

$$e_l : E(F_k)[l] \times E(F_q)[l] \rightarrow F_{q^k} \tag{4}$$

So that $e_l(P, Q) = e_l(P, \varphi(Q))$ for any $P, Q \in E(F_q)[l]$ .The order of $E(F_q)$ is $\# E(F_q) = q + 1$. This curve is named as A1 curve in the PBC software library [7].

Finite Field Size versus Security Level. The security of pairing-based cryptosystems generally rely on two hard problems, elliptic curve discrete logarithm problem (ECDLP) in G and logarithm problem in the extension field F_{q^k}, that is, G_T. When a pairing-based cryptosystem requires 1024 bits security, the size of the extension field F_{q^k} should at least be 1024 bits long and the group order of G should at least be 160 bits long [9]. Besides, the security of most composite-order pairing-based cryptographic constructions also relies on the intractability of a problem called Subgroup Decisional Problem (SDP) [2]: for a bilinear map $e : G \times G \rightarrow G_T$ of composite order l, without knowing the factorization of the group order l, the SDP is to decide if an element x is in a subgroup of G or in G. For the intractability of SDP, the group order l of G should be at least 1024 bits long. As $l|q + 1$, q should also be at least 1024 bits long. As the embedding degree k is 2, the size of the extension field F_{q^2} is at least 2048 bits long [11].

3 Bilinear Pairing Algorithm

The arithmetic operations involved in our algorithms consists of the operations in the extension field F_{q^2} and the elliptic curve $E(F_q)$ which are based on the base field operations in F_q. We mainly construct the operations in base field in RNS using the RNS Montgomery multiplication algorithm. The concrete algorithms are described in [11], reader can refer it for the details.

In our paper, we use Barreto et al.'s algorithm [1] based on the composite-order bilinear pairing in F_q to realize the bilinear pairing, because the computation flow of this algorithm relies on the system parameters instead of the input values, which is suitable for SIMD on GPU. In this algorithm, all operations are feasible on GPU, such as double operations in $E(F_q)$ and multiplication operations in F_{q^2} discussed in [11].

We propose an improved algorithm, which transforms the calculation of $g_{U,V\varphi}$ [5] into separate steps. We take two lists of the elements in elliptic curve field $in1$ and $in2$ as the input values and construct $cacheVP$ list to store the results in the algorithm. $CacheVP$ is a defined structure which contains the followings: V in elliptic curve field, VV in elliptic curve field, and a boolean $hasVP$ indicating which branch the flow in line 9 of Barreto et al.'s algorithm [1] will go into. When $n_1 = 0$, $hasVP$ will be set to true, otherwise false. In line 11–15 of $computeVP$, we record the information of the parameters instead of computing the value of $g_{U,V\varphi}$ directly.

In $g_{U,V\varphi}$ algorithm [5], the input values become the list $cacheVP$, which is conducted in the $computeVP$ $algorithm$. From the tuple in $cacheVP$, we can get the input value of U and V, and whether we should compute the result once or twice. Traversing all tuples in $cacheVP$, the computing of $g_{U,V\varphi}$ can be done, and get all the results stored in valuable cacheG that are needed in the subsequent steps. We call the phase above preprocessing.

Improved Composite Order Bilinear

Algorithm 1: computeVP

Require: $heVP * cacheVP$, elliptic curve field $in1[], in2[]$

Ensure: $E : y^2 = x^3 + x, q > 3$ and $q \equiv 3 \pmod 4$

1:   $(x, y) \in E(F_q)[n], \ i \in F_{q^2}(i^2 = -1), \phi(x, iy) = (-x, y) \in E(F_{q^2})[n]$

2:   $n = (n_t, \ldots, n_0), \ n_i \in \{0, 1\}, \ n_t = 1$

3:   $V \leftarrow P$

4:   for $i \in [t-1, 0]$ do

5:     $idx = kernel\_index * len_n + i$

6:     $V \leftarrow 2V$

7:     if $i = 0$ then

8:       break;

9:     end if

10:     if $n_i = 1$ then

11:       $cacheVP[idx].VV \leftarrow V$;

12:       $cacheVP[idx].hasVP \leftarrow true$;

13:       $V \leftarrow V + P$

14:     else

15:       $cacheVP[idx].hasVP = false$;

16:     end if

17:   end for

18:   return

4 Implementation

We implement our solution using CUDA. We use a block of 67 (or 131) threads to represent an element in F_q. The thread number on one block is decided on the security level and the word length of GPU. For example, for the 1024/2048 bit composite order and word length of 32 bits, $1024/32 = 32(64)$ bases are the least needed to represent a number (actually 33(65)). Considering another set of bases for the extension operation, $33 + 33 + 1(65 + 65 + 1)$ are used to represented one element in F_q. Elements of the extension field and elliptic curve are represented based on base field F_q. For example, the element in extension field can be presented by a 2D tuple as (x, y), with x and y be the elements in base field. Zhang et al. [11] discusses computation between many pairings at one time which is called basic scheme in our paper.

Global memory is time-consuming to load and communicate. Since the reduction operations will be computed over and over again, we utilize the texture memory to store the intermediate values. Texture memory is better for some specific data structures and we set it as 2D in our scheme. We bind some frequently-used data precomputed for the reduction operation to the 2D texture memory called *tex_ref*. Besides, shared memory is suitable for sharing information in the same block, so we allocate some share memory to store the intermediate information among threads in one block.

The basic scheme in [11] only considers the parallelism among many pairings and neglects the parallelism within the operations of extension field and elliptic curve and within the bilinear algorithm. The computation on one single pairing is still sequential. Through further analysis of the flow, we propose a method to improve the efficiency of the parallelism within one single pairing. According to the algorithm, $n = E(F_q) = q + 1$. We assume that $n = (n_t \ldots n_0), n_i \in \{0, 1\}$ and $n_t = 1$, then the main loop will go over t times. In each round of the computation, about 20 times of multiplication should be done plus to the reduction operations. It is computational expensive. We describe the solution to this problem in the following.

As presented in the computation details of $g_{U, V}(\varphi(Q))$, the computation of a, b and c only relies on the input value of U and V and independent on the results of the previous step. It is suitable for parallelism model. We apply each computation to a different thread and then read the data and compute the return value of $(c - ax_3) + by_3 i$, which will be stored in a global array. With the help of this preprocessing, we only need to read the pre-computing result according to the index i when calculating the value of $g_{U, V}(\varphi(q))$. In terms of the communication of different blocks, we use the global and shared memory to store the shared results. It is feasible to obtain the index i according to the data structures of CUDA (which are *threadID, blockId, blockDim and gridDim*). By applying this improvement, the computation efficiency of composite-order pairing will be greatly improved.

5 Experimental Result and Analysis

We compare the efficiency of the CPU scheme, the basic scheme [11] and our improved GPU scheme. The security level is 1024 bits. The CPU scheme was run on Intel Core 2 E8300 CPU at 2.83 GHz and 3 GB memory. We chose NVIDIA GTX 480 as the GPU running environment. When implementing the experiment on CPU, we adopt the Pairing Based Cryptography (PBC [7]) library.

Analysis and Evaluation As shown in Table 1, the basic-scheme is slower (49.9 ms) than the CPU version (38.4 ms) when the number of pairings is small. Parallel computation does not make significant contribution to the efficiency. It is time-consuming to do the pre-computing. However, as the number of pairings increases, the result reveals the advantage of using GPU. In the case of 100 pairings, the average time used by the improved scheme is only 16.23 % of the CPU version. When the number of pairings increases (from 20 pairings), the improved scheme starts to win over the basic scheme. We also analyze the loading time of memory for schemes based on GPU. Though the first access of the memory in GPU is expensive, the average time is quite short as the number of pairings

Improved Composite Order Bilinear

Table 1 Results on CPU, basic and improved scheme

Time(ms)	Total-time			Average-time		
Num	CPU	Basic	Improved	CPU	Basic	Improved
1	38.4	49.9	52.3	38.4	49.9	52.3
20	783.1	201.8	128.8	39.2	10.1	6.4
100	3884.9	859.9	631.2	38.8	8.6	6.3
200	7698.7	1716.0	1262.3	38.5	8.5	6.3

increases. When the number of pairings grows large enough, the average time will become comparatively small. It can easily be deduced from the experimental results. In the basic scheme, when the pairing number is small, the time used is larger than that of the CPU version because the average loading time is large while it is nearly zero in the CPU scheme. Since we add some preprocessing in our improved scheme, the loading time will be longer than the basic scheme. However, according to the experimental result, the advantage becomes obvious when the number of pairing becomes large.

Acknowledgment This project is partially supported by the Small Project Funding of HKU (201109176091).

References

1. Barreto PSLM, Kim HY, Lynn B, Scott M (2002) Efficient algorithms for pairing-based cryptosystems. In: CRYPTO, pp 354–368
2. Boneh D, Goh E-J, Nissim K (2005) Evaluating 2-dnf formulas on ciphertexts. In: TCC, Lecture notes in computer science, vol 3378. Springer, Heidelberg, pp 325–341
3. Fleissner S (2007) Gpu-accelerated montgomery exponentiation. In: International conference on computational science, pp 213–220
4. Freeman DM (2010) Converting pairing-based cryptosystems from composite-order groups to prime-order groups. In: EUROCRYPT, pp 44–61
5. Guillermin N (2010) A high speed coprocessor for elliptic curve scalar multiplications over Fp. In: CHES, Lecture Notes in Computer Science, vol 6225. Springer, Berlin, pp 48–64
6. Lewko AB, Rouselakis Y, Waters B (2011) Achieving leakage resilience through dual system encryption. In: TCC, pp 70–88
7. Lynn B Pbc: the pairing-based cryptography library. http://crypto.stanford.edu/pbc/
8. Meiklejohn S, Shacham H, Freeman DM (2010) Limitations on transformations from composite-order to prime-order groups: the case of round-optimal blind signatures. In: ASIACRYPT, pp 519–538
9. NIST (2007) Recommendation for key management
10. NVIDIA Corporation (2010) Nvidia CUDA C programming guide
11. Zhang Y, Xue CJ, Wong DS, Mamoulis N, Yiu SM (2012) Accelerating bilinear pairing on graphics hardware. In: ICICS, pp 341–348

Deployment and Management of Multimedia Contents Distribution Networks Using an Autonomous Agent Service

Kilhung Lee

Abstract This paper introduces an agent application service and shows its application through the implementation of a multimedia data distribution service. This service is broadly comprised of agents, agent systems, an agent master and agent manager components. A software component of relaying multimedia date traffic is developed as an agent application, and is created dynamically with the distribution of the content client in the network. By using this service, it is possible to deploy a new multimedia data distribution with greater speed, efficiency and convenience.

Keywords Management · Multimedia content distribution · Autonomous agent service

1 Introduction

With the development and adoption of the Internet, many new and hitherto unimagined application services have been introduced through this revolutionary medium. Multimedia data service will be the major application of the future Internet and Intranet environments. The peer-to-peer (P2P) method has strong merits given its scalability, and has the potential to serve as an applicable means of tackling errors more easily. Likewise, mobile agent technologies that can conduct critical operations at precise locations are of significant importance in the overall software technology and development progress [1]. An agent is an autonomous programming object that can perform by itself those functions that have been entrusted to it by the software user [2]. This operational method is an innovative

---

K. Lee (✉)
Department of Computer Science and Engineering, Seoul National University of Science and Technology, 172 Gongnung2-dong, Nowon-gu, Seoul 139-743, Korea
e-mail: khlee@seoultech.ac.kr

J. J. (Jong Hyuk) Park et al. (eds.), *Multimedia and Ubiquitous Engineering*, Lecture Notes in Electrical Engineering 240, DOI: 10.1007/978-94-007-6738-6_91, © Springer Science+Business Media Dordrecht(Outside the USA) 2013

way of implementing the traditional mobile and distributed computing system and constitutes a new computing paradigm [3]. Through proper utilization of the functions of an agent, which can be used to operate in specific places at precise times, we can achieve significant reductions in network traffic and contribute to increased operational efficiencies [4, 5].

2 Component of Autonomous Agent Service

Our Autonomous Agent Service is an agent system framework developed Java language and consists of several components. Each component is a package of classes that has a role in agent application environment. The agent framework component is composed of the agent, agent system, agent master system, and agent manager [6].

Agent: An agent is a software component and is a small program that is designed to fulfill certain operations. An agent is also capable of moving around the different systems and once it arrives at its desired location it begins specific operations. An agent is one of the classes devised for certain interface in Java.

Agent System: An agent system manages the agent operation system and provides the place (i.e. the operation part of the agent). After receiving the service code, the agent system creates the agent object and provides the operation environment using services such as begin, stop, restart, and termination.

Agent Master: An agent master controls the local agent system, intercedes with clients who are service users for the provided agent service, and provide management interface to the agent manager. The agent system signals the start of its service operations by beginning it at the same time it registers in the agent master. When an agent from the client connects with the service to be used, the agent master returns a system list containing available agents.

Agent Manager: An agent manager manages the environments of the overall agent application framework. An agent manager can help achieve management efficiency, flexibility and scalability with fewer complications by managing the agent master. The agent master embodies agent management information that is necessary to manage how agents behave. The agent manager can carry out monitoring activities by reading a Management Information Base (MIB) and service management functions by creating needed values.

Client: A client request and control the service and can offer service codes through the agent master. A client can either provide the requested service operation code directly or the agent system can remove the address that the client provides to the service storage. The client can designate the agent system directly or help the agent master choose the appropriate agent to carry out the service. In the case of P2P applications, the agent system and the client can be combined and operated together [7].

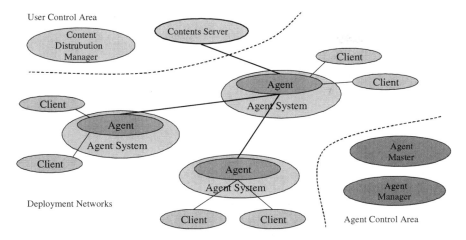

Fig. 1 Content distribution application environment

3 Management of Multimedia Contents Distribution Network

The content distribution network in Fig. 1 is controlled by the distribution manager. The distribution manager is one type of agent client in our agent application framework. Through the agent master system, the distribution manager initiates the service agent in the necessary agent system after searching the profile of the agent system. When agents commence new operations, they transmit the message to the distribution manager and register themselves. The distribution manager, according to its location, locates the new agents into their proper position within the distribution tree. When the content clients near the new agents request a content service, the distribution manager makes it possible to provide the service by connecting new agents to the proper agents. The functions of the components in the content service model are as follows.

Content Server: The content server is the data source that provides the information to the content client. The services provided by the content server include broadcasting, chatting, Video on Demand (VOD), video conferencing, stock information services, real-time news services, and more. In addition, depending on the types of services provided, there are various kinds of servers to be used, such as a push server where the client gathers information, and pull servers where the server provides information [8].

Content Distribution Agent: A distribution agent takes the contents from the content server and distributes them again to the client or other distribution agents. Distribution agents, depending on the content types served, take different forms. Therefore, the agents locate themselves where the service providers want them to place and are served in specific manner depending on the particular services and type of server. Like servers, agents also provide services with the client list and depending on the quality and forms of service, one or more threads are involved in the operations. Agents monitor the quality of service while at the same time providing services.

Content Distribution Manager: A content distribution manager is the component that controls the data service between the server and the client. A content distribution manager possesses all of the information regarding the server, distribution agent, and the client served, and manages the types of services provided by coordinating the distribution tree. In addition, the content distribution manager is involved in the preparation of the list and location of new agent through the service quality that is monitored and reported, and readjusts the distribution tree. This system is the component that performs the function of a client in the agent application framework environments.

Content Client: A content client receives the content from the agent distributor and sends it to users in the proper forms. In general, the communication handling and information handling portions constitute the client system.

Figure 2 is the snapshot of the content distribution manager application. After multimedia data source server activates, content distribution manager initiates relaying agents in some important points of networks. Thereafter, new clients are then attached to the appropriate relaying agent that meets the traffic requirements, and receives data services from them. When client are increased and the quality of service is decreased in specific sectors, content distribution manager detects and

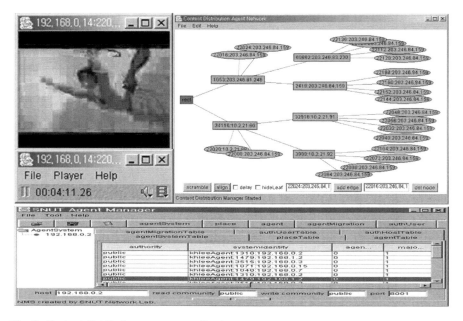

Fig. 2 Content distribution manager application

Deployment and Management of Multimedia Contents Distribution Networks

Table 1 Transmission characteristic of the content distribution service

Hop count	2	3	4	5
Average delay (ms)	32	48	62	81
Jitter (ms)	+58/−14	+64/−21	+68/−25	+71/−28

deploy a new relaying agent to an agent management system resides in that sectors [9].

Table 1 shows the property of the data traffic delivered in a multicast tree design using the agent application framework. The transmission property in the Intranet is very satisfactory and is subject to significant property changes depending on the status of number of client and host load. As the transmission hop of the tree is increased, delays are also proportionally increased. Despite this delay increasing property, the average error in delays tends not to be dramatically changed. Jitters also show slight increasing properties.

Between autonomous systems, the sending and receiving transmission property had 2973 ms in average delay and an average jitter of +349/−572 ms. The delays in the Intranet and the value of the jitter are not subject to significant changes.

4 Summary

This paper has introduced the management of an agent application service and the example development of content distribution network as an application. The multimedia delivery performance of the agent application framework is acceptable in this test. A multimedia data service network can be created and controlled by this framework with simplicity, efficiency, accessibility, increased controllability and manageability properties.

References

1. Yang X, Zhang Y, Niu Q, Tao X, Wu L (2007) A mobile-agent-based application model design of pervasive mobile devices. In: 2nd International conference on pervasive computing and applications, pp 1–6
2. Karmouch A, Pham VA (1998) Mobile software agents: an overview. IEEE Commun Mag 36(7) 26–37
3. Wang YH, Keh HC, Hu TC, Liao CH (2005) A hierarchical dynamic monitoring mechanism for mobile agent location. In: 19th International conference on advanced information networking and applications, vol 1, pp 351–356
4. Glitho RH, Olougouna E, Pierre S (2002) Mobile agents and their use for information retrieval: a brief overview and an elaborate case study. IEEE Network 14:34–41
5. Baek JW, Yeom HY (2003) d-Agent: an approach to mobile agent planning for distributed information retrieval. IEEE Trans Consum Electron 49(1):115–122
6. DC00087C (2002) Mobile agent facility specification support for mobility specification

7. Stolarz D (2001) Peer-to-peer streaming media delivery. In: Proceedings of the first international conference on peer-to-peer computing, pp 48–52
8. Yang S, Yang H, Yang Y (2003) Architecture of high capacity VOD server and the implementation of its prototype. IEEE Trans Consum Electron 49(4):1169–1177
9. Chen JS, Shi HD, Chen CM, Hong ZW, Zhong PL (2008) An efficient forward and backward fault-tolerant mobile agent system. In: Eighth international conference on intelligent systems design and applications, vol 2, pp 61–66

Part XIII
Advanced Mechanical and Industrial Engineering, and Control II

Design Optimization of the Assembly Process Structure Based on Complexity Criterion

Vladimir Modrak, Slavomir Bednar and David Marton

Abstract This paper focuses on configuration design optimization of the assembly supply chain network. It is intended to use this approach to select an optimal assembly process structure in early stages of manufacturing/assembly process design. For the purpose of optimization, structural complexity measures as optimality criteria are considered. In order to compare alternatives in terms of their complexity, a method for creating comparable process structures is outlined. Subsequently, relevant comparable process structures are assessed to determine their structural complexity.

Keywords Complexity indicator · Assembly model · Vertex · Arc · Tier

1 Introduction

The design optimization of assembly networks is one of the challenging issues for the practitioners and researchers in order to get high performances with low prices. To react to the trend of agile manufacturing, companies are endeavoring to provide a wide variety of modular products. A major advantage of this strategy is larger quantity of standard modules, which contribute to cost reduction and reduce total cycle time. However, the assembly process can become quite complex as the

---

V. Modrak (✉) · S. Bednar · D. Marton
Faculty of Manufacturing Technologies with Seat in Presov, Technical University of Kosice, Bayerova 1, Presov, Slovakia
e-mail: vladimir.modrak@tuke.sk

S. Bednar
e-mail: slavomir.bednar@tuke.sk

D. Marton
e-mail: david.marton@tuke.sk

J. J. (Jong Hyuk) Park et al. (eds.), *Multimedia and Ubiquitous Engineering*, Lecture Notes in Electrical Engineering 240, DOI: 10.1007/978-94-007-6738-6_92, © Springer Science+Business Media Dordrecht(Outside the USA) 2013

product variety increases. Therefore, by optimizing the assembly process structure it is aimed to reduce complexity and, thereby, reduce assembly times. This work applies network complexity indicators in order to compute the structural complexity measures of a simple case of the assembly process. To frame this problem we apply here a method for exact generating of all possible assembly process structures based on a number of initial nodes.

Related works

Probably the most important optimization criterion of assembly lines as parts of manufacturing processes has long been focused on reducing costs in each phase from product development to market achievement [1]. Many such efforts have been focused on use of modern managerial tools with aim to ensure high product quality standards, volume and mix flexibility, and delivery speed and reliability [2–5]. In configuration design, there is a number opportunities to adopt methods and tools for the evaluation of assembly process structures. Undoubtedly, novel metrics for assessing the structural complexity of manufacturing system configurations are a demanding challenge [6]. In general, the complexity of assembly supply chain networks can be characterized in terms of several interconnected aspects. Some of these aspects were described by, e.g. [7–10]. Three basic dimensions of structural complexity that links the uncertainty with performance were identified in the work presented by Milgate [11]. An innovative complexity measure for assembly supply chains has been proposed by Hu et al. [12]. This complexity measure is based on Shannon's information entropy and is closely related to Index of vertex degree [13] that was used in presented research.

2 Problem and Method Description

In this work, we are interested in optimization of the assembly process structure using minimal complexity level as criterion. For this purpose we have selected a simple real assembly process, model of which is shown in Fig. 1a.

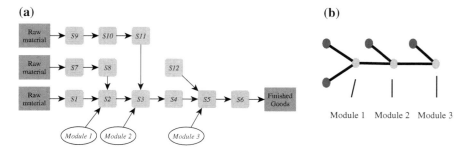

Fig. 1 a Original assembly process structure [14], b Simplified process structure

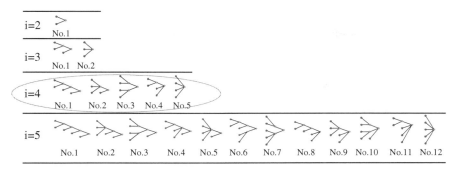

Fig. 2 Fragment of assembly process structure classes [15]

In order to compare this process structure with alternative ones we firstly transformed this model into simplest network (see Fig. 1b). Then we can formally describe this assembly network as a structure with the following elementary parameters: Number of initial nodes-4; number of all nodes-6, number of modules -3, number of tiers-4. A fragment of this classification is shown in Fig. 2. In this stage we can compare structural complexity level of all relevant process networks. For this purpose we used set of different complexity metrics to be able to objectively determine complexity differences among alternative structures of given class. These complexity indicators are described in the next paragraph.

3 Description of Complexity Indicators

3.1 Index of Vertex Degree I_{vd}

Bonchev and Buck [13] by adopting Shannon's information theory propose the following indicator to characterize complexity of a network:

$$I_{vd} = \sum_{i=1}^{V} \deg(v)_i \log_2 \deg(v)_i \qquad (1)$$

where deg(v)$_i$ represents number of the nearest-neighbors of a vertex i.

3.2 Modified Flow Complexity MFC

Modified flow complexity indicator [16] combines FC together with Multi-Tier ratio (MTR) and index (MTI), and Multi-Link ratio (MLR). Using MTI, MTR and MLR we can determine α, β and γ coefficients. MFC basically counts all Tiers (including Tier 0), Nodes and Links and adds all these counts, weighted with determined α, β and γ coefficients. In MFC indicator, Nodes and Links are counted only once, even if they are repeated in graph. Presence of Nodes and Links repetition is included in coefficients. In mathematically term, the MFC indicator can be expressed as follows:

$$MFC = \alpha \cdot T + \beta \cdot N + \gamma \cdot L, \tag{2}$$

$$\alpha = MTI = \frac{TN - N}{(T - 1) \cdot N}, \tag{3}$$

$$\beta = MTR = \frac{TN}{N}, \tag{4}$$

$$\gamma = MLR = \frac{LK}{L}. \tag{5}$$

where: N—Number of Nodes, TN—Number of Nodes per i-th Tier Level, L—Number of Links, LK—Number of Links per i-th Tier Level, T—Number of Tiers.

3.3 Supply Chain Length LSC

Németh and Foldesi [17] described Supply Chain Length (LSC) indicator and its extended definition. The LSC indicator takes besides number of nodes also number of links weighted by the complexity of links into consideration. It is mainly focused on material flows. The equation formula of LSC is expressed by the equation:

$$LSC = c_1 \cdot \sum_{i \in P} w_s \cdot V_i + c_2 \cdot \sum_{(i,j) \in P} f(D_{i,j}) \cdot A_{i,j} \tag{6}$$

where: c_1—constant represent the technical and managerial level of vertices, c_2—constant represent the technical and managerial level of edges, w_S—weight corresponding the nature of node, P—path from the origin to the destination, Vi—the vertices (nodes) in the path, A_{ij}—the arcs (edges) in the path, D_{ij}—distance in logistic terms (in this study it equals 1), $f(D_{ij})$—the weight determined by the distance in logistic terms.

3.4 Links Tiers Index LTI

When comparing two or more structures with the same number of tiers "t" and nodes "n" but with different number of links "l" the following argument can be constructed. The structure with the smallest number of links is topologically less complex than other one(s). Then, it is proposed to measure structural complexity by formula Links/Tiers Index [15]:

Design Optimization of the Assembly Process Structure

$$LTI = \sum_{j=1}^{p} \sum_{l=1}^{m} l_j \cdot t_l \, 0,1 \tag{7}$$

3.5 Flow Complexity FC

The FC is proposed by Crippa [18]. It can be expressed by Eq. 8 and it counts all Tiers (including Tier 0), Nodes and Links and adds all these counts, weighted with arbitrary chosen α, β and γ coefficients. Nodes are counted only once, even if they are repeated in Tiers. Presence of repetition is included in Links count.

$$FC = \alpha \cdot \sum_{i=1}^{n} T_i + \beta \cdot \sum_{s=1}^{m} N_s + \gamma \cdot \sum_{i=1}^{n} \sum_{j=1}^{k} LK_{ij} \tag{8}$$

where: T_i—ith Tier, N_s—sth Node, LK—ith and jth Link.

3.6 Complexity Degree κ

Maksimovic and Petrovic [19] described a Complexity degree (κ) indicator and its extended definition based on two fundamental constituents of each structure. The κ indicator takes besides number of elements and the interrelation between elements within the structure. It is mainly focused on flows in a system. Formally κ is expressed by the formula:

$$\kappa = \frac{\sum_{i=1}^{i=m} m_i}{m} \tag{9}$$

where: m_i—number of links, m—number of nodes.

3.7 Average Shortest Length ASP

The ASP is a network indicator which is applicable for determination distance of network between every pairs of nodes. Alex and Efstathiou [20] used it for interpretation of robustness complex networks as fragmentation of network. Formally can be described as follows:

$$ASP = \frac{1}{N.(N-1)} \cdot \sum \sum d_{ij}. \tag{10}$$

where: d_{ij}—is the shortest path in the network for all nodes from i till j.

Table 1 Computational results of individual indicators

Graph No.	Indicators						
	Ivd	MFC	LSC	LTI	FC	K	ASP
No. 5	8	9	9	0,8	11	0,8	0,2
No. 2	9,51	11	11	1,5	14	0,83	0,36
No. 4	10	11	11	1,5	14	0,83	0,3
No. 3	11,51	13	13	1,8	16	0,86	0,33
No. 1	11,51	13	13	2,4	17	0,86	0,48

4 Testing of Alternative Process Structures

Using the above mentioned indicators for complexity levels computation of alternative assembly structures we have obtained values summarized in Table 1. As we can see individual indicators assign an approximate complexity values to individual structures. It means that even if individual indicators use different complexity calculation methods they still show comparable results. For us it is important to know that the process we want to optimize can theoretically be replaced by the other four structures except for structure No. 3, where topology is not transferable into structure No. 1 and vice versa. The simplest transformation way of structure No. 1 is provided by structures No. 2 and 4. It is because the transformation only needs a single integration of two modules into one. For the replacement of structure No. 1 by structure No. 5 it is necessary to integrate 3 modules into one. This type of reduction is no effective from our perspective. For that reason we consider structure No. 5 as irrelevant for purpose of substitution. From our perspective structures No. 2 and 4 are comparable. Taking in mind the specifics of the structure we want to optimize we only need to integrate two modules to obtain structure No. 2 and we would have to integrate two modules together with one operation to obtain structure No. 4.

5 Conclusions

Presented approach showed that complexity reduction of the assembly process structure can be effectively achieved through fusion of only two modules into one. All other attempts leading to integration of modules would give less effective results. As described above, principally, there exist only three possible ways of structural complexity reduction of given process structure. Such a method can be used as a supportive tool for designers in optimal process designing of any assembly networks.

References

1. Patterson KA, Grimm CM, Corsi TM (2003) Adopting new technologies for supply chain management. Transp Res E-Log 39:95–121
2. Thomas DJ, Griffin PM (1996) Coordinated supply chain management. Eur J Oper Res 94:1–15
3. Holmes G (1995) Supply chain management: Europe's new competitive battleground. EIU Research report
4. Gots I, Zajac J, Vojtko I (1995) Equipment for measuring the degree of wear to cutting tools. Tech Mes 1:8–11
5. Cuma M, Zajac J (2012) The impact analysis of cutting fluids aerosols on working environment and contamination of reservoirs. Tech Gaz 19:443–446
6. Kuzgunkaya O, ElMaraghy HA (2006) Assessing the structural complexity of manufacturing systems configurations. Int J Flex Manuf Sys 18:145–171
7. Deshmukh AV, Talavage JJ, Barash MM (1998) Complexity in manufacturing systems. Part 1: analysis of static complexity. IIE Trans 30:35–44
8. Modrak V (2006) Evaluation of structural properties for business processes. In: 6th international conference of enterprise information systems ICEIS, Porto, pp 619–622
9. Calinescu A, Efstathiou J, Schirn J, Bermejo J (1998) Applying and assessing two methods for measuring complexity in manufacturing. J Oper Res Soc 49:723–733
10. Modrak V (2007) On the conceptual development of virtual corporations and logistics. In: Symposium on logistics and industrial informatics, Wildau, pp 121–125
11. Milgate M (2011) Supply chain complexity and delivery performance: an international exploratory study. Sup Ch Manag Int J 6:106–118
12. Hu SJ, Zhu XW, Wang H, Koren Y (2008) Product variety and manufacturing complexity in assembly systems and supply chains. CIRP Ann Manuf Technol 57:45–58
13. Bonchev D, Buck GA (2005) Quantitative measures of network complexity. In: Bonchev D (eds) Complexity in chemistry, biology and ecology, Springer, pp 191–235
14. Wang S, Bhaba RS (2005) An assembly-type supply chain system controlled by kanbans under a just-in-time delivery policy. Eur J Oper Res 162:153–172
15. Modrak V, Marton D, Kulpa W, Hricova R (2012) Unraveling complexity in assembly supply chain networks. In: 4th IEEE international symposium on logistic and industrial informatics LINDI, Smolenice, pp 151–155
16. Modrak V, Marton D (2012) Modelling and complexity assessment of assembly supply chain systems. Proc Eng 48:428–435
17. Németh P, Foldesi P (2009) Efficient control of logistic processes using multi-criteria performance measurement. Act Tech Jaur Log 2:353–360
18. Crippa R, Bertacci N, Larghi L (2006) Representing and measuring flow complexity in the extended enterprise: the D4G approach. In: RIRL international congress for research in logistics
19. Maximovic R, Petrovic S (2009) Complexity of production structures. Fact Univer Mech Eng 7:119–136
20. Alex KSNg, Efstathiou J (2006) Structural robustness of complex networks. Phys Rev (3):175–188

Kinematics Modelling for Omnidirectional Rolling Robot

Soo-Yeong Yi

Abstract A ball-shaped mobile robot, called a ballbot, has a single point of contact with the ground. Thus, it has low energy consumption for motion because of the reduced friction. This paper presents the systematic kinematics modelling for a type of ballbot with omnidirectional motion capability. This kinematics modelling describes the velocity relationship between the driving motors and the robot body for the motion control of the robot.

1 Introduction

A ball-shaped robot has a single point of contact with the ground, which reduces its friction with the ground. A ball-shaped robot is generally called a ballbot. In comparison with a conventional wheeled mobile robot, a ballbot consumes less energy for motion because of the reduced friction [1]. There are two types of ballbots, as illustrated in Fig. 1. The ballbot shown in Fig. 1a has a cylindrical body on the top of a ball [2, 3]. This cylindrical body has driving motor and wheel assemblies in contact with the exterior of the ball to exert a driving force. In contrast, the ballbot shown in Fig. 1b has a pure spherical shape and contains a driving mechanism inside the ball [4]. The driving mechanisms can be classified into two types: (i) the wheeled platform type (Fig. 1b) [4] and (ii) the pendulum type (Fig. 2) [1].

The ballbot shown in Fig. 1a has many motion control difficulties because its posture is essentially unstable. In contrast, the pure ball-shaped robots shown in Fig. 1b and Fig. 2 are inherently stable, so the motion control is relatively stable.

S.-Y. Yi (✉)
Department of Electrical and Information Engineering, Seoul National University
of Science and Technology, Seoul, Republic of Korea
e-mail: suylee@seoultech.ac.kr

J. J. (Jong Hyuk) Park et al. (eds.), *Multimedia and Ubiquitous Engineering*,
Lecture Notes in Electrical Engineering 240, DOI: 10.1007/978-94-007-6738-6_93,
© Springer Science+Business Media Dordrecht(Outside the USA) 2013

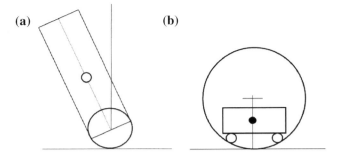

Fig. 1 Types of the ballbot. **a** Cylindrical robot body. **b** Spherical robot body

Fig. 2 Pendulum type driving mechanism

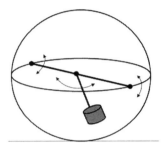

However, these ballbots still have a motion control problem because the driving mechanism cannot provide omnidirectional motion capability.

In this paper, the systematic kinematics modelling for a ballbot with a wheel-type driving mechanism inside a ball is addressed for the motion control of the ballbot. More specifically, the driving mechanism is a platform with three Swedish wheels, so the ballbot has omnidirectional motion capability without nonholonomic constraints. Thus, the motion control of the ballbot becomes comparatively simple.

2 Differential Motion Between Coordinate Frames

The kinematics modelling can be described by the velocity relationship between the active driving motor and the robot body. When the transformation between two coordinate frames, **B** and **C**, is given as T_B^C, the relationship of the differential motions between the coordinate frames shown in Fig. 3 is described as

$$\Delta^C = T_B^{C-1} \cdot \Delta^B \cdot T_B^C, \tag{1}$$

Kinematics Modelling for Omnidirectional Rolling Robot

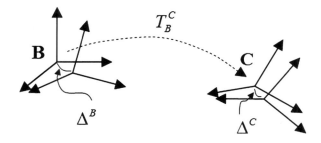

Fig. 3 Relationship of differential motions between coordinates frames

where Δ^B and Δ^C denote the differential motions in the corresponding coordinate frames [5]. The differential motion (Δ) implies a velocity transform if the motion occurs in a small time interval, δt, and can be written as

$$\Delta = \begin{bmatrix} 0 & -\delta_z & \delta_y & d_x \\ \delta_z & 0 & -\delta_x & d_y \\ -\delta_y & \delta_x & 0 & d_z \\ 0 & 0 & 0 & 0 \end{bmatrix}, \qquad (2)$$

where $\bar{\boldsymbol{\delta}} = [\delta_x \ \delta_y \ \delta_z]^t$ and $\bar{\mathbf{d}} = [d_x \ d_y \ d_z]^t$ denote the rotational and translational differential motions, respectively.

The transformation T_B^C is represented by column vectors as (3).

$$T_B^C = [\bar{\mathbf{n}} \ \bar{\mathbf{o}} \ \bar{\mathbf{a}} \ \bar{\mathbf{p}}]. \qquad (3)$$

Then, from (1) and (2), each component of the differential motions in coordinate system **C** becomes (4-1) and (4-1).

$$\bar{\boldsymbol{\delta}}^C = \begin{bmatrix} \delta_x^C & \delta_y^C & \delta_z^C \end{bmatrix}^t = \begin{bmatrix} \overline{\boldsymbol{\delta}^B} \cdot \bar{\mathbf{n}} & \overline{\boldsymbol{\delta}^B} \cdot \bar{\mathbf{o}} & \overline{\boldsymbol{\delta}^B} \cdot \bar{\mathbf{a}} \end{bmatrix}^t, \qquad (4-1)$$

$$\bar{\mathbf{d}}^C = [d_x^C \ d_y^C \ d_z^C]^t$$
$$= \begin{bmatrix} \bar{\mathbf{n}} \cdot \left(\overline{\boldsymbol{\delta}^B} \times \bar{\mathbf{p}} + \overline{\mathbf{d}^B} \right) & \bar{\mathbf{o}} \cdot \left(\overline{\boldsymbol{\delta}^B} \times \bar{\mathbf{p}} + \overline{\mathbf{d}^B} \right) & \bar{\mathbf{a}} \cdot \left(\overline{\boldsymbol{\delta}^B} \times \bar{\mathbf{p}} + \overline{\mathbf{d}^B} \right) \end{bmatrix}^t. \qquad (4-2)$$

In the above equations, "·" and "×" denote the inner product and outer product, respectively, of the vectors.

3 Structure of Ballbot and Assignment of Coordinate Frames

The structure of the ballbot considered in this paper is shown in Fig. 4a. The driving system inside the ball is an omnidirectional mobile platform having three Swedish wheels with 120° spacing. Each wheel is normal to the interior tangential

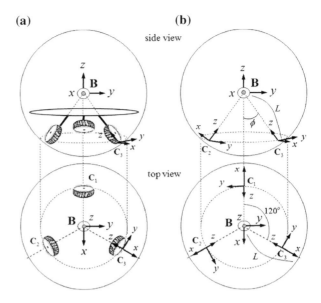

Fig. 4 Structure and coordinate assignment of proposed ballbot. **a** Structure of ballbot. **b** Coordinate frames

plane of the ball at the point of contact. The driving force of the ball comes from the friction between the wheel and the interior surface of the ball. The coordinate frame assignment of the ballbot is depicted in Fig. 4b. In this figure, the coordinate frame at the centre of the ball is denoted as **B**, which is the inertial coordinate frame attached to the driving platform. The coordinate frames at the contact points of the wheels on the inside surface of the ball are represented as $\mathbf{C}_i$, $i = 1, 2, 3$.

To derive the kinematics model, it is assumed that the motion of the ballbot is quasi-static. This quasi-static motion implies that the motion has a constant velocity and, as a consequence, the driving platform inside the ballbot maintains level always when in motion. This assumption simplifies the motion of the ballbot by disregarding the dynamics effects and gives the velocity relationship between each driving wheel and the robot body. The motion of the ballbot can be described by the translational and rotational velocities at ground contact **G**. Here, the translational velocity implies the differential motion on the horizontal x–y plane, and the rotational velocity denotes the differential motion about the vertical z axis at the ground contact (Fig. 5). It should be noted that the translational and rotational velocities (v_{xy}^G, ω_z^G) of the ballbot are the same as the velocities of the inertial coordinate frame, **B**, of the platform (v_{xy}^B, ω_z^B). Thus, the motion kinematics of the ballbot can be represented by the velocity relationship between coordinate frame **B** and each wheel coordinate frame $\mathbf{C}_i$, $i = 1, 2, 3$.

Fig. 5 Motion of ballbot (v_{xy}, ω_z): **a** motion at ground contact **G** and at centre of ball **B**; **b** motion at centre of ball **B** and at each driving wheel **C**$_i$

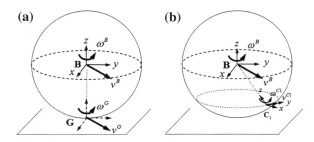

4 Velocity Relationship Between Wheel and Robot Body

From Fig. 4b, coordinate transformations from **B** to **C**$_i$, $i = 1, 2, 3$ are given as follows:

$$T_B^{C_1} = Rot(z, \pi) \cdot Rot(y, -\phi) \cdot Trans(z, -L)$$
$$= \begin{bmatrix} -c\phi & 0 & -s\phi & -Ls\phi \\ 0 & -1 & 0 & 0 \\ s\phi & 0 & c\phi & -Lc\phi \\ 0 & 0 & 0 & 1 \end{bmatrix}, \quad (5-1)$$

$$T_B^{C_2} = Rot\left(z, -\frac{\pi}{3}\right) \cdot Rot(y, -\phi) \cdot Trans(z, -L)$$
$$= \begin{bmatrix} -\frac{1}{2}c\phi & \frac{\sqrt{3}}{2} & -\frac{1}{2}s\phi & \frac{1}{2}Ls\phi \\ -\frac{\sqrt{3}}{2}c\phi & \frac{1}{2} & \frac{\sqrt{3}}{2}s\phi & -\frac{\sqrt{3}}{2}Ls\phi \\ s\phi & 0 & c\phi & -Lc\phi \\ 0 & 0 & 0 & 1 \end{bmatrix}, \quad (5-2)$$

$$T_B^{C_3} = Rot\left(z, \frac{\pi}{3}\right) \cdot Rot(y, -\phi) \cdot Trans(z, -L)$$
$$= \begin{bmatrix} \frac{1}{2}c\phi & -\frac{\sqrt{3}}{2} & -\frac{1}{2}s\phi & \frac{1}{2}Ls\phi \\ \frac{\sqrt{3}}{2}c\phi & \frac{1}{2} & -\frac{\sqrt{3}}{2}s\phi & \frac{\sqrt{3}}{2}Ls\phi \\ s\phi & 0 & c\phi & -Lc\phi \\ 0 & 0 & 0 & 1 \end{bmatrix}. \quad (5-3)$$

In (5-1)–(5-3), L denotes the radius of the ball, ϕ is the zenith angle as shown in Fig. 4b, and $s\phi$ and $c\phi$ represent $\sin(\phi)$ and $\cos(\phi)$, respectively. The transformations in (5-1)–(5-3) can be written by column vectors as (3). In the case of $i = 1$ for example, the relationship between each component of the differential motion of (2) can be obtained as follows. It should be noted that $d_z^B = 0$ and $\delta_x^B = \delta_y^B = 0$ at first. From (1), (2), (3), and (5-1) through (5-3), the differential motion components are given as follows by equating both sides of (1):

$$\bar{\delta}^{C_1} = \begin{bmatrix} \delta_x^{C_1} & \delta_y^{C_1} & \delta_z^{C_1} \end{bmatrix}^t = \begin{bmatrix} s\phi\,\delta_z^B & 0 & c\phi\,\delta_z^B \end{bmatrix}^t, \tag{6-1}$$

$$\bar{d}^{C_1} = \begin{bmatrix} d_x^{C_1} & d_y^{C_1} & d_z^{C_1} \end{bmatrix}^t = \begin{bmatrix} -c\phi\,d_x^B & Ls\phi\,\delta_z^B - d_x^B & -s\phi\,d_x^B \end{bmatrix}^t. \tag{6-2}$$

It should be noted that the active motion at each wheel is only $v_y^{C_i}$ according to the coordinate assignment in Fig. 4b, which can be generated by the driving motor of the wheel. From (6-2), $v_y^{C_1}$ is given as follows:

$$v_y^{C_1} = L\,s\phi\,\omega_z^B - v_y^B. \tag{7}$$

Equation (7) represents the velocity relationship between a wheel and the robot body.

Similarly, from (1)–(6-2), the relationship between the velocity motion at **B** and the active motion of each wheel can be obtained as

$$v_y^{C_2} = Ls\phi\omega_z^B + \frac{\sqrt{3}}{2}v_x^B + \frac{1}{2}v_y^B, \tag{8}$$

$$v_y^{C_3} = Ls\phi\omega_z^B - \frac{\sqrt{3}}{2}v_x^B + \frac{1}{2}v_y^B. \tag{9}$$

Finally, from (7)–(9), the velocity kinematics of the ballbot in matrix form is described as

$$\begin{aligned} \begin{bmatrix} v_y^{C_1} \\ v_y^{C_2} \\ v_y^{C_2} \end{bmatrix} &= \begin{bmatrix} 0 & -1 & Ls\phi \\ \frac{\sqrt{3}}{2} & \frac{1}{2} & Ls\phi \\ -\frac{\sqrt{3}}{2} & \frac{1}{2} & Ls\phi \end{bmatrix} \begin{bmatrix} v_x^B \\ v_y^B \\ \omega_z^B \end{bmatrix} \rightarrow \\[2ex] \begin{bmatrix} v_x^B \\ v_y^B \\ \omega_z^B \end{bmatrix} &= \begin{bmatrix} 0 & -1 & Ls\phi \\ \frac{\sqrt{3}}{2} & \frac{1}{2} & Ls\phi \\ -\frac{\sqrt{3}}{2} & \frac{1}{2} & Ls\phi \end{bmatrix}^{-1} \begin{bmatrix} v_y^{C_1} \\ v_y^{C_2} \\ v_y^{C_3} \end{bmatrix}. \end{aligned} \tag{10}$$

5 Conclusions

The ballbot considered in this paper uses an omnidirectional mobile platform with three Swedish wheels inside it as a driving mechanism and has several advantages over conventional ballbots: free motion without nonholonomic constraints, an inherently stable posture, and low energy consumption for motion. Kinematics modelling as an equation of motion is an essential prerequisite for motion control. Systematic kinematics modelling was addressed in this paper, which described the velocity relationship between the driving motors and the robot body for the motion control.

References

1. Kim J, Kwon H, Lee J (2009) A rolling robot: design and implementation. In: Proceedings of 7th Asian control conference, pp 1474–1479
2. Lauwers T, Kantor G, Hollis R (2006) A dynamically stable single-wheeled mobile robot with inverse mouse-ball drive. In: Proceedings of IEEE international conference on robotics and automation
3. Kumagai M, Ochiai T (2008) Development of a robot balancing on a ball. In Proceedings of international conference on control, automation and systems, pp 433–438
4. Bicchi A, Balluchi A, Prattichizzo D, Gorelli A (1997) Introducing the sphericle: an experimental testbed for research and teaching in nonholonomy. In: Proceedings of IEEE international conference on robotics and automation, pp 2620–2625
5. McKerrow P (1990) Introduction to robotics. Addison-Wesley, Reading

Design of Device Sociality Database for Zero-Configured Device Interaction

Jinyoung Moon, Dong-oh Kang and Changseok Bae

Abstract Nowadays people connected to the Internet are using more than six mobile connected personal devices, such as smartphones, smart pads, and laptops. To provide a simple way to share multiple devices owned by themselves or by their family and friends without configuration, this research aims at building and managing social relationships of personal devices by using human relationships obtained by social networking services. This paper proposes the design of device sociality database on the basis of ER diagrams, which is a critical step to store and manage data and information required for zero-configured device interaction. The database design is made up of the resource specification of personal devices and device sociality including device ownership, human relationships, and access permission of device resources.

Keywords Database design · Device sociality · Human relationship · Device collaboration

1 Introduction

Nowadays people own multiple connected mobile personal devices, such as laptops, smart pads, and smartphones. According to the white paper of Cisco [1], the number of mobile devices connected to the Internet was 12.5 billion in 2010 and

---

J. Moon (✉) · D. Kang · C. Bae
Eelectronics and Telecommunications Research Institute, 218 Gajeong-ro,
Yuseong-gu, Daejeon 305-700, South Korea
e-mail: jymoon@etri.re.kr

D. Kang
e-mail: dongoh@etri.re.kr

C. Bae
e-mail: csbae@etri.re.kr

J. J. (Jong Hyuk) Park et al. (eds.), *Multimedia and Ubiquitous Engineering*,
Lecture Notes in Electrical Engineering 240, DOI: 10.1007/978-94-007-6738-6_94,
© Springer Science+Business Media Dordrecht(Outside the USA) 2013

will be 25 billion by 2015. In addition, the number of connected devices per person was 1.84 and will be 3.47 by 2015. Because about 2 billion people among world population use the Internet actually, the real number of connected devices per person can be regarded as 6.25 in 2010 instead of 1.84. Therefore people need a simple and convenient way to share resources of devices owned by them or by their family and friends without explicit setting.

The human relationships are disclosing online through Social Networking Services (SNS) [2], such as Twitter, Facebook, and MySpace. Now the 65 % of adult Internet users use a social networking service, which increased dramatically compared to 5 % of adult Internet users in 2005 according to the statistics of SNS usage in [3]. The prevalence on SNS usage has speeded up online human relationships. The categorized human relationships and list of friends included in each relationship group can be retrieved by open Application Programming Interfaces (APIs) provided by commercial SNSs. If there are no explicit relationship groups provided by the SNSs, affinity-based group can be generated by analyzing SNS activities between friends, such as replying, commenting or representing emotions on a post, photo, or video, and sending and receiving private messages. Therefore, the SNS can be the feasible source for obtaining human relationships.

The purpose of our study is to build and manage social relationships between the personal devices with zero-configuration of device interaction by either extracting human relationship groups from commercial SNSs or by inferring human relationship groups from the history of device interactions [4], as shown in Fig. 1. The social relationship between devices is called device sociality and enables the devices to interact each other without manual setting of their device resources. To store and manage data required for the zero-configured device interaction, we design a device sociality database including specification of each device, human relationships between device owners, ownerships of devices, access permission of device resources.

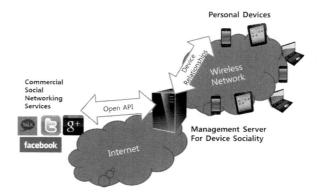

Fig. 1 Concept of device sociality on the basis of human sociality extracted from commercial SNS

Design of Device Sociality Database

2 ER Modeling

We generate an Entity Relationship (ER) model for the device sociality by using the ER modeling, which is the primary method for database design. The ER model includes entities with their attributes and relationships between the entities optionally with their attributes [5]. The model is made up of the part for device specification and the part for device sociality on the basis of human relationships. Each part is shown by an ER diagram.

Figure 2 shows the ER diagram describing the part for device specification. The ER model for the device specification describes device resources, which are a hardware device, an operating system, and other components including a CPU, storage, display, input interface, power, and sensors.

In the ER model, the device entity has two many-to-one relationships with device hardware and operating system entities because a device has one device hardware and one operating system and device hardware and an operating system can be employed to multiple devices. In addition, the device entity has a one-to-many relationship with the location log entity because a device can file zero or many location logs. The device entity has two dependent relationships with data storage and service entities because a device can share its data storage or provides its service for device interaction.

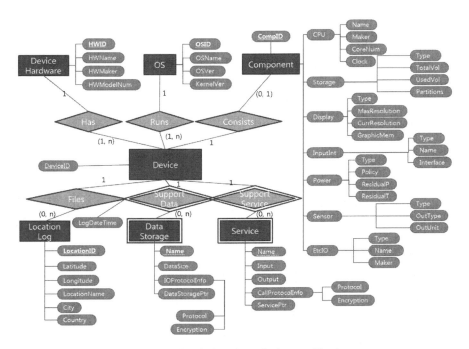

Fig. 2 ER diagram of device sociality design about device specification

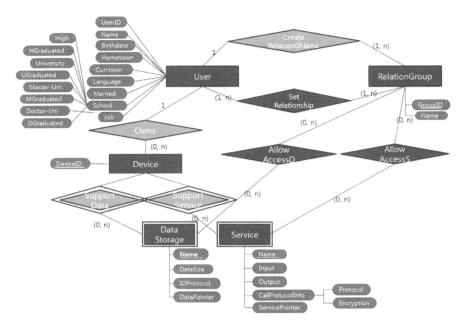

Fig. 3 ER diagram of device sociality design about device relationships

Figure 3 shows the ER diagram for the part for device sociality on the basis of human sociality. In the ER model, the user entity describes all the data related to users, such as name, birthdate, hometown, main language, schools, and current job. The user entity has a one-to-many relationship with the relation group entity. A relation group corresponds to either an extracted human relationship groups from a commercial SNS or a self-group for the user. The user entity has a many-to-many relationship with the relation group entity. The set relationship lists all the users included in each relation groups. In the ER model, the relation group can have access permission to shared resources like data storages or services. The relation group entity has many-to-many relationships with the data storage and service entities because a data storage or service can be allowed to be accessed by multiple relation groups and a relationship group can have access permission to multiple data storages and services.

3 Schema Design and Database Implementation

By using the proposed ER model, we obtained a schema design for device sociality database according to mapping rules of ER modeling [6]. All the entities were mapped into corresponding tables in the database schema. The attributes of an entity were mapped into columns of a corresponding table. An identifying attribute

Design of Device Sociality Database 767

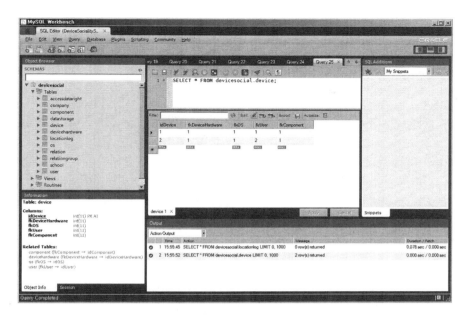

Fig. 4 Screenshot of database implementation in a commercial database system

of an entity was mapped to the primary key of the table. A one-to-many relationship between two entities was represented by a foreign key from one table, which is the primary key of the other table. A binary many-to-many relationship between two entities is mapped into a table with two foreign keys from tables corresponding to the entities.

After build the database schema from the proposed ER models, we implemented the database for device sociality by using one of commercial relational database systems, as shown in Fig. 4. By inserting data into the implemented database, we validated the proposed design for device sociality.

4 Conclusion

Because people connected to the Internet are nowadays using more than six mobile connected personal devices, they need a simple and easy way to share multiple devices owned by them or by their family and friends with zero-configuration. Our research aims at building and managing social relationships of personal devices by using human relationships obtained by open APIs from commercial SNSs. Therefore, this paper proposes the design of device sociality database by using ER modeling, which is a critical step to store and manage data for zero-configured device interaction.

The proposed ER model for device sociality includes the part for device ownership, human relationships, and access permission of resources as well as the part for the device specification. To validate the proposed ER model, we obtained the database schema by using the mapping rules and implemented the database in a commercial database system.

For the future work, we are going to collect and analyze real data for device sociality database from device sociality management servers and associated devices.

Acknowledgments This work was supported by the IT R&D program of MKE/KEIT (K10041801, Zero Configuration Type Device Interaction Technology using Device Sociality between Heterogeneous Devices).

References

1. Dave E (2011) The internet of things: how the next evolution of the internet is changing everything. White paper, CISCO Internet Business Solutions Group (IBSG), April 2011
2. Wikipedia, Social Networking Service, http://en.wikipedia.org/wiki/Social_networking_service
3. Mary M, Kathryn Z (2011) 65% of online adults use social networking sites. The Pew Research Center's Internet & American Life Project surveys, 26 Aug 2011
4. Kang K, Kang D, Bae C (2013) Novel approach of device collaboration based on device social network, consumer electronics (ICCE). In: Proceedings of IEEE international conference on 11–14 Jan 2013, pp 248–249
5. Wikipedia, Entity-relationship model, http://en.wikipedia.org/wiki/Entity-relationship_model
6. Elmasri R, Navathe S (2010) Fundamentals of database systems, 6th edn. Addison Wesley, Boston

Image Processing Based a Wireless Charging System with Two Mobile Robots

Jae-O Kim, Chan-Woo Moon and Hyun-Sik Ahn

Abstract This paper presents the image processing algorithm for wireless charging between each mobile robot. The image processing algorithm converts Red Green Blue (RGB) format of inputted image to detect edge. It calculates a specific area using Hough Transformation (HT) in detected edge and judges correct charging antenna using Speeded-Up Robust Features (SURF). Accordingly, the image processing algorithm can control position and direction of mobile robot and antenna for wireless charging. The image processing algorithm is implemented wireless charging systems, which are set up on each two mobile robot and it is verified with experiment.

Keywords Wireless charging · HT · SURF · Mobile robot · Color edge

1 Introduction

A typical mobile robot has actuators which are operated by battery. As the robot moves for a long time, the battery is exhausted, and recharging is needed. Currently, automatic recharging is general trend for an intelligent robot. And usually, contact type of recharging with a guide mechanism is widely used. But in a circumstance on which guide mechanism is unavailable, for example, recharging between different type of robots, recharging outdoor and recharging when a robot is still moving, a wireless power transmission can be considered.

J.-O. Kim · C.-W. Moon (✉) · H.-S. Ahn
Department of Electronics Engineering, Kookmin University, Seoul, Korea
e-mail: mcwnt@kookmin.ac.kr

J.-O. Kim
e-mail: futurejo@paran.com

H.-S. Ahn
e-mail: ahs@kookmin.ac.kr

J. J. (Jong Hyuk) Park et al. (eds.), *Multimedia and Ubiquitous Engineering*, Lecture Notes in Electrical Engineering 240, DOI: 10.1007/978-94-007-6738-6_95, © Crown Copyright 2013

In this paper, wireless power transmission method for two mobile robots is investigated. To obtain the maximum efficiency of power transmission, vision based position control of a transmission antenna is implemented. To recognize the antenna, Color Edge, Hough Transform and Speeded-Up Robust Features (SURF) methods are used. The information of pose of antenna is used to control the mobile robot.

2 Inductive Power Transmission

The basic principle of an inductively coupled power transfer system consist of a transmitter coil and a receiver coil. Both coils form a system of magnetically coupled inductors. An alternating current in the transmitter coil generates a magnetic field which induces a voltage in the receiver coil. This voltage can be used to power a mobile device or charge a battery. In this research, near range direct induction method is used.

3 Color Edge Detection

Common image processing is the processing of the input image is converted to grayscale and binary coded. At this time, a lot of ambient lighting, and the impact on the environment, it is difficult to extract the straight edge. As a way to solve this problem, changing the format of the input images from RGB to HSV is widely used, which changes only the size of the color difference image after edge detection.

4 Hough Transform (HT)

Hough Transform is widely used for searching a figure such as line and circle in an image pixel data. Equation 1 denotes a basic equation of line and Eq. 2 is another formulation of line which is represented in the parameter space. If a line becomes vertical, the slope becomes infinity, then, Eq. 3 is used alternatively [2, 5].

$$y_i = ax_i + b \tag{1}$$

$$b = -x_i a + y_i \tag{2}$$

$$x \cos \theta + y \sin \theta = \rho \tag{3}$$

5 Speeded-Up Robust Features (SURF)

SURF is faster than the video from one of the algorithm to find the feature points invariant to scale, lighting, point to changes in the environment such as SIFT and similarity, however, faster than SIFT when compared with key points finds. We base our detector on the Hessian matrix because of its good performance in computation time and accuracy. However, rather than using a different measure for selecting the location and the scale (as was done in the Hessian-Laplace detector), we rely on the determinant of the Hessian for both. Given a point x = (x, y) in an image I, the Hessian matrix H(x, σ) in x at scale. σ is defined as follows

$$H(x,\sigma) = \begin{bmatrix} L_{xx}(x,\ \sigma) & L_{xy}(x,\ \sigma) \\ L_{xy}(x,\ \sigma) & L_{yy}(x,\ \sigma) \end{bmatrix} \quad (4)$$

6 The Structure of Experimental System

Figure 1 shows block diagram of the system used in the experiment.

The image processing algorithm for wireless charging converts inputted data format of camera. Converted image is used to detect edge by difference of color signal of image. Position of Antenna is judged by HT method and image is cropped. Cropped image is confirmed by SURF and KNN for alignment with target. The wireless charging control system decides target and direction using this process. And it is implemented on mobile robot task for motion control of mobile robot.

7 Experimental Result

Experimental environment consists of a mobile robot for wireless charging, a mobile robot for transmission power, test antenna for wireless charging and main control program. It is as shown in Fig. 2.

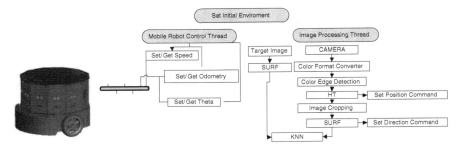

Fig. 1 Block diagram

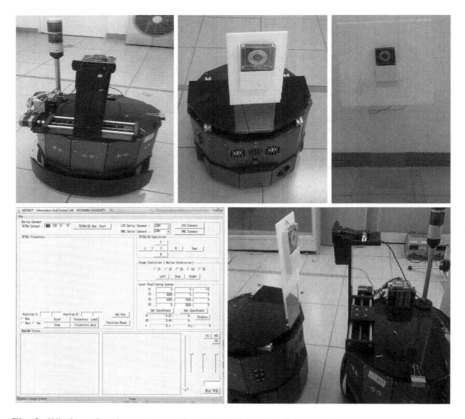

Fig. 2 Wireless charging system and mobile robot and main control program

Figure 3 is typed as it seems the main routine for processing the data.

Experimental result of edge detection in inputted image using color format is as shown in Fig. 3. Compare image with result of edge detection of binary image is as shown in Fig. 4.

The system detects straight line in input image using HT method, extracts feature and compares with target. Position of antenna is confirmed by this process. It is shown by Fig. 5.

Image Processing Based a Wireless Charging System 773

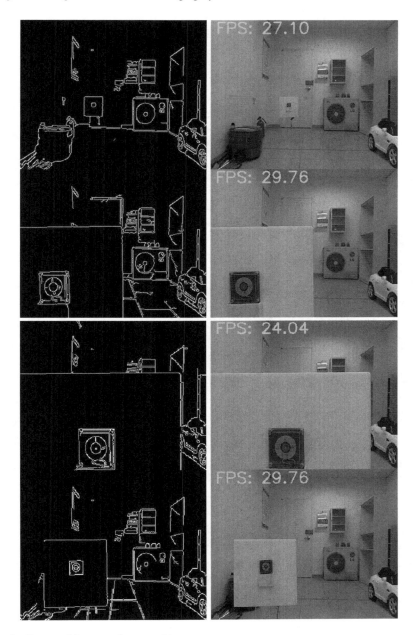

Fig. 3 Converted image and captured image

Fig. 4 The result of edge detection image used grayscale and binary data

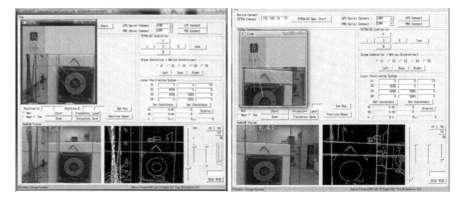

Fig. 5 The detected image

8 Conclusion

In this paper, we implemented a wireless charging system between two robots using image processing. Image processing algorithm is implemented with color edge method to detect position and direction of antenna, HT method for extraction of an interested area and SURF to judge correct feature of antenna. Thus wireless charging system could control position and direction of mobile robot.

9 Summary

A wireless charging system between two mobile robots using image processing methods in which difference of color for edge detection, calculation of antenna position and direction using HT, SURF is used is implemented.

Acknowledgments This research was supported by the The Ministry of Knowledge Economy (MKE), Korea, under the Information Technology Research Center (ITRC) support program (NIPA-2012-H0301-12-2007) supervised by the National IT Industry Promotion Agency (NIPA).

References

1. Wireless Power Consortium. http://www.wirelesspowerconsortium.com
2. Hough PVC (1962) Methods and means for recognizing complex patterns. US Patent 3,069,654
3. Bay H, Ess A, Tuytelaars T, Van Gool L (2008) SURF speeded up robust features. Comput Vis Imag Underst 110(3):346–359
4. TETRA-DS DasaRobot, DasaRobot (2009)
5. Kim J-O, Rho S, Moon C-W, Ahn H-S (2012) Imaging processing based a wireless charging system with a mobile robot. Computer applications for database, education, and ubiquitous computing. Communications in computer and information science, vol 352. Springer, Heidelberg, pp 298–301

Design of a Reliable In-Vehicle Network Using ZigBee Communication

Sunny Ro, Kyung-Jung Lee and Hyun-Sik Ahn

Abstract This paper presents a new configuration for in-vehicle networks to increase the reliability of the communication between electronic control units (ECU) and to improve the safety level of a vehicle. Basically, the CAN (Controller Area Network) protocol is assumed for the data communication in vehicles but more reliable communication can be guaranteed by adding the ZigBee communication function to each ECU. To show the validity and the performance of the presented network configuration for some network faults, the Electronic Stability Control (ESC) operation is analyzed by using an ECU-In-the-Loop Simulation (EILS). The experimental set-up for EILS of ESC system consists of two 32-bit microcontroller boards which can be communicated with the CAN or the ZigBee protocol. A 7-DOF (Degrees Of Freedom) vehicle model and ESC algorithm is implemented on each microcontroller. It is shown by the experimental results that ESC using the high reliability CAN system can achieve the same performance as using only CAN protocol without disconnected CAN bus.

Keywords In-vehicle network · Controller area network · Reliability · Fault tolerance · Electronic stability control

---

S. Ro · K.-J. Lee · H.-S. Ahn (✉)
Department of Electronics Engineering, Kookmin University, Jeongneung-dong, Seongbuk-gu, Seoul, Korea
e-mail: ahs@kookmin.ac.kr

S. Ro
e-mail: sunyda88@nate.com

K.-J. Lee
e-mail: streizin@nate.com

J. J. (Jong Hyuk) Park et al. (eds.), *Multimedia and Ubiquitous Engineering*, Lecture Notes in Electrical Engineering 240, DOI: 10.1007/978-94-007-6738-6_96, © Crown Copyright 2013

1 Introduction

To enhance the handling performance and the safety of vehicles, many active chassis control systems to ensure the vehicle stability have been consistently studied. ESC is a stability enhancement system designed to electronically detect and assist drivers in critical driving situation and under adverse conditions automatically [1].

In addition, ECUs are distributed throughout the vehicle to perform a variety of different vehicle functions. Accordingly, In order to transmit data between distributed ECUs and software correctly, safely, performance improvement in in-vehicle network is essentially needed [2]. Controller Area Network (CAN) is an asynchronous serial communication protocol which follows ISO 11898 standards and is widely accepted in automobiles due to its real time performance, reliability and compatibility with wide range of devices. The main features of CAN protocol are high-speed data transmission up to 1 Mbps, bus access control depending on a multi-master principle, and bus off function in the event of transmission abnormalities [3]. However, if a critical fault occurs in CAN (e.g. disconnection), ECU must transfer the data to another ECU by replacement of CAN protocol.

In this paper, we present the new method for improvement reliability of CAN for in-vehicle network. To improve reliability, CAN is replaced with ZigBee when fault is generated by a disconnection of CAN bus. Also, in order to verify the proposed CAN system, the ESC operation is analyzed by using an EILS. The experimental set-up for EILS of ESC system consists of two 32-bit microcontroller boards which can be communicated with the CAN or the ZigBee protocol.

2 The High Reliability CAN System

Typically, CAN is used in a wide range of industrial automation and an important element in a protocol of distributed real-time control. The distributed real-time control functionalities have been studied for reliable communication network systems when separate ECUs are connected with each other through a CAN protocol [4]. To prevent a critical fault such as a disconnection, CAN protocol must be replaced with another protocol. Accordingly, we propose a fault-tolerant CAN controller system called the high reliability CAN system, which consists of CAN and ZigBee to avoid a fault such as a disconnection. This proposed CAN system for improvement of reliability is based on ZigBee to tolerate any single permanent fault in one CAN controller [5]. In this study, the proposed high reliability CAN system using ZigBee is as shown in Fig. 1.

When CAN is disconnected in this system, disconnected ECU transfers information on disconnected CAN bus to another ECUs by using ZigBee. Then, the high reliability CAN system operates as a normal system in fault. The process to prove the performance of this system is as follows [6, 7]. The transmission time C_m

Design of a Reliable In-Vehicle Network

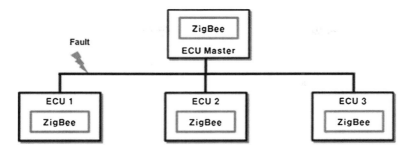

Fig. 1 The high reliability CAN network

of CAN messages containing s_m data byte, a 47 bit overhead per message and the transmission time for a single bit τ_{bit} is given by:

$$C_m = (47 + 8s_m)\tau_{bit} \tag{1}$$

The transmission bit N_{CAN} for a single message of 8 byte CAN is given by:

$$N_{CAN} = (47 + 8s_m) = 111 \text{ bit} \tag{2}$$

If CAN transmits at 500 kbps, the number of transmission messages per second are 4505. Accordingly, the transmission period of CAN T_{CAN_P} is given by:

$$T_{CAN\_P} = \frac{N_{CAN}}{4505} = 220 \text{ us} \tag{3}$$

In the same Eqs. (1) and (2), the transmission bit N_{ZigBee} for a single message of 1 byte ZigBee is given by:

$$N_{ZigBee} = (208 + 8s_m) = 216 \text{ bit} \tag{4}$$

At this time, ZigBee transmits at 250 kbps and the number of transmission messages per second is 1157. The transmission period of ZigBee T_{ZigBee_P} is given by:

$$T_{ZigBee\_P} = \frac{N_{CAN}}{1157} = 864 \text{ us} \tag{5}$$

Considering the data length of 8 byte CAN, the control period multiplies T_{ZigBee_P} by 8. And then, the control period T_{ZigBee_P} is 6.91 ms.

We implement the high reliability CAN system in ESC for reliability verification. ESC is generally controlled by the control period at 10 ms because the bandwidth is considered for control, self-diagnosis and so on of a particular vehicle. The transmission period should not affect to the performance of control because the transmission messages are completed within 10 ms. In the two control period from Eqs. (3) and (5), T_{ZigBee_P} and T_{CAN_P} are included in the control period of ESC. Therefore, the high reliability CAN system is predicted to show a performance as well without a fault.

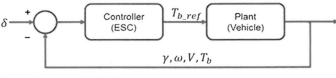

δ	Steering angle [rad]
ω	Wheel angular velocity [rad/sec]
V	Vehicle speed [m/sec]
γ	Yaw rate [red/sec]
T_b	Brake Troqe [N]

Fig. 2 The control scheme of ESC

3 Experimental Environment and Results

The experimental environment for EILS of ESC system consists of two 32-bit microcontroller connected by CAN protocol. ESC system consists of the controller and the plant as shown in Fig. 2. A 7-DOF vehicle model applied to Plant is developed to obtain the longitudinal, lateral and yaw motions of vehicle dynamics and the other four degrees of motion representing 4 wheel dynamics [8, 9].

In this paper, it is assumed that a fault occurs when CAN to transfer braking reference torque T_{b_ref} from ESC to a vehicle model. In each microcontroller, a 7-DOF vehicle model and ESC algorithm are respectively implemented on ECU. Also, CAN monitoring device is used to confirm EILS as shown in Fig. 3.

When the vehicle drives at 80 km/h, the driver rapidly changes the lane as shown in Fig. 4. The behavior of vehicle by the steering input is shown in Figs. 5 and 6. The vehicle model without CAN fault similarly tracks the reference yaw rate. However, the vehicle model with CAN fault appears unstable condition at 2 s by compared with the reference yaw rate as shown in Fig. 5a and b.

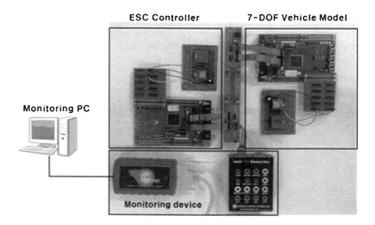

Fig. 3 EILS environment for ESC system

Design of a Reliable In-Vehicle Network 781

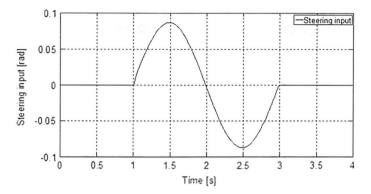

Fig. 4 The steering input for single lane change

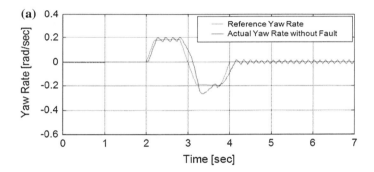

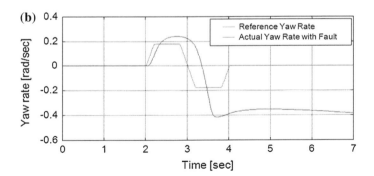

Fig. 5 EILS results with and without CAN fault. **a** Yaw rate in EILS without fault. **b** Yaw rate in EILS with fault

The experiment assumes that ESC system has a fault. The experimental result of the proposed high reliability CAN system using ZigBee is similar to the reference yaw rate as shown in Fig. 6.

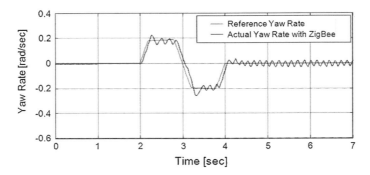

Fig. 6 EILS results of high reliability CAN system using ZigBee

Therefore, the performance of the high reliability CAN system to tolerate CAN fault is verified through the experimentation of EILS. Also, it is confirmed that all transmission messages are exactly processed by the calculated T_{CAN_P} and T_{ZigBee_P} within the control period of ESC.

4 Conclusion

In this paper, we proposed the high reliability CAN system based on ZigBee for tolerance of a CAN fault in ESC system. The main objects of this system were avoidance of a critical CAN fault and the performance as ESC system without a CAN fault. The efficiency of the high reliability CAN system was inferred from the transmission period calculation formula of CAN and ZigBee, and verified by the experiment of EILS of ESC system. EILS consists of a 7-DOF vehicle model and ESC algorithm. The experiment of EILS had been implemented on two 32-bit microcontroller and verified the performance of ESC system when ESC occurred a CAN fault. The experimental result of the proposed high reliability CAN system only using ZigBee was shown as a without CAN fault.

References

1. Neuhaus D, Willms J (2005) Vehicle dynamics-continuous improvements in vehicle safety from abs to electronic stability control. SAE Technical paper 2005-26-065, pp 729–736
2. Chaavan K, Leserf P (2009) Simulation of a steer-by-wire system using flexray-based ecu network. International conference on ACTEA, pp 21–26
3. Chen H, Tian J (2009) Research on the controller area network. International conference on networking and digital society, vol 2, pp 251–254
4. Palai D (2012) Design methods to optimize the performance of controller area network. SAE Technical paper 2012-01-0194, pp 1–11

5. Guerrero C, Rodriguez-Navas G, Proenza J (2002) Hardware support for fault tolerance in triple redundant can controllers. International conference on electronics, circuits and systems, vol 2, pp 457–460
6. Mary GI, Alex ZC, Jenkins L (2012) Response time analysis of messages in controller area network: a review. J Comput Netw Commun, vol 2013, Article ID 14805
7. Johnstone MN, Jarvis JA (2011) Penetration of zigbee-based wireless sensor networks. The 12th Australian information warfare and security conference, pp 16–23
8. Zhao C, Xiang W, Richardson P (2006) Vehicle lateral control and yaw stability control through differential braking. IEEE international symposium on industrial electronics, vol 5, pp 384–389
9. Zhao C, Xiang W, Richardson P (2011) Monitoring system design for lateral vehicle motion. J IEEE Trans Vehicular Technol 6(4):1394–4103

Wireless Positioning Techniques and Location-Based Services: A Literature Review

Pantea Keikhosrokiani, Norlia Mustaffa, Nasriah Zakaria and Muhammad Imran Sarwar

Abstract With advent of satellite positioning system and availability of wireless communication network, it is possible for an end user to navigate even in the scarce location where there are fewer inhabitants. With affordable cost and vast coverage, millions of users can access location co-ordinates from any part of the world due to wireless positioning techniques. In this study, we highlight different positioning methods, location-based services and vast variety of applications benefited from these methods and services. This paper covers brief mathematical models used among all the wireless positioning systems along with their comparison. In today's fast pace information era, location-based services are not only used for hotspot navigation but, also used for marketing strategy and so on. In addition, this article includes location-based services that access mobile network, and utilized the current location of the mobile device appropriately. Finally we classify the location-based solutions that have been used in variety of models such health services, marketing, tourism, entertainment and advertisement, and so forth. The study concludes that with evolution of technological advancement, wireless positioning system will be more improved and will be used in every part of our daily life in an effective manner.

Keywords Location positioning methods · Wireless technologies · Location-based services · Global positioning system (GPS) · Cellular network

---

P. Keikhosrokiani (✉) · N. Mustaffa · N. Zakaria
School of Computer Sciences, Universiti Sains Malaysia, Minden 11800, Penang, Malaysia
e-mail: pantea.kia@ieee.org

N. Mustaffa
e-mail: norlia@cs.usm.my

N. Zakaria
e-mail: nasriah@cs.usm.my

M. I. Sarwar
National Advance IPv6 Centre (NAv6), Universiti Sains Malaysia, Penang, Malaysia
e-mail: imrans@live.com.my

J. J. (Jong Hyuk) Park et al. (eds.), *Multimedia and Ubiquitous Engineering*,
Lecture Notes in Electrical Engineering 240, DOI: 10.1007/978-94-007-6738-6_97,
© Springer Science+Business Media Dordrecht(Outside the USA) 2013

1 Introduction

Rapid development of wireless communication technology and wide usage of wireless networks had a strong effect on the possibility of location-based services. One of the fundamental elements of many applications such as e-commerce, emergency and medical, advertisement, navigation and other location-based services (LBS) is object location positioning. Different functionalities and events are required to approximate the location of a node of interest. These functionalities involve coordinates in two or three dimensions as well as some information such as latitude, longitude, and altitude of the nodes. These information are available everywhere, outside environment, inside building, in the airplane, in the sea, etc. but different methods along with various mathematical principles are needed to track the location in each environment [1]. Location-based services (LBS) refer to services provided based geographical position of mobile device. Such services consist of commerce, emergency and medical, advertisement, navigation and routing, social networking, finding friends and so forth. LBS require location positioning techniques such as satellite and wireless technologies as well as geographical information system in order to map the location of a node of interest [2–4]. The main purpose of this paper is to overview the main mathematical principles of location positioning, different location positioning methods as well as emerging class of location-based services. Such information assists us to be familiar with various location-based services and recognize the right positioning methods and technologies for each service.

To understand how wireless positioning techniques and location-based services work, this paper first introduces basic mathematical positioning methods and their comparison. Furthermore, this paper presents an overview of existing positioning solutions using different wireless communication systems along with their weaknesses and strengths. Different location-based services and their technology is illustrated in next section followed by concluding remarks.

2 Positioning Technologies and Methods

Mathematical Positioning Methods. There are three basic mathematical principles that support every location positioning technique used today. Some positioning methods such as angle of arrival (AOA), time of arrival (TOA), time difference of arrival (TDOA), and received signal strength (RSS) are based on geometric principles to calculate the position of a node. These principles consist of Triangulation, multilateration and hyperbolic used lines and angles in order to calculate the position. Table 1 illustrates these basic mathematical principles in detail [1, 5].

Table 2 illustrates mathematical formulation of basic mathematical principles that are used in position location. The table shows three different columns for

Wireless Positioning Techniques and Location-Based Services

Table 1 Basic mathematical principles for location positioning [1, 5]

Method	Basic Approach	Figure
Triangulation	– Angles of two sides, φ_1 and φ_2 are calculated to get desired location when the distance of points is unknown – Importantly the desired point has to be intersection of two lines from two sides	
Multilateration	– An extension of Triangulation with three reference points – Three point of intersection will give calculated distance value from reference point to object T	
Hyperbolic principle	– It is a set of points that have constant difference values from two fixed points – Hyperbola's focus is represented by each point where focus is an anchor node or reference point – Position can be calculated when the target resides between two foci of hyperbola curve – Curve's distance to each hyperbola focus are fixed	

Triangulation, Multilateration and Hyperbolic principle. The purpose of the table is to explain the mathematical concept of getting final formulae used to calculate the location based positioning system.

Types of Location positioning. Location tracking of the handset devices becomes important among various fields such as advertisement, healthcare, tourism, navigation and routing, entertainment, observation and so forth. There are different ways and techniques in order to calculate the location of devices. Location positioning system consists of several components such as hardware, measuring unit and signal transmitter. Wireless location positioning systems can be classified based on the functionality of these components and their interactions. Hence, positioning systems are categorized into three groups: (1) handset-based positioning, (2) network-based positioning, and (3) hybrid-positioning system [1, 5]. In handset-based positioning, the handset calculates its own location while

Table 2 The formulation of three mathematical principle in position location

Triangulation	Multilateration	Hyperbolic
The intersection point T, we get the 2 angles φ_1 and φ_2	To calculate the coordinates of the target T: first, the distances between reference nodes and the target T	The equations of hyperbola are:
$R = \dfrac{d}{\tan\varphi_1} + \dfrac{d}{\tan\varphi_2}$	$d_1 = (t_1 - t_0) \cdot c$	$1 = \dfrac{x^2}{a^2} - \dfrac{y^2}{b^2}$
R is line between reference point N1 and N2	$d_2 = (t_2 - t_0) \cdot c$	$a^2 = \left(\dfrac{\Delta d}{2}\right)^2$
$d = \dfrac{R \cdot \sin\varphi 1 \cdot \sin\varphi 2}{\sin(\varphi_1 + \varphi_2)}$	$d_3 = (t_3 - t_0) \cdot c$	$b^2 = \left(\dfrac{D}{2}\right)^2 - a^2$
d is perpendicular line between target point T and line R	where c is the speed of light, t_0 is time of a signal sent from T,	Where a and b can be obtained from quantities d and D
$\varphi_1 = \tan^{-1}\left(\dfrac{h - y}{g - x}\right)$	d_1 is distance between N_1 and T,	$\Delta d = d_2 - d_1 = c(t_1 - t_1)$
$\varphi_2 = \tan^{-1}\left(\dfrac{b - Y}{a - X}\right)$	d_2 distance between N_2 and T,	Where,
The coordinates of the target (X, Y) can be calculated as mentioned below	d_3 distance between N_3 and T,	c is the speed of light
$Y = x\tan\varphi_2 + (b - a\tan\varphi_1)$	t_1 time of arrival of signal T to N_1	t1 is the time of node N_1
$X = \dfrac{b - h - a\tan\varphi_2 + g\tan\varphi_1}{\tan\varphi_1 - \tan\varphi_2}$	t_2 time of arrival of signal T to N_2	t_2 is the time of node N_2
distances between reference points N1 and N2, and the target point T	t_3 time of arrival of signal T to N_3	
	Equation of 3 intersecting circle with centers at the reference point	
$d_1 = \|g - X\| = \sqrt{(g - X)^2 - (h - Y)^2}$	$d_1^2 = x^2 + y^2$	
	$d_2^2 = (x - x_2)^2 + y^2$	
$d_2 = \|a - X\| = \sqrt{(a - X)^2 - (b - Y)^2}$	$d_3^2 = (x - x_3)^2 + (y - y_3)^2$	
	Solving the above equations:	
	$x = \dfrac{x_2^2 + d_1^2 - d_2^2}{2 \cdot x_2}$	
	$y = \dfrac{x_3^2 + y_3^2 + d_1^2 - d_3^2 - 2 \cdot x \cdot x_3}{2 \cdot y_3}$	

in network-based positioning the network calculates the handset's location. In hybrid-positioning method, there is collaboration between the network and handset in order to measure and calculate the device's position. GPS is one of the examples of handset-based positioning in which position estimation will be done by handset and GPS based on signals received from at least four satellites [6]. There are some examples for network-based positioning systems such as the cellular networks and Airborne Early Warning and Control System (AWACS) as stated in [1, 7]. Lastly, Assisted GPS(A-GPS) is a good example for hybrid-positioning system [8]. The most important concept of positioning technology is locating users in outdoor and indoor environment and it can be divided into three categories as mentioned above. The fundamental attributes of those approaches are reviewed in Table 3.

Each method illustrated in Table 3 has some weaknesses and strengths; thus, they must be used in the proper situation. For instance, from handset-based category, Global Positioning System (GPS) is appropriate to use for outdoor environment and it does not have indoor services. GPS coverage is poor in urban canyons, it has delay in calculating the location, and a modern handset along with power is required. On the other hand, GPS does not required new network infrastructures, and it is accurate with improved privacy for the user. The next method of handset-based positioning is Enhanced Observed Time Difference (E-OTD) that has enhanced privacy; whereas, some modification must be done in handset and network investment is needed. The next method is Forward Link Triangulation (FLT) that decrease complexity and cost for handset but same as E-OTD some handset modification and network investment is needed. FLT consists of two categories of Advanced Forward Link Trilateration (A-FLT) and Enhanced Forward Link Trilateration (E-FLT). These two categories have various accuracies as shown in Table 3. The first method from network-based category is Cell-ID, Cell of Origin (COO). COO is available now, no handset modification is required for this method and the cost is lower in compare with other methods. COO has lower accuracy in compare with other positioning methods especially in rural cells. Another weakness of COO is low privacy for users. Time of Arrival (TOA) and Uplink Time Difference of Arrival (U-TDOA) have better accuracy in compare with COO and in addition to position, it can determine velocity and heading. Moreover, TOA does not need any modification in handset while TDOA requires some handset modification. TDOA has lower accuracy for TDOA in analog and narrowband digital systems. One of the weaknesses of TOA and TDOA is new equipment is required for base stations. In addition, they have less privacy for users. Angle of Arrival (AOA), another network-based positioning method needs some special equipments for base stations such as special antennas and receivers. It has low privacy for users same as TOA, TDOA and COO but it does not require any handset modification. Received Signal Strength (RSS) is competitive in terms of simplicity and cost in compare with other methods. It is valuable to merge different positioning methods. For instance, combining AOA with RSS concludes better accuracy in compare with using one of the methods alone. Nevertheless, combination methods will increase cost of the network infrastructure. Fingerprint method overcomes many problems by using RSS at the

Table 3 Positioning technologies for location-based services [1, 5]

Type	Positioning Method	Basic Approach	Technology	Accuracy
Handset-Based	Global Positioning System (GPS)	Triangulation method by using timing signals from at least 4 satellites	Satellite	50–100 m
	Enhanced Observed Time Difference (E-OTD)	Triangulation calculation is used to determine location	Cellular Network	60–200 m
	Advanced Forward Link Trilateration (A-FLT)	Measure the time difference of signals from nearby cellular base stations (BS) to triangulate location	Cellular Network	50–200 m
	Enhanced Forward Link Trilateration (E-FLT)	Existing pilot signal measurement message (PSMM) is used from mobile device to BS	Cellular Network	250–300 m
Network-Based	Cell-ID, Cell of Origin (COO)	Location of base station is used to illustrate subscribers location	Cellular Network	10–35 km
	Time of Arrival (TOA)	Uses timing of signals sent by mobile device to triangulate the location	Cellular Network	100–400 m
	Uplink Time Difference of Arrival (U-TDOA)	Uses differences in arrival time between the received signals to identify the location	Cellular Network	50–150 m
	Angle of Arrival (AOA)	It is based on the angle of the received signal of a mobile device into two or more base stations	Cellular Network, WLAN	50–150 m
	Received Signal Strength (RSS)	The energy of the received signal at one end is used to estimate the distance between two nodes	Cellular Network, WLAN	
	(Multipath-) Fingerprint	Measured fingerprints at the existing position location of the nodes will be compared with the fingerprints of diverse positioning locations that are stored in a database	Cellular Network, WLAN	
Hybrid	Timing Advance (TA)	The length of time a signal takes to reach the base station from a mobile phone	GSM	100–550 m
	Assisted Global Positioning System (A-GPS)	GPS receivers are embedded in the cellular network which assist a partial GPS receiver in the handset, reducing the calculation burden	GPS Satellite, Cellular Network	3–20 m

modeling location. Fingerprint method does not require any handset modification but some receiving equipment is needed for base stations. Furthermore, updating and development of database is required in fingerprint and users have less privacy. Assisted Global Positioning System (A-GPS) has some strength in compare with GPS. For example, it reduced the cost imposed by GPS handset and the handset can be smaller with better battery life. Moreover, A-GPS reduced delay in calculating the location. The only problem of A-GPS is new handset requirement as well as indoor positioning accuracy.

3 Location-Based Services

Location-based services can be defined as services that can be accessed by mobile network, and utilized the current location of the mobile device appropriately. Many industries used GPS to enhance their products and services such as automotive industries that used navigation systems for their produced cars. Location-based services assist user to access to the information regarding to the current geographic area of the user [2–4]. Additionally, location-based services make possible two way communication and interaction between customers and businesses. In this way, users will get information according to their needs and requirements. Location based services are a combination of information and telecommunication technologies including Web GIS, Mobile GIS, Mobile Internet, Spatial Database, Internet and mobile devices [9]. Generally, location based services are composed of some components: a mobile device, a communication network, a positioning component, a service and application provider, and a data and content provider.

4 Classification of Location-Based Services

Nowadays, a wide range of services has been offered by relying on user's location information. By accessing to different types of geographical information services (GIS), the location information can be provided simply. As mentioned before, there are several ways to exploit the location in order to provide new services. It will be more effective when the location information merged with other user profile information to offer new location-based services [10]. There are many classifications for location-based services. For instance, Levijoki (2000) categorized location-based services into billing, safety, information, tracking and proximity awareness [11]. Moreover, Kar and Bouwman (2001) grouped location-based services into different services such as information, entertainment, communication, transaction, mobile office and business process support [12]. Additionally, the classification that has been done by Steinfield et al. (2004) is: Emergency, Safety and Medical/Health, Information, Navigation/Routing,

Transactions and Billing, Asset Tracking and Fleet Management, Mobile Office, Entertainment, Proximity Services [13]. After reviewing different location-based services, we classified it into Emergency/Medical/Health Services, Tourism, Navigation/routing/Tracking, Proximity Services, E-Commerce, Vehicular Services, Entertainment, and Advertisement as shown in Table 4.

Rapid technological growth in mobile communications in the last decade has led to innovative and unique mobile healthcare systems. Adding location-based services into healthcare systems will assist patient in terms of searching nearby doctors and healthcare centers. Doctors can check patients remotely and current patient's location can be tracked in case of emergency. References [14–16] are examples of location-based Emergency/Medical/Health Services. Moreover, location-based services can provide the wide range of information related to the Points of Interest (POI) such as hotels, restaurant, tourism attraction and so on. This information can be offered based on the current location of the tourists who are looking for an appropriate POI. References [17–19] proposed tourism location-based services. In addition to using location-based services in healthcare and tourism areas, location-based services can guide users in order to find the best routes. Navigation/routing/Tracking services can be used for tracking the friends, patients, etc. It can be used in order to direct users to their destination and find a way with less traffic. References [20–22] offered navigation, localization and monitoring of patients, and pedestrian navigation respectively. Proximity is another type of the location-based services. Proximity services can notify users while they are within the certain distance of other people, businesses, and so forth. On the other hand, businesses can be informed while users are in their proximity; therefore, they can send advertisement to those users and attract them to their businesses. References [23–25] are good cases for proximity services. Rapid technological evaluation affects electronic commerce (e-commerce) by changing the way of shopping, booking and marketing. Location-aware shopping has been developed by [26] in order to provide information of the customer's preferred vendors that are in their neighborhood.

Furthermore, [27] provides a dynamic service discovery mechanism that enables mobile users in a given coverage area to easily access available services that are provided by suppliers. In addition, [28] designed an intelligent agent based hotel search and booking system. The system is agent-based to perform hotel-booking activities. The agent will check all of the hotels in terms of available facilities, price, customer experience, transportation etc. and forward this information back to the user's mobile phone. Another category of location-based services is vehicular services. It can be either related to traffic information sharing like [29], or it can be used for vehicle location prediction such as [30] and [31]. Entertainment is the next category of location-based services. There are many location-based services such as [32–34] that can be used for the purpose of language learning specially [32]. The last location-based services is advertisement. Businesses can detect near-by customers and send them their promotion and new product in order to attract them to their businesses. On the other hand, this service will benefit customers who are looking for their favorite product. While the

Table 4 Classification of location-based services

Application Type	Years	Framework	Technology	Application Type	Years	Framework	Technology
Emergency/ Medical/ Health Services	2012	Location-Based Mobile Cardiac Emergency System (LMCES) [14]	GPS/GPRS	E-Commerce	2008	Location-aware recommender system for mobile shopping environments [26]	Internet, Cellular Network
	2010	Location Application for Healthcare System [15]	Internet, GPS		2007	Location-based M-Commerce [27]	Cellular Network
	2011	Android-based emergency alarm and healthcare management system [16]	GSM, GPS		2007	Hotel Search and Booking System [28]	Internet, Web Application
Tourism	2009	Personalized Tourism Information System in Mobile Commerce [17]	RFID	Vehicular Services	2006	Sharing Traffic Jam Information using Inter-Vehicle Communication [29]	GPS, WLAN
	2012	MyTourGuide.com [18]	GPS		2012	Vehicular location prediction based on mobility patterns for routing in urban VANET [30]	VANET
	2012	A Trajectory-Based Recommender System for Tourism [19]	GPS		2009	Wireless LAN-Based Vehicular Location Information Processing [31]	WLAN
Navigation/ routing/ Tracking	2012	RFID Assisted Navigation Systems for VANETs [20]	RFID	Entertainment	2009	Location-based Game for Supporting Effective English Learning [32]	WLAN
	2010	Localization and monitoring of patients [21]	Zigbee		2009	Mobile Game Based Learning [33]	GPS
	2010	Pedestrian Navigation System [22]	GPS		2011	iDetective [34]	GPS

(continued)

Table 4 (continued)

Application Type	Years	Framework	Technology	Application Type	Years	Framework	Technology
Proximity Services	2011	Framework for quantifying the system performance of proximity-based services (PBS) [23]	Cellular Network	Advertisement	2005	Location Based Information/Advertising [35]	Bluetooth
	2008	Proximity-based peer selection [24]	GPS, Cellular Network		2011	e-Brochure [36]	GPS
	2000	Context-aware Electronic Tourist Guide [25]	Cell-based wireless communications		2012	Targeted mobile advertising system (TMAS) [37]	Internet

customer is confused between different products, sending promotion and product information can draw their attention to the special business close to their current location. References [35–37] are proposed for location-based advertisement services.

5 Concluding Remarks

Accuracy is one of the important parameters of positioning methods used in location-based services. Accuracy of different positioning methods along with their pros and cons are discussed in this paper. As mentioned in this paper, the accuracy of different positioning methods varies about 10 m–35 km. Therefore the developer must be careful to decide which method is best fitting, depending on the needs of the location based service. Another important parameter of selecting positioning method is the environment of positioning method. It is important to select appropriate method for indoor and outdoor environment. For instance, GPS is a reliable outdoor positioning method while it cannot be used for indoor positioning. TOA, Cell ID, and E-OTD are suitable for indoor positioning; whereas, GPS and A-GPS are not recommended to locate the indoor position of any node. Location positioning is the key features of location-based services and existing location-based services prove the significance of location positioning method. Although LBS represent promising services for user, privacy, concerns, quality of service problems, fair access to location information, and the lack of standards for technology and service providers may hinder market development and represent critical policy issues to be resolved.

References

1. Khalel AMH (2010) Position location techniques in wireless communication systems. Master electrical engineering emphasis on telecommunications. Blekinge Institute of Technology Karlskrona, Sweden
2. Schiller J, Voisard A (2004) Location-based services. Morgan Kaufmann Publishers-Elsevier, San Francisco
3. Virrantaus K et al (2001) Developing GIS-supported location-based services. In: Web information systems engineering, pp 66–75
4. Espinoza F et al (2001) GeoNotes: social and navigational aspects of location-based information systems. In: Abowd GD, Brumitt B, Shafer S (eds) UbiComp 2001. LNCS. vol 2201. Springer, Heidelberg, pp 2–17
5. Willaredt J (2010) WiFi and Cell-ID based positioning-protocols, standards and solutions. Presented at the 2nd international conference on Computer and Network Technology
6. Rappaport TS et al (1996) Position location using wireless communications on highways of the future. Commun Mag IEEE 34:33–41
7. Kayton M, Fried WR (1997) Avionics navigation systems, 2nd edn. Wiley-Interscience, New York

8. Ficco M, Russo S (2009) A hybrid positioning system for technology-independent location-aware computing. Softw Pract Experience 39:1095–1125
9. Brimicombe AJ (2002) GIS-Where are the frontiers now? In: Proceedings GIS 2002, Bahrain
10. Searby S (2003) Personalisation-an overview of its use and potential. BT Technol J 21:13–19
11. S. Levijoki "Title", unpublished
12. van de Kar E, Bouwman H (2001) The development of location based mobile services. Edispuut conference
13. Steinfield C et al (2004) The development of location based services in mobile commerce. In: Preissl B, Bouwman H, Steinfield C (eds) Elife after the dot.com bust. Springer, Berlin, pp 177–197
14. Keikhosrokiani P et al (2012) A proposal to design a location-based mobile cardiac emergency system (LMCES). Stud Health Technol Inform 182:83–92
15. Sobh T et al (2010) Mobile application for healthcare system-location based. In: Sobh T (ed) Innovations and advances in computer sciences and engineering. Springer, Netherlands, pp 297–302
16. Yuanyuan D et al (2011) An android-based emergency alarm and healthcare management system. In: International Symposium on IT in Medicine and Education (ITME), 2011, pp 375–379
17. Zheng W (2009) Personalized tourism information system in mobile commerce. In: International conference on management of e-Commerce and e-Government, 2009. ICMECG '09, pp 387–391
18. Husain W et al (2012) MyTourGuide.com: a framework of a location-based services for tourism industry. In: International conference on Computer and Information Science (ICCIS), 2012, pp 184–189
19. Huang R et al (2012) A trajectory-based recommender system for Tourism, in Active Media Technology. Springer Berlin Heidelberg, pp 196–205
20. Wei C et al (2012) On the design and deployment of RFID assisted navigation systems for VANETs. Parallel Distrib Syst IEEE Trans 23:1267–1274
21. Redondi A et al (2010) LAURA-LocAlization and Ubiquitous monitoRing of pAtients for health care support. In: IEEE 21st international symposium on personal, indoor and Mobile Radio Communications Workshops (PIMRC Workshops), 2010, pp 218–222
22. Popa M et al (2010) Car finding with a pedestrian navigation system. In: 3rd conference on Human System Interactions (HSI), 2010, pp 406–411
23. Günes Karabulut K (2011) On the performance of proximity-based services. Wirel Commun Mobile Comput
24. El-Nahas A, Helmy D (2008) Proximity-based peer selection for service lookup in areas of sudden dense population. In: IET 4th international conference on intelligent environments, 2008, pp 1–7
25. Cheverst K et al (2000) Developing a context-aware electronic tourist guide: some issues and experiences. Presented at the proceedings of the SIGCHI conference on Human Factors in computing systems, The Hague, The Netherlands
26. Yang W-S et al (2008) A location-aware recommender system for mobile shopping environments. Expert Syst Appl 34:437–445
27. Mzila PD et al (2007) Service supplier infrastructure for location-based M-commerce. In: Second international conference on Internet Monitoring and Protection, ICIMP 2007, pp 35–35
28. McTavish C, Sankaranarayanan S (2010) Intelligent agent based hotel search and booking system. In: IEEE international conference on Electro/Information Technology (EIT), 2010, pp 1–6
29. Shibata N et al (2006) A method for sharing traffic jam information using inter-vehicle communication. IEEE
30. Xue G et al (2012) A novel vehicular location prediction based on mobility patterns for routing in urban vanet. EURASIP J Wirel Commun Netw C7–222(2012):1–14

31. Takeda K et al (2009) Wireless lan-based vehicular location information processing. In: Takeda K et al (eds) In-Vehicle corpus and signal processing for driver behavior. Springer, US, pp 69–82
32. Chih-Ming C, Yen-Nung T (2009) Interactive location-based game for supporting effective english learning. In: Environmental Science and Information Application Technology, 2009, ESIAT 2009. International Conference, pp 523–526
33. Schadenbauer S (2009) Mobile game based learning: designing a mobile location based game. In: Bruck P (ed) Multimedia and e-content trends. Vieweg Teubner, Wiesbaden, pp 73–88
34. Yoshii A et al (2011) iDetective: a location based game to persuade users unconsciously. In: Embedded and Real-Time Computing Systems and Applications (RTCSA), 2011 IEEE 17th International Conference, pp 115–120
35. Rashid O et al (2005) Implementing location based information/advertising for existing mobile phone users in indoor/urban environments. In: Mobile Business, 2005. ICMB 2005. International Conference, pp 377–383
36. Keikhosrokiani P et al (2011) A study towards proposing GPS-based mobile advertisement service. Commun Comput Inf Sci 252:527–544
37. Li K, Du TC (2012) Building a targeted mobile advertising system for location-based services. Decis Support Syst 54:1–8

Part XIV
Green and Human Information Technology

Performance Analysis of Digital Retrodirective Array Antenna System in Presence of Frequency Offset

Junyeong Bok and Heung-Gyoon Ryu

Abstract In this paper, we design and analyze a digital retrodirective array antenna (RDA) system based on bandpass sampling for wireless communication. The proposed system has low power consumption thanks to increased signal to interference noise ratio (SINR) because digital RDA can automatically make beam toward source with no information about the direction of incoming signal. Also, this paper presents a robust communications system to frequency offset due to digital PLL. Digital PLL can automatically compensate for frequency offset which occurs because of different frequency between transceivers. Simulation results show that the proposed scheme has better BER performance about 5 dB than that of without phase conjugation when the array elements are three.

Keywords Retrodirective array antenna · Phase detection · Phase conjugation · Phase lock loop

1 Introduction

Retrodirective array technique is able to transmit signal toward source without a priori knowledge of the arrival direction [1]. Retrodirective array has more simple structure than smart antenna technique and it is possible to do automatically beam-tracking. Also, retrodirective system has merit such as high link gain, easy interference elimination, and high energy efficiency. The design of phase conjugation is important in order to develop a retro-directive system. Various schemes

---

J. Bok · H.-G. Ryu (✉)
Department of Electronic Engineering, Chungbuk National University, Cheongju, Korea
e-mail: ecomm@cbu.ac.kr

J. Bok
e-mail: bjy84@nate.com

J. J. (Jong Hyuk) Park et al. (eds.), *Multimedia and Ubiquitous Engineering*,
Lecture Notes in Electrical Engineering 240, DOI: 10.1007/978-94-007-6738-6_98,
© Springer Science+Business Media Dordrecht(Outside the USA) 2013

have been studied to design efficient phase conjugation. Above all, Corner reflector is well known as analog phase conjugation scheme [2]. The Corner reflector scheme is contributed by placing two intersecting flat reflectors perpendicular to each other. Passive retrodirective arrays such as these methods are easy to be implemented. But, it is cannot be update or modify of whole system. Another method heterodyne mixing is proposed to design phase conjugation as another schemes. The phase conjugation is achieved by mixing the received signal of know frequency with double frequency of RF frequency. It is very big disadvantage to need double frequency of RF.

Retrodirective array system using direct down-conversion scheme is proposed for resolving these problem [3]. Direct down conversion method is sensitivity to DC offset and frequency offset. Recently, retrodirective array system using under-sampling (bandpass sampling) based on SDR (software-defined-radio) is studied to solve the problem of direct conversation methods [4, 5].

In this paper, we analyze the BER performance of digital retro-directive array system based on bandpass sampling considering noise at retro-directive array antenna system.

2 Digital Retrodirective Array Antenna

The each received signal has different phase lags $(0, \Delta\varphi, \Delta 2\varphi, \Delta 3\varphi)$ when incident wave is θ in shown Fig. 1. Frequency offset is defined as $\Delta\gamma$. For example, first array element has phase lag (0), and adjacent second array element has phase lag $(\Delta\varphi)$ when frequency offset effect does not exist. In presence of frequency offset,

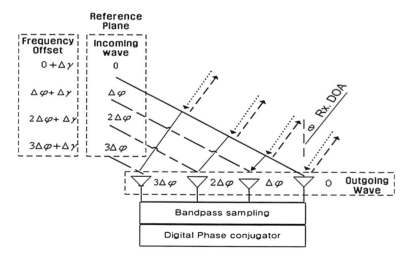

Fig. 1 Digital RDA consider frequency offset

Fig. 2 Block diagram of digital PLL

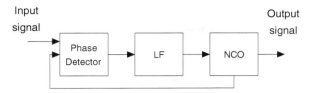

first array element and second array element have phase lag such as $0 + \Delta\gamma$ and $\Delta\varphi + \Delta\gamma$. In order to compensate for the frequency offset, we design the digital PLL.

2.1 Digital Phase Lock Loop (PLL)

Figure 2 shows the block diagram of digital PLL. Digital PLL is a control system that generates an output signal whose phase is related to the phase of an input "Reference" signal. "Reference" signal is hard decision value of received signal. The proposed scheme has three step process to make that reference phase is locked as 0 degrees. Firstly, different phase between input signal and feedback signal was detected using the phase detector block. Secondly, output signal of phase detector is passing through loop filter (LP) for stabilizing the signal. Lastly, the output signal of LF is passing through numerical control oscillator (NCO). NCO is a digital signal generator which creates a signal with amplitude 1 and with conjugation about the phase of output of phase detector.

Transfer function of NCO and LF given by

$$H_{NCO}(z) = \frac{z^{-1}}{1 - z^{-1}} \tag{1}$$

$$H_{LP}(z) = \frac{(G_2 - G_1)z^{-1} + G_1}{1 - z^{-1}} \tag{2}$$

2.2 Phase Detector

In the case of QPSK modulation, phase difference is calculated by using equation (5). The phase difference can be expressed as follows equation.

$$\begin{aligned}e^{j\varphi} &= e^{(\varphi_a - \varphi_b)} = \cos(\varphi_a - \varphi_b) + j\sin(\varphi_a - \varphi_b) \\ &= \frac{I_a}{\sqrt{I_a^2 + Q_a^2}}\frac{I_b}{\sqrt{I_b^2 + Q_b^2}} + \frac{Q_a}{\sqrt{I_a^2 + Q_a^2}}\frac{Q_b}{\sqrt{I_b^2 + Q_b^2}} \\ &+ j(\frac{Q_a}{\sqrt{I_a^2 + Q_a^2}}\frac{I_b}{\sqrt{I_b^2 + Q_b^2}} - \frac{Q_b}{\sqrt{I_b^2 + Q_b^2}}\frac{I_a}{\sqrt{I_a^2 + Q_a^2}})\end{aligned} \tag{3}$$

where, Q_a, I_a are the quadrature and in phase component of received signal. Q_b, I_b are the quadrature and in phase component of hard decision signal.

The amplitude of QPSK signal has $\sqrt{I_b^2 + Q_b^2} = \sqrt{2}$, we assume$(|\varphi| < 20)$, phase difference φ can be approximated as $(\varphi \simeq \sin \varphi)$

(3) can be reformed as

$$\varphi \simeq \sin \varphi = \frac{1}{\sqrt{2(I_a^2 + Q_a^2)}} \cdot (I_b Q_a - I_a Q_b) \tag{4}$$

Finally, only phase information is given by

$$\varphi = (I_b Q_a - I_a Q_b) \tag{5}$$

We can detect the phase difference by using Eq. (5).

3 Simulation Results

Table 1 shows simulation parameters. We assume that transmitter has one antenna, receiver is array antenna (the number of array is 1, 2, and 3). Each received signal has different phase delay. The phase of received signal of first array element is fixed to 0 degrees by using digital PLL.

Figure 3 is constellation of received signal at receiver. Figure 3a–c show that phase delays are 10, 35, and 55 degrees respectively when phase offset is 10.

The phase delay of the received signal is 0 by digital PLL as shown in Fig. 4a. We ensure that digital PLL can compensate for phase delay (10 degrees) by frequency offset. The phase delay of 1st element is 0 degrees after passing through the digital PLL. Phase delays of 2nd element and 3rd element are 20 and 40 after passing through the digital PLL as shown in Fig. 4b, c.

Figure 5 shows comparison of bit error ratio (BER) performance according to the number of array elements at digital RDA. The BER performance of receiver is improved by array gain in the case of increasing the number of array elements. The proposed system can efficiently retransmit the data signal using digital RDA. When the number of array elements is 2, the propose system need 3 dB lower SNR compare to that array element with 1.

Figure 6 shows comparison of BER performance by with and without phase conjugation schemes. Simulation results show that the propose system with phase

Table 1 Simulation parameters	Parameters	Values
	Symbol rate	1 Mbps
	Modulation	QPSK
	Channel	AWGN
	# of array elements	1, 2, 3

Performance Analysis of Digital Retrodirective Array Antenna System 805

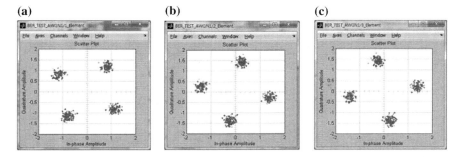

Fig. 3 Constellation of received signal of digital RDA. **a** 1st element, **b** 2nd element, **c** 3rd element

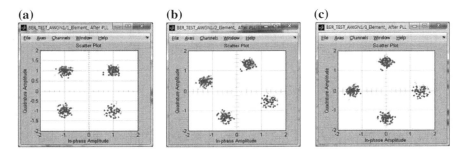

Fig. 4 Constellation of received signal by digital PLL. **a** 1st element, **b** 2nd element, **c** 3rd element

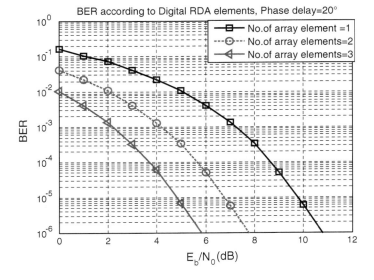

Fig. 5 Comparison of performance by the number of elements

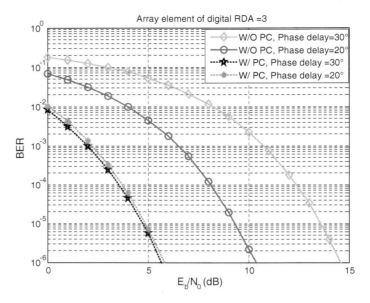

Fig. 6 Comparison of BER performance by w/ and w/o phase conjugation (PC)

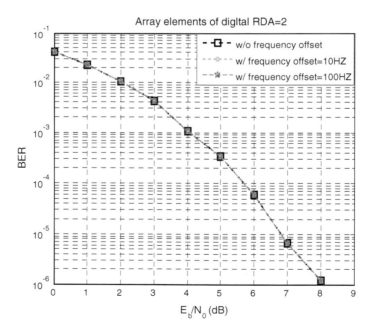

Fig. 7 Comparison of BER performance in presence of frequency offset

conjugation has better BER performance than that without phase conjugation. When we do not use phase conjugation technique, BER performance at receiver is poor because transmitter send data signal toward different direction at source. Simulation results show that the proposed scheme has better BER performance about 5 dB than that of without phase conjugation when the array elements are three.

Figure 7 shows comparison of BER performance by frequency offset. The received signal is shifted by frequency offset. The proposed system has not changing BER performance when same frequency offset occurs at receiver. The proposed system is robustness to frequency offset by digital PLL.

4 Conclusion

We study wireless communication using digital retrodirective array antenna for low power consumption and high quality. The proposed system can efficiently communicate compare to using omni-directional antenna because digital RDA can make automatically beam without prior information about source position. Simulation results that the proposed scheme has better BER performance about 5 dB than that of without phase conjugation when the array elements are three. We ensure that the proposed scheme is an energy efficiency system and robustness to frequency offset through designing the digital PLL.

Acknowledgments This research was supported by the Korea Communications Commission (KCC), Korea, under the R&D program supervised by the Korea Communications Agency (KCA) (KCA-2012-11-921-04-001).

References

1. Sharp ED, Diab MA (1960) Van atta reflector array. IEEE Trans Antennas Propag 8:436–438
2. Pon C (1964) Retrodirective array using the heterodyne technique. IEEE Trans Antennas Propag 2:176–180
3. Miyamoto RY, Qian Y, Itoh T (2001) A reconfigurable active retrodirective direct conversion receiver array for wireless sensor systems. In: Proceedings of IEEE MTT-S international microwave symposium, Phoenix, pp 1119–1122
4. Sun J (2007) A bandpass sampling retrodirective antenna array for time division duplex communications. M.A.Sc. thesis, Dalhousie University, Halifax, NS, Canada
5. Sun J, Zeng X, Chen Z (2008) A direct RF-undersampling retrodirective array system. In: Proceedings of IEEE radio and wireless symposium, Orlando, pp 631–634

A Novel Low Profile Multi-Band Antenna for LTE Handset

Bao Ngoc Nguyen, Dinh Uyen Nguyen, Tran Van Su, Binh Duong Nguyen and Mai Linh

Abstract A low profile antenna, using coupled-fed, meandered, and folded Planar Inverted–F antenna (PIFA), is proposed to cover multi-band Wireless Wide Area Network (WWAN) operations. The proposed antenna, which is suitable for modern 4G mobile phones, requires only a small foot print of 44×20 mm^2. The antenna covers band 14 of LTE-700, GSM-850, GSM-900, DCS-1800, PCS-1900, WCDMA-2100, and band 41 of LTE-2500. In addition to the common WWAN frequencies for mobile communication, the antenna also covers the IEEE802.11b band. The proposed antenna has three simple structures comprising a coupled-fed strip, a meandered shorted-patch, and a folded-patch. These elements are capacitively coupled to each other to form resonant regions in the low bands and the high bands. Modifications to the ground plane are added to achieve a good operation in LTE low band. In the scope of this paper, simulation results, using CST software, are presented to show the effectiveness of the proposed antenna.

Keywords PIFA · Low profile · Broadband · Folded patch · Meandered-line · LTE · Modified ground

1 Introduction

Currently, mobile communication devices require antennas to have the ability to operate in multi-frequency bands, such as GSM 850/900, DSC 1800, PCS 1900, UMTS 2100, and new LTE bands. In addition, mobile devices are becoming

---

B. N. Nguyen (✉) · D. U. Nguyen · T. Van Su
B. D. Nguyen · M. Linh
School of Electrical Engineering, International University—Vietnam National University, Hochiminh City, Vietnam
e-mail: baongocvt1@gmail.com

D. U. Nguyen
e-mail: nduyen@hcmiu.edu.vn

J. J. (Jong Hyuk) Park et al. (eds.), *Multimedia and Ubiquitous Engineering*, Lecture Notes in Electrical Engineering 240, DOI: 10.1007/978-94-007-6738-6_99, © Springer Science+Business Media Dordrecht(Outside the USA) 2013

smaller, slimmer and multi-function integrated, forcing the size of the internal antenna to be smaller. Thus, many researches in designing compact and broad band handset antenna have been published recently [1–9].

In the near future the demanding of the 4th generation of communications is expected to grow. Therefore, the mobile communication antenna will have to incorporate the LTE bands as well. Although LTE standard can be used with many different frequency bands, LTE 700 is of interest because of its availability and low cost deployment. Designing an antenna for a low frequency, specifically the LTE-700 band, with a low profile poses a difficulty on the limited dimensions of typical modern mobile phones. The typical size of the antenna at the low frequency band is usually much larger than the typical size of the mobile devices.

Recently, planar inverted-F antenna (PIFA) has been employed widely as internal antenna for mobile handsets thanks to its low profile, light weight, low cost, versatile characteristics. Nonetheless, PIFA in its original shape has disadvantage of narrow bandwidth. Thus, some techniques to make PIFA broadband and compact such as inserting slots, matching network, meandering and folding must be applied to PIFA [10, 11].

In this paper, we will propose a novel low profile multiband antenna which has the potential to cover 8 frequency bands of Long Term Evolution US band 14 (LTE 758–798 MHz), Global System for Mobile communications (GSM850/900 824–960 MHz), Digital Communication System (DCS 1710–1880 MHz), Personal Communication Services (PCS 1850–1990 MHz), Universal Mobile Telecommunications System (UMTS 1920–2170 MHz), IEEE802.11b (2400–2495 MHz), and LTE band 41 (2496–2690 MHz).

The proposed antenna is a coupled-fed, meandered, and folded PIFA which is printable on a FR4 substrate circuit board of the mobile devices, making it easy to fabricate at low cost and attractive for slim mobile phone applications.

2 Proposed Antenna Geometry

The proposed coupled-fed folded PIFA antenna with modified ground for eight-band LTE/GSM/UMTS/WLAN operation in the mobile phone is shown in Fig. 1a. The antenna is printed on a no-ground space of 20×44 mm^2 and occupies the bottom of the system circuit board which is a 0.8-mm thick FR4 substrate of relative permittivity 4.3, loss tangent 0.025, length 100 mm and width 44 mm.

The detailed dimensions of the antenna are described, in various types of view, in Fig. 1b, c, d, e, respectively. The proposed antenna is composed of a monopole antenna acting as a coupled-fed strip and a shorted folded-patch antenna. The monopole antenna is a quarter-wavelength inverted-L shape antenna centered around 2436 MHz. The monopole antenna is used as the coupled-fed structure to the shorted folded-patch antenna. The coupled-fed structure was proven to enhance considerably impedance bandwidth of the antenna in both high bands and low bands [2, 4, 6, 7]. To obtain the low bands for the antenna, the folded antenna is

A Novel Low Profile Multi-Band Antenna for LTE Handset

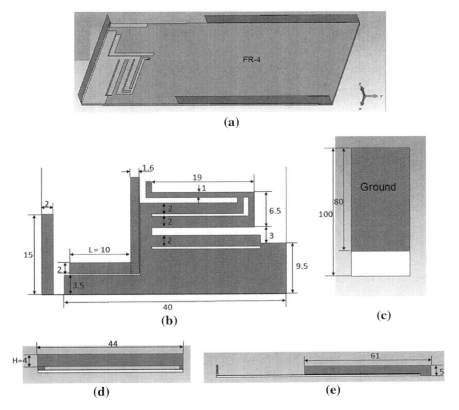

Fig. 1 Geometry and detailed dimensions of the proposed antenna; **a** The overall antenna on the circuit board, **b** Front view, **c** Back view, **d** Bottom view, **e** Right side view

meandered with two strip lines and a shorting pin. The shorting pin here plays important roles not only in the making the antenna physically smaller but also improving impedance matching [10]. To extend the covering of the low band, the radiator is lengthened with a perpendicular patch (size 4 × 44 mm^2) to the main radiator. The perpendicular patch is connected to the radiator by two connectors; one is a metal piece (size 1 × 1 mm^2) connected at one end and the other is a strip line (size 2 × 15 mm^2) connected at other end.

The shorted meandered strip lines and the perpendicular patch together (hence the shorted folded-patch antenna) attribute in generating a resonant mode to form the antenna's lower band to cover LTE band 14, GSM850 and GSM900. Moreover, two slots are created in order to gain more impedance matching at both low band (758–1107 MHz) and high band (1.68–2.18 GHz).

Even the antenna is designed to resonate at the low frequency bands; the impedance matching is still a problem due to the limitation of the ground length [3]. To achieve resonance at LTE band 14 which has the longest wavelength, the ground is at least 100 mm in length, thus the total circuit board must be 120 mm; which is considerable larger than the typical length of the mobile phone. To make the ground larger without increasing the antenna size, folded arms are attached to the ground [3]. Our proposed antenna has two arms with optimized length of 61 mm are attached to the ground (Fig. 1e).

3 Simulation Results and Discussions

Figure 2 shows the simulated return loss, S11 parameter, of the proposed antenna. The simulated results were obtained using Microwave Studio software from Computer Simulation Technology (CST). The proposed antenna has three simple structures, the coupled-fed strip, a meandered shorted-patch, and a folded-patch. By varying each of the structure individually, the optimum combination was used to achieve the desired bandwidth at the selected frequency bands with better return loss less than −6 dB. The −6 dB threshold is generally the acceptable level for broadband internal mobile phone antenna. The optimum combination includes the coupled-fed strip length L = 10 mm, the folded-patch with the height of H = 4 mm, and the two meandered lines with the total length of 46.5 mm. In Fig. 2, there are two frequency areas of interest that satisfy the −6 dB recommendation; ranging from 0.756–1.042 GHz and 1.709–2.725 GHz. Within these two areas of interest, all the desired frequency bands have the return loss below −6 dB threshold. In other words, the proposed antenna can operate in LTE 700 MHz bands 14 758–798 MHz, GSM 824–960 MHz, DCS 1710–1880 MHz, PCS 1850–1990 MHz, UMTS 1920–2170 MHz, and LTE band 41 2496–2690 MHz. In addition to the mobile communication bands, the result shows the covering of the WLAN as well, specifically the IEEE802.11b 2400–2495 MHz frequency band.

The maximum simulated radiation gains and efficiency at the centered frequencies are shown in Table 1.

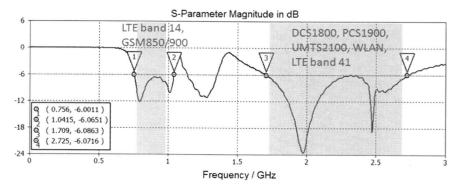

Fig. 2 Return loss simulated result of the proposed antenna

Table 1 Maximum simulation gain of the proposed antenna

Application	Centre frequencies (MHz)	Peak gain (dBi)	Total efficiency (%)
LTE band 14 (758–798 MHz)	780	1.9	90.4
GSM850 (824–894 MHz)	860	2.0	82.5
GSM900 (880–960 MHz)	920	1.8	76.6
DCS (1710–1880 MHz)	1795	4.0	84.2

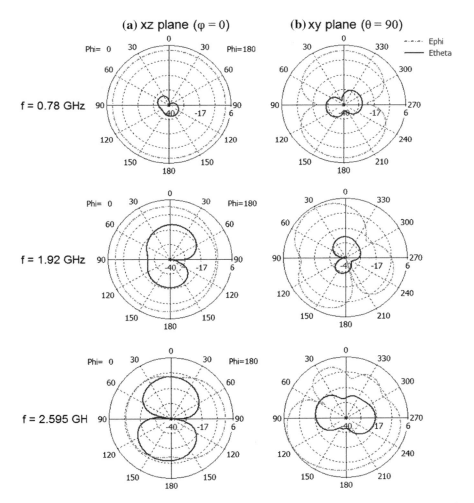

Fig. 3 Polar view of Eφ and Eθ radiation pattern of the proposed antenna in xz-plane and xy-plane

The polar views of the radiation patterns of the proposed antenna are shown in Fig. 3. At low bands from 758–960 MHz, antenna's patterns are similar to a dipole's pattern. On the other hand, at the high bands, the radiation patterns do not follow any specific patterns.

4 Conclusion

The novel wide bands LTE-700 band 14, GSM-850, GSM-900, DCS-1800, PCS-1900, WCDMA-2100, LTE-2500 band 41 and WLAN 2400 has been proposed for mobile phone application. Coupling the feeding strip with the microstrip antenna was proven to enhance considerably the bandwidth of the antenna in both high bands and low bands. Adding the two meandered strip lines with the PIFA was proven to improve the impedance matching at the lower band. The shorted meandered strip line and the perpendicular patch together attribute in generating a resonant mode at the lower band to cover LTE band 14, GSM850 and GSM900. Moreover, to improve the antenna operating in LTE low band, the ground is lengthened by attached two arms. The proposed antenna, simulated on a FR4 substrate, has shown the simulation results with acceptable of return loss (S11), radiation patterns, gains and efficiencies.

References

1. Lee WY, Jeong YS, Lee SH, Oh JR, Hwang KS, Yoon YJ (2010) Internal mobile antenna for LTE/DCN/US-PCS. In: IEEE microwave Conference Proceedings (APMC), 2010 Asia-Pacific, pp 2240–2243, 7–10 Dec 2010
2. Wong KL, Chen WY, Kang TW (2011) On-board printed coupled-fed loop antenna in close proximity to the surrounding ground plane for penta-band WWAN mobile phone. IEEE Trans Antennas Propag 59(3):751–757
3. Jeong YS, Lee SH, Yoon JH, Lee WY, Choi WY, Yoon YJ (2010) Internal mobile antenna for LTE, GSM850, GSM900, PCS1900, WiMAX, WLAN. In: IEEE conference publications on Radio and Wireless Symposium (RWS) 2010, pp 559–562 (Jan 2010)
4. Ying LJ, Ban YL, Chen JH (2011) Low-profile coupled-fed printed PIFA for internal seven-band LTE, GSM, UMTS mobile phone antenna. In: IEEE conference publications on cross strait quad-regional radio science and wireless technology conference (CSQRWC), Vol. 1, pp 418–421 (July 2011)
5. Ying Z (2012) Antennas in cellular phones for mobile communications. IEEE J Mag, Proc IEEE 100(7):2286–2296
6. Yang CW, Jung YB, Jung CW (2011) Octaband internal antenna for 4G mobile handset. IEEE J Mag, Antennas Wirel Propag Lett 10:817–819
7. Kim MH, Lee WS, Yoon YJ (2011) Wideband antenna for mobile terminals using a coupled feeding structure. In: IEEE international symposium on antennas and propagation (APSURSI) 2011, pp 1910–1913, July 2011
8. Zheng M, Wang H, Hao Y (2012) Internal hexa-band folded monopole, dipole, loop antenna with four resonances for mobile device. IEEE Trans Antennas Propag 60(6):2880–2885

9. Tsai PC, Lin DB, Lin HP, Chen PS, Tang IT (2011) Printed inverted-f monopole antenna for internal multi-band mobile phone antenna. In: Vehicular technology conference (VTC Spring), 2011 IEEE 73rd, 15–18 May 2011
10. Wong KL (2002) Compact and broadband microstrip antennas. John Wiley & Sons, Inc., New York
11. Chen ZN, Chia MYN (2006) Broadband planar antennas design and applications. John Wiley & Sons, Inc., New York

Digital Signature Schemes from Two Hard Problems

Binh V. Do, Minh H. Nguyen and Nikolay A. Moldovyan

Abstract In this paper, we propose two new signature schemes and a novel short signature scheme from two hard problems. The proposed schemes have two prominent advantages. Firstly, they are developed from some signature schemes where the security and efficiency have been proven. Therefore, they inherit these properties from the previous schemes. Secondly, the security of the proposed schemes is based on two hard problems. Therefore, they are still safe even when cryptanalysis has an effective algorithm to solve one of these problems, but not both. Moreover, we also propose a method for reducing signatures and this is the first attempt to reduce signatures based on two hard problems. Therefore, our proposed schemes are suitable for the applications requiring long-term security in resource limited systems.

Keywords Cryptographic protocol · Digital signature · Factorization problem · Discrete logarithm problem · Short signature scheme

1 Introduction

One of the vital objectives of a information security systems is providing authentication of the electronic documents and messages. Usually this problem is solved with digital signature schemes (DSSes) [1]. There were many proposals for

---

B. V. Do (✉)
Military Information Technology Institute, Hanoi, Vietnam
e-mail: binhdv@gmail.com

M. H. Nguyen
Le Qui Don Technical University, Hanoi, Vietnam
e-mail: hieuminhmta@ymail.com

N. A. Moldovyan
St. Petersburg Institute for Informatics and Automation of Russian Academy of Sciences, 14 Liniya, 39, St., Petersburg, Russia199178,

---

J. J. (Jong Hyuk) Park et al. (eds.), *Multimedia and Ubiquitous Engineering*, Lecture Notes in Electrical Engineering 240, DOI: 10.1007/978-94-007-6738-6_100, © Springer Science+Business Media Dordrecht(Outside the USA) 2013

signature schemes published based on a single hard problem such as factoring (FAC), discrete logarithm (DL) or elliptic curve discrete logarithm (ECDL) problems [1, 2]. However, these schemes only guarantee short-term security. In order to enhance the security of signature schemes, it is desirable that the signature schemes are developed based on multiple hard problems. This makes it much harder to attack these schemes sine it requires solving multiple problems simultaneously. Some schemes based on two problems, FAC and DL, have been published [3–5]. However, designing these schemes is not easy. Moreover, most of them have been proven that they are not secure [6–8]. Therefore, it is necessary to develop new safe signature schemes based on two hard problems.

In bandwidth and resource limited systems, it is important that the signature schemes have a short signature length. So far, the problem of signature reducing is only investigated for the schemes with single hard problem [9, 10]. We can easily implement a combination of two or more hard mathematical problems in a unified DSS. Breaking such schemes requires simultaneously solving all hard problems. Such implementations require increasing signature length, because the signatures must be present elements belonging to different mathematical problems. It is therefore of interest to develop DSSes, that provide an acceptable signature length. The rest of this paper is organized as follows. In Sect. 2, describes the DSSes based on two hard problems (FAC and DL). Section 3, presents the design of two new DSSes, which requires the simultaneous breaking of FAC and DL problems. Section 4 proposes a novel and efficient short signature scheme. Section 5, describes the security analysis of our schemes. Section 6, describes the performance analysis of our schemes. In the last section, the conclusion of our research is presented.

2 Signature Schemes Based on Factoring and Discrete Logarithms

Previously, DSSes were proposed based on the difficulty in solving the factorization and discrete logarithm problems. For example, the scheme in [11] used a prime modulo p with a special structure $p = 2n + 1$, where $n = q'q$, q' and q are large prime numbers with at least 512 bits. We use the following notations to describe these signature schemes. H is a hash value computed from the signed document M. F is a one-way function, for which can be used to calculate the value of $H = F_H(M)$. α is a primitive element in Z_p^* with order q satisfying $\alpha^q \equiv 1 \bmod p$. The value of λ is a bit length of q, where q is a prime divisor value of n.

The public key is a triple of (p, α, λ). The private key is q.

Signature generation procedure:

(1) Compute $r = F_H(\alpha^k \bmod p)$, where k is a secret random number, $1 < k \leq q - 1$.
(2) The equation generating the parameter S is given by the following equation:
$S = k(Hr)^{-1} \bmod q$.

The signature is a pair of values (r, S), in which the length of the second value is equal to $|S| \leq \lambda$;

When using 1024-bit prime p and a compression function F whose output is a t-bit length and assuming $t = 160$ bits, the length of the digital signature is $|F| + |q| \approx 160 + 512 \approx 672$ bits.

Signature verification procedure:

The verification equation is as follow: $r = F_H(\alpha^{HSr} \mod p)$.

An important part of the verification procedure is to verify the authenticity of a digital signature with the condition $|S| \leq \lambda$, because signature (r, S') with second element of which has the size $|S'| \approx 1023$ bits (if $|p| \approx 1024$ bits) can be easily generated without knowing of the secret parameter q. Such signature (r, S') will satisfy the verification equation. However the signatures (r, S') do not satisfy the condition $|S'| \leq \lambda$. Computing the forged signature (r, S') satisfying both the verification equation and the condition $|S'| \leq \lambda$ without knowing the private key q is not easier than factoring the number $n = (p - 1)/2$ [11]. Security of the considered DSS is based on the difficulty of solving any of the following two problems, factorization and discrete logarithm. Indeed, it is easy to show that solving the factorization problem or solving the discrete logarithm problem allows one to compute the private key and to forge the signature.

In the Schnorr signature in [1], we can use a prime module with the structure of $p = 2n + 1$. This leads to the DSS with public key in the form of four values (p, α, λ, y), where the first three parameters are defined as in the scheme [11] and y is calculated by the formula $y = \alpha^x \mod p$, where x is one element of the secret key.

Signature generation procedure:

(1) Compute $R = \alpha^k \mod p$, where k is a secret random number, $1 < k \leq q - 1$.
(2) Compute $E = F_H(M||R)$.
(3) Compute $S = k - xE \mod q$, such that $R = \alpha^S y^E \mod p$.

The signature is the pair (R, S).

Signature verification procedure:

(1) If $|S| \leq \lambda$, then calculating the value of $R^* = \alpha^S y^E \mod p$. Otherwise, the signature is rejected as invalid.
(2) Compute $E^* = F_H(M||R^*)$.
(3) Compare the values E^* and E. If $E^* = E$, then signature is valid.

Breaking the last signature scheme can be done by simultaneously solving the discrete logarithm problem, which allows to find the secret key x and the factorization problem, which allows to find the value of q, required to compute the value of signature S, whose size will not exceed the value of $\lambda|q|$.

However, the simultaneously solving of these two independent hard problems is not necessary to break this scheme. Indeed, the secret parameters of the scheme can be calculated by solving only the discrete logarithm problem.

This can be done as follow:

We choose an arbitrary number t, the bit length does not exceed the value $\lambda - 1$. Then calculate the value of $Z = \alpha^t \mod p$. After that we find the logarithm of Z on the basis of α, using the index calculus algorithm [1]. This gives a value of T, calculated modulo $n = (p - 1)/2$. With a probability close to 1, the size of this value is equal to $|T| \approx |n| > |t|$. Because α is number with order q over Z_p^* then we have $t = T \bmod q$, so q evenly divides the difference between $T - t$. This means that by following the factorization of $T - t$, we can find the secret parameter q. The probability that a factorization of $T - t$ will have a relatively low complexity is quite high. This means that following the above procedure several times, we will find the value of $T - t$, which can be easily factored.

Thus, for breaking of the two DSSes in this section, we only need to solve discrete logarithm problem modulo a prime. In order to design the DSS, which requires simultaneous solving both the factorization problem and the discrete logarithm problem to break, the last signature scheme should be modified. For example, one can use the value α having order equal to n and introduce a new mechanism for calculating the value S, which will require knowledge of the factors of n while computing S.

3 New Signature Schemes Based on Difficulty of Solving Simultaneously Two Hard Problems

In this section, we propose two new signature schemes from two hard problems. Breaking the modified signature schemes described below requires simultaneous solving two different hard problems, computing discrete logarithm in the ground field $GF(p)$ and factoring n.

3.1 The First Scheme

The following modifications have been introduced in the first signature scheme: (i) as parameter α it is used a value having order equal to n modulo p; (ii) instead of the value S in the signature verification equation it is introduced the value S^2.

Key generation:

(1) Choose large distinct primes q' and q in the form $4r + 3$, and compute $n = q'q$.
(2) Choose randomly a secret key x with $x \in Z_p^*$.
(3) Compute $y = \alpha^x \bmod p$.

The public key is (p, α, y). The secret key is (x, q', q).
Signature generation procedure:

(1) Compute $R = \alpha^k \bmod p$, where k is a secret random number, $1 < k \leq n - 1$.
(2) Compute $E = F_H(M||R)$.

Digital Signature Schemes from Two Hard Problems 821

(3) Calculate the value S, such that $S^2 = k - xE \bmod n$.

The signature is the pair (E, S).
Signature verification procedure:

(1) Compute $R^* = \alpha^{S2} y^E \bmod p$
(2) Compute $E^* = F_H(M||R^*)$.
(3) Compare the values E^* and E. If $E^* = E$, then signature is valid.

It is easy to see that, the advantage of using this exponent 2 (calculate the value S) is computational load smaller compared to larger exponents. The disadvantage is if $S^2 = k - xE \bmod n$ has no solution, the signature cannot be directly generated [1].

3.2 The Second Scheme

The following modifications have been introduced in the second signature scheme: (i) as parameter α it is used a value having order equal to n modulo p; (ii) it is used one additional element e of the public key; (iii) it is used one additional element d of the private key; (iv) instead of the value S in the signature verification equation it is introduced the value S^e. The values e and d are generated like in the RSA cryptosystem [1].
Key generation:

(1) Choose randomly an integer $e \in Z_n$ such that gcd $(e, n) = 1$.
(2) Calculate a secret d such that $ed \equiv 1 \bmod \phi(n)$.
(3) Choose randomly a secret key x with $x \in Z_p^*$.
(4) Compute $y = \alpha^x \bmod p$.

The public key is (e, α, y). The secret key is (x, d).
Signature generation procedure:

(1) Compute $R = \alpha^k \bmod p$, where k is a secret random number.
(2) Compute $E = F_H(M||R)$.
(3) Calculate the value S, such that $S^e = k - xE \bmod n$, i.e. $S = (k - xE)^d \bmod n$ such that $R = \alpha^{Se} y^E \bmod p$.

The signature is the pair (E, S). It is easy to see that the length of signature is $|E| + |S| \geq 1184$ bits.
Signature verification procedure:

(1) Compute $R^* = \alpha^{Se} y^E \bmod p$.
(2) Compute $E^* = F_H(M||R^*)$.
(3) Compare the values E^* and E. If $E^* = E$, then signature is valid.

4 Novel Short Signature Scheme

One of important problems is developing digital signature schemes with short signature length [9]. To reduce the signature length in the case of DSSes from two hard problems we use signature formation mechanism, which is based on solving a system of equations [10].

We use the signature formation mechanism that can be applied while developing DSSes with three-element signature denoted as (k, g, v).

The mechanism is characterized in using a three element public key with the structure (y, α, β), where $y = \alpha^x \bmod p$; α is the δ order element modulo p, i.e. $\alpha^\delta \bmod p = 1$; β is the γ order element modulo n, i.e. $\beta^\gamma \bmod n = 1$ ($p = 2n + 1$, where $n = q'q$) and in solving a system of three equations while generating signature. The secret key is γ.

In this scheme, q and q' are strong primes and easy to generate using Gordon's algorithm [1]. The prime q and q' are supposed to be of large size $|q| \approx |q'| \geq 512$ bits. Gordon's algorithm allows to generate strong primes q and q' for which the numbers $q - 1$ and $q' - 1$ contain different prime devisors γ' and γ'', respectively.

Some internal relation between the β and n values provides potentially some additional possibilities to factorize modulus n. This defines special requirements to the β element of the public key [10]. One should use composite γ, i.e. $\gamma = \gamma'\gamma''$, where $\gamma'|q - 1$, $\gamma''|q' - 1$, $\gamma''\backslash q - 1$ and $\gamma'\backslash q' - 1$. To choose the size of the γ value we should take into account that the β value can be used to factorize the n modulus calculating $\gcd(\beta^i \bmod n - 1, n)$ for $i = 1, 2,\dots \min\{\gamma', \gamma''\}$. Therefore we should use the 80-bit values γ' and γ''. Thus, for γ we get the following required length: $|\gamma| = 160$ bits.

A secure variant of the DSS with the 480-bit signature length is described by the following verification equation: $k = \left(y^k \alpha^{gH} \bmod p + \beta^{kgv+H} \bmod n\right) \bmod \delta$, where δ is a specified prime number and H is the hash value of the signed message. *The signature generation is performed as follows*:

(1) Generate two random number u_1 and u_2 calculate $z_1 = \alpha^{u1} \bmod p$ and $z_2 = \beta^{u2} \bmod n$.
(2) Solve simultaneously three equations:

$$k = (z_1 + z_2) \bmod \delta; \quad g = (u_1 - kx)H^{-1} \bmod \delta; \quad v = (u_2 - H)k^{-1}g^{-1} \bmod \gamma.$$

Breaking this scheme requires the simultaneously solving of the factorization the modulus n and the discrete logarithm modulo p.

In this scheme the signature length is compared for different DSSes in the case of minimum security level that can be estimated at present as 2^{80} operations [1]. The minimum level of security provided under the following size parameters: $|p| \geq 1024$ bits, $|n| \geq 1024$ bits, $|\delta| \geq 160$ bits and $|\gamma| \geq 160$ bits. It is easy to see that the size of a digital signature is $|k| + |g| + |v| \geq 480$ bits.

5 Security Analysis

This section presents an analysis on the security of the proposed signature schemes. The results show that the new schemes are only broken when two hard problems, DL and FAC, are solved simultaneously.

The first scheme: In this scheme, solving the DL problem in $GF(p)$ is not sufficient for breaking the modified scheme. The solution of the DL problem leads to the computation of the secret key x and to the possibility to calculate the value $S^* = (k - xE) \bmod n$. However, calculating the signature S requires to extract the square root modulo n from the value S^*. The last represent a hard problem until the value n is factorized.

The second scheme: Similar to the first scheme, solving the DL problem in $GF(p)$ is not sufficient for breaking the modified scheme. To break this signature scheme it is required to know the factorization of n. The solution of the DL problem leads to the computation of the secret key x and to the possibility to calculate the value $S^* = (k - xE) \bmod n$. However, to calculate the signature S, it is required to extract the eth root modulo n from the value S^*. This requires factoring the modulus n.

Theorem 1 *If an ORACLE O can solve DL and FAC problems, then it can break the proposed schemes.*

In other words, if an ORACLE O has the prime factors (q', q) of n and (x, k) by solving FAC and DL problems, then (E, S) will be the eligible sign of document M generated by the proposed methods.

We indicate that the following attacks can be used to break the proposed schemes.

Attack 1: In order to break these schemes, the adversary needs to calculate all secrete elements in the systems. In this case, the adversary needs to solve DL problem to calculate values (x, k). Moreover, the adversary also have to solve FAC problem. It means that the adversary have to solve both DL and FAC problems in order to break the proposed schemes.

Attack 2: The adversary may receive values (R, E, S). By selecting S arbitrarily and computing $E = F_H(M||R)$, the adversary try to find S satisfying equation $R = \alpha^{Se} y^E \bmod p$. In order to solve this equation, the adversary also needs to solve both DL and FAC problems.

Attack 3: All attacks on RSA, Rabin, Schnorr [1] can not be successful on the proposed schemes, because these schemes are the combination of two fundamental algorithms.

Table1 Time complexity comparison of the proposed schemes and the scheme of [5]

	Time complexity (our first scheme)	Time complexity (our second scheme)	Time complexity [5]
Key generation	T_{EXP}	$T_{EXP} + T_{INV}$	$T_{EXP} + T_{INV}$
Signature generation	$T_{EXP} + T_{MUL} + T_{SR} + T_H$	$2T_{EXP} + T_{MUL} + T_H$	$3T_{EXP} + 3T_{MUL} + 2T_{SR} + T_H$
Signature verification	$3T_{EXP} + T_{MUL} + T_H$	$3T_{EXP} + T_{MUL} + T_H$	$4T_{EXP} + 2T_{MUL} + T_H$

6 Performance Analysis

The performance of the proposed algorithms is evaluated based on the complexity of the following procedures: key generation, signing generation and verification. For the sack of comparison, we use the following notations. T_{EXP} denotes Time complexity for executing the modular exponentiation. T_{MUL} denotes Time complexity for executing the modular multiplication. T_H denotes Time complexity for performing hash function. T_{SR} denotes Time complexity for executing the modular square root computation. T_{INV} denotes Time complexity for executing the modular inverse computation.

The results in Table 1 show that the proposed scheme have better performance than the previous scheme in [5].

7 Conclusion

This paper presents the ability to efficiently develop signature schemes based on the widely used fundamental schemes. Based on some well-know schemes, RSA, Rabin and Schnorr, we proposed two new signature schemes. The proposed schemes possess the higher security than well-know schemes because they are based on two independently difficult problems.

The paper also introduces a new method for reducing signature length. This leads to the proposed signature schemes have the shortest signature length in comparison with the other schemes based on two hard problems.

References

1. Menezes AJ, van Oorschot PC, Vanstone SA (1996) Handbook of applied cryptography. CRC Press, Boca Raton
2. Pieprzyk J, Hardjono T, Seberry J (2003) Fundamentals of computer security. Springer, New York
3. Harn L (1994) Public-key cryptosystem design based on factoring and discrete logarithms. IEEE Proc Comput Digit Tech 141(3):193–195

4. Tzeng SF, Yang CY, Hwang MS (2004) A new digital signature scheme based on factoring and discrete logarithms. Int J Comput Math 81(1):9–14
5. Ismail ES, Tahat NMF (2011) A new signature scheme based on multiple hard number theoretic problems. ISRN Commun Netw
6. Li J, Xiao G (1998) Remarks on new signature scheme based on two hard problems. Electron Lett 34(25):2401–2402
7. Chen T-H, Lee W-B, Horng G (2005) Remarks on some signature schemes based on factoring and discrete logarithms. Appl Math Comput 169:1070–1075
8. Buchmann J, May A, Vollmer U (2006) Perspectives for cryptographic long term security. Commun ACM 49(9):50–55
9. Boneh D, Lynn B, Shacham H (2001) Short signatures from the Weil pairing. In: ASIACRYPT '01, vol 2248. LNCS, pp 514–532
10. Moldovyan NA (2009) Short signatures from difficulty of factorization problem. Int J Netw Secur 8(1):90–95
11. Dernova ES (2009) Information authentication protocols based on two hard problems. PhD Dissertation, St.Petersburg State Electrotechnical University. St. Petersburg, Russia

Performance Improvements Using Upgrading Precedences in MIL-STD-188-220 Standard

Sewon Han and Byung-Seo Kim

Abstract Deterministic Adaptable Priority Net Access Delay, one of the channel access methods defined in MIL-STD-188-220 standard, is designed focusing on the reliable transmissions of the highest priority packets. Therefore, it degrades the performances of the low priority traffics. This paper proposes a method to improve the performances of the low priority traffics while maintaining the performances of the highest priority traffics. The method upgrades intentionally the priority of the traffics that stays in a queue for a certain time period, so that it gives an opportunity for even lower priority traffic to be transmitted. The proposed method is simulated and the results show it improves the network performances.

Keywords MIL-STD-188-220 · Tactical networks · DAP-NAD · Precedence

1 Introduction

MIL-STD-188-220 standard specifies physical, data link, and intranet protocol layers for narrowband-based tactical Digital Message Transfer Devices (DMTD) [1] used for remote access to automated Command, Control, Communication, Computer and Intelligence (C4I) systems and to other DMTDs. DMTD is a portable tactical communication devices having limited data generation and processing capability. The standard has been versioned up from MIL-STD-188-220A to MIL-STD-188-220D with Change 1. MIL-STD-188-220 standard defines 6

S. Han
Korea Radio Promotion Association, Seoul, Korea
e-mail: swhan@rapa.or.kr

B.-S. Kim
Department of Computer and Information Communications Engineering, Hongik University, 2639, Sejong-ro, Sejong, Sejong-si, Korea

J. J. (Jong Hyuk) Park et al. (eds.), *Multimedia and Ubiquitous Engineering*, Lecture Notes in Electrical Engineering 240, DOI: 10.1007/978-94-007-6738-6_101, © Springer Science+Business Media Dordrecht(Outside the USA) 2013

different protocols for data link layer which are Random-Network Access Delay (R-NAD), Prioritized-NAD (P-NAD), Hybrid-NAD (H-NAD), Radio Embedded-NAD (RE-NAD), Deterministic Adaptable Priority NAD (DAP-NAD), and Data And Voice NAD (DAV-NAD). Moreover, MIL-STD-188-220 standard defines three precedences for the traffics: Urgent, Priority, and Routine. Urgent precedence is the highest precedence and Routine precedence has the lowest precedence. P-NAD, DAP-NAD, and DAV-NAD provide methods for priority-based transmissions.

While conventional tactical communication devices process mainly voice traffic and short messages over the narrowband systems, recent tactical communication devices have been designed to deal with various types of traffics including voice, short/long messages, pictures and video over wideband systems as shown in [2] because sufficient information on the battle field leads to the better command and control. As a part of the movements on the tactical communications, MIL-STD-188-220 standard also needs to be modified to deal with various traffic types over wideband systems. Even though there are some studies as shown in [3–11], all studies are based on the narrowband systems. Studies in [3–9] have focused on the old version of MIL-STD-188-220 standard. Recently, MIL-STD-188-220 standard has been evaluated over wideband channel environments in [12–14]. In [12], R-NAD and DAP-NAD are evaluated comparing with IEEE802.11-based Wireless Local Area Networks (WLAN) over wideband network scenarios and the study concludes using the standard over wideband systems is feasible. In [13], the utility of bump-slot, that is one of time slot used in DAP-NAD, is evaluated over wideband systems and it is concluded that the bump-slot is not useful over wideband systems. In [14], a method to provide enhancements on voice packet transmission over MIL-STD-188-220-based standard system is proposed. The method utilizes the bump slots to give higher opportunity to voice traffics.

In this paper, a method to improve the performances of the lower precedence traffics using DAP-NAD is proposed. While as shown in [12], DAP-NAD gives the best performances on Urgent traffics. Routine traffic may not have any transmission opportunity. The proposed method upgrades the precedences of the frames if the lower frame stays more than certain time period, so that it give an opportunity of transmission to even the lower precedence frames.

In this paper, Sect. 2 introduces the specifications on DAP-NAD specified in MIL-STD-188-220 standard. In Sect. 3, the proposed method is introduced and the extensive simulation results are provided in Sect. 4. Finally, the conclusions and future works are made in Sect. 5.

2 DAP-NAD

DAP-NAD is Time-Division Multiple Access (TDMA)-based medium access protocol and the transmissions sequence among the participating nodes is pre-scheduled based on the assigned unique node number and the sequence is repeated whenever the last ordered node's turn is completed. Each node transmits its own frames during its own time slot whose length is variable depending on the length of frame if the Network Precedence (NP) has to be same as Message Precedence (MP) of a node's frame. NP indicates the precedence that is allowed to be transmitted during one transmission sequence. NP might be changed after one transmission sequence is completed. Each transmission sequence is set to one of three precedences. The first NP is set to Urgent, so that during a whole one-sequence, only a node with Urgent pending frame is allowed to transmit. If a node does not have Urgent pending frame when the NP is Urgent, it give up is turn and the next node have opportunity to transmit. If no node transmits during one sequence, then the NP is downgraded to Priority which means nodes having higher precedence frame than Priority are allowed to transmit. If any node transmits a frame during this sequence with Priority NP, the NP is upgraded to Urgent. Otherwise, the NP is downgraded to Routine which means nodes having higher precedence frame than Routine are allowed to transmit. That is, only a node having a frame whose MP is same as NP can transmit. If a node transmits its frame no matter what NP is, the next sequence automatically upgraded to Urgent and start over again. Each node has to keep tracking the NP by itself. For synchronizing NP over all participating nodes, the header of frame contains the MP, so that nodes enable to adjust their NPs by overhearing the header.

3 Proposed Method

As mentioned in Sect. 2, DAP-NAD provides the highest transmission opportunity on the Urgent traffics. While it gives the high reliability on the Urgent traffic, the performances of the lower priority traffics are degraded. In certain case, the transmission opportunities for the lower priority traffics are totally prohibited even though there are some opportunities for the lower priority traffics. For example, If a node has periodic Urgent traffic and others have lower traffics, the node with Urgent traffic keep transmitting and the time slots for other nodes is passed without any transmission. This is because only one transmission make NP upgrade to Urgent. In this scenario, some transmissions of the lower MP frames may not degrade the performances of Urgent MP traffic. To resolve this problem, we proposes a method to intentionally upgrade the higher MP of the lower MP to provides transmission opportunities even for the nodes with the lower MP frames. The process of the proposed method is shown in Fig. 1. The proposed method is reclusively performed for the frames in a queue. When a frame is arrived in the queue, the queuing time of the frame, T_q, is recorded. In every a unit time, the T_p is evaluated with T_{MAX} and Precedence-Critical-Time (PCT). T_{MAX} is the life time of the frame and PCT is time threshold to upgrade the frame's precedence. In every T_u, if T_p is larger than T_{MAX}, the frame is dropped. Otherwise, T_p of the frame is compared with PCT. If the T_q is larger than PCT and $Flag_{up}$ is not set, then the precedence of the frame is upgraded to next higher level. $Flag_{up}$ indicates

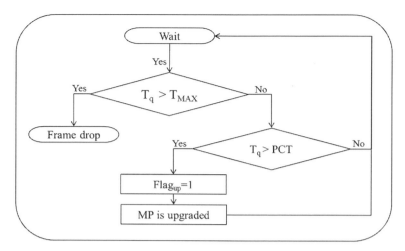

Fig. 1 Flow chart of the proposed method

if the frame's precedence has been upgraded. Upgrading the precedence is done by changing the values in the *Data Link Precedence* subfield in the transmission header. '1 0' in *T-Bits* indicates the network uses DAP-NAD method and *First Station Number* indicate the node number that is owner of the first time slot in the new sequence. Based on the standard, the values for Urgent, Priority, and Routine are 00, 10, and 11, respectively. Therefore, 10 and 11 in the current value in the Data Link Precedence subfield is changed to 10 and 11, respectively. In order to prevent from over-crowded in Urgent traffics, the precedence level is upgraded once. That is, the precedence is upgraded one level higher, not more than two level. As the standard defines, when the next time slot for the node is arrived, the current Network Precedence (NP) is compared with the Frame Precedence (FP) of the pending frame in queue and if both precedences are equal, the pending frame is transmitted.

4 Performance Evaluations

The proposed method is simulated using the simulator used in [13, 14]. Because we are targeting to the wideband systems and the latest commercial tactical communication system in [2] adopts Orthogonal Frequency Division Multiplexing (OFDM)-based system, IEEE802.11a-based OFDM system [15] is used as a physical layer for the evaluation of the proposed method. Because based on [2], the channel bandwidth is 10 MHz, the system parameters of IEEE802.11a is redefined as Table 1 in order to corresponding to 10 MHz channel bandwidth. The traffic type for this simulation is Type 3 connectionless and coupled acknowledgement operation mode defined in [1] which is similar to unicast transmission in

Table 1 Simulation parameters

Parameter	Value
Data rate	3 Mbps
Preamble	32 us
Physical layer header	8 us
MAC header	272 bits
Default slot time	13 us
ACK packet	88 us
SIFS	32 us

IEEE802.11-based system. The channel error is not considered. In each simulation case, the number of nodes with Urgent traffic is fixed to 6 and the nodes with Priority and Routine traffics are randomly chosen between 1 and 10. The frame sizes of all traffics are fixed to 256-byte and the packet inter-arrival times of Urgent, Priority, Routine traffics are set to 0.005, 0.01, 0.05 s, respectively. As the part of evaluations, the proposed method is applied to only Routine traffics. That is, only Routine traffic is upgraded to Priority if the waiting time in queue is larger than PCT. This is to evaluate if there are some opportunities for Routine traffic even though only upgrading Routine traffic to Priority traffic.

Figure 2 shows the transmission rates as a function of PCT and the precedences of traffics. The transmission rate is defined as the ratio of successfully transmitted

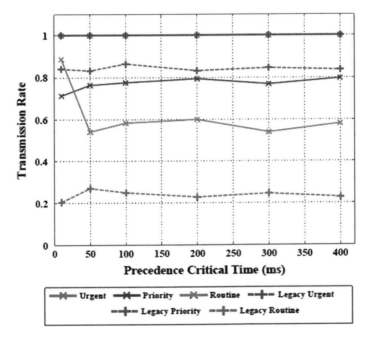

Fig. 2 Average transmission rates as a function of PCT

frames to totally generated frames. In the figure, Legacy means conventional MIL-STD-188-220-standard-based system. As shown in Fig. 3, the performances of Routine traffic using the proposed method is improved up to 3 times comparing to that using conventional DAP-NAD. On the other hand, the performances of Priority traffic is degraded up to 9 %.

5 Conclusion and Future Works

In this paper, a method to improve the performances of the lowest precedence traffic is proposed while minimizing the impacts on the performances of the higher precedence traffic. The method upgrades MPs of Priority and Routine frames if the waiting–time of the frames in queue is larger than a certain time, and as a consequence, the lower MP frames have the more opportunities for their transmissions. Through the simulation studies, it is proved that the proposed method improves the performances of Routine frames up to three times while the performance of Priority frame is degraded 9 % and the performance of Urgent frames is not degraded. However, the scenario might be different with different environments. Therefore, as the future works, the proposed method will be evaluated with the actual tactical traffic model and various network environments such as erroneous channel.

Acknowledgments This research was supported in part by the National Research Foundation of Korea (NRF) grant funded by the Korea government (MEST) (2012-0003609) and in part by the MKE (The Ministry of Knowledge Economy), Korea, under the ITRC (Information Technology Research Center) support program supervised by the NIPA (National IT Industry Promotion Agency) (NIPA-2012-(H0301-12-2003)).

References

1. MIL-STD-188-220D with Change1 (2008) Digital message transfer device subsystem. 23 June 2008
2. Abacus Programming Corporation. http://www.abacuscorp.com/se.htm
3. Thuente DJ, Borchelt TE (1997) Simulation studies of MAC algorithms for combat net radio. In: 16th military communications conference, pp. 193–199. IEEE Press, New York
4. Thuente DJ, Borchelt TE (1998) Simulation model and studies of MIL-STD-188-220A. In: 17th military communications conference, pp. 198–204. IEEE Press, New York
5. Thuente DJ, Borchelt TE (2000) Efficient data and voice media access control algorithm for MIL-STD-188-220B. In: 19th military communications conference, pp. 115–121. IEEE Press, New York
6. Thuente DJ, Whiteman JK (2001) Modified CSMA/implicit token passing algorithm for MIL-STD-188-220B. In: 20th military communication conference, pp. 838–844. IEEE Press, New York
7. Thuente DJ (2002) Improving quality of service for MIL-STD-188-220C. In: 21st military communication conference, pp. 1194–1200. IEEE Press, New York

8. Yang J, Liu Y (2006) An improved implicit token passing algorithm for DAP-NAD in MIL-STD-188-220C. In: 2nd international conference on wireless communications, pp. 1–4
9. Liu Y, An J, Liu H (2008) The modified DAP-NAD-CJ algorithm for multicast applications. In: 4th international conference on wireless communications, networking and mobile computing, pp. 1–4
10. You J, Baek I, Kang H, Choi J (2010) Effective traffic control for military tactical wireless mobile ad-hoc network. In: 6th IEEE international conference on wireless and mobile computing, networking and communications, pp. 1–8. IEEE Press, New York
11. Kim J, Kim D, Lim J, Choi J, Kim H (2011) Effective packet transmission scheme for real-time situational awareness based on MIL-STD-188-220 tactical ad-hoc networks. In: Military communication conference, pp. 956–960. IEEE Press, New York
12. Kim B-S (2010) Comparative study of MIL-STD-188-220D standard over IEEE802.11 standard. SK Telecommun Review 20:256–264
13. Han S, Kim B-S (2011) Evaluations on effectiveness of bump-slot over DAP-NAD-based tactical wideband wireless networks. In: Altma E, Shi W NPC 2011. LNCS, vol. 6985, pp. 341–350. Springer, Heidelberg
14. Han S, Kim B-S (2012) Efficient voice transmissions for MIL-STD-188-220-based wideband tactical systems. IEICE Trans Commun. E95-B, 264-2967
15. Part 11 (2007) Wireless LAN medium access control (MAC) and physical layer (PHY) specifications, IEEE Std. 802.11, 12 June 2007

Blind Beamforming Using the MCMA and SAG-MCMA Algorithm with MUSIC Algorithm

Yongguk Kim and Heung-Gyoon Ryu

Abstract Satellite communication system does not use training sequence because the satellite communication channel is similar to the additive white gaussian noise (AWGN). But, in the mobile satellite communication environment, inter-symbol-interference (ISI) seriously occurs due to movement of receiver. We must use the blind equalization for remove the ISI in mobile satellite communication. Blind equalization techniques such as MCMA and SAG-MCMA is suitable for channel equalization in the mobile satellite environment. But equalization performance of blind equalizer were not as satisfactory as expected. In this paper, we propose a blind equalization technique based on coordinate change and beamforming method in order to improve the BER performance of receiver in mobile satellite communication. The simulation results show that the proposed scheme with coordinate change need to less SNR about 1 dB to satisfy BER performance (10-5).

Keywords Blind equalizer · Coordinate change · MUSIC algorithm · MCMA · SAG-MCMA

1 Introduction

In digital communication system, it's important to transmit more information data. According to the given power, the amount of information is limited based on information theory. Channel noise and inter symbol interference (ISI) are main factors to limit amount of information. Conventional adaptive equalizations are

---

Y. Kim · H.-G. Ryu (✉)
Department of Electronic Engineering, Chungbuk National University, Cheongju, Korea
e-mail: ecomm@cbu.ac.kr

Y. Kim
e-mail: coolfeelyg@naver.com

J. J. (Jong Hyuk) Park et al. (eds.), *Multimedia and Ubiquitous Engineering*,
Lecture Notes in Electrical Engineering 240, DOI: 10.1007/978-94-007-6738-6_102,
© Springer Science+Business Media Dordrecht(Outside the USA) 2013

using the training sequence to estimate the channel characteristic. Through the channel characteristic, we estimate the characteristic coefficient of reverse channel. After that, transmit signals are passed, have a characteristic coefficient of reverse channel, the filter. Using this method, we reduce the ISI and random phase rotation influence. Therefore, the communication system can improve overall performance. Training sequence is promised signal between transmitter and receiver. In other words, training sequence is additional information. So, Bandwidth efficiency is decreased.

In the blind equalization, using the cumulative rate of received signal and using the modulus constant modulus algorithm (CMA) is represented in a way [1, 2]. Inter symbol interference (ISI) and the phase rotation can be restored at the cumulative rate method. However, it requires high-level operation. So, high speed transmission may have a problem as equalization. In the CMA, ISI and phase rotation compensate is impossible at a time. However, this method has the advantage of reduces the amount of computation. CMA equalization method for updating the equalizer coefficients, using the LMS adaptive filtering algorithm the actual implementation is very simple. LMS method the Eigen value distribution of the correlation matrix of the input signal is large; the rate of convergence is slow. CMA blind equalization algorithm is one of the most used techniques [3, 4].

MCMA(modified CMA) can compensate phase rotation problem. The MCMA accomplishes the correction of phase error and frequency offset with the modified cost functions. But, the MCMA does not judge whether the adjustment of tap coefficients is correct or not. Picchi and Prati was define the SAG (stop and go) algorithm. SAG algorithm is comparing the received signal decision error with the Sato algorithm error. If two error sign is equal, tap coefficient is updated (go). Another case, tap coefficient is not updated (stop) [5].

In this paper, MCMA blind equalization system using the beamforming, MUSIC algorithm and coordinate change method. Receive SNR is increased through the beamforming and the MUSIC algorithm. The propose method improve BER performance because through the coordinate change reduces the modulus and error function.

2 Coordinate Change for MCMA and SAG-MCMA

2.1 Proposed MCMA Algorithm

Figure 1 shows block diagram of the MCMA with coordinate change. We explain the coordinate change scheme in 16APSK for improving the performance of equalizer. The ratio of the inner circle and outer circle is expressed as follows.

$$\gamma = \frac{R_2}{R_1} \tag{1}$$

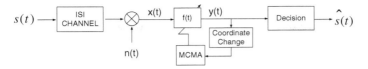

Fig. 1 Block diagram of the proposed scheme in MCMA

Table 1 Coordinate change for 16-APSK

Original coordinates	New coordinates	Original coordinates	New coordinates
1 + i	1 + i	1 − i	1 − i
2.0153 + 2.0153i	1 + i	2.0153 − 2.0153i	1 − i
2.7529 + 0.7376i	1 − i	2.7529 − 0.7376i	1 + i
0.7376 + 2.7529i	−1 + i	0.7376 − 2.7529i	−1 − i
−1 + i	−1 + i	−1 − i	−1 − i
−2.0153 + 2.0153i	−1 + i	−2.0153 − 2.0153i	−1 − i
−2.7529 + 0.7376i	−i − i	−2.7529 − 0.7376i	−1 + i
−0.7376 + 2.7529i	1 + i	−0.7376 − 2.7529i	1 − i

γ of 16-APSK signal has a value of 2.85, each symbol has a value of $\{\pm 1 \pm i, \pm 2.0153 \pm 2.0153i, \pm 2.7529 \pm 0.7376i, \pm 0.7376 \pm 2.7529i\}$. Coordinate Change can be seen in Table 1.

We can get the new coordinate value using Table 1.

Coordinate change of R'_2 is defined as follows.

$$R'_{2,R} = \frac{E[|[a_R'(t)|^4]}{E[|[a_R'(t)|^2]}, R'_{2,I} = \frac{E[|[a_I'(t)|^4]}{E[|[a_I'(t)|^2]} \quad (2)$$

In the case of coordinate change, constant modulus values are calculates by using new coordinate values in shown Table 1.

The tap coefficients are updated through the following equation.

$$f(t+1) = f(t) - mu(fr_R(t)e'_r(t) + jfi(t)e'_i(t))x(t) \quad (3)$$

where m is the step size.

2.2 Proposed SAG MCMA Algorithm

Figure 2 shows block diagram of the SAG-MCMA with coordinate change.

Coordinate change of R'^2 of SAG-MCMA defined as follows.

$$R_{2,R}' = \frac{E[|[s_R'(t)|^4]}{E[|[s_R'(t)|^2]}, R_{2,I}' = \frac{E[|[s_I'(t)|^4]}{E[|[s_I'(t)|^2]} \quad (4)$$

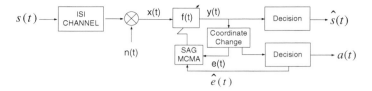

Fig. 2 Block diagram of the propose scheme in SAG-MCMA

The output signal of the equalizer is

$$y(t) = f^T(t)x(t) \tag{5}$$

The proposed error function is

$$\widehat{e_R}(t) = y_R'(t)(y_R'(t)^2 - a_R(t)^2) \\ \widehat{e_I}(t) = y_I'(t)(y_I'(t)^2 - a_I(t)^2) \tag{6}$$

Cost function of the proposed CMA is as follows.

$$J'_{CMA}(f) = E[\{e'(t)\}^2] \tag{7}$$

The tap coefficients are updated through the following equation.

$$f(t+1) = f(t) - mu(fr_R(t)e'_r(t) + jfi(t)e'_i(t))x(t) \tag{8}$$

3 Blind Beamforming System

This Fig. 3 shows the block diagram of blind beamforming system. Each element has weighting factor for beamforming. Transmit signal is passed ISI channel. After then, we detect direction of Passed signal using MUSIC algorithm. Through the detected direction, receiver elements have weighting factor for receive beamforming. Received signals are coordinate changed. Finally, signals are equalized using the SAG MCMA and MCMA algorithm.

4 Simulation Results

In this paper, we like to compare the proposed method with the conventional MCMA. We can find a better BER performance by blind beamforming system. We consider Table 2 for analyzing the improvement of BER. In the simulation, the ISI channel was used. SNR is 30 dB. Equalizer has 21 tabs.

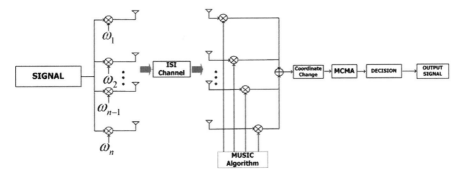

Fig. 3 Block diagram of propose system

Table 2 Simulation parameters

Modulation	16-APSK
Channel	ISI Channel [0.8, 0.3, 0, 0.2 + j0.2, 0, 0]

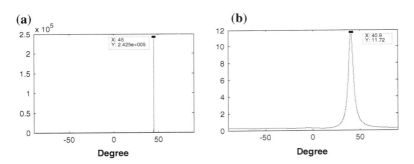

Fig. 4 MUSIC spectrum(desired angle $= 45°$). (**a**) AWGN channel (**b**) ISI channel

Figure 4 shows the results the result of tracking the direction of the signal in the AWGN and ISI channels. In the ISI channel, detection result has lower accuracy than in the AWGN channel. But, MUSIC algorithm can accurately detect arrival direction of receiver.

Figure 5 is shows the BER performance of the SAG MCMA and MCMA. Both MCMA and SAG MCMA, we confirm that the BER performance is improved approximate 1 dB when using coordinate change scheme. We can see that the BER performance change in MCMA is greater than in the SAG MCMA.

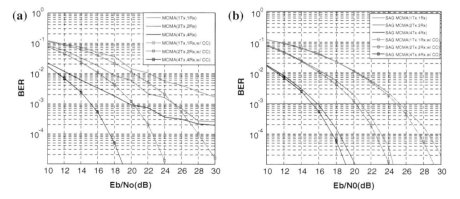

Fig. 5 BER performance of the SAG MCMA and MCMA using coordinate change. (**a**) BER performance of MCMA (**b**) BER performance of SAG MCMA

5 Conclusion

In this paper, we propose the stop-and-go MCMA algorithm and MCMA based on the coordinate change using MUSIC algorithm for beamforming. The proposed scheme has better BER performance than that of general system in 16-APSK in shown the Fig. 5. To improve receive performance, we using the beamforming and the MUSIC algorithm. The case of the MUSIC algorithm in AWGN channel environment, detecting very accurately the direction of the signal. However, In ISI channel environment, less accurate than in the AWGN channel environment in shown Fig. 4.

Acknowledgments This research was supported by Basic Science Research Program through the National Research Foundation of Korea(NRF) funded by the Ministry of Education, Science and Technology(No. 2012017339).

References

1. Rao W, Yuan K-M, Guo Y-C, Yang C (2008) A simple constant modulus algorithm for blind equalization suitable for 16-QAM signal. In: The 9th international conference on signal processing, vol. 2, pp 1963–1966
2. Godard D (1980) Self-recovering equalization and carrier tracking in two dimensional data communication systems. In: IEEE Trans Commun COM-28: 1867–1875
3. Johnson CR Jr, Schniter P, Endres JT et al (1998) Blind equalization using the constant modulus criterion: a review. Proc IEEE 86(10):1927–1949
4. Rao W, Guo, Y-C (2006) New constant modulus blind equalization algorithm based on variable segment error function. J Syst Simul 19(12): 2686–2689
5. Suzuki Y, Hashimoto A, Kojima M, Sujikai H, Tanaka S, Kimura T, Shogen K (2009) A study of adaptive equalizer for APSK in the advanced satellite broadcasting system. In: Global telecommunications conference. GLOBECOM 2009. IEEE, pp 1–6

Performance Evaluation of EPON-Based Communication Network Architectures for Large-Scale Offshore Wind Power Farms

Mohamed A. Ahmed, Won-Hyuk Yang and Young-Chon Kim

Abstract In order to meet the growing demand of large-scale wind power farms (WPF), integration of high reliability, high speed, cost effectiveness and secure communication networks are needed. This paper proposes the Ethernet passive optical network (EPON) as one of promising candidates for next generation WPF. Critical communication network characteristics such as reliability, mean downtime, optical power budget, path loss and network cost are evaluated and compared with conventional switched-based architectures. The results show that our proposed EPON-based network architectures are superior to conventional switched-based architectures.

Keywords Ethernet network · SCADA · Communication network · EPON · Reliability · Power budget · Path loss · Cost · Wind power farm

1 Introduction

There is a rapid development in wind farm industry around the world. Many large-scale projects are scheduled for construction in the coming years with a huge number of wind turbines. The communication networks are considered a fundamental infrastructure that enable transmission of measured information and control signals

---

M. A. Ahmed · W.-H. Yang · Y.-C. Kim
Department of Computer Engineering, Chonbuk National University, Jeonju, Korea
e-mail: mohamed@jbnu.ac.kr

W.-H. Yang
e-mail: whyang@jbnu.ac.kr

Y.-C. Kim (✉)
Smart Grid Research Center, Jeonju 561-756, Korea
e-mail: yckim@jbnu.ac.kr

J. J. (Jong Hyuk) Park et al. (eds.), *Multimedia and Ubiquitous Engineering*,
Lecture Notes in Electrical Engineering 240, DOI: 10.1007/978-94-007-6738-6_103,
© Springer Science+Business Media Dordrecht(Outside the USA) 2013

between the wind turbines and control center. Traditional communication infrastructures for monitoring the wind power farms (WPF) are based on Ethernet communication, and consist of an independent set of network switches and communication links in every wind turbine [1]. In case of network failure, serious problems may interrupt the operation, generation and control of the whole wind farm. To consider a new infrastructure for WPF communication network, critical characteristics need to be evaluated such as reliability, scalability and network cost [2].

EPON technology provides high performance data communication with a high bandwidth, flexibility, high reliability, low maintenance costs and compatibility with existing Ethernet networks. EPON could be configured with different topologies includes star, bus, tree and ring. It consists of optical line terminal (OLT), passive optical splitter (POS) and optical network unit (ONU) [3]. This paper proposes the Ethernet passive optical network (EPON) as one of promising candidates for next generation WPF.

2 Related Work

2.1 Offshore Wind Farm Layout

The offshore wind power farm consists of wind turbines, local wind turbine grid, collecting point and transmission system. The electrical layout can be designed with different configurations depending on the wind farm size and redundancy, such as radial, ring and star. Figure 1a shows an offshore wind farm consists of five radials, each of them with five turbines, connected to an offshore platform via a circuit breaker and switch. Cables with different cross section areas are used to connect the turbines. The voltage is stepped up using an offshore transformer and the transmission system transmits the total output power from the 25 turbines to shore at the point of common coupling (PCC) [4].

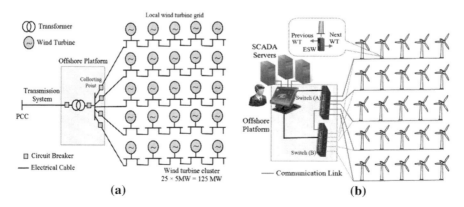

Fig. 1 a Layout of wind power farm. b Conventional communication network of wind farm

2.2 Conventional Wind Farm Communication Network

The communication network for wind power farm defines the SCADA communication between the control center and wind turbines. This configuration usually follows the electrical topologies because the optical fibers are integrated with the submarine medium voltage cables. Furthermore, a wireless network or radio link can be incorporated into the design to increase the reliability [5].

3 Proposed EPON-Based Architectures

This section describes the proposed EPON-based communication network architectures for offshore wind power farms based on electrical topologies of Ref. [4]. There are four different cases designed with different number of feeders. The spacing between individual turbines is equivalent to 1.12 km within a row and between rows.

The proposed network model consists of an ONU deployed on the wind turbine side. All ONUs from different wind turbines are connected to a central OLT, placed in the control center. We considered each wind turbine have two devices; one ONU device and one POS (1 × 2). The ONU collects data from different devices (turbine controller, video cameras, internet telephones, etc.). The POS has two output ports as shown in Fig. 2, one port is connected to the ONU unit and the other is connected to next wind turbine. At offshore platform side, the OLT unit is installed, and connected to WTs-ONUs using feeder fiber (FF), distributed fiber (DF) and POS.

Architecture of case (1) begins with an OLT unit located at offshore platform, connected with five cables with different length of feeder fiber, (FF1 → FF5 to WT-A01 → WT-E01). Cascade splitters are used to reduce the amount of deployed fiber in the network. For all wind turbines (from WT-A01 → WT-E05), each WT has only one POS (1 × 2); one port is connected using DF to the next WT, while the other port is connected to the WT ONU unit. All DFs in all architectures are of 1.12 km length (the distance between WTs). Note that, POS (1 × 2) is used at the end of the feeder in order to help extending the network, in case of installing new WTs (one port is used, while the other is left free).

Architectures of case (2) and case (3) differ with configuration (A) with respect to number of feeder fibers, based on the electric system layout. All POS are (1 × 2), the same like case (1) with some interconnection between turbines such as WT-B03 to WT-C03 and WT-D01 to WT-E01. In Architecture of case (4), the WT-B01 and WT-D01 have two identical POS (primary and secondary); where POS is (1 × 2), with different insertion loss. Table 1 shows the details of network elements for the proposed EPON-based network architectures.

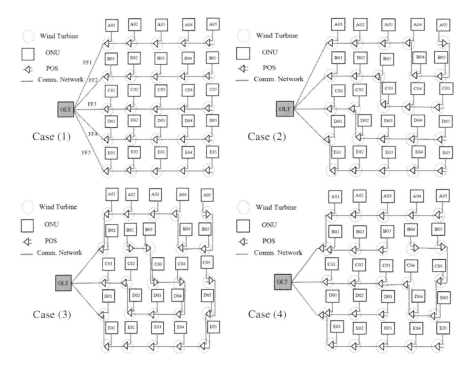

Fig. 2 Proposed wind farm communication network based on EPON

Table 1 Network elements of EPON-based large-scale WPF (unprotected)

Architecture	# OLT	FF (Km)	# POS	FF (Km)	# ONU
Case (1)	1	9.28	25 (1 × 2)	22.4	25
Case (2)	1	6.78	25 (1 × 2)	23.52	25
Case (3)	1	4.28	25 (1 × 2)	24.64	25
Case (4)	1	4.28	27 (1 × 2)	24.64	25

4 Performance Evaluation

4.1 Reliability

We studied the connection availability between the OLT located at offshore platform and each wind turbine. We consider TDM-PON architecture defined by ITU-T, with unprotected architecture in [6]. Using the failure rate of a communication network component, unavailability of a component (U_x) is derived from its failure rate in FIT (1 FIT = 1 failure/10E09 h) and the mean repair time (MTTR) in hours. We considered that MTTR is 24 h, as all network elements located offshore and the only way

Performance Evaluation of EPON-Based Communication Network Architectures 845

Table 2 Component reliability of EPON-based WPF

Components	Failure rate (FIT)	MTTR (h)	Unavailability
OLT	256	24	6.144E-06
ONU	256	24	6.144E-06
Splitter (1 × 2)	50	24	1.20E-06
ESW	1250	24	3.00E-05
Fibe (/Km)	570/Km	24	1.368E-05

to access is by boat or helicopter. The expression for the connection unavailability (U_{EPON}) for EPON-based architecture is given as follow:

$$U_{EPON} = U_{OLT} + U_{FF} + KU_{POS} + U_{DF} + U_{ONU} \qquad (1)$$

where, U_{OLT}, U_{FF}, U_{POS}, U_{DF} and U_{ONU} are the unavailabilities of OLT, FF, POS, DF and ONU, respectively. K is the number of passive optical splitters. The calculations of unavailability for different network elements are shown in Table 2.

Figure 3a shows the expected downtime of EPON-based and switched-based architectures. As we can see, switched-based architectures have the highest downtime. The lowest MDT is 309 min/year for EPON-based architectures compared with 618 min/year for switched-based architectures (Table 3).

4.2 Network Cost

The communication network cost can be divided into active devices cost (OLT, ONU and Ethernet switch) and passive components cost (POS and fiber). Table 4 detailed the components cost used in our network model [6]. The total network cost for EPON-based and switched-based architectures can be represented as follow:

$$Cost_{EPON} = C_{OLT} + C_{FF} + C_{POS} + C_{DF} + N_{ONU} \cdot C_{ONU} \qquad (2)$$

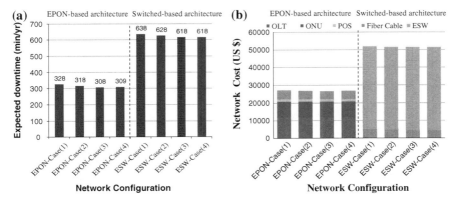

Fig. 3 a Mean down time for WPF architectures. b Network cost for WPF architectures

Table 3 Component cost (US $)

OLT	ONU	Ethernet switch	Splitter (1×2)	Splitter (1×16)	Fiber (/Km)
12100	350	1800	50	800	160

Table 4 Component insertion loss

Component	Fiber	Connector	Splitter	
Attenuation	0.4 dB/Km	0.2 dB	1×2 (5 %:95 %) 0.4 dB	1×2 (50 %:50 %) 0.4 dB

$$\text{Cost}_{\text{Ethernet}} = C_{\text{ESW}} + C_{\text{FF}} + C_{\text{DF}} \tag{3}$$

where $C_{\text{OLT}}, C_{\text{POS}}, C_{\text{ESW}}$ and C_{ONU} represent the component cost of OLT, POS, Ethernet switch and ONU, respectively. C_{FF} and C_{DF} represent the costs of optical fiber cable of feeder fiber and distributed fiber, respectively. N_{ONU} represents the number of WTs-ONUs.

Figure 3b shows the network cost for EPON-based and switched-based architectures in US$. EPON-based architectures have the lowest costs by about 52 % which represent the most economic deployment solution compared with switched-based architectures which have the dominant cost of Ethernet switches.

4.3 Power Budget

The optical power budget is analyzed to ensure that received signal power is enough to maintain acceptable performance. The power budget for EPON specified in IEEE 802.3ah standard is 26.0 dB in case of 1000Base-PX20 for both upstream and downstream traffic. The optical budget [dB] is defined as the difference between the minimum transmitter launch power (Ptx, dBm) at the input of the optical link, and the minimum sensitivity of the receiver (Prx, dBm) at the output of optical links [7].

$$\text{Power Budget} = P_{\text{tx}} - P_{\text{rx}} \tag{4}$$

There are many sources of attenuation including splitters (Loss_{POS}), connections ($\text{Loss}_{\text{conn}}$) and the fiber cable itself ($\text{Loss}_{\text{fiber}}$). The optical path loss is equal to:

$$\text{Loss}_{\text{epon}} = \sum \text{Loss}_{\text{POS}} + \sum \text{Loss}_{\text{conn}} + \sum \text{Loss}_{\text{fiber}} \tag{5}$$

The total insertion loss must be less than the value of power budget. Table 3 shows the network elements insertion loss. A safety margin should be considered in total optical power loss calculation due to factors of aging the Tx/Rx elements and the effect of temperature. The network component insertion loss is shown in Table 3.

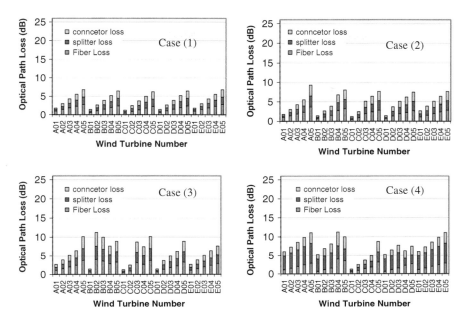

Fig. 4 Total optical path loss for WPF architectures

Figure 4 shows the total optical path loss calculation for four different configurations. The highest optical path loss value represents the farthest turbine, while the lowest value represents the nearest turbine. For example, the highest optical path loss value for WT-B02 in case (3) is about 11.23 dB, while the lowest value for WT-C01 is about 1.25 dB. Considering the IEEE 802.3 std. requirements, all EPON-based architectures satisfy the standard requirements.

5 Conclusion

In this paper, we proposed EPON-based network architectures for large-scale wind power farm. We evaluated the network performance in view of reliability, path loss, optical power budget and network cost. The results show that EPON-based architectures have superior performance to conventional switched-based networks. This work proves the applicability and robustness of EPON-based communication network architectures for next generation WPF.

Acknowledgments This work was supported by the National Research Foundation of Korea (NRF) funded by the Korea government (MEST) (2012-0009152).

References

1. Carolsfeld R (2011) Practical experience from design and implementation of IEC 61850 based communications network in large offshore wind installation. In: Technical Meeting, CIGRE-AORC, Thailand
2. Yu R, Zhang P, Xiao W, Choudhury P (2011) Communication systems for grid integration of renewable energy resources. IEEE Netw 25(5):22–29
3. Kramer G (2005) Ethernet passive optical networks. McGraw-Hill, New Jersey
4. Shin J-S, Cha S-T, Wu Q, Kim J-O (2012) Reliability evaluation considering structures of a large scale wind farm. In: European power electronics (EPE) wind energy and T&D chapter seminar, Aalborg
5. Ahmed MA, Kim Y-C (2012) Network modeling and simulation of wind power farm with switched gigabit ethernet. In: 12th international symposium on communications and information technologies, Australia
6. Wosinska L, Chen J, Larsen CP (2009) Fiber access networks: reliability analysis and Swedish broadband market. IEICE Trans Commun E92-B(10):3006–3014
7. Huang H, Zhang H (2011) Application and analysis of long-distance EPON in transmission lines monitoring system. Adv Mater Res 317–319:1583–1589

A User-Data Division Multiple Access Scheme

P. Niroopan, K. Bandara and Yeon-ho Chung

Abstract The conventional Interleave Division Multiple Access (IDMA) employs interleavers to separate users, while the conventional Code Division Multiple Access (CDMA) uses user specific spreading sequences for user separation. In this paper, we propose a User-Data Division Multiple Access (UDMA) scheme that employs user data as the spreading sequence for user separation with chip-by-chip iterative multiuser detection strategy. As such, this spreading sequence is not only as random as user data and independent of current symbols, but also dynamically changes from one symbol to another according to the user data. Therefore, this spreading sequence makes unwanted detection of the data by unintended receivers practically impossible. Also, in UDMA, identical interleavers are used and thus do not require to store all interleaving patterns. The simulation results show that the proposed scheme is superior to the bit error rate (BER) performance of the system in flat fading channel.

Keywords IDMA · Interleaver · Multiple access · Spreading

P. Niroopan · K. Bandara · Y. Chung (✉)
Department of Information and Communications Engineering, Pukyong National University, Busan, Korea
e-mail: yhchung@pknu.ac.kr

P. Niroopan
e-mail: niroopan86@gmail.com

K. Bandara
e-mail: kassae6@gmail.com

J. J. (Jong Hyuk) Park et al. (eds.), *Multimedia and Ubiquitous Engineering*,
Lecture Notes in Electrical Engineering 240, DOI: 10.1007/978-94-007-6738-6_104,
© Springer Science+Business Media Dordrecht(Outside the USA) 2013

1 Introduction

Wireless technology has been gaining rapid popularity for some years. Multiple access schemes are the major concern for researchers that can support high data rate and high reliability for next generation wireless communications systems.

Interleave Division Multiple Access (IDMA) is a new multiple access technique [1], where interleaver that gives the name to IDMA has an important role in the system architecture. Every single user has its own interleaver which differs from others. IDMA is a special case of the CDMA system [2]. A user specific spreading sequence is used in the CDMA system where the spreading sequence must be orthogonal and also needs to maintain synchronization. It is shown that the IDMA system has the edge over the CDMA system [3].

In this paper, we propose User-Data Division Multiple Access (UDMA) systems that use user data for user separation instead of interleavers in the IDMA system. As the randomly generated user data are used as a spreading sequence, it will be more secure and less probable to be intercepted [4, 5]. In fact, this dynamically changing spreading sequence makes unwanted detection of the data by unintended receivers practically impossible. Also, we remove the repetition code that serves as a spreading sequence in the IDMA system. This repetition code is not only unsophisticated but also bandwidth inefficient. Thus, the UDMA system provides more secure and efficient communications than CDMA and IDMA systems. For the error checking in the despreading sequence at the receiver of the UDMA system, we use genetic search algorithm and Markov chain analysis [6].

2 IDMA System

The transmitter and receiver structures of an IDMA system with K-simultaneous users are shown in Fig. 1. At the transmitter, the block size of N-length information bits from each user-k is denoted as $d_k = [d_k(0),\ldots\ldots, d_k(N-1)]^{\mathrm{T}}$, $k = 1, 2, \ldots, K$. The data sequence is encoded using a convolutional code into $b_k = [b_k(0),\ldots\ldots, b_k(N_C - 1)]^{\mathrm{T}}$. That is, the code rate is defined as $R_1 = N/N_C$. Then each bit of b_k is again encoded using a low rate code such as a spread encoder with a rate of $R_2 = 1/S_k$, where S_k is a spreading factor. Thus, the overall code rate is $R_1 R_2$, which produces a chip signal. The second encoder output is fed into the user specific interleaver $(\pi_1, \pi_2,\ldots\ldots, \pi_K)$ for user separation, which generates $x_k(j)$, $j = 1, 2, \ldots\ldots, J$, where J is the user frame length. The resultant signal is then transmitted through the multiple access channel. In the receiver, the received signal is given by

$$r(j) = \sum_{k=1}^{K} h_k x_k(j) + n(j), \quad j = 1, 2, \ldots, J \tag{1}$$

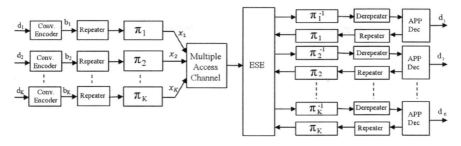

Fig. 1 The transmitter and receiver structure of the IDMA system

where h_k is the channel gain for user-k, x_k is the corresponding transmitted signal and n is the additive white Gaussian noise (AWGN) process with zero mean and variance, $\sigma^2 = N_0/2$. It is assumed that the channel coefficients $\{h_k\}$ are known a priori at the receiver.

This received signal is passed to a multi-user detection (MUD) receiver that consists of an elementary signal estimator (ESE) and K a posteriori probability (APP) decoders (DECs), one for each user. The ESE performs chip-by-chip detection to roughly remove the interference among users. The outputs of the ESE and DECs are extrinsic log-likelihood ratios (LLRs) about $\{x_k\}$ defined as

$$e(x_k(j)) = \log\left[\frac{p(y|x_k(j) = +1)}{p(y|x_k(j) = -1)}\right] \quad (2)$$

Those LLRs are further distinguished by $e_{ESE}(x_k(j))$ and $e_{DEC}(x_k(j))$, depending on whether they are generated by the ESE or DECs. For the ESE section, y in (2) denotes the received channel output while for the DECs, y in (2) is formed by the deinterleaved version of the outputs of the ESE block. These results are then combined using a turbo-type iterative process for a pre-defined number of iterations. Finally the DECs produce hard decisions on information bits for each user.

3 UDMA Scheme

We have developed a practical method to generate spreading sequences from the user data and regenerate them for data detection in the intended receiver. This user-data based spreader not only spreads the information bits but also separates from individual users. Note that the user data based spreading sequences are not only random but also changing dynamically from symbol to symbol.

The system model is shown in Fig. 2. In the transmitter, the user data are first encoded using the convolutional code. Then, this coded data enter the spreading sequence generator. We use shift registers for the spreading sequence. Initially, all shift registers are initialized with '1'. The all '1' sequence is multiplied by the first coded data. After outputting the first sequence from the generator, the shift

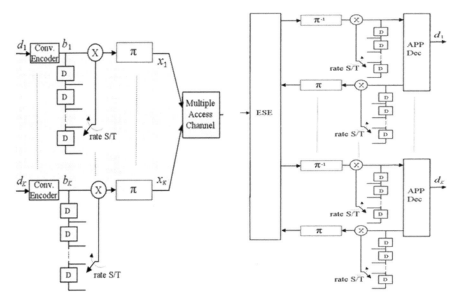

Fig. 2 The transmitter and receiver structure of the UDMA scheme

registers shift one bit right. This process continues for all data. Thus, the spreading sequences are changing dynamically according to the user data. This spread data is fed into the interleaver (π) that is identical to all users, instead of user specific interleavers. The resultant signal is then transmitted through the multiple access channel.

In IDMA, user specific interleavers are used for user separation and thus the specific interleaving pattern needs to be known to the receiver. User specific spreading sequences are employed in CDMA and the spreading sequences must be orthogonal between users and have to be synchronous. In UDMA, however, all identical interleavers are used and thus do not require to store all interleaving patterns at the uplink transmission. In addition, the UDMA sequences do not require orthogonality and synchronization.

The UDMA scheme employs a chip by chip detection. Initially, we assign the same spreading sequences for spreading and despreading in the transmitter and the receiver. In the detection process, despreading sequences are provided by decoded data for each user. Severe multiple access interference and error propagation in the receiver may cause the despreading sequences to be mismatched with the spreading sequences. We consider the recovery of the spreading sequence at the receiver without a priori knowledge. The received signal strength would determine the integrity of the recovered spreading sequence and would thus affect bit decisions subsequently. For the error checking in the despreading sequence, we use genetic search algorithm and Markov chain analysis. Those algorithms help to refine despreading sequences and updates with optimization efficiently.

Table 1 Parameters used in the simulation over flat fading channel

Parameters	Specifications
Number of users	5, 10
Data length	1024 bits
Encoder	Convolutional code $(23, 35)_8$
Spreader length	16, 24
Modulation	BPSK
Iteration	10
Interleaver	Random interleaver

4 Simulation Results

To verify the performance of the UDMA scheme, we have conducted performance evaluation and comparative study. For a comparison, we have used the IDMA system. It is assumed that all users use the same energy level. The simulation parameters used in the simulation are given in the Table 1.

Figure 3 shows the BER performance comparison between UDMA and IDMA with 5 simultaneous users. The spreading sequence lengths used for the simulation are 16 and 24. When the spreading length is 16, the performance of the UDMA scheme shows comparable performance to the IDMA system. However, when the spreading length increases, the UDMA scheme gives better performance than the IDMA system. In Fig. 4, we further performed the evaluation with 10 simultaneous users with the spreading lengths remained unchanged. Compared with IDMA, the BER performance of the UDMA scheme is better than the IDMA system. It is important to note that the performance gain of the UDMA over the IDMA increases as the number of the user increases. Likewise, when the spreading sequence length increases, the BER performance further improves. This performance improvement stems from more sophisticated spreading and efficient detection process.

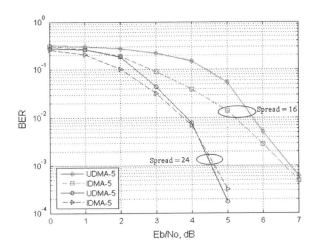

Fig. 3 Performance comparison between UDMA and IDMA with K = 5

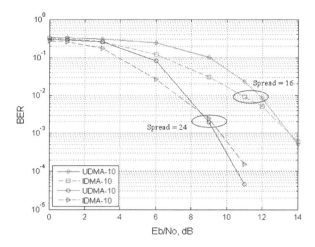

Fig. 4 Performance comparison between UDMA and IDMA with K = 10

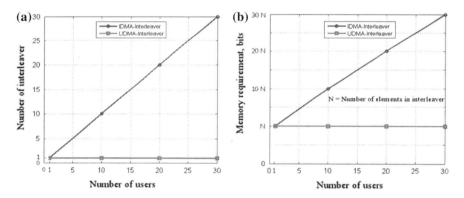

Fig. 5 Comparison of interleaver between UDMA and IDMA (**a**) Number of interleaver (**b**) Memory requirement of interleaver

Figure 5 shows comparison of interleaver between UDMA and IDMA. When the number of simultaneous user increases, number of interleaver also increases linearly in the IDMA system which requires the storage of the entire interleaving pattern for each user. This can be expensive or infeasible for applications that have limited storages when the number of users is large. But, the UDMA scheme uses identical interleaver. Thus, it uses a same interleaving pattern for all users.

5 Conclusion

We have proposed a User-Data Division Multiple Access (UDMA) scheme with user data based spreading sequences. Unlike IDMA and CDMA, the proposed scheme does not require the storage of all interleaving patterns and orthogonality and synchronization of spreading sequences. In addition, these spreading sequences vary dynamically from symbol to symbol according to the user data. As a result, the UDMA scheme provides enhanced security and privacy. For a BER performance comparison, the proposed scheme improves the BER performance as the spreading sequence length and the number of users increase.

References

1. Li P, Liu L, Wu KY, Leung WK (2006) Interleave-division multiple-access. IEEE Trans Wirel Commun 5:938–947
2. Prasad R, Ojanpera T (1998) A survey on CDMA: evolution towards wideband CDMA. In: Proceedings, IEEE 5th international symposium on spread spectrum techniques and applications, vol. 1, pp 323–331 (IEEE)
3. Li P, Liu L, Wu KY, Leung WK (2003) Interleave division multiple access (IDMA) communication systems. In: Proceedings of 3rd international symposium on turbo codes & related topics, pp 173–180
4. Kim YS, Jang WM, Nguyen L (2006) Self-encoded TH-PPM UWB system with iterative detection. In: The 8th international conference on advanced communication technology, pp 710–714 (ICACT)
5. Jang WM, Nguyen L (2012) Distributed and centralized iterative detection of self-encoded spread spectrum in multi-channel communication. J Commun Netw 14(3):280–285 (IEEE)
6. Nguyen L, Jang WM (2008) Self-encoded spread spectrum synchronization with genetic algorithm and Markov chain analysis. In: 42nd annual conference on information sciences and systems, pp 324–329 (CISS)

On Channel Capacity of Two-Way Multiple-hop MIMO Relay System with Specific Access Control

Pham Thanh Hiep, Nguyen Huy Hoang and Ryuji Kohno

Abstract For the high end-to-end channel capacity, the amplify-and-forward (AF) scheme multiple-hop MIMO relays system is considered. The distance between each transceiver and the transmit power of each relay node are optimized to prevent some relays from being the bottleneck and guarantee the high end-to-end channel capacity. However, when the system has no control on Mac layer, the interference signal should be taken in account and then the performance of system is deteriorated. Therefore, the specific access control on MAC layer is proposed to obtain the higher end-to-end channel capacity. The optimum number of relays for the highest channel capacity is obtained for each access method. However, there is the trade-off of channel capacity and delay time.

Keywords Multiple-hop MIMO relays system · MAC-PHY cross layer · Optimization distance · Optimization transmit power · Specific access control · Channel capacity-delay time tradeoff · Outdated channel state information

1 Introduction

In order to achieve the high performance, the multiple-hop relays system is considered. [1–3]. However, in these papers the SNR at receiver(s) is assumed to be fixed and the location as well as the transmit power of each transmitter(s) are not dealt. In the multiple-hop MIMO relay system, when the distance between the source (Tx) and the destination (Rx) is fixed, the distance between the Tx to a relay

---

P. T. Hiep (✉) · R. Kohno
Graduate School of Engineering, Yokohama National University, Yokohama, Japan
e-mail: hiep@kohnolab.dnj.ynu.ac.jp

N. H. Hoang
Le Quy Don Technical University, Ha Noi, Viet Nam

J. J. (Jong Hyuk) Park et al. (eds.), *Multimedia and Ubiquitous Engineering*,
Lecture Notes in Electrical Engineering 240, DOI: 10.1007/978-94-007-6738-6_105,
© Springer Science+Business Media Dordrecht(Outside the USA) 2013

(RS), RS to RS, RS to the Rx called the distances between transceivers, is shorten. Consequently, according to the number of relay and the location of the relay, the SNR and the capacity are changed. Hence, to achieve the high end-to-end channel capacity, the location of each relay meaning the distance between each transceiver needs to be optimized. We have analyzed the one-way AF scheme multiple-hop MIMO relay system (MMRS) in case the interference is taken in account and optimized distance between each transceiver to obtain the high end-to-end channel capacity [4]. However, in order to achieve the higher end-to-end channel capacity when the interference is taken in account, the specific access control on Mac layer for multiple-hop MIMO relay system needs to be analyzed. In this paper, we propose the specific access control on MAC layer for one-way multiple-hop relay system and apply this method into two-way multiple-hop relay system. The proposed access control is compared to the existing method using network coding technology [5, 6]. The end-to-end channel capacity, the delay time and the relation of them is analyzed. Note that the channel capacity which is analyzed in this paper is the ergodic channel capacity. The rest of the paper is organized as follows. We introduce the concept of MMRS in Sect. 2. Section 3 shows specific access control on MAC layer. The two-way MMRS is described in Sect. 4. Finally, Sect. 5 concludes the paper.

2 Multiple-hop MIMO Relays System

The MMRS is described in details in [4]. However we choose some important parts to help the reader understand easier.

2.1 Channel Model

Let M, N and K_i ($i = 1,...,m$) denote the number of the antenna at the Tx, Rx and RS_i, respectively. The distance between each transceivers is denoted by d_i ($i = 0,...,m$). The distance between the Tx and the Rx is fixed as d. The Tx and all the relays employ amplify-and-forward strategy. Mathematical notations used in this paper are as follows. x and X are scalar variable, $\boldsymbol{x}$ and $\boldsymbol{X}$ are vector variable or matrix variable $(\bullet)^H$ is conjugate transpose. In order to easily describe, the Tx, Rx are also be denoted as the RS_0 and RS_{m+1}, respectively. Since the path loss is taken into consideration, channel matrix is a composite matrix and we model as $\sqrt{l_i}H_i, i = 0, \ldots, m$, of which l_i and H_i represent the path loss and the channel matrix between the RS_i and the RS_{i+1}, respectively. H_i is a matrix with independent and identical distribution (i.i.d.), zero mean, unit variance, circularly symmetric complex Gaussian entries. We assume that the transmit power of the Tx (E_{tx}), the Rx (E_{rx}) and the total transmit power of relays (E_{rs}) are fixed and are not affected by the change in the number of relays and antennas at each relay. In order to simplify the composition of relay and demonstrate the effect of optimizing the distance and the transmit power of each relay, we assume that the transmit power of each relay is equally divided into each antenna and the number of antenna in each relay is the same. Moreover, the perfect channel state information is assumed to be available and the zero forcing algorithms is applied to both the transmitter and the receiver.

3 Specific Access Control on MAC Layer

3.1 Multiple-Phases Transmission

The transmission of each relay in the system can be divided into the multiple-phases. The relays in the same phases transmit the signal in the same time and the allocation time (t_i). In the other phases, the relay keeps the silence or receives the signal. Since the neighbor relay transmits the signal in different phases, the interference signal is weaker than that of the system without control.

Figure 1 shows 2 phases and 3 phases transmission protocol. The 2 phases transmission protocol is explained as follows. The even-number relays and the odd-number relays transmit the signal in phase 1 and phase 2, respectively. The system has no control on MAC layer can be seem as the system with 1 phase transmission protocol. Therefore, the end-to-end channel capacity of the system with n phases can be written as

$$C = \log_2 \left(\det \left(I_M + \frac{HH^H \left(\sum_{i=0}^{m} l_i p_i \right)}{\sigma^2 + \sum_{i=1}^{m+1-n} l_{i-1 i+n} p_{i-1} + \sum_{i=0}^{m-1-n} l_{ii+1+n} p_{i+1+n}} \right) \right) \quad (1)$$

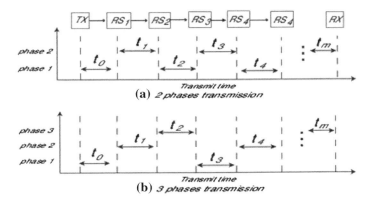

Fig. 1 2 phases and 3 phases transmission protocol

Compare the interference component of system has no control to that of the system with n phases transmission protocol (1), the distance from interference relay is longer and the number of interference relay is also larger. Hence, we can say that according to the control on MAC layer, the power of interference is decreased, thus the end-to-end channel capacity is expected to be higher.

3.2 Comparing to the existing method

The access control for one-way was proposed in Sec. 3.1. For two-way transmission, the transmission of downlink and uplink is assumed to alternate. Therefore, although the delay time increases 2 times, this transmission protocol can be extended for two-way transmission. The uplink end-to-end channel capacity is the same as the channel capacity of downlink in (1). We compare the proposed access method with existing once.The access method for two-way have being considered. There are some methods using the network coding technologies [5, 6]. In case of interference from 2d (2 times of distance), the transmission of all transmitters is divided into 3 phases. It means the delay time in this case is 3 s if we assume that the transmission time on each phase is 1 s. Moreover, in the proposed access method, the 1 phase method with MIMO beamforming to cancel the interference from uplink has the interference from 2d. However, it needs only 2 phases for two-way. It means the delay time is 2 s, smaller than the delay time of network coding method. Similarly, in case of interference from 3d, the delay time is 4 s for network coding method. In the proposed method, the 2 phases has the same distance of interference and the same delay time for two-way.

3.3 Numerical Evaluation for Proposed Access Method

In order to obtain the high end-to-end channel capacity, the distance and the transmit power should be optimized. The mathematical optimization method is explained in [4]. However, the mathematical method is complicated in case the channel model between each transceiver is different. Hence, the particle filter method is applied to optimize the distance and the transmit power simultaneously. The system parameter is summarized in Table 1.

Table 1 Numerical parameters

Antennas at TX, RX, RS	4
Transmit power of TX (mW)	100
Transmit power of RX (mW)	10
Total transmit power of RS [mW]	100
Noise power (mW)	6.12e-011
Reflection factor	0.38
Distance between TX-RX (m)	3000

On Channel Capacity of Two-Way Multiple-hop MIMO Relay System

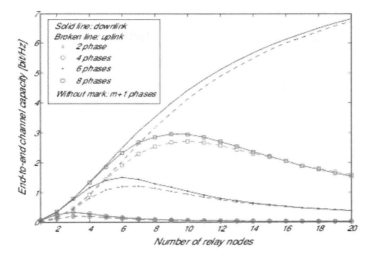

Fig. 2 The end-to-end channel capacity of two-way transmission under access control on MAC layer, the interference from both uplink and downlink

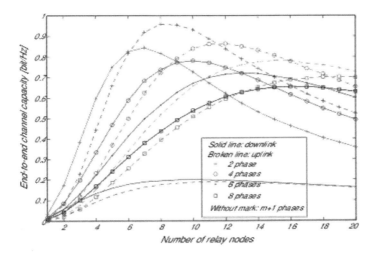

Fig. 3 The end-to-end channel capacity in case the transmission time is normalized

The end-to-end channel capacity of two-way under control on MAC layer is shown in Fig. 2. There is the optimum number of relays that has the maximum end-to-end channel capacity. In addition, the end-to-end channel capacity of the high number of phases is higher than that of the low number of phases. However, according to the transmission environment and the access method, the optimum number of relays is changed. Moreover, the allocation time for each phase was assumed as 1 s. Thus, the delay time of each access method increases when the number of phases increases. It means that there is the trade-off between channel capacity and delay time. In case the transmission time of the system is normalized meaning the transmission time from the Tx to the Rx is 1 s, the end-to-end channel capacity is shown in Fig. 3. According to the channel model (the transmission environment, the transmit power and so on), the optimum number of relays and the number of phases is changed for the highest end-to-end channel capacity

4 Conclusion

The access control method on MAC layer for two-way MMRS is proposed based on the access method of one-way and compared to the existing method. There are the trade-off of channel capacity-delay time and the optimum number of relays for highest end-to-end channel capacity. According to the channel model and the number of phases, the optimum number of relays is different. In this paper, we have optimized the distance and the transmit power for each transmission protocol on MAC layer to obtain the highest end-to-end channel capacity. However, the combination of physical layer and MAC layer is not optimized. Additionally, the perfect channel state information is assumed and the ergodic channel capacity is analyzed. In the future, the system with the imperfect channel state information and the instantaneous channel capacity will be analyzed.

References

1. Gastpar M, Vetterli M (2005) On the capacity of large Gaussian relay networks. IEEE Trans Inf Theor 51(3):765–779
2. Levin G, Loyka S (2010) On the outage capacity distribution of correlated keyhole MIMO channels. IEEE Trans Inf Theor 54(7):3232–3245
3. Peyman R, Yu W (2009) Parity forwarding for multiple-relay networks. IEEE Trans Inf Theor 55(1):158–173
4. Hiep PT, Kohno R (2010) Optimizing position of repeaters in distributed MIMO repeater system for large capacity. IEICE Trans Commun E93-B(12):3616–3623
5. Katti S, Rahul H, Hu W, Katabi D, Mdard M, Crowcroft J (2006) Xors in the air: practical wireless network coding, Proceedings of the 2006 conference on applications, technologies, architectures, and protocols for computer communications, vol 36, no 4, pp 243–254
6. Popovski P, Yomo H (2007) Physical network coding in two-way wire- less relay channels, IEEE International Conference on Communications (ICC07), pp 707–712
7. Kita N, Yamada W, Sato A (2006) Path loss prediction model for the over-rooftop propagation environment of microwave band in suburban areas (in Japanese) IEICE Trans Commun J89-B(2):115–125
8. Edwards HM (1997) Graduate texts in mathematics: Galois theory, Springer, New York

Single-Feed Wideband Circularly Polarized Antenna for UHF RFID Reader

Pham HuuTo, B. D. Nguyen, Van-Su Tran, Tram Van and Kien T. Pham

Abstract In this paper, a single-feed wideband circularly polarized antenna has been proposed for UHF RFID reader. This antenna is designed to cover the frequency range from 860 to 960 MHz. In this antenna, the main patch is a modified form of the conventional E-shaped patch to obtain a circular polarization. A parasite patch is placed at the same layer of main patch to enhance axial ratio bandwidth. A short-circuited cylinder is also added in main patch to broaden the impedance bandwidth. The 3 dB axial ratio bandwidth is over 11 %, from 850 to 960 MHz. The impedance bandwidth is of 17 % (850–1000 MHz). Thus, The impedance bandwidth and 3 dB axial ratio bandwidth totally covers the universal UHF RFID band (860–960 MHz). The simulated and measured results indicate that the proposed antenna will be a good candidate for UHF RFID reader system.

Keywords Wideband · UHF · RFID · Circularly polarized antenna · LHCP

1 Introduction

Basic RFID systems are based on wireless communication between a reader and a tag. Antenna is one of the most important components; it will affect the performance of the whole RFID system. The operating frequencies authorized for UHF RFID

P. HuuTo (✉) · B. D. Nguyen · V.-S. Tran · T. Van · K. T. Pham
School of Electrical Engineering, International University, Ho chi minh, Vietnam
e-mail: phto@hcmiu.edu.vn

B. D. Nguyen
e-mail: nbduong@hcmiu.edu.vn

V.-S. Tran
e-mail: tvsu@hcmiu.edu.vn

K. T. Pham
e-mail: ptkien@hcmiu.edu.vn

J. J. (Jong Hyuk) Park et al. (eds.), *Multimedia and Ubiquitous Engineering*,
Lecture Notes in Electrical Engineering 240, DOI: 10.1007/978-94-007-6738-6_106,
© Springer Science+Business Media Dordrecht(Outside the USA) 2013

applications are varied in different countries and regions (866–869 MHz in Europe, 902–928 MHz band in North and South of America, 866–869 and 920–925 MHz in Singapore, and 952–955 MHz in Japan...). Hence, a universal UHF RFID antenna for reader is necessary to cover all UHF RFID frequency range. Since the RFID tags are always arbitrarily oriented, circularly polarized (CP) antenna is the best solution for RFID system to ensure the reliability of communications between readers and tags. It can increase orientation diversity and reduce the loss caused by the multipath effects between the reader antenna and the tag antenna. In the literature, there are several configurations proposed to create the circular polarization. The commonly used methods are rectangular patch with truncated corners and selecting suitable feed position [1], or a power splitting network to excite two orthogonal patch modes in phase quadrature [2]. However, these antennas have inherent narrow impedance bandwidth and axial ratio (AR) bandwidth (typically 1–4 %). Some designed structures allow a wide axial ratio (AR) bandwidth covering the entire ultra-high frequency (UHF) band (860–960 MHz) [3–6], while some of them show complex feeding networks.

In this paper, we propose a new geometry of antenna to achieve a wide band of circular polarization. The antenna which is built on the low-cost FR-4 substrate, is a single layer with main patch and parasite patch, and fed directly by a single coaxial probe. The main patch is a modification from conventional E-antenna [7] to generate the circular polarization. A parasite patch is added on the same layer with main patch to broaden the circular polarization band. A slot and short-circuited cylinder is also added to the main patch antenna in order to widen impedance bandwidth. After this model has been designed, the return loss was measured to validate the simulation results.

2 RFID Antenna Design

In order to obtain a circular polarization, two conditions must be satisfied. One of them is to have two degenerated orthogonal modes. The other is that the difference of phase of two orthogonal modes is 90°. By this way, a conventional E-patch antenna is modified to become dissymmetric to provide a circular polarization. Fig. 1 shows the geometry of the proposed antenna for RFID. This antenna is printed on the FR-4 substrate with relative permittivity $\varepsilon_r = 4.4$, loss tangent $\tan \delta = 0.02$ and the substrate thickness is 1.2 mm. The medium between the FR-4 substrate and the ground is an air-layer with height of 13.8 mm. The designed antenna structure is simulated and optimized by using HFSS simulation software. The optimized geometric parameters are listed in Table 1.

The proposed geometry of this antenna is shown in Fig. 1. The main patch antenna is modified from a rectangular patch antenna (198 mm, 138 mm). The antenna is fed by a coaxial probe at position F (−37 mm, 0 mm) at the middle line of the patch. Basically, the dissymmetric dimensions of slot No. 1 (g1, h1) and slot No. 2 (g2, h2) leads to two orthogonal currents on the patch; hence, circularly

Fig. 1 Geometry of the proposed UHF-RFID antenna **a** Top view of the antenna. **b** Cut-plane view of the antenna

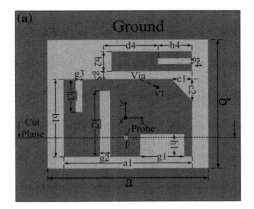

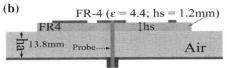

Table 1 Optimized parameters of the antenna

Main patch	Unit: mm	Sub-patch	Unit: mm
a1	198	g4	5
b1	138	h4	65
c1	28.5	b2	35
c2	28.5	g5	3
g1	71	d4	58
h1	36	**Ground plan**	**Unit: mm**
g2	14	a	270
h2	120	b	270
g3	8		
h3	56		

polarized fields are excited. The resonant lengths of the x and y orthogonal currents are thus dependent on the width (g) and length (h) of slots. The patch is also connected to the ground via a short-circuited cylinder with diameter of 2 mm to improve the impedance bandwidth; it is placed at the position V1 (53 mm, −37 mm). This antenna is fabricated as shown in Fig. 2 and measured by E5071CAgilent Network Analyzer.

In order to widen and tune the axial ratio band, a parasitic patch is added on the same layer with main patch. On the surface of parasite patch, a slot is also created to improve and tune the axial ratio band. The antenna is truncated at one corner with equal side lengths. The purpose of truncating is to enhance the 3 dB axial ratio band.

Fig. 2 The fabricated UHF-RFID antenna

Figure 3 shows the current-vector distribution on the antenna at 860 MHz for different phase states. As shown in Fig. 3, at time instant of 0°, the current at left side of the main patch flows in the negative x-axis, while current at right side of

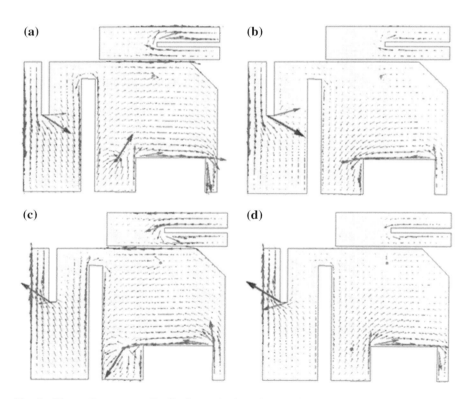

Fig. 3 The surface current distribution and orientation on the patches for the designed UHF-RFID antenna at frequency of 860 MHz at four time instants; **a** Time instant of 0°. **b** Time instant of 90°. **c** Time instant of 180°. **d** Time instant of 270°

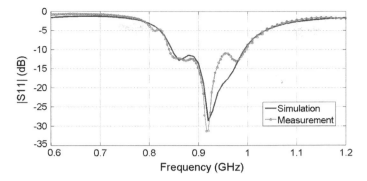

Fig. 4 The simulated and measured return loss of the proposed antenna

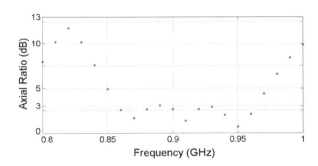

Fig. 5 The simulation axial ratio with respect to frequency

that flows in the negative y-axis. Then, at time instant of 90°, the current direction at right side of the patch changes to +y axis while at left side of the patch still remains its own direction. Similarly, at 180° and 270°, both currents are in opposite directions with that in case of 0° and 90° respectively. This implies a quadrature phase between the x- and y-directed currents. As a result, the current flows turning the x-axis into y-axis like a left-handed circularly polarized (LHCP). Note that the surface currents at other frequencies within the 3 dB axial ratio band are varied as functions of time in a similar manner.

3 Results and Discussion

Figure 4 shows the simulated and measured return loss of antenna. It is observed that the measured return loss is less than −10 dB over the frequency range of 850–1000 MHz (17 %), which can easily cover the entire universal UHF RFID frequency band of 860–960 MHz.

Figure 5 illustrates the simulated axial ratio of antenna. As can be seen from it, the simulated 3 dB AR bandwidth is of 860–960 MHz or 11 %. It is able to cover the entire universal UHF RFID frequency band of 860–960 MHz.

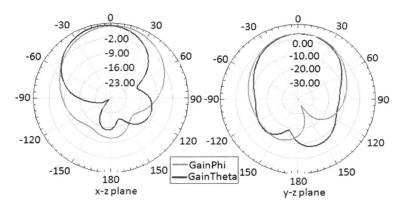

Fig. 6 Simulation radiation pattern of the UHF-RFID antenna at 860 MHz

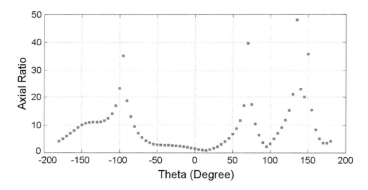

Fig. 7 The simulated axial ratio with respect to theta of the implemented antenna simulated at 860 MHz

The radiation patterns at 860 MHz are illustrated in Fig. 6 in x–z and y–z planes. In both planes, wide-angle axial ratio characteristics have been examined.

As shown in Fig. 7, the simulated 3 dB AR beam width of the implemented antenna is about 90° (from −50° to 40°).

The performance of simulation gain with respect to frequency is illustrated in Fig. 8. The value of gain varies from 6.0 to 7.6 dBic along the universal UHF RFID bandwidth, with a peak gain of 7.6 dBic at 930 MHz.

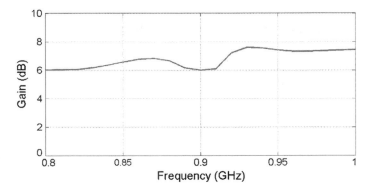

Fig. 8 Simulation gain of the implemented UHF antenna

4 Conclusion

In this paper, a broadband circularly polarized antenna has been presented for UHF RFID applications. The antenna structure is fed by a single coaxial probe. By combining several techniques, the implemented antenna has achieved the desired performance over the UHF band: the return loss of 17 % or 850–1000 MHz, 3 dB axial ratio bandwidth of 11 % or 860–960 MHz, the gain of more than 6dBic. A prototype is fabricated to validate the simulation results. The measured results show that the designed antenna can provide broad impedance bandwidth of 17 % (850–1000 MHz) and prove that it is a good candidate for UHF RFID reader system.

Acknowledgments The authors would like to thank Advantech Jsc Company for their support of Ansoft Designer HFSS used as simulation tool to obtain these results in this paper.

References

1. Chen W-S, Wu CK, Wong K-L (1998) Single feed square-ring microstrip antenna with truncated corner for compact circular polarization operation. Electron Lett 34:1045–1047
2. Targonski SD, Pozar DM (1993) Design of wideband circularly polarized aperture-coupled microstrip antenna. IEEE Trans Antennas Propag 41:214–220
3. Chen ZN, Qing X, Chung HL (2009) A universal UHF RFID reader antenna. IEEE Trans Microw Theory Tech 57(5):1275–1282
4. Lau P-Y, Yung KK-O, Yung EK-N (2010) A low-cost printed CP patch antenna for RFID smart bookshelf in library. IEEE Trans Industr Electron 57(5):1583–1589
5. Wang P, Wen G, Li J, Huang Y, Yang L, Zhang Q (2012) Wideband circularly polarized UHF RFID reader antenna with high gain and wide axial ration beam widths. Prog Electromagn Res 129:365–385

6. Kwa HW, Qing X, Chen ZN (2008) Broadband single-fed single-patch circularly polarized antenna for UHF RFID applications. In: IEEE AP-S International Symposium on Antennas and Propagation, San Diego, pp 1–4
7. Yang F, Zhang XX, Ye X, Ramat-Samii Y (2001) Wide band E-shaped patch antenna for wireless communications. IEEE Trans Antennas Propag 49(7):1094–1100

Experimental Evaluation of WBAN Antenna Performance for FCC Common Frequency Band with Human Body

Musleemin Noitubtim, Chairak Deepunya and Sathaporn Promwong

Abstract Wireless communication systems have become important in daily life such as wireless body area network (WBAN). An ultra-wideband (UWB) technology is chosen to be used for short-range communication scenarios, low-power and high data rate technology which accommodates appropriate technology in WBAN. In this paper, we design a printed circular monopole with coplanar waveguide (CPW) fed for WBAN with common frequency band following FCC common band (7.25–8.5 GHz). The antenna structure is simple, using FR4 circuit board (PCB) with overall size of $18 \times 20 \times 1.6 \text{ mm}^3$. The simulation and experiment results show that the proposed antenna achieves good impedance matching and stable radiation patterns over operating bandwidth of 6.8–12.34 GHz. Moreover, the authors evaluate UWB antenna in two scenarios are free-space and with human-body to consider the antenna performance in time domain.

Keywords WBAN · BAN · UWB · Wideband antenna · Human body

M. Noitubtim (✉) · S. Promwong
Department of Telecommunication Engineering, King Mongkut's Institute of Technology Ladkrabang, Bangkok, Thailand
e-mail: m_leemin@hotmail.com

S. Promwong
e-mail: kpsathap@kmitl.ac.th

C. Deepunya
Department of Electrical Engineering, King Mongkut's Institute of Technology Ladkrabang, Bangkok, Thailand
e-mail: kdchaira@kmitl.ac.th

J. J. (Jong Hyuk) Park et al. (eds.), *Multimedia and Ubiquitous Engineering*, Lecture Notes in Electrical Engineering 240, DOI: 10.1007/978-94-007-6738-6_107, © Springer Science+Business Media Dordrecht(Outside the USA) 2013

1 Introduction

There are increasing interests in ultra wideband (UWB) communication systems because of a radio technology with high data rate, anti-multipath interference and simple transceiver structures that is possible to make it as potentially powerful technology for low complexity, low cost communications. The power density of the UWB signal is considered to be noise for other communication systems because its power spectrum is below the noise level or part 15 limited. Therefore, UWB technology can exist with other RF technologies and can use any applications. The Federal Communications Commission (FCC) [1] in the United States allocated the fractional bandwidth ≥ 0.2 and having occupied bandwidth ≥ 500 MHz.

The regulation for UWB indoor devices defined the frequency of high band UWB and the common band is ranged from 7.25 to 8.5 GHz with power spectral density (PSD) is -41.3 dBm/MHz for the UWB applications [2] indoor communication systems such as wireless body area network (WBAN). WBAN is wireless communication system that enable communications between electronic devices, that place on and/or into the human body. The systems are of great interest for various applications such as sport, multimedia, health care, and military applications [3].

The antenna is an important component for ultra wide band system. The coplanar waveguide (CPW) is very suited for patch antenna design and has been widely used [4, 5]. It also has wide bandwidth characteristic and can be easily integrated with microwave monolithic integrated circuits (MMICs). The printed circular monopole with CPW fed antenna used in WBAN fundamental requirements such as optimized characteristics in frequency and time domains small size, low profile and good on-body propagation. The proposed antenna has good return loss frequency range 6.88–12.23 GHz. It cover FCC common band (7.25–8.5 GHz).

In this paper, printed circular monopole with coplanar waveguide (CPW) fed is proposed for WBAN applications. The common parameters of the proposed antenna such as return loss, radiation patterns are shown. Moreover, the authors evaluate UWB antenna in two scenarios are free-space and with human-body to consider the antenna performance in time domain.

2 Common Parameter of the Proposed Antenna

All the simulations were carried out with CST Microwave Studio. The proposed antenna optimal parameter values are listed in Table 1. The printed circular monopole with CPW fed antenna verify in this paper depicted in Fig. 1a. The proposed antenna fabricated was show in Fig. 1b. The antenna was printed on one side of a FR4-Epoxy (PCB) substrate which has dielectric constant of 4.3, thickness of 1.6 mm and size 18×20 mm^2. The CPW transmission line is designed with 50 ohm and terminated with SMA connector for measurement purpose in this paper. In the Fig. 2 is shown simulated compare with measured return loss (S_{11}).

Experimental Evaluation of WBAN Antenna Performance

Table 1 The optimized parameters of the printed circular monopole with CPW fed antenna

Parameter	Value (mm)	Parameter	Value (mm)
W	20	L	18
x	3	y	7
z	6	g	0.35
s	2	r	5

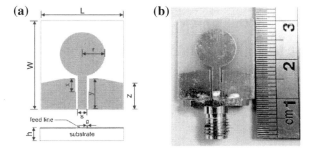

Fig. 1 The printed circular monopole with CPW fed antenna. **a** Geometric and dimensions. **b** Phototype

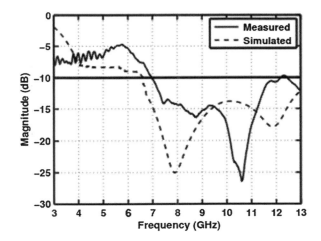

Fig. 2 Characteristic of the printed circular monopole with CPW fed antenna |S11|

3 UWB-IR Transmission

3.1 System Transfer Function

Friis' transmission formula cannot be directly applied to the UWB radio as the bandwidth of the pulse is extremely wide. The complex form Friis' Transmission Formula is extended for estimating the link budget of UWB transmission System [6–9]. Transfer function can be defined as the ratio between the voltages received

(Rx-antenna) and the voltage at the Transmitter antenna (Tx-antenna), the transfer function $H(f)$ can simply expressed

$$H(f) = H_{tx}(f)H_f(f)H_{rx}(f) \tag{1}$$

H_{tx} is transfer function of the transmitter antenna, H_{rx} is transfer function of the receiver antenna, and $H(f)$ is free space transfer function, can be written as

$$H_f(f,d) = \frac{c}{4\pi fd}\exp(-j\frac{2\pi f}{c}d) \tag{2}$$

The transfer function $H(f)$ can be directly obtained from measurement, and the system response can be completely determined when the transfer function is known. In this paper we used the rectangular passband transmitted waveform, which is in time domain and its spectral density as the model which is given by

$$v_t(t) = \frac{A}{f_b}[f_H \sin c(2f_H t) - f_L \sin c(2f_L t)] \tag{3}$$

$$V_t(f) = \begin{cases} \frac{A}{2f_b}, ||f| - f_c| \le \frac{f_b}{2} \\ 0, ||f| - f_c| \ge \frac{f_b}{2} \end{cases} \tag{4}$$

where is the maximum amplitude, f_b is the occupied bandwidth, f_c is the center frequency, $f_L = f_c - f_b/2$ and $f_H = f_c + f_b/2$ are the minimum and maximum frequencies. Investigate the waveform occupying the entire UWB band, $f_L = 7.25$ GHz and $f_H = 8.5$ GHz. After knowing the channel transfer function and determining the transmitted waveforms, the receiver antenna output waveform $v_r(t)$ is given by

$$v_r(t) = \int_{-\infty}^{\infty} H_c(f)V_t(f)\exp(j2\pi ft)df \tag{5}$$

And in case of using isotropic antennas on both sides, the receiver output waveform $v_{r-iso}(t)$ can be written

$$v_{r-iso}(t) = \int_{-\infty}^{\infty} H_f(f)V_t(f)\exp(j2\pi ft)df \tag{6}$$

3.2 Power Delay Profile

The mean relative power of the taps are specified by the power delay profile (PDP) of the channel, defined as the variation of mean power in the channel and $h(\tau)$ is the channel impulse response.

$$PDP_\tau = |h(\tau)^2| \tag{7}$$

4 Experimental Setup

In Fig. 3 shows the sketch of the experimental setup. The UWB radio channel transfer function was measured as S_{21} in frequency domain by using the vector network analyzer. Measurement in frequency range from 7–11 GHz, number of frequency points are 801 and dynamic range is 80 dB. In this study the printed coplanar waveguide (CPW) antennas was used as Tx and Rx antennas. Tx-antenna was rotate start from 0° to 350° for 10° step. Two scenarios shows in this experiment, first take data in azimuth plane without body and second in this case takes S_{21} when port 1 or Tx-antenna place on body. The Tx-antenna was fixed on human bodies (chest) and Rx-antenna for receive signal height 1.3 m. The distance between Tx-antenna and Rx-antenna are 1 m.

5 Measured Results and Discussion

The radiation patterns of the E-plane and H-plane which is obtained at 7.25 GHz, 8.5 GHz and 10.25 GHz was shown Fig. 4. From the figure, note that the antenna has Omni-directional radiation pattern. The return loss from measurement is less than −10 dB from 6.8–12.34 GHz covering the entire common band 7.25–8.5 GHz. The power delay profiles in free-space and with human-body are shown in Fig. 5. However, power delay profile has worse case when antenna place on-body in angle from 75° to 260°.

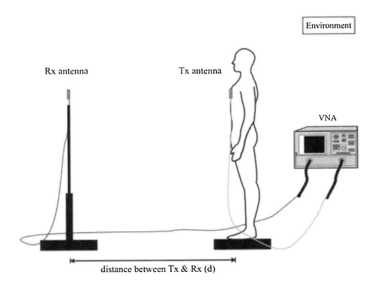

Fig. 3 Experimental setup

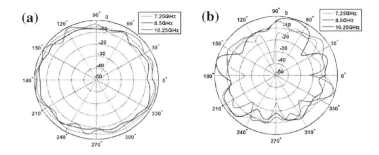

Fig. 4 Radiation pattern of the printed circular monopole with CPW fed antenna. **a** E-plane. **b** H-plane

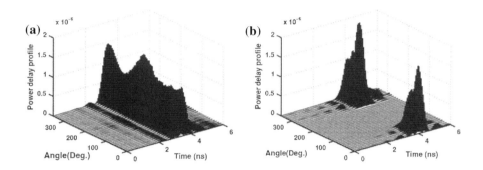

Fig. 5 Power delay profile. **a** Free-space. **b** With human-body

6 Conclusion

The printed circular monopole with CPW fed antenna for common frequency band with human body is proposed antenna. The proposed antenna has structure simple and compact size covering bandwidth from 6.8 to 12.34 GHz. The result from simulations and measured show the proposed antenna has good return loss has achieved good impedance matching, Omni-directional for common frequency band with human body (7.25–8.5 GHz) WBAN applications. However, transmission performance for on-body shows not good quality in case NLOS.

References

1. Federal communications commission: revision of part 15 of the commission's rules regarding ultra-wideband transmission systems, First report, FCC 0248, Apr (2002)
2. Ian Oppermann (2004) UWB Theory and Applications, Ian Oppermann, Matti Hamalainen and Jari Iinatti
3. Chen ZN (2006) Broadband planar antennas design and applications. Wiley, London
4. IEEE P802.15 (2010) Wireless personal area Networks (WPANs). Channel model for body area network (BAN), 10 Nov
5. Natarajamani S(2009) CPW-fed octagon shape slot antenna for UWB application, International conference on microwaves. Antenna propagation and remote sensing
6. Promwong S, Supanakoon P, Takada J (2010) Waveform distortion and transmission gain due to antennas on ultra wideband impulse radio. IEICE Trans Commun E93-B:2644–2650
7. Promwong S (2008) Experimental evaluation of complex form Friis transmission formula with indoor/outdoor for ultra wideband impulse radio. Computer and Communication Engineering ICCCE 13–15 May 2008
8. Promwong S, Hachitani W, Ching GS, Takada J (2004) Characterization of ultra-wideband antenna with human body. In: International symposium on communication and information technologies, Sapporo, Japan, 21–24 Oct 2004
9. Promwong S (2005) Experimental study of UWB transmission antennas for short range wireless system. In: International symposium communications and information technology, 12–14 Oct 2005

Performance Evaluation of UWB-BAN with Friis's Formula and CLEAN Algorithm

Krisada Koonchiang, Dissakan Arpasilp and Sathaporn Promwong

Abstract An ultra wideband impulse radio (UWB-IR) are developing to use in communication system and medical application because it has been an increase interest in using on body for health monitoring and body area networks (BAN). This research want to improving UWB channel propagation on body by using CLEAN algorithm for eliminate noise in channel propagation. In addition to, we use result from previous work for easier to compare performance before system. Moreover, in this paper us analysis performance of system when using CLEAN algorithm and compare without use CLEAN Algorithm by us will show BER in each position on body for analysis performance of CLEAN algorithm can be reduce noise or effect on body when without CLEAN Algorithm.

Keywords BAN · UWB · Impulse radio · Friis's transmission formula · CLEAN Algorithm

1 Introduction

Current demand in the connection network electronics device with other electronics device for convenience of use has increased. Whether the network connections within buildings. Or a network connection to the entertrainment

---

K. Koonchiang (✉) · D. Arpasilp · S. Promwong
Department of Telecommunication Engineering, Faculty of Engineering,
King Mongkut's Institute of Technology Ladkrabang, Chalongkrung Rd,
Ladkrabang, Bangkok 10520, Thailand
e-mail: Boatwi@hotmail.com

D. Arpasilp
e-mail: pangdumjungka@hotmail.com

S. Promwong
e-mail: kpsathap@kmitl.ac.th

J. J. (Jong Hyuk) Park et al. (eds.), *Multimedia and Ubiquitous Engineering*,
Lecture Notes in Electrical Engineering 240, DOI: 10.1007/978-94-007-6738-6_108,
© Springer Science+Business Media Dordrecht(Outside the USA) 2013

within the housing. The popular wireless technology used to connect such devices include WiFi, Bluetooth and shortwave technology, however, present a wide range of interesting trends about technology, Ultra wideband (UWB). Which is expected to change the data communication system with high efficacy over a WiFi or Bluetooth technology was an obvious, Ultra wideband, or UWB is likely to increase and become a standard wireless network in indoor environment. Thus, this re-search sees a critical need to consider and study the technology, Ultra wideband, which this project was to study the impact of the human body that affect the signal propagation radio impulse for Ultra wide band technology. There are a variety of modeling to analyze and compare.

The CLEAN algorithm (de-convolution algorithm) was first used to enhance the radio astronomical imaging of the sky and microwave communication. Whereat, it has been widely use in both narrowband [1] and UWB [1, 2] communication in localization and UWB biomedical imaging applications [1]. In general, the algorithm processes data by serially cancelling (i.e., cleans) the similarity between a dirty map (e.g., the measurement) and the a priori information (e.g., the template), and reconstructs the clean map (i.e., CIR) based on these detected similarities. However, CLEAN inherently assumes the channel to be non-dispersive, and that the resultant CIR is simply a summation of amplitude scaled and time-shifted versions of the a priori information. For time domain UWB channel sounding, this assumption must be considered with care when it involves probing the channel with sub-nanosecond impulses. Because of the wide spectral occupancy of these pulses, and a significant number of objects in the channel, the received signal is always severely distorted due to the frequency selectivity of the propagation phenomena, which often arise due to object's material, orientation and shape, especially for non line-of-sight (NLOS) and long-range line-of-sight (LOS) measurements.

The purpose of this paper is first to using CLEAN algorithm for improving UWB channel propagation on body and compare without use CLEAN Algorithm. So, this research has been using CLEAN algorithm from UWB-localization to UWB-BANs by implement apply CLEAN algorithm for application on body because communication system or medical want to performance of channel propagation more than directional of signal or estimate channel propagation. Where fore, we must be using CLEAN algorithm for reduce noise in UWB channel propagation (in time domain) by building new channel impulse response (CIR) or CLEAN map of the original. Specifically, we have used the results of previous research [3] but in this research change to use CLEAN for improving CIR for easier to compare in previous work. Finally, we use type of CLEAN algorithm is single-template [1] because it basic CLEAN algorithm and easy to analysis.

The rest of paper is organized as follow: in Sect. 2, describes about measurement setup from previous work, Sect. 3 we present some background of CLEAN algorithm, theory and analysis in channel propagation, Sect. 4 result and discussion we give some show BER and receive signal wave form from match filter, Sect. 5 conclusion.

2 Measurement Setup

This research will be implementing to experiment the UWB channel propagation on body by using vector network analyzer (VNA) in indoor-environment and we install Tx and Rx Antenna on body as well as standing position in room of experiment and antenna under test (AUT), It has been shown in previous work [3]. However, this research uses correlator in receiver side for signal distortion analysis, it shown in Fig 1.

In UWB channel propagation characterization, when we get data in frequency-domain (UWB channel) from VNA by we sent frequency range between 3–11 GHz (sampling frequency 801 point) and we processing data in matlab program for converse to time-domain by use Inverse Fourier Transform (IFT) because CLEAN technique must be process in time domain [2] or channel impulse response (CIR). In this research we generate pulse wave form from programming by generate sinc function (rectangular wave form in frequency-domain) and sent thought UWB channel, so it can be analysis power delay profile and bit error rate (BER) of channel on body.

3 Theory and Analysis

3.1 UWB Channel Impulse Response (UWB-CIR)

In this research we get data from VNA in frequency-domain (UWB channel) and we multiply by transmit signal for we want to get receive signal, the after that must be use Inverse Fourier Transform (IFT) for converse from frequency-domain to time-domain includes receive signal in time-domain show in equation.

$$V_r(f) = V_t(f)H_c(f) \tag{1}$$

$$v_r(t) = F^{-1}\{V_r(f)\} \tag{2}$$

Where is receive signal in frequency-domain, is UWB channel propagation in frequency (UWB-CIR) and is receive signal in time-domain from inverse fourier transform technique.

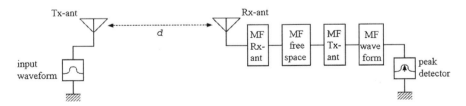

Fig. 1 Block diagram of extention Friis's transmission formula for UWB system [4]

3.2 CLEAN Algorithm

In this topic, a CLEAN algorithm for UWB-CIR characterization is proposed. Although, CLEAN algorithm were used in localization [1, 2] for estimate signal or decrease of signal distortion but this paper present CLEAN algorithm for UWB-BANs application, so object of this algorithm was Improved can be reduce noise and decrease signal distortion.

The basic algorithm to process the narrowband channel was introduced in [2] from image processing to estimate details of the time of arrival (ToA) but This paper implementation for UWB-BAN by involves the computation of the correlation coefficient function and the removal and the reconstruction of detected Fig. 2.

Similarity on both the dirty and the clean maps, respectively, for each iteration, as follows:

1. Initialize normalized cross-correlation between $v_r(t)$ and $v_t(t)$ normalized autocorrelation of $v_t(t)$ as $C_{cc}(\tau) = v_r(t) \odot v_t(t)$ and $C_{au}(\tau)$ respectively, and define the dirty and clean maps as $d_o(\tau) = C_{cc}(\tau)$ and $c_o(\tau) = 0$.
2. Compute $a_k = \max|C_{cc}(\tau)|$
3. If all $a_k <$ threshold, go to step 7.
4. Clean the dirty map by $d_t = d_{t-1} - (a_k \times C_{au}(\tau))$.
5. Update the clean map by $c_t = c_{t-1} - (a_k \times \delta(t_0))$.
6. Go to step 2.
7. The CIR is then $c_t = h_{clean}(t)$.

The above algorithms assume to be independent of the generator output, the measurement system, and the antennas used. Despite accurate estimation, the aforementioned algorithms are still based on a modeled approach. Therefore, their outputs (i.e., the CIR) must be carefully interpreted.

3.3 Bit Error Rate

This parameter will show performance when using CLEAN algorithm for improve channel impulse response on body that show in the equation by this paper using correlation coefficient for analysis about performance of signal when sent on body, so this research will show equation as:

$$H_{clean}(f) = F\{H_{clean}(t)\} \tag{3}$$

$$V_{rc}(f) = V_t(f)H_{clean}(f) \tag{4}$$

$$v_{rc}(t) = F^{-1}\{V_{rc}(f)\} \tag{5}$$

Performance Evaluation of UWB-BAN

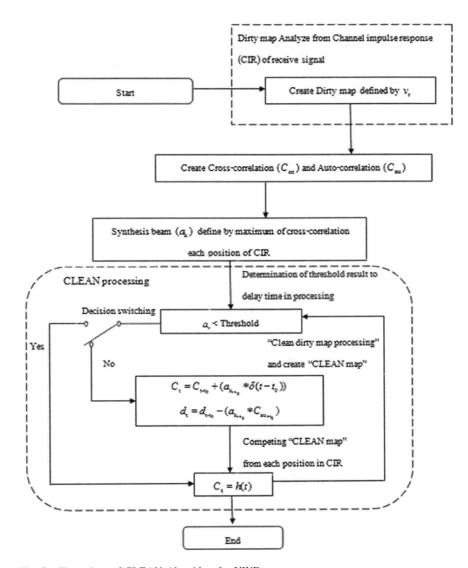

Fig. 2 Flow chart of CLEAN Algorithm for UWB system

The equation above shows $V_{rc}(f)$ is receive signal in frequency-domain when through UWB channel from processing by using CLEAN algorithm or "CLEAN map" and $v_{rc}(t)$ is receive signal in frequency-domain by using inverse fourier transform technique for analysis in correlation coefficient and that can be analysis bit error rate by use correlation coefficient in error function [4].

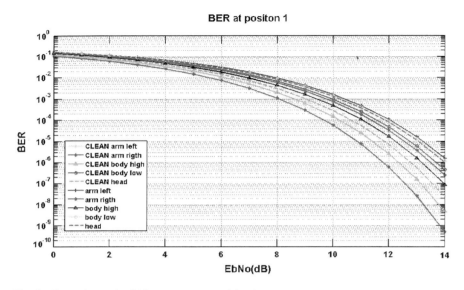

Fig. 3 Example result of bit error rate at position 1

4 Result and Discussion

This resulted will be show performance of system by using CLEAN technique for improve channel propagation. Figure 3 shows BER at standing position 1, it observe BER of signal when it through channel propagation from create by CLEAN technique or "CLEAN map" has BER better than in [1], Since performance of CLEAN technique can be decrease signal distortion from effect on body or multipath this figure position on body has lowest BER at arm right and position has worst at around stomach area when Observed Fig. 4 shows BER at standing position 4, so farthest between Tx and Rx antenna. The best of position on body (lowest BER) at head and arm right position has highest BER while previous work [3] the position of worst at head cause we known CLEAN can be improve channel to make reduce signal distortion.

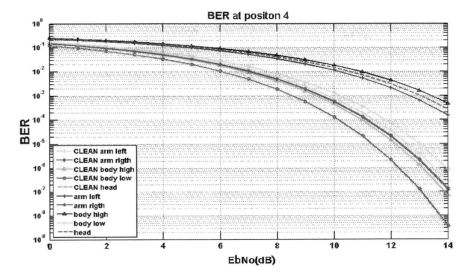

Fig. 4 Example result of bit error rate at position 4

5 Conclusion

This paper present improves UWB channel propagation by using CLEAN technique for solve MPC, signal distortion, noise from free space and effect on body. The resulted shows that CLEAN algorithm can be reduce noise and effect of multipath by show in PDP and solve problem of signal distortion in receiver side by show in BER when compare [3]. Specifically, CLEAN algorithm can be implementing about time of arrival (ToA) by time of receive has reduce in receiver side. However issue of CLEAN is amplitude of signal very low when compare using Extension Friis's formula and this algorithm modified from single-template CLEAN technique may be considered further.

Finally, for solve problem about amplitude of receive signal. That will present new receiver by using RAKE receiver for increase amplitude or power of signal at receiver side. Although, CLEAN algorithm has problem about amplitude of signal as very low but CLEAN technique still benefit for using on body and waiting develop.

References

1. Liu TCK, Kim DI, Vaughan RG (2007) A high-resolution, multi-template deconvolution algorithm for time-domain UWB channel characterization. Can J Elect Comput Eng 32(4):207–213
2. Yang W, Naitong Z (2006) A new multi-template CLEAN algorithm for UWB channel impulse response characterization, This project was supported by the key program of national natural science foundation of China, 27–30 Nov 2006

3. Arpasin D, Narongsak M, Promwong S (2011) Experimental characterization of UWB channel model for body area networks. In: ISPACs, 7–9 Dec 2011
4. Takada J, Promwong S, Hachitani W (2003) Extension of Friis' transmission formula for ultra-wideband systems. IEICE Tech Rep WBS2003-8/MW2003-20
5. Molisch AF (2005) Ultrawideband propagation channels: theory measurement and modeling. IEEE Trans Veh Technol 54(5):1528–1545
6. Xia L, Redfield S, Chiang P (2011) Experimental characterization of a UWB channel for body area networks. EURASIP J Wirel Commun Netw 2011(703239):11 (Hindawi publishing corporation)

A Study of Algorithm Comparison Simulator for Energy Consumption Prediction in Indoor Space

Do-Hyeun Kim and Nan Chen

Abstract In last couple of years many research have been done to develop the technology for minimizing the energy consumption, security and maintaining a comfortable living environment in smart buildings. In this paper, we propose comparison simulator to analyze algorithms such as averaging method, moving averaging, Low-pass filter, Kalman filter and Gray model for predicting energy consumption in indoor space. Additionally, we evaluate energy prediction algorithms in order to facilitate the testing. Our propose comparison simulator support to verify the performance of the prediction algorithms and effective estimation of energy usage in indoor environment.

Keywords Indoor space · Energy consumption prediction · Simulator

1 Introduction

Recent patterns of economic growth, worldwide energy consumption of resources, its conservation and agreements to minimize the global carbon emission and environments friendly consumption patterns are converged. Research in this connection still continues to develop the efficient technology for minimizing the energy consumption, security and maintaining a comfortable living environment in smart buildings. Such kind of building energy management systems provides energy saving effects through optimal operation of different equipment's by getting energy consumption information and comprehensively analyzing operation information of various equipment's connected to building automation system.

D.-H. Kim (✉) · N. Chen
Department of Computer Engineering, Jeju National University, Jeju, Korea
e-mail: kimdh@jejunu.ac.kr

N. Chen
e-mail: xuehu001@gmail.com

J. J. (Jong Hyuk) Park et al. (eds.), *Multimedia and Ubiquitous Engineering*,
Lecture Notes in Electrical Engineering 240, DOI: 10.1007/978-94-007-6738-6_109,
© Springer Science+Business Media Dordrecht(Outside the USA) 2013

Energy management and consumption in future buildings is predicted to analyze space–time form of energy consumption patterns or configures converted meaningful information in step on simple storing energy collected base of the interior or extracting statistical data. In particular, by drawing relation between energy data collected by importing ontology concept to building energy management system and space–time, user, subject etc., it is predicted on reflecting decision-making or policy for indoor energy savings or efficiency. The interior space of the building consists of floors, rooms, and hallways. Here we focus on displaying real-time energy data and demand on the map by showing room and the object in the room centrally. Demand energy expected value calculated by using predicted model and real-time energy data.

In this paper, we present algorithm comparison simulator for energy consumption prediction in indoor space. This is a part of energy information collector of indoor energy monitoring system and we focus on short-term energy consumption information using prediction algorithm for providing meaningful information based on real-time energy data.

The rest of this paper is structured as follows. In Sect. 2, we will describe prediction algorithms of indoor energy consumption in detail. In Sect. 3, we describe our proposed simulator and show how our design addresses the problems. Finally we conclude in Sect. 4.

2 Prediction Algorithms of Indoor Energy Consumption

In order to show the characteristics of the set of measured values usually we use the averaging method. Averaging method is the numerical summation observations. Therefore, the averaging method considers the mid value of all the observations. All the other values of averaging method are fluctuating around this midpoint. We apply the averaging method to predict the indoor energy consumption. Equation 1 below can be used to calculate the mid value using averaging method. In this equation "x_k" is the measurement value and 'i' is the accumulative value from 1 to "k". The accumulated value is divided by the number "k", the result is the predicted value in time "$k + 1$", "k" is the current time and "x_i" is the energy consumption value at time "i".

$$X_{k+1} = \frac{1}{k} \sum_{i=1}^{k} X_i \tag{1}$$

Figure 1 shows the sequence diagram of averaging method for energy prediction in indoor space. Initially import the simulated energy consumption value and then we get first stage prediction values and error values through the averaging method, prediction module and finally we correct the predicted value of the first stage through the Kalman filter to get the second stage prediction value and error values.

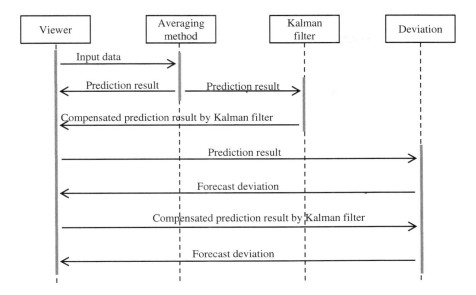

Fig. 1 Sequence diagram of energy prediction using averaging method

Disadvantage of prediction using averaging method is that, it is unable to reflect the dynamic changes in the system. In order to solve this problem the moving averaging method is introduce. The moving averaging algorithm is the modification of the averaging algorithm. The characteristic of the moving averaging method is that, it does not take all the measured value as the object and just select couple of values as objects. When it receives the latest data, the previous data will be replaces with this new data automatically. This method can be maintained at a certain number of sample data.

We calculate the moving averaging method for prediction of indoor energy consumption. At first we set "S" as move set. Then obtain a measured value "x_k" at the present stage. Second we accumulate values from the time of k to S. 'k' is the sum value and it is divided by move set value "S".

$$x_{k+1} = \frac{1}{n} \sum_{i=k-n}^{k} x_i \qquad (2)$$

Equation 2 is the moving averaging method for prediction. Here "k" represents the current stage. The value of "x_i" is the energy consumption value at time "i". The value of "n" is move set of the moving averaging method. First, moving averaging method energy prediction import the simulated energy consumption value and we get first stage prediction values and error values through the moving averaging method prediction module and correct the predicted value of the first stage through the Kalman filter to get the second stage prediction value and error values. Moving average method algorithm disadvantage is that, the new and old values use the equivalent specific gravity. Therefore, the Low-pass filter algorithm can adjust the proportion of value came into existence. High-frequency signals will be filtered out when the signal go through the low-frequency pass filter. The moving set value "S" and weight value "W" are adjusted. Then input the load value "x_k" of the current time. Accumulate all values from "k-s", moment measured value to a measured value of the present time k. Then we get moving average method value by using the result to divide moving set value and obtain the prediction value from the proportion of adjusting this value and the current time.

$$x_{k+1} = \alpha \bar{x}_{k-1} + (1 - \alpha)x_k \tag{3}$$

Equation 3 is the Low-pass filter prediction of the prediction formula. "x_k" is the energy consumption value at current time and "α" is the weight value.

First, Low-pass filter prediction algorithm import the simulate energy consumption value and then get first stage prediction values and error values through the Low-pass filter prediction module and correct the predicted value of the first stage through the Kalman filter to get the second stage prediction value and error values.

Kalman filter correct prediction error to get the right estimated value. Therefore, it reduces the error using the correction function of Kalman filter. As shown in Fig. 2, the Kalman filter comprises five computation phases. In Kalman model, input has the measurement value "Z_k" and after taking a series of internal calculations automatically we get the predicted value "x_k" as a output. In addition to the first time we run the initialization steps, while the remaining four are calculation steps.

The short-term indoor energy consumption predictions also use gray prediction algorithm. The gray model is proposed by Deng Julong who is from China Huazhong University of Science and Engineering in 1982. This model is used to predict the unknown information by the known part of the information. The Gray model can find out the law of development of the data by the collation of the raw data. The advantage of this model is that it does not require much of specimen data that is needed in case of moving averaging method. The model generates small specimen of data. Then new sequence is generated by accumulating and near averaging method sequence, generated by calculating the averaging method of adjacent elements. Figure 3 is prediction model of energy consumption using Gray model.

3 Simulator for Prediction Algorithms Comparison in Indoor

In order to predict the energy, load data must analyze the raw energy consumption data. We assume that the data in second interval determined by the room where energy consumption sensors are located.

A Study of Algorithm Comparison Simulator

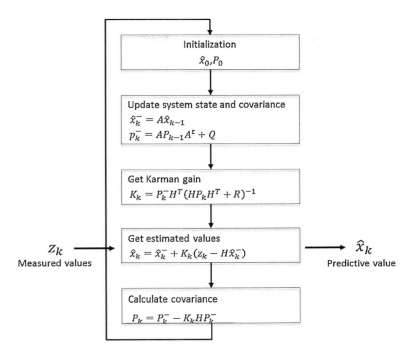

Fig. 2 Kalman filter prediction model

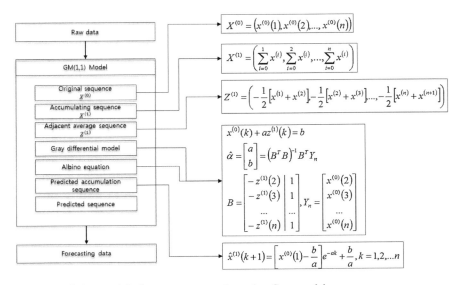

Fig. 3 Prediction model of energy consumption using Gray model

A room for the object and hour interval record load of energy 3600 s an hour, we must generate 3600 simulation energy consumption values. Room energy consumption is 5474 kW. So every second load will be 5474 kW centered fluctuated. We assume that this fluctuation is to meet the normal distribution.

$$f(x) = \frac{1}{\sqrt{2\Pi}\sigma} e^{-\frac{(x-\mu)^2}{2\sigma^2}} \qquad (4)$$

Equation 4 is a normal distribution equation. $f(x)$ is the probability of occurrence of the "x" events. μ is the overall averaging method of the event and σ is the offset value of the specimen. The following procedure describes how the program automatically generates simulated load data.

Random energy consumption data for simulation used various prediction algorithms (averaging method, moving average method, Low-pass filtering, Kalman filter and Gray model, compensated methods by Kalman filter) and it results in a linear graph display. Following picture shows a linear plot about 5474 kW as center for 3600 data that is generated by simulating data generator.

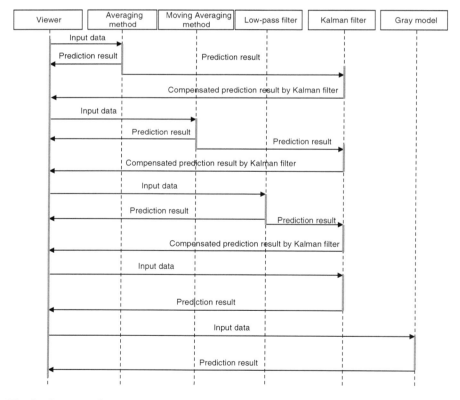

Fig. 4 Sequence diagram of the simulation of energy consumption prediction

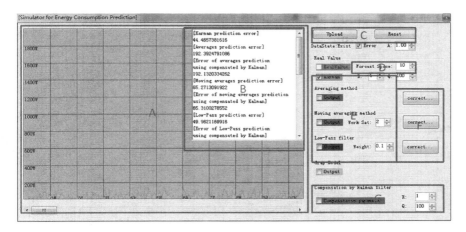

Fig. 5 Simulator for evaluating energy consumption prediction in indoor space

Figure 4 shows the data processing sequence diagram of the simulator. The simulator imports simulated data and display it in the form of a linear graph. It predicts indoor energy consumption values through the averaging method, moving average method, Low-pass filter, Kalman filter, Gray model prediction module. As the .Net environment does not have class library for dealing with the matrix. So we need to implement a matrix calculation module for matrix addition, subtraction, division and multiplication.

Figure 5 shows the indoor energy consumption performance evaluation simulator. This simulator is divided into seven areas. 'A' area is used to show linear diagram. Simulated data, prediction data, prediction error values are shown in an intuitive way through converting the values to a linear plot. 'B' area is used to display the error value of the predicted results of the prediction module. 'C' area is used to control the import simulation data through the "Upload" button to import a text file to get simulated data and clear it by "Reset" button. 'D' area is used to adjust the moving set value of the moving averaging method. 'E' area is used to set the parameters of each prediction model. 'F' area is used to correct the various modules of the prediction results through the Kalman filter. 'G' area is used to set the parameter "R" and "Q" of the Kalman filter.

4 Conclusions

In this paper, we present the simulator for comparing prediction methods of energy consumption in indoor space. Additionally, we evaluate energy prediction algorithms in order to facilitate the testing. We compare statistical prediction algorithms such as the averaging method, moving averaging method, Low-pass filter, Kalman filter, Gray model. Then we verified that prediction by using Low-pass filter and Kalman filter demonstrate the best performance. Additionally, we have developed energy prediction simulator and energy data generation tools based on normal distribution in order to facilitate the testing.

Acknowledgments This work was supported by the Industrial Strategic Technology Development Program funded by the Ministry of Knowledge Economy (MKE, Korea). [10038653, Development of Semantic based Open USN Service Platform]. (No. 2011-0015009). This work was supported by the National Research Foundation of Korea (NRF) grant funded by the Korea government (MEST) (No. 2011-0015009).

References

1. Jia J, Niu D (2008) Application of improved gray markov model in power load forecasting. Electr Util Deregul Restruct Power Technol 1:1488–1492
2. Crossley F (2007) Advanced metering for energy supply in Australia. Energy Future Australia Pty Ltd, Australia July
3. Gu T, Pung HK, Zhang DQ (2005) Service-oriented middleware for building context-aware services. J Netw Comput Appl 28:1–18

Energy Efficient Wireless Sensor Network Design and Simulation for Water Environment Monitoring

Nguyen Thi Hong Doanh and Nguyen Tuan Duc

Abstract In recent years, flooding in Ho Chi Minh city, Vietnam has become more and more serious, it affects much the lives and economics of the citizens. There are some reasons which caused this situation; among them the unfavorable natural conditions like rainfall, tide, are the crucial one. It is the reason why to build an intelligent network which is suitable for data collection precisely, rapidly and timely for water environment monitoring is essential. While traditional networks spend much money and effort to ensure continuous information in specific regions, Wireless Sensor Network is best known its advantages such as low cost, reliable and accurate over long term and required no real maintenance. Our contribution is that to build a network topology with suitable hardware and Low Energy Adaptive Clustering Hierarchy (LEACH) routing protocol in order to reduce energy consumption of whole network and prolong network life time. Fact has shown that LEACH protocol achieves a factor of 8–15 % energy improvement compared to direct transmissions.

Keywords Wireless sensor network · LEACH protocol · Energy consumption · Network lifetime

1 Introduction

A wireless sensor network consists of sensor nodes deployed over a geographical area for monitoring physical phenomena like water level, temperature. For a wireless sensor network design, to select a network topology and hardware are two

---

N. T. H. Doanh (✉) · N. T. Duc
School of Electrical Engineering, Vietnam National University—International University, Vietnam, China
e-mail: doanhnth04@hcmiu.edu.vn

N. T. Duc
e-mail: ntduc@hcmiu.edu.vn

J. J. (Jong Hyuk) Park et al. (eds.), *Multimedia and Ubiquitous Engineering*, Lecture Notes in Electrical Engineering 240, DOI: 10.1007/978-94-007-6738-6_110, © Springer Science+Business Media Dordrecht(Outside the USA) 2013

main points need to be considered for purpose of network energy efficient [1]. In traditional design, a network is designed with distance between nodes about 50–100 m. This means we have to use hundreds even thousands nodes to monitor a stage of a river. This, actually, is challenge because it required much money and time to maintain a huge system. Hence, firstly, a suitable hardware is chosen to guarantee transmission range at least from 2000 to 2500 m, saving energy and increasing network lifetime. After a long time study and test existed hardware in cutting edge technology, a table below makes a comparison some hardware [2].

	MRF24J40 MB (Microchip ZigBee)	CC1110 (Texas Instrument)	CC1120 (Texas Instrument)
Standard	IEEE 802.15.4	TI proprietary	IEEE 802.15.4 g
Data rate	250 Kbps	1.2–500 Kbps	200 Kbps
Output power	+20 dBm	10–12 dBm	+16 dBm
Received sensitivity	−102 dBm (at 250 Kbps)	−110 dBm (at 1.2 kBaud) −94 dBm (at 250 kBaud)	−123 dBm (at 1.2 kbps) −110 dBm (at 50 kbps)
Frequency band	ISM band (industrial, scientific and medical band)	868/915 MHz ISM band	868 MHz ISM band
Max. current consumption (RX/TX)	25 mA/130 mA	20.4 mA/ 36.2 mA	22 mA/45 mA

It can be concluded that the SmartRF Transceiver CC1120 evaluation board for Low Power RF transceiver devices from Texas Instrument is the best choice for environment applications.

1.1 LEACH for Energy Constrained Wireless Sensor Network Hardware

LEACH is traditional topology that helps to reduce energy consumption of whole network because of its energy efficient and simplicity. LEACH divides nodes into clusters with one node from each cluster serving as a cluster-head (CH). It randomly selects some predetermined number of nodes as cluster heads. CHs then advertise themselves and other nodes join one of those cluster heads whose signal they found strongest (i.e. the CH which is nearest to them). In this way, a cluster is formed. The CHs collect the data from their clusters and aggregate it before sending it to the other CHs or base station (BS) [3].

1.2 Network Topology for Water Environment Monitoring

In reliability, to build a wireless sensor network for water environment monitoring meets many difficult because of its complex geography. In addition, HCMC is locating in the lower basin of the Saigon-Dongnai river, HCMC has an area of 2093.7 km^2 and at 0.5–32.0 m above mean sea level. In our simulation, we assumed that there are 31 sensor nodes organized into 3 clusters. Distance between each nodes is about from 500 to 2500 m (Fig. 1).

2 Simulation Materials

2.1 Simulation Environment

I integrated LEACH code by making all of the changes as specified in the uAMPS changes package to the ns-allinone-2.34.tar.gz package, which latest version because the MIT uAMPS created by Massachusetts Institute of Technology on ns-2.e1b5 release, a very old version of the program. In order to change some parameters that are suitable for TI CC1120 hardware, we make some changes in physical layer. In MAC protocol, it is a combination of a carrier-sense multiple access (CSMA), Time division multiple access (TDMA), and a simple model of direct-sequence spread spectrum (DS-SS) to send data messages to avoid inter-cluster interference.

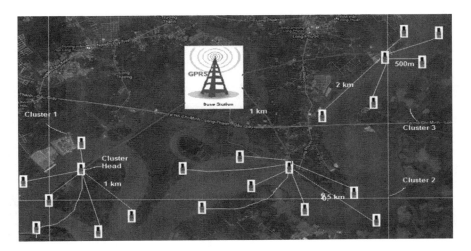

Fig. 1 Network topology for HCMC water monitoring

2.2 Channel Propagation Model

There are many propagation models that can describe comprehensively environment between transmitter and receiver, but we use two-ray ground propagation model is for simple observation [5]. Received power is calculated as:

$$P_r = \frac{P_t * G_t * G_r * h_t^2 * h_r^2}{d^4}$$

where P_r is the received power at distance d, P_t is transmitted power, G_t is gain of the transmitting antenna and G_r is gain of the receiving antenna.

In order the packet is successfully detected and to avoid the collision, there are three factors should to be considered. They are: Receive threshold value (RXthresh), Carrier Sense Threshold (CSthresh) and Capture Threshold (CPthresh). The CC1120 transceiver is built to be very sensitive. It is expected to decode signals with power as low as -123 dBm.

$$10 \times \log\left(\frac{P_{RXthresh}}{1 \text{ mW}}\right) = -123$$

Hence, $P_{RXthresh} = 5.0118e - 16$ W and CPthresh $> = 10$.

3 Simulation Analysis

In order to evaluate performance of this system, we make a comparison between direct transmission and LEACH in some parameters like energy consumption average, number of alive nodes and total end to end delay.

3.1 Simulation Parameters

Parameters	Value
Number of nodes	31
Number of clusters	3
Size of network	10000 m $\times$ 10000 m
BS location	6000 m $\times$ 8000 m
Initial energy of node	10 J
Propagation model	Tworayground
Time simulation	1000 s
Mac	CSMA/CA
Distance between nodes	500–2500 m

(continued)

(continued)

Parameters	Value
Frequency	868 MHz/139 MHz
CSThresh	1e−15 W (−12 dBm)
RXThresh	5.1594e−15 W (−12.3 dBm)

3.2 Simulation Performance and Evaluation

3.2.1 Total Energy Consumption

Energy consumption is the most importance factor that considered in our research. In order to dominate LEACH's advantage, I simulate scenarios that consisted 31 nodes (Fig. 2).

As can be seen, when the time increases, the energy consumption rises up significally. Energy consumption in direct transmission is always higher than LEACH from 0 to nearly 800 s. After that, energy consumption in former is constant because all nodes are dead while the later is continuously goes up due to LEACH's nodes still alive. In this situation, LEACH protocol actually saved energy of whole network than other routing protocol.

3.2.2 Lifetime of Network

There is a reduction in number of alive nodes in both routing protocol due to the far distance from nodes to base station. In LEACH routing protocol, sensor nodes witnessed a slight decrease while other protocol goes down strongly and ends at nearly 800 s (Fig. 3).

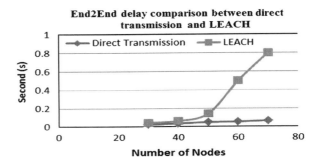

Fig. 2 Energy consumption average between direct transmission and LEACH

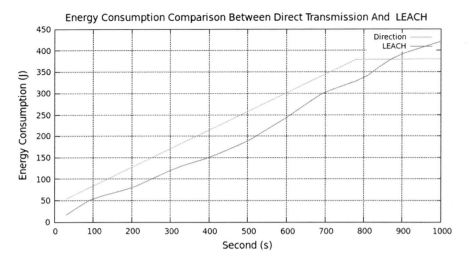

Fig. 3 Number of alive nodes comparison between direct transmission and LEACH

3.2.3 End to End Delay

Although LEACH protocol gives us the optimal energy consumption, end2end is not minimum in comparison with direct transmission. This is explained because LEACH has to undergo many steps before sending data to base station while direct transmission sees opposite trend. We cannot fix cluster head position because in case of far cluster head, it cannot send message to base station successfully (Fig. 4).

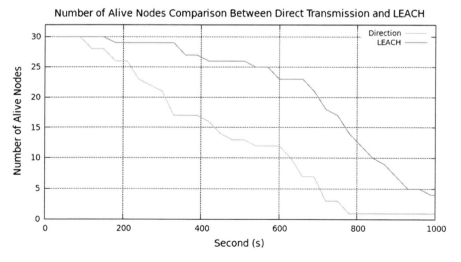

Fig. 4 End to end delay comparison between direct transmission and LEACH

4 Conclusion

In this paper, we integrated the wireless sensor network hardware for water environment monitoring in a large geography. For this model, LEACH protocol is also studied in order to reduce the energy consumption and prolong network lifetime of whole network.

References

1. Padmavathy TV, Gayathri V, Indhumathi V, Karthiga G (2012) Energy constrained reliable routing optimized cluster head protocol for multihop under water acoustic sensor networks. Int J Netw Secur Appl (IJNSA) 4(3):57
2. Texas Instruments. High performance RF transceiver for narrowband systems, CC1110, CC1120. Datasheet
3. Tong H, Zheng J (2011) An energy and distance based clustering protocol for wireless sensor networks. 978-1-61284-307-0/11/$26.00 ©2011IEEE
4. The LEACH code is only compatible with ns-2.1b5. It was originally developed by WendiB.Heinzelman and is no longer being updated. http://www.mtl.mit.edu/uamps/research/cad.shtmlresearchgroups/icsystems/
5. Rao GS, Vallikumari G (2012) A beneficial analysis of node deployment schemes for wireless sensor networks. Int J Advanced Smart Sensor Netw Syst (IJASSN) 2(2):33–43

An Energy Efficient Reliability Scheme for Event Driven Service in Wireless Sensor Actuator Networks

Seungcheon Kim

Abstract Wireless sensor network has been evolving to wireless sensor actuator networks (WSAN), which is supposed to provide more dynamic services based on events from sensor nodes. WSANs require event reliability to support more accurate services. The optimum number of event notification messages from sensor nodes is the prerequisite for the further service reliability in WSANs. In this paper, we provide the analysis about how the number of event notification affects the energy consumption of actuator node in wireless sensor actuator networks.

Keywords WSAN · Event reliability · Service reliability · Energy consumption · Actuator

1 Introduction

Sensor network usually refers to the network that provides information services by using sensing information from sensor nodes [1]. To provide sensing information, it should have an infrastructure of sensor nodes and sink node, which is dealing with how to deliver and process the sensing information. Those efforts resulted in making Zigbee, IEEE 802.15.4, as PHY and MAC of wireless sensor network (WSN). And lots of works are concentrated in providing Internet services in WSNs with IPv6 [2].

But now WSN is evolving toward the wireless sensor actuator network (WSAN) that can react promptly to the event of an interest based on the sensing information from sensor nodes. In WSANs, actuator nodes are required to do specific actions or services after exchanging query or response messages with

---

S. Kim (✉)
Department of Information and Communication Engineering, Hansung University,
Seoul, Korea
e-mail: kimsc@hansung.ac.kr

J. J. (Jong Hyuk) Park et al. (eds.), *Multimedia and Ubiquitous Engineering*,
Lecture Notes in Electrical Engineering 240, DOI: 10.1007/978-94-007-6738-6_111,
© Springer Science+Business Media Dordrecht(Outside the USA) 2013

sensor nodes [3]. The reliability matters a lot especially in WSAN, which can be categorized into the network reliability, data transmission reliability and the service reliability. Among those, service reliability means how the sink or actuator node can verify if the wanted event happens. Usually the number of event notification message is used to verify and confirm that the wanted event happens [4].

As time goes, QoS of WSN has been issued depend on the services of WSN. One of the main concerns in WSN is how to reduce the energy consumption when it exchanges sensing information. Since the energy consumption is directly connected with the life time of sensor network, nearly all the searches have been done from MAC to Application Layers.

When it comes to WSAN, the energy consumption in WSAN is also a crucially important matter especially in aspect of network life time. Usually actuator node uses more energy than normal sensor node since it has to perform a specific action as a reaction to the events. The more the actuator node reacts to the unwanted event, therefore, the shorter the network life time becomes [5]. We need to find out the ways to reduce the energy consumption in WSAN.

This paper is focused in investigating the proper number of event notification messages that satisfies the service reliability with minimum energy consumption of actuator node in WSAN and provides the scheme to reduce the energy consumption in WSAN.

2 Wireless Sensor Actuator Network

2.1 Reliability in WSAN

Reliability in WSN/WSAN can be categorized as Network reliability, Data transmission reliability and Service reliability.

First, the network reliability of WSAN means the network stability itself. WSN/WSAN can change its shape or configurations when its purpose of service has been changed or modified. Even though those changes happen, WSN/WSAN should be operated stably.

Second, the data transmission means to deliver the sensing data safely and reliably from sensor node to sink node. For this, we need to reconsider the way the data is transmitted in wireless sensor network environments. Typically there are three methods for data delivery: End-to-End Delivery, Hop-to-Hop Delivery and Cache Mode Delivery. Those are probably compared in aspect of reliability and efficiency. And also it should be revised to deal with the transmission errors for the data transmission reliability. Considerations on the use of ACK, NAK or Timer should be done for better transmission reliability.

Lastly, the service reliability or event reliability is very far from the data transmission reliability in WSAN. The event is the most important interest in WSAN, where the specific reactions are related to the specific events. The service reliability, therefore, means that the actuator node can verify whether it receives proper events from sensor nodes. Usually WSANs are composed of lots of sensor nodes. The event reliability depends on the number of event notification messages and confirming messages from sensor nodes. Sometimes sensor nodes can send a notification message on a wrong event. Therefore actuator node is recommended to react when it receives a certain number of notification messages from sensor node to identify the event.

2.2 Sensor–Actuator Coordination

The first thing we should think about in WSAN is how to assign an actuator node to the corresponding sensor nodes considering the reacting pattern of actuator node over the event. To determine the mutual relationship between sensor node and actuator node, we should look into how many actuator nodes are required to respond to the event notification and which actuator node are to be selected in the end. And also we need to categorize the event according to the task that is assigned to the actuator in WSANs.

2.2.1 Single-Actuator Task Versus Multi-Actuator Task

The most general example of WSANs is the single actuator task (SAT) that is performed by single actuator on an event. In this case, the actuator normally performs a reaction towards the event after exchanging some messages with sensor nodes about the event.

On the contrary, multi actuator task (MAT) requires actuators to do a more complicated task than SAT. In this case, task for actuators might be able to be assigned according to the mutual communication between sensor node and actuator nodes and the proper reaction plan for a specific event can be changed after communication among actuator nodes.

2.2.2 Centralized Decision Versus Distributed Decision

The decision making method in WSANs can be divided into centralized decision (CD) and distributed decision (DD). In CD, the event is notified to the server in the information center through sink node and the server determines how the actuator reacts to the event as shown in Fig. 1.

In DD, the actuator has the right to decide whether it will perform a reaction or not when it receives the event notification from sensor nodes. The result of reaction is reported to the server through sink node by the actuator node.

In DD, the reaction is much prompter than in CD but the actuator node takes the responsibility of wrong decision of the event, which result in shortening the life

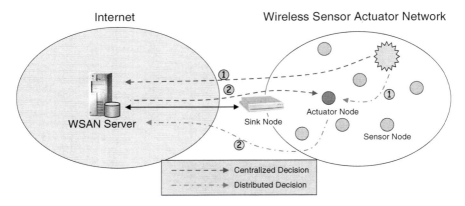

Fig. 1 Decision making in WSAN

time of actuator and WSAN since the energy consumption of reaction is much larger than the energy consumption needed for communication.

3 Analysis of the Event Reliability

In WSAN, the number of notification messages for an event is definitely related to the energy consumption of actuator, which can affect the whole WSAN. Here we are going to investigate the relationship between the number of notification message and the energy consumption of actuator node.

3.1 Probability of Receiving k Messages During τ in Actuator

Assuming that every sensor node can send only one notification message when if detects an event, the probability of receiving k-messages from n-sensor nodes is possibly calculated as follows.

$P_e(k) = \binom{n}{k} p^k (1-p)^{n-k}$, where p is the event detection probability in sensor node.

If we assume that the average event rate is λ and the interval time of each event is independent, we can use exponential distribution for the probability density function of the interval between two events in WSANs. Then, the probability of event during a specific time τ can be described like this.

$$P_\tau = \int_0^\tau \frac{1}{\lambda} e^{-\lambda t} dt = 1 - e^{-\lambda \tau} \tag{1}$$

Therefore, the probability that the actuator could receive k messages from n-sensor nodes during τ is finally described like this.

$$P_e(k) = \binom{n}{k} p_\tau^k (1 - p_\tau)^{n-k}, \text{ where } P_\tau = 1 - e^{-\lambda t} \tag{2}$$

3.2 Power Consumption in Actuator

The power consumption in actuator (E_{av}) is composed of the energy (E_{act}) that is needed for action for specific service and the energy (E_c) for exchanging data with sensor nodes.

$$E_{av} = E_{act} + E_c$$

E_T requires actuator node to receive more than the threshold (T) number of messages from sensor nodes to verify the event in WSANs. If the number of messages in actuator node is less than T, actuator is supposed not to react on the event notification messages from sensor node and thus it consumes only communication energies needed to exchange only k messages with sensor nodes. Usually the energy for actuator to perform a service in WSANs is more than 10 times as big as the one needed for communication. Therefore, the average energy consumption in actuator is possibly described like this.

$$\begin{aligned} E_{av} &= Prob[k < T] \times E_c + Prob[k \geq T] \times (E_{av} + E_{act}) \\ &= \sum_{i=1}^{T-1} p_e(i) \times E_c + \sum_{i=T}^{n} p_e(i) \times (E_c + E_{act}) \end{aligned} \tag{3}$$

When the number of sensor nodes is set to 20 and the average arrival rate of notification message is set to 0.01/s and the duration time of accepting notification messages in actuator is set to 2 s, the result is shown in Fig. 2.

4 Proposed Scheme

Consequently we found that the number of the notification messages is connected to the energy consumption of the actuator in WSAN. Based on the information we found between the number of the notification messages and the energy consumption of actuator in WSAN, we propose a method that can regulate the threshold number of notification messages for the event to minimize the energy

Fig. 2 Variation of energy consumption

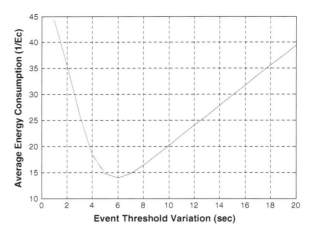

consumption of the actuator. The idea is described in Fig. 3. As shown in the Fig. 3, after receiving the notification messages from sensor nodes, actuator needs to start the timer for event checking and count the number of event notification message to see if it can be considered as a real event. Upon seeing the number of messages going over the threshold, it would perform the scheduled action for the

Fig. 3 Proposed algorithm

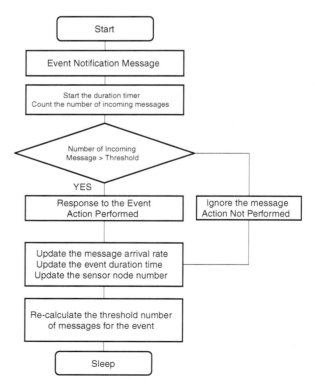

corresponding event. Otherwise, it would ignore the notification messages from sensor nodes and continue to count the number of the messages and finally update the information required to calculate the threshold of notification messages for an event.

With the updated information about arrival rate of event notification message, event checking duration and the number of sensor nodes, the proper number of event notification messages for minimum energy consumption is calculated with the Eq. (3). This is also updated in an actuator for later detection of an event.

5 Conclusions

In WSAN, the number of notification messages for an event is definitely related to the energy consumption of actuator, which can affect the whole WSAN. Here we investigated the relationship between the number of notification message and the energy consumption of actuator node. And we propose a new scheme that can regulate the threshold of event notification messages for an event and reduce the energy consumption of an actuator in WSAN. The proposed scheme is only used in an actuator and does not affect the other sensor nodes in WSAN. And also it is considered as a self-adaptive scheme for better performance. With the minor improvements, the proposed scheme is expected to contribute in alleviate the burden of energy consumption problem in WSAN/WSN.

Acknowledgments This Research was financially supported by Hansung University.

References

1. Akyildiz IF, Su W, Sankarasubramaniam Y, Cayirci E (2002) A survey on sensor networks. IEEE Commun Mag, August 2002
2. Akyildiz IF, Kasimoglu IH (2004) Wireless sensor and actor networks: research challenges. Ad Hoc Netw 2(4):351–367
3. Yick J, Mukherjee B, Ghosal D (2008) Wireless sensor network survey. Comput Netw 52(12):2292–2330
4. Xi F (2008) QoS challenges and opportunities in wireless sensor/actuator networks. Sensors 8:1099–1110
5. Xia F, Tian Y-C, Li Y, Sun Y (2007) Wireless sensor/actuator network design for mobile control applications. Sensors 7:2157–2173

Efficient and Reliable GPS-Based Wireless Ad Hoc for Marine Search Rescue System

Ta Duc-Tuyen, Tran Duc-Tan and Do Duc Dung

Abstract Based on work in wireless ad hoc network for Marine Search and Rescue Sys-tem (MSnR) system, this paper presents an improving GPS-based wireless ad hoc network capable of providing location and emergency service to small fishing boats by improving the weak sea-to-land wireless radio link from small fishing boats to the central in-land stations. The proposed approach provides continuous report and monitoring of all boats and its locations for searching and rescuing process during emergencies. The message priority assignment allows the system to operate more efficient and reliable when one or several boats boat in distress. System model, communication mechanism and network simulation results are presented.

Keywords Ad hoc network · Global positioning system (GPS) · Medium access control (MAC)

1 Introduction

According to our recent study in real-time location monitoring of small fishing [1], we propose an improve model of GPS-based mobile ad hoc network which will pro-vide more efficient and reliable sea-to-land communication link from small fishing boats to central base-stations. The proposed network combines the Global Positioning System (GPS) service (positioning) with a wireless ad hoc network (wireless communication). In addition, the message priority assignment allows the system to operate more efficient and reliable when one or several boats boat in distress. For the land-to-sea communication link, the proposed network simply

---

T. Duc-Tuyen (✉) · T. Duc-Tan
VNU University of Engineering and Technology, 144 Xuan Thuy,
Cau GiayHa Noi, Viet Nam
e-mail: tuyentd@vnu.edu.vn

D. Duc Dung
Samsung Mobile R&D Center, Samsung Viet Nam, Yen Phong, Bac Ninh, Viet Nam

---

J. J. (Jong Hyuk) Park et al. (eds.), *Multimedia and Ubiquitous Engineering*,
Lecture Notes in Electrical Engineering 240, DOI: 10.1007/978-94-007-6738-6_112,
© Springer Science+Business Media Dordrecht(Outside the USA) 2013

utilizes the existing coastal radio network; hereby greatly reduces cost and simplifies network design.

The paper is organized as follows: In Sect. 2 we introduce model of Marine Search and Rescue System and benefits of the proposed system. Section 3 present the communication mechanism of proposed system, included packet structure, medium access control and routing protocol. The simulation scenarios and results is shown in Sect. 4. Finally, in Sect. 5 we will conclude the paper and discuss future possibilities for this research.

2 Monitoring, Searching and Rescuing System

Operates of the proposed GPS-based wireless ad hoc network for marine monitoring, searching and rescuing system as follows. The forward link from land to sea uses the current coastal radio system with high-power base-stations along the coastal line to transmit weather and relevant information to all fishing. The forward link can reach all of fishing boats based on the high power transmission. In contrast to the forward link, the return link from sea to land is limited by a low transmit power and a low antenna height mounted on small fishing boats. This problem is remedied by the proposed wireless ad hoc network, which enables the return link to be first established between fishing boats and then finally connected from the closest boats to the central base-station. The routing protocol of the proposed system is location-based routing. A data packet being routed in these links contain a boats identification number (boat ID), its current GPS location, and a short message. In order to improve the reliability and efficiently of proposed system in case of emergency mode, a message priority assignment is added. It means that messages are divided into three kinds: emergency message, the node's message and normal forwarded message with descending priority. In addition, to identify boats position and establish a wireless connection, a commercially off-the-shelf integrated GPS receiver (location determination) and a programmable digital signal processing (DSP) board (wireless routing and medium access control) are added to the existing low-power radio on fishing boats. Finally, packets are transmitted over existing radio system. Compare with the previous system, the priority mode ensure the probability of successful reception of emergency message from the boat in distress to the base station.

3 Communication Mechanism

3.1 Medium Access Control (MAC)

In this work, we propose a hybrid MAC solution. The core idea of the hybrid MAC protocol is a combination and smart selection of time division medium access (TDMA) and carrier sense multiple access (CSMA). The operating principle is as

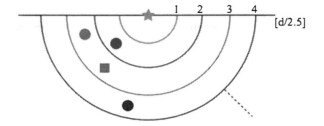

Fig. 1 A graphical representation of a coverage area

follows. A coverage area is divided into an equal number of concentric circles (red, blue, green, black in Fig. 1, which share the same central point at the local base station (red star). These circles have a radius equal to a multiple of 2.5 km (r = n × 2.5) where n is an integer number (n = 1; 2; 3…). Nodes residing on different concentric circles [i.e. circles with n = 2 (blue) and n = 4 (black)] are given two different time slots to transmit/receive. This is similar to TDMA. When there are two or more nodes within the same concentric circle [i.e. circle with n = 3 (green)], these nodes employ CSMA protocol with a carrier detection capability to avoid transmit collision. In addition, handshaking is implemented in conjunction with the CSMA/CA scheme in order to reduce the hidden node problem in the same concentric circle. When node A wants to transmit data, it will broadcast the Ready-to-Send (RTS) message to check the medium idle or not. The received node forward this message to all node in its radio cover-age. If no other nodes in the same concentric circle with A makes a transmission at-tempt, the medium is idle. A Clear-to-Send (CLS) message is replied to node A. Else, the medium is busy and A must be wait for a random time and then transmits again.

3.2 Data Packet Format

Figure 2 shows a format of each data packet used in the proposed network. A SNI is the source node identifier or the ID of a transmit node. A DNI is the destination node identifier or the ID of a receive node. When DNI is set to all 0s, the data are sent to base station nodes (default mode). A RNI is the relay node identifier or the ID of a message forwarded node. It set same the SNI field at message's source. At immediately node, which the message will be, relay, the RNI is the node identifier. A source position (POS) is the position of node at the time a packet is sent. It includes latitude (SLA) and longitude (SLO) of the source nodes at that time. An initial time (IT) field is the original time when a packet is first transmitted. It contains two sub-fields: UDate and UTime, which are Coordinated Universal Time (UTC) date and time of day of the GPS signal. A hop count (HC) indicates the number of nodes that a packet has been traversed. It is set to 0 at the

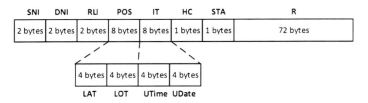

Fig. 2 Format of a data packet used in the proposed GPS-based wireless ad hoc network

source node and incremented by 1 at each subsequent forwarding node. A status (STA) field indicates the status of the source node. It is set to all 0s when the source node is in its normal mode of operation, and set to all 1s if it is in an emergency mode, i.e., when the source node seeks help from other ships or the base-station. Reserved field a reserved area for other purposes if any.

3.3 Routing Protocol

The main use of the proposed ad hoc network is monitoring and reporting ships location, therefore it does not require a high data rate and can support a large network delay tolerance [2, 3]. Based on these requirements, we adopt a modified hybrid proactive-passive, location-based routing protocol similar to the DREAM protocol first proposed by Basagni in [4]. In the proposed hybrid routing protocol, a lookup table (LUT) at each node is updated when a node receives a data packet. Unlike DREAM or other passive protocol, the LUT only contains locations of one-hoop neighboring nodes. As a result, the LUT is significantly smaller and hence less time is required updating the LUT entries.

In this protocol, each node in our ad hoc network broadcasts its packet m to all of its one-hoop neighbors regardless of their directions and locations. Default that the emergency message will be broadcast immediately at receiving node. It means the emergency packet is served with the highest priority. In other hand, when the neighbor receives a normal message, it decides to relay or drop the packet based on the relative position of the neighbors to the base station. In cases the neighbor decides to relay the packet, it first switches from receive to transmit mode, then re-transmits the packet m and updates its forward table (FT). Other node repeats the same procedure, until the final destination (base station) is eventually reached. The routing algorithm is described next. Algorithm 1 shows operating of a node in it is receive mode when Algorithm 2 shows two different transmit scenarios for a given node.

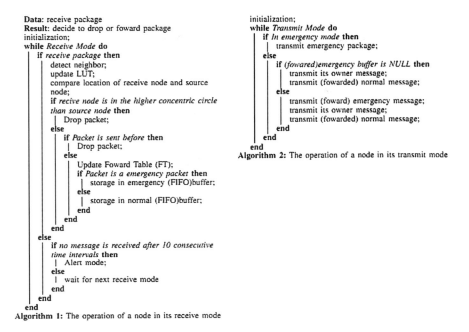

Algorithm 1: The operation of a node in its receive mode

Algorithm 2: The operation of a node in its transmit mode

4 Network Simulation and Results

The proposed routing algorithm and MAC are implemented and tested in OMNET++ tool with INETMANET framework, a discrete event simulation tool for Mobile Ad Hoc Networks. The simulation environment is a Mobile Ad Hoc Network consists of 20–100 nodes in a 1000 × 1000 area. We assume that each node will be active by a 8.1 MHz radio frequency with bandwidth of 100 kHz and support data rate of 200 Kbps. The correspondence convergence circle area of each node with radio radius of 100. The simulated area is considered as a two dimensional square and nodes movement freely throughout the area. The movement of node has been simulated according to Random Way-point model with maximum movement speed is 10 m per second (36 km per hour or 19.4 nautical mile per hour).

In order to evaluate the performance and the efficiently of the proposed system, a set of simulation were operated with duration of 2000 s. We select a set of parameters to show the efficiency of our algorithm in two case: (1) no emergency message is generated and (2) at least one emergency message is generated in network. These parameters include:

Fig. 3 Average packet latency for different number of nodes (N) and different number of emergency message

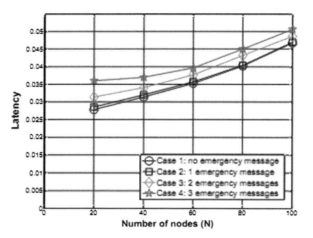

4.1 Average Packet Latency

Figure 3 is a plot of the average packet latency versus the number of network nodes. That is the average time taken by any node to receive the message. In our case, this is the time taken by base station to receive the message from boat. In case of no emergency message is generated, the packet latency increases proportionally with the number of network nodes. This is expected since a large network necessary causes a long delay because packets must be relayed through more number of nodes. In other hand, when at least one emergency packet is generated, the average packet latency has slightly decreased. However, when the number of emergency message at the same time is increased, the average packet latency has increased. This is due to fact that the crashed node only broadcast the emergency message and does not participate in forwarding the message from other

Fig. 4 Average latency of emergency packet for different number of emergency packets at the same time

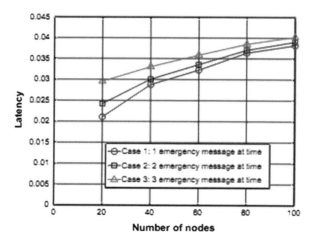

Fig. 5 The average packet delivery ratio (PDR) for different number of nodes

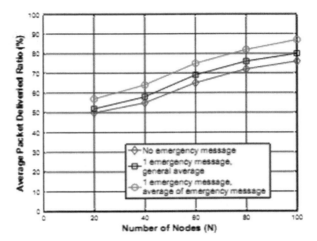

node in network. In worst case, the network will be collapse when the number of network nodes in an emergency increased.

The average latency of emergency packet is shown in Fig. 4. We can see that latency increases faster in small-scale network than it in large-scale network. This can be explained by the increase in emergency nodes reduce the ability to forward messages on the network, especially for small networks.

4.2 Average Delivery Ratio

The average packet latency increases when increase size of network. In addition, a larger ad hoc network increases the possibility of packet delivery, too. Figure 5 indicates the average packet delivery ratio for normal message in case of normal mode and emergency and normal message in case of emergency mode with 1 emergency message is generated.

5 Conclusions

A wireless ad hoc network capable of providing location service to small fishing boats and improving the weak sea-to-land wireless radio link from small fishing boats to the central in-land stations has been proposed. The proposed location-based routing protocol (DREAM) and hybrid MAC have been verified using OMNET++ tool with INETMANET Framework. The simulation results show that the message priority assignment will be help to improve the reliability and the great potential of the proposed concept for marine monitoring, searching, and rescuing applications.

Acknowledgments The authors want to thank the project CN 12.04 of VNU University of Engineering and Technology for the financial support of this work.

References

1. Do DD, Nguyen HV, Tran NX, Ta TD, Tran TD, Vu YV (2011) Wireless ad hoc network based on global positioning system for marine monitoring, searching and rescuing (MSnR). In: APMC 2011: Asia-Pacific microwave conference, Melbourne, pp 1510–1513
2. Karl H, Willig A (2005) Protocols and architectures for wireless sensor networks. Wiley
3. Zhang Z (2009) Routing protocols in intermittently connected mobile ad hoc networks and delay-tolerant networks. In: Boukerche a (ed) Algorithms and protocols for wireless and mobile ad hoc networks. Wiley
4. Basagni S, Chlamtac I, Syrotiuk VR (1998) A distance routing effect algorithm for mobility (DREAM). In: Proceeding of the annual international conference on mobile computing and networking, USA

Improved Relay Selection for MIMO-SDM Cooperative Communications

Duc Hiep Vu, Quoc Trinh Do, Xuan Nam Tran and Vo Nguyen Quoc Bao

Abstract In this paper, we propose two relay selection algorithms based on the signal-to-noise ratio (SNR) and the eigenvalue which achieve improved bit error rate (BER) performance compared with the previous one based on the mean square error (MSE) at the same complexity order.

Keywords Cooperative communication · Relay selection · MIMO · SDM

1 Introduction

The modern wireless communication is developing very fast in order to meet the human demand for high-speed data access. The last decade has witnessed various successful developments in the air interface technology. The most important development is probably the multiple-input multiple-output (MIMO) transmission [1]. MIMO transmission systems can be implemented in the form of either the transmit diversity [2] or spatial division multiplexing (SDM) [3]. The aim of the transmit diversity is to achieve diversity gain in order to reduce the bit error rate (BER) and thus increasing the link reliability. This transmit diversity scheme is also known as the space–time block code [2]. The MIMO-SDM systems, on the other hand, aim at achieving multiplexing gain in order to increase the spectral efficiency. For a centralized MIMO system where multiple antennas are placed at the transmitter and the receiver, it was shown in [4] that there is a trade-off between the diversity and multiplexing gain. This means that the centralized

D. H. Vu (✉) · Q. T. Do · X. N. Tran
Le Quy Don Technical University, 236 Hoang Quoc Viet, Cau Giay, Ha Noi, Vietnam

V. N. Q. Bao
Post and Telecommunications Institute of Technology, 11 Nguyen Dinh Chieu Str., District 1, Ho Chi Minh City, Vietnam

J. J. (Jong Hyuk) Park et al. (eds.), *Multimedia and Ubiquitous Engineering*, Lecture Notes in Electrical Engineering 240, DOI: 10.1007/978-94-007-6738-6_113, © Springer Science+Business Media Dordrecht(Outside the USA) 2013

MIMO systems do not achieve full diversity and multiplexing gain at the same time. In order to achieve both diversity and multiplexing gain, a so-called MIMO-SDM cooperative communication system was proposed in [5]. In this work the authors proposed three distributed relay selection schemes and a linear minimum mean square error (MMSE) combining scheme which achieve full diversity and full multiplexing gain at the same time. Among the three proposed selection algorithms based on maximum channel matrix norm, maximum channel harmonic mean, and minimum MSE, the MSE-based algorithm was shown to achieve the best BER performance [5]. In this paper, based on the idea of [6] for the case of MIMO-SDM, we developed two relay selection algorithms based on the signal to noise ratio (SNR) and eigenvalue. The two proposed algorithms have improved BER performance over the MSE-based algorithm while requiring the same complexity order.

2 System Model

We consider a MIMO-SDM cooperative communication network similar to [5] as illustrated in Fig. 1. The network consists of a source and a destination communicating with each other with the help of a relay node via a relaying path. Without loss of generality, we assume that all nodes (including source, destination and intermediate) are equipped with $N = 2$ antennas for both transmission and reception. There are K capable intermediate nodes $k = 1, 2, \ldots, K$ between the source and the destination. Based on a distributed relay selection protocol [5, 7] the K intermediate nodes will interact with one another to select the best one to act as the relay (denoted by the index r). The channels between nodes are assumed flat uncorrelated Rayleigh fading and unvarying during a transmission period. We denote $\mathbf{H}^{sd}$, $\mathbf{H}^{sk}$, $\mathbf{H}^{kd}$ the channel matrices between the source and the destination, the source and the intermediate node k and the destination, respectively. The channel between a node a and a node b is denoted as the matrix $\mathbf{H}^{ab} =$

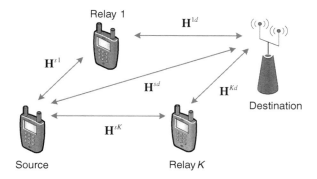

Fig. 1 System model of the MIMO cooperative communications

$[h_{11}^{ab}, h_{12}^{ab}; h_{21}^{ab} h_{22}^{ab}]$ where h_{mn}^{ab} is the channel between the mth antenna of node b to the nth antenna of a.

The communication between the source and the destination involves two phases: relay selection and signal transmission. The relay selection is done using the distributed protocol as mentioned above while the signal transmission uses two time slots. In the first slot, the source transmits a signal vector $s = [s_1, s_2]^T$ consisting of two symbols s_1 and s_2 from the two antennas to both the destination and the relay. Here the superscript T denotes the matrix transpose. The received signal vector at the destination and the relay is given by $\mathbf{y}_1 = \mathbf{H}^{sd}\mathbf{s} + \mathbf{z}_1, \mathbf{x}_r = \mathbf{H}^{sr}\mathbf{s} + \mathbf{z}_r$, where $\mathbf{z}_1$ and $\mathbf{z}_r$ are the noise vector at the destination and relay r, respectively. In the second time slot, the relay performs amplifying-and-forwarding (AF) the received signal $\mathbf{x}_r$ to the destination. The amplification matrix $\mathbf{G}_r$ is a diagonal matrix with the amplification factor used for the ith branch given by [5]: $g_i^r = \sqrt{E_s/N(E_s/N\|\mathbf{h}_i^{sr}\|^2 + 1)}$, where E_s is the transmit symbol energy and $\mathbf{h}_i$ is the ith row of the appropriate channel matrix. The received signal at the destination during the second time slot is given by [5]

$$\mathbf{y}_2 = \mathbf{H}^{rd}\mathbf{G}_r\mathbf{x}_r + \mathbf{z}_2 = \mathbf{H}^{rd}\mathbf{G}_r\mathbf{H}^{sr}\mathbf{s} + \mathbf{H}^{rd}\mathbf{G}_r\mathbf{z}_r + \mathbf{z}_2 \tag{1}$$

where $\mathbf{z}_2$ is the noise vector at the destination in the second time slot. The destination will combine the received signal vectors $\mathbf{y}_1$ and $\mathbf{y}_2$ to obtain the received signal vector $\mathbf{y}$. Define $\mathbf{y} = [\mathbf{y}_1^T, \mathbf{y}_2^T]^T$, $\mathbf{H} = [(\mathbf{H}^{sd})^T, (\mathbf{H}^{srd})^T]^T$ and $\mathbf{z} = [\mathbf{z}_1^T, (\mathbf{H}^{rd}\mathbf{G}_r\mathbf{z}_r + \mathbf{z}_2)^T]^T$, the system equation is given by

$$\mathbf{y} = \mathbf{Hs} + \mathbf{z}. \tag{2}$$

3 Proposed Relay Selection Algorithms

The proposed selection algorithms are performed in a distributed manner as described in [5, 7]. The intermediate nodes $k = 1, 2, \ldots, K$ are assumed to know the forward channel from itself to the destination $\mathbf{H}^{kd}$ and the backward channel from it back to the source $\mathbf{H}^{ks}$. Due to reciprocity the channel $\mathbf{H}^{sk}$ is assumed to be the same as $\mathbf{H}^{ks}$. Each node k will calculate the channel quality index (CQI) of the relaying path via itself. The node with the largest CQI, denoted by κ be selected as the relay.

Table 1 Eigenvalue-based relay selection algorithm

Input: $K, \mathbf{H}^{sk}, \mathbf{H}^{kd}$
For $k = 1$ to K
Calculate $\lambda_{1,2}^{sk}, \lambda_{1,2}^{kd}$ using (3)
Select $\lambda^k = \min\{\lambda_{1,2}^{sk}, \lambda_{1,2}^{kd}\}$
$CQI_k = \lambda^k$
$\kappa = \arg\max_{k}\{CQI_k\}$
End
Output: node κ as relay r

3.1 Eigenvalue-Based Relay Selection

The idea of selecting relay based on eigenvalues comes from the fact that in the MIMO systems eigenvalue of the channel matrix is considered the power gain of the channel [8]. As a result, the channel which has larger eigenvalues will have better power gain. The eigenvalues of the channel $\mathbf{H}^{ab}$ is the solutions to the following characteristic equation $\det\left(\mathbf{H}^{ab} - \lambda\mathbf{I}\right) = 0$, where $\mathbf{I}$ is an $N \times N$ identity matrix and $\det(\cdot)$ represents the determinant of the matrix formed by $\left(\mathbf{H}^{ab} - \lambda\mathbf{I}\right)$. For an $N \times N$ complex matrix $\mathbf{H}^{ab}$ there are at most N distinct eigenvalues. For the 2×2 channel matrix $\mathbf{H}^{ab}$ considered here there are two eigenvalues given by

$$\lambda_{1,2}^{ab} = \frac{\left(h_{11}^{ab} + h_{22}^{ab}\right) \pm \sqrt{\left(h_{11}^{ab} + h_{22}^{ab}\right)^2 - 4\left(h_{11}^{ab}h_{22}^{ab} - h_{12}^{ab}h_{21}^{ab}\right)}}{2} \tag{3}$$

In order to obtain the associated CQI each intermediate node k will first select $\lambda^k = \min\left\{\lambda_{1,2}^{sk}, \lambda_{1,2}^{kd}\right\}$ and then calculate $CQI^k = \lambda^k$. The max–min selection algorithm based on eigenvalues is summarized as pseudocodes in Table 1. It is worth noting that for the case of using a larger number of antennas, i.e. $N > 2$, the calculation of the eigenvalues as used in (3) is not straightforward and a more complicated calculation algorithm should be used.

3.2 SNR-Based Selection Algorithm

As SNR is inversely proportional to BER, selecting a relaying path with better SNR promises lower BER. In order to perform SNR-based relay selection, we assume that the destination uses the linear MMSE detector proposed in [5]. Based on this assumption intermediate nodes will calculate the received SNR at the destination via its relaying path. The CQI associated with each path will be assigned based on the calculated SNR. From [5] we can write the combining

Table 2 SNR-based algorithm

Input: $K, \mathbf{H}^{sk}, \mathbf{H}^{kd}, \mathbf{G}_k$
For $k = 1$ to K
Calculate SNR_n^k using (6)
$\text{SNR}^k = \min\{\text{SNR}_1^k, \text{SNR}_2^k\}$
$\text{CQI}_k = \text{SNR}^k$
$\kappa = \arg\max_k\{\text{CQI}_k\}$
End
Output: node κ as relay r

weight matrix that the destination would use to combine the relaying signal with that from the direct path as follows

$$\mathbf{W}_2^k = \left[\frac{E_s}{N}\mathbf{H}^{skd}\left(\mathbf{H}^{skd}\right)^H + \sigma_{z_k}^2\mathbf{H}^{kd}\mathbf{G}_k^2\left(\mathbf{H}^{kd}\right)^H + \sigma_{z_2}^2\mathbf{I}_2\right]^{-1}\frac{E_s}{N}\mathbf{H}^{skd}, \qquad (4)$$

where $\mathbf{H}^{skd} = \mathbf{H}^{sk}\mathbf{G}_k\mathbf{H}^{kd}$, $\sigma_{z_k}^2$ and $\sigma_{z_2}^2$ are the variance of the noise induced at node k and at the destination during the second time slot, respectively. For simplicity, we assume that $\sigma_{z_d}^2 = \sigma_{z_1}^2 = \sigma_{z_2}^2$ and that $\sigma_{z_k}^2 = \sigma_{z_d}^2$. Since we use the assumption that the source sends two parallel streams, the estimated symbol $\tilde{s}_n$ of s_n if node k acts as the relay would be

$$\tilde{s}_n = \left(\mathbf{w}_{2,n}^k\right)^H\mathbf{y}_2 = \left(\mathbf{w}_{2,n}^k\right)^H\mathbf{H}^{kd}\mathbf{G}_k\mathbf{H}^{sk}\mathbf{s} + \left(\mathbf{w}_{2,n}^k\right)^H\left(\mathbf{G}_k\mathbf{H}^{kd}\mathbf{z}_k + \mathbf{z}_2\right) \qquad (5)$$

where $\mathbf{w}_{2,n}^k$ is the nth column of $\mathbf{W}_2^k$ and $(\bullet)^H$ denotes the Hermitian operation. The received SNR at the destination is defined as follows

$$\text{SNR}_n^k = \frac{E_s\left\|\left(\mathbf{H}^{sk}\right)^H\mathbf{G}_k^H\left(\mathbf{H}^{kd}\right)^H\mathbf{w}_{2,n}^k\right\|^2}{N\left(\sigma_{z_k}^2\left\|\mathbf{G}_k^H\left(\mathbf{H}^{kd}\right)^H\mathbf{w}_{2,n}^k\right\|^2 + \sigma_{z_2}^2\left\|\mathbf{w}_{2,n}^k\right\|^2\right)} \qquad (6)$$

From this equation the SNR-based relay algorithm as summarized in Table 2.

4 Complexity Analysis

In order to compare the complexity of the proposed algorithms with that based on the MSE, we perform detailed calculation of the number of addition/subtraction, multiplication and division for all the case of complex–complex, complex–real, and real–real operations. These computational operations will be then converted into floating points (flops) for comparison. The complexity of the SNR-based algorithm involves mainly with calculating (5) and (7). The approximated complexity of the SNR-based algorithm is given by $C_{\text{SNR}} = 36N^3 + 34N^2 + 28N + 5$ [flops]. The main complexity of the eigenvalue-based algorithm is used for computing the eigenvalues of the two square matrices $\mathbf{H}^{sk}$, $\mathbf{H}^{kd}$ both of the same size $N \times N$. The complexity for calculating the eigenvalues using singular value decomposition is $72N^3$ [flops]. The complexity of the MSE-based relay selection algorithm mainly involves with calculating equations (30), (34) and (35) in [5]. The number of computational operations required by the MSE-based algorithm is $C_{MSE} = 20N^3 + 26N^2 + 4N + 3$ [flops]. Therefore, it is clear that all the algorithms have the same complexity order $O(N^3)$.

5 Simulation Results

In order to demonstrate the advantage of the proposed algorithms, we have performed Monte-Carlo simulations to obtain the average BER. In the first simulation, we use a simple model with three nodes, i.e., source, relay, and destination. In order to select the relay, we assume that there are two intermediate nodes within the coverage area of the source and destination. The proposed algorithms will be used to select the better node as the relay. The channels between the source to intermediate nodes and from the intermediate nodes to the destination are all assumed to undergo flat uncorrelated Rayleigh fading. All nodes are equipped with two antennas and transmit BPSK signal over the two parallel branches. The average symbol energy of each node is normalized to E_S. The destination employs the MMSE detector in [5] to estimate the transmit signal. In the second simulation, we use a similar model but the ratio E_b/N_0 is fixed while the number of intermediate nodes is increased to analyze the effect of selecting a relay from a large number of nodes. In all simulations, BER of the MSE-based algorithm is also plotted for comparison. The average BER curves obtained using the proposed algorithms and the MSE based are shown in Fig. 2. It can be seen clearly from the figure that both the proposed algorithms have the same BER performance at the low E_b/N_0 region but outperform the MSE-based for large E_b/N_0. Specifically, at BER $= 10^{-5}$, the proposed eigenvalue based algorithm has about 0.5 dB better E_b/N_0 while the SNR based achieves up to 2.5 dB improvement. It is also clear that the gap between the SNR-based algorithm and the MSE-based is much larger than that of the eigenvalue-based. Figure 3 illustrates the BER performance of the three algorithms obtained at $E_b/N_0 = 10, 20$ dB for the case of $N = 2, 3, 4, 5, 6$. It still can be seen that the two proposed algorithms achieve better BER performance than the MSE-based, particularly at high E_b/N_0. However, similar to [5] it is interesting to note that increasing the number of intermediate nodes does not achieve better improvement. This is the inherent property of the MIMO-SDM cooperative communication as explained in [5].

Fig. 2 BER performance of different selection algorithm, $N = 2$; 2 select 1

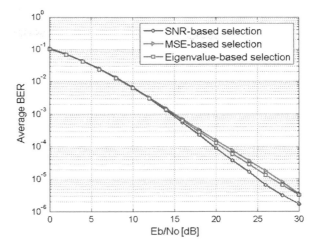

Fig. 3 BER performance versus the number of candidate nodes

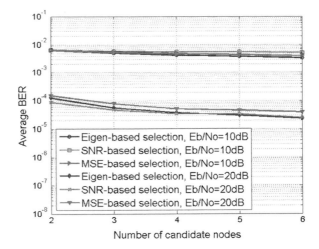

6 Conclusions

In this paper, we have proposed two relay selection algorithms based on SNR and eigenvalue for the MIMO-SDM cooperative communication networks. Both the proposed algorithms have better BER performance over the previous MSE-based algorithm. We have also carried out detailed complexity analysis to show that both the proposed algorithms and the MSE-based have the same complex order $O(N^3)$. The proposed SNR-based algorithm was shown to be the best candidate in terms of both BER performance and required complexity.

Acknowledgments This work is sponsored by National Foundation for Science and Technology Development (Nafosted) under project number 102.03-2012.18.

References

1. Foschini GJ, Gans MJ (1998) On limits of wireless communications in a fading environment when using multiple antennas. Wireless Pers Commun 6(3):311–335
2. Alamouti SM (1998) A simple transmit diversity technique for wireless communications. IEEE J Sel Areas Commun 16(8):1451–1458
3. Wolniansky P, Foschini GJ, Golden GD, Valenzuela RA (1998) V-BLAST: an architecture for realizing very high data rates over the rich-scattering wireless channel. In: The URSI international symposium on signals, systems, and electronics, Italy, pp 295–300
4. Zheng L, Tse D (2003) Diversity and multiplexing: a fundamental tradeoff in multiple antenna channels. IEEE Trans Inf Theory 49(5):1073–1096
5. Tran XN, Nguyen VH, Bui TT, Dinh TC (2012) Distributed relay selection for MIMO-SDM cooperative networks. In: IEICE Trans Commun E95-B:1170–1179
6. Paulraj A, Heath RW (2001) Antenna selection for spatial multiplexing systems with linear receivers. IEEE Commun Lett 5(4):142–144
7. Bletsas A (2003) A simple cooperative diversity method based on network path selection. IEEE J Sel Areas Commun 24(3):659–672
8. Andersen JB (2000) Array gain and capacity for known random channels with multiple element arrays at both ends. IEEE J Sel Areas Commun 18(11):2172–2178

Freshness Preserving Hierarchical Key Agreement Protocol Over Hierarchical MANETs

Hyunsung Kim

Abstract Recently, Guo et al. proposed an efficient and non-interactive hierarchical key agreement protocol applicable to mobile ad-hoc networks, which is a try to solve the open question: How can secrets be established if an adversary can eavesdrop on every message exchange? However, their protocol does not support freshness of the established session key that key agreement protocols should have. Thereby, we propose a freshness preserving hierarchical key agreement protocol over the hierarchical MANETs. Compared with other existing protocols, the proposed protocol offers much better performance on the bandwidth consumption, the computational cost, and the storage cost.

Keywords Hierarchical key agreement · Security protocol · Mobile ad-hoc network · Information security · Cryptography

1 Introduction

Mobile Ad hoc Network (MANET) is a collection of mobile nodes that are dynamically and arbitrarily located in such a manner that communication between nodes does not rely on any fixed network infrastructure. The absence of static infrastructure and centralized administration makes MANETs to be self organized and relying on the cooperation of neighboring nodes in order to find the routes between the nodes for reliable communication. Hence, the performance of MANETs is highly dependent on collaboration of all the participating nodes. The more the number of nodes that participate in packet routing, greater aggregate bandwidth and shorter routing paths can be realized. It will further minimize network partition in the case of failures. The MANET has a wide range of applications in diverse fields

H. Kim (✉)
Department of Cyber Security, Kyungil University, Kyungsansi, Kyungpook, Korea
e-mail: kim@kiu.ac.kr

ranging from low power military wireless sensor networks to large scale civilian applications, emergency search and rescue operations [1, 2].

The major challenges in providing secure authenticated communication for MANETs come from the following unique features of such networks [3]: lack of a fixed reliable public key infrastructure, dynamic network topology due to high mobility and joining/leaving devices, energy and resource constrained nodes with limited storage, communication and computation power, lack pre-distributed symmetric keys shared between nodes, high-level of self-organization, vulnerable multi-hop wireless links, etc.

Key agreement is a fundamental tool for secure communication, which lets two nodes in a MANET agree on a shared key that is known only to them, thus allowing them to use that key for secure communication [4–10]. In environments where bandwidth is at a premium, there is a significant advantage to non-interactive schemes, where two nodes can compute their shared key without any interaction. Daffie-Hellman key agreement protocol is an example of a non-interactive scheme [6]. But the nodes in the Diffie-Hellman protocol must still get each other's public keys, which require coordination. To minimize the required coordination, one may use identity-based key-agreement that provides each node with a secret key that corresponds to that node's name [7]. In this setting, the non-interactive identity-based scheme of Sakai et al. in [8] is based on bilinear maps. Recently, Guo et al. proposed an efficient and non-interactive hierarchical key agreement protocol, named as HNAKA, applicable to mobile ad-hoc networks [9]. However, the HNAKA could not support freshness for the established session key to support non-interactive. Thereby, the purpose of this paper is to remedy the HNAKA to preserve freshness. The proposed revision, named as $HNAKA_{fresh}$, establishes a secure channel by setting up a fresh session key between any two nodes in the MANETs. The $HNAKA_{fresh}$ could support security and robustness over the hierarchical MANETs.

2 Related Works

This section reviews Guo et al.'s hierarchical non-interactive identity-based authenticated key agreement protocol (HNAKA) based on the pairing [8, 9]. Furthermore, we show that the HNAKA does not provide session freshness, which is one of important features for the key agreement protocol.

2.1 Guo et al.'s Hierarchical Non-interactive Authenticated Key Agreement

Similarly to other identity-based authenticated key agreement protocols, Guo et al.'s HNAKA requires a private key generator (PKG) and consists of three phases: system setup, private key generation, and key agreement [9]. Let k be the

Freshness Preserving Hierarchical Key Agreement Protocol

security parameter, G and G_T be two cyclic groups of prime order q, and $\hat{e}: G \times G \to G_T$ be a bilinear pairing. They denote by G^* the non-identity elements set of G. They assume that public keys (identities or IDs) at depth l are vectors of elements in $(G^*)^l$. The jth component corresponds to the identity at level j. They later extend the construction to public keys over $\{0, 1\}^*$ by first hashing each component I_j using a collision resistant hash $H_1: \{0, 1\}^* \to G$.

Setup To generate system parameters for our scheme of maximum depth l,

- Select a random generator $P_0 \in G$ and choose master keys s_i $(i = 1, \ldots, l)$ uniformly at random from Z_q and compute the PKG's public key as $s_i P_0$

The resultant public parameters and the master key are $params = \{q, G, G_T, P_0, \hat{e}, H_1, s_1 P_0, \ldots, s_l P_0\}$ and master—key $= \{s_1, \ldots, s_l\}$.

KeyGen For user A with the identity tuple $(ID_1, \ldots, ID_t)$, given the master secret key $\{s_1, \ldots, s_l\}$ and the system public parameters, compute $P_i = H_1(ID_i)$ and $d_i = s_i P_i$ $(i = 1, \ldots, t)$. Send $D_A = (s_1 P_1, \ldots, s_t P_t, s_{t+1}, \ldots, s_l)$ to user A via an authenticated and private channel. Indeed, given user A with the identity tuple $(ID_1, \ldots, ID_t)$ and his parent C with the identity tuple $(ID_1, \ldots, ID_{t-1})$, C can compute the private key D_A for user A using his own private key $D_C = (s_1 P_1, \ldots, s_{t-1} P_{t-1}, s_t, \ldots, s_l)$ as follows:

- C computes $P_t = H_1(ID_t)$ and $s_t P_t$ using his private key.
- C sends $D_A = (s_1 P_1, \ldots, s_t P_t, s_{t+1}, \ldots, s_l)$ to user A via an authenticated and private channel.

Suppose user A and user B want to establish a shared session secret key: A has the identity $(ID_1, \ldots, ID_m)$ and B has the identity $(ID'_1, \ldots, ID'_n)$, where $m > n$.

Key agreement To establish a shared session secret key, A and B conduct the following tasks:

- A computes $P'_i = H_1(ID'_i)$, where $i = 1, \ldots, n$, and B computes $P'_j = H_1(ID'_j)$, where $j = 1, \ldots, m$.
- A computes

$$sk_{AB} = \hat{e}\left(s_1 P_1, P'_1\right) \cdots \hat{e}\left(s_n P_n, P'_n\right) \cdot \hat{e}\left(s_{n+1} P_{n+1}, P'_n\right) \cdots \hat{e}\left(s_m P_m, P'_n\right) \quad (1)$$

- B computes $sk_{BA} = \hat{e}\left(P_1, s_1 P'_1\right) \cdots \hat{e}\left(P_n, s_n P'_n\right) \cdot \hat{e}\left(P_{n+1}, P'_n\right)^{s_{n+1}} \cdots \hat{e}\left(P_m, P'_n\right)^{s_m}$

2.2 No Key Freshness Support Problem in HNAKA

A key establishment/agreement process among the participants should guarantee that each shared session key is fresh, i.e. has not been reused by one of the participants. This also means that a key used in one cryptographic association has not been used in another association. Thus, the session key needs to be changed over time since a key may be compromised during pre-deployment or operational phases of communication networks. In Guo et al.'s HNAKA, each party computes sk_{AB} via the Eq. (1), which depends on both of their own private key and identity tuples but not on the session dependent random value. Thereby, the HNAKA does not provide key freshness. No freshness support means that the established session keys in different sessions are always the same, which could provide some means or useful information to attacker. One of serious effects is traffic analysis attack, which is focused on traffic flow identification, traffic flow tracking, or disclosing application-level information.

3 Freshness Preserving Hierarchical Key Agreement Protocol

This section proposes a freshness preserving hierarchical key agreement protocol, named as HNAKA$_{fresh}$ over hierarchical MANETs, which is a remedy of Guo et al.'s HNAKA. This protocol falls into the same phases with the HNAKA: system setup, private key generation, and key agreement. The first two phases are the same with the HNAKA but the last one is different due to support key freshness to the protocol. Figure 1 shows the hierarchical key generation model for the HNAKA$_{fresh}$.

The assumptions in the HNAKA$_{fresh}$ is the same as in Guo et al.'s protocol, which are user A with the identity tuple $(ID_1, \ldots, ID_m)$ and private key tuple $D_A = (s_1P_1, \ldots, s_mP_m, s_{m+1}, \ldots, s_l)$ at level m and user B with the identity tuple

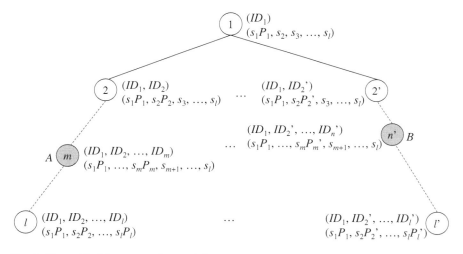

Fig. 1 Hierarchical key generation model

Freshness Preserving Hierarchical Key Agreement Protocol 931

$(ID'_1, \ldots, ID'_n)$ and private key tuple $D_B = (s_1P'_1, \ldots, s_nP'_n s_{n+1}, \ldots, s_l)$ at level n, where $m > n$ and l is the depth of the hierarchy as shown in Fig. 1.

Key agreement To establish a shared session secret key, A and B conduct the following tasks:

- A chooses a random number r_1, computes $R_1 = r_1P_m$, $P'_i = H_1(ID'_i)$, where $i = 1, \ldots, n$, $sk_{AB} = \hat{e}(s_1P_1, P'_1)\cdots\hat{e}(s_nP_n, P'_n)\cdot\hat{e}(s_{n+1}P_{n+1}, P'_n)\cdots\hat{e}(s_mP_m, P'_n)^{r1}$ and $MAC_1 = H_1(sk_{AB}, R_1)$, and sends $\{R_1, MAC_1\}$ to B.
- B computes $P'_j = H_1(ID'_j)$, where $j = 1, \ldots, m$, and $sk_{BA} = \hat{e}(P_1, s_1P'_1)\cdots\hat{e}(P_n, s_nP'_n)\cdot\hat{e}(P_{n+1}, P'_n)^{sn+1}\cdots\hat{e}(R_1, P'_n)^{sm}$.
- B assures the correctness of the established session key only if the validity check of MAC_1 is successful by comparing it with $H_1(sk_{BA}, R_1)$.

4 Security Analysis

This section provides security analysis of the $\text{HNAKA}_{\text{fresh}}$. Although it is important to provide a formal security proof on any cryptographic protocols, the formal security proof of the protocols remains one of the most challenging issues for cryptography research. Therefore, we follow the approaches used in [10].

4.1 Computational Problems

Bilinear map captures an important cryptographic problem, i.e., the Biliniear Diffie-Hellman (BDH) problem, which was introduced by Boneh and Franklin in [7]. The security of the $\text{HNAKA}_{\text{fresh}}$ relies on a variant of the BDH assumption.

Let G and G_T be two groups of a prime order q. Suppose that there exists a bilinear map $\hat{e}: G \times G \to G_T$. We consider the following computational assumptions

- Bilinear Diffie-Hellman (BDH): For a, b, and $c \in_R Z_q^*$ and given aP, bP and cP, computing $\hat{e}(P, P)^{abc}$ is hard
- Decisional Bilinear Diffie-Hellman (DBDH): For a, b, c and $r \in_R Z_q^*$, differentiating $(aP, bP, cP, \hat{e}(P, P)^{abc})$ and $(aP, bP, cP, \hat{e}(P, P)^r)$ is hard

4.2 Security Analyses

Our security analysis is focused on verifying the overall security requirements for the $\text{HNAKA}_{\text{fresh}}$ including passive and active attacks as follows.

Proposition 1 *The HNAKA$_{fresh}$ is secure against passive attack.*

Proof We assume that an adversary is success if the adversary could learn some useful information from the intercepted messages. We show that probability to succeed in learning them is negligible due to the difficulty of the underlying cryptosystem, the BDH problem and the DBDH problem.

1. A completeness of the key agreement protocol is already proven by describing the run of the protocol in Sect. 3.
2. If the adversary is passive adversary, all the adversary can gather are as follows: the identity tuples $(ID_1, \ldots, ID_m)$ of A and $(ID'_1, \ldots, ID'_n)$ of B and the message $\{R_1, MAC_1\}$. However, it is negligible to find the key related information from them due to the difficulty of the underlying cryptosystem, the keyed hash function.

Finally, we could say our protocol is secure against passive attack.

Proposition 2 *The HNAKA$_{fresh}$ is secure against active attack.*

Proof We assume that an adversary is success if the adversary finds the session key sk (from now on, we use sk instead of using sk_{AB} or sk_{BA} for simplicity) or the session key related private key information $\{s_1, s_2, \ldots, s_l\}$. Therefore, we show that probability to succeed in finding them is negligible due to the difficulty of the underlying cryptosystem, the BDH problem and the DBDH problem.

1. The acceptance by all entities means that MAC_1 in the corresponding message is successfully verified. That is, MAC_1 is verified successfully by using the correct session key sk and the session dependent random related value R_1. We show that if it is the case that entities accept the message and continue the session, then the probability that the adversary have modified the message being transmitted is negligible. And the only way for the adversary to find the session key or security related information is to solve the difficulty of the underlying cryptosystem, the BDH problem and the DBDH problem.
2. Now, we consider the active adversary with following cases.

   (a) There is no way that an adversary could get the secret information $\{s_1, s_2, \ldots, s_l\}$ due to the difficulty of the BDH problem and the DBDH problem.
   (b) An adversary cannot impersonate A or B to cheat the others in the hierarchy. That is the attacker cannot generate valid message without deriving the correct session key sk, since the attacker cannot pass the verification of MAC_1 in the protocol.
   (c) An adversary cannot compute session key from any useful information from outside of the hierarchy nodes or gathered information from the network, which are the identity tuples $(ID_1, \ldots, ID_m)$ of A and $(ID'_1, \ldots, ID'_n)$ of B and the message $\{R_1, MAC_1\}$ due to the difficulty of the underlying cryptosystem, the BDH problem and the DBDH problem.

Finally, we could say the HNAKA$_{fresh}$ is secure against active attack.

5 Conclusion

Recently, Guo et al. proposed an efficient and non-interactive hierarchical key agreement protocol applicable to mobile ad-hoc networks, which is a try to solve the open question: How can secrets be established if an adversary can eavesdrop on every message exchange? However, their protocol could not support freshness of the session key. Thereby, we proposed a freshness preserving hierarchical key agreement protocol, named $HNAKA_{fresh}$, over the hierarchical MANETs. Compared with the other existing protocols, the $HNAKA_{fresh}$ offers much better performance on the bandwidth consumption, the computational cost, and the storage cost.

Acknowledgment This work was supported by the Kyungil University Research Fund and was also partially supported by the National Research Foundation of Korea Grant funded by the Korean Government (MEST) (NRF-2010-0021575).

References

1. Conti M, Giordano S (2007) Multihop ad hoc networking: the theory. J IEEE Commun Mag 45(4):78–86
2. Gopalakrishnan K, Uthariaraj VR (2012) Collaborative polling based routing security scheme to mitigate the colluding misbehaving nodes in mobile ad hoc networks. Wireless Pers Commun 67:829–857
3. Dutta R, Dowling T (2011) Provably secure hybrid key agreement protocols in cluster-based wireless ad hoc networks. Ad Hoc Netw 9:767–787
4. Yang H, Luo H, Ye F, Lu S, Zhang L (2004) Security in mobile ad hoc networks: challenges and solutions. IEEE Wireless Commun 11(1):38–47
5. Anjum F, Mouchtaris P (2007) Security for wireless ad hoc networks. Wiley, Hoboken
6. Diffie W, Hellman ME (1976) New directions in cryptography. IEEE Trans Inf Theory 22(6):644–654
7. Boneh D, Franklin M (2001) Identity-based encryption from the weil pairing. Lect Notes Comput Sci 2139:213–229
8. Sakai R, Ohgishi K, Kasahara M (2000) Cryptosystems based on Pairings. In: Proceedings of the symposium on cryptography and information security 2000
9. Guo H, Mu Y, Lin Z, Zhang X (2011) An efficient and non-interactive hierarchical key agreement protocol. Comput Secur 30:28–34
10. Kim H (2011) Location-based authentication protocol for first cognitive radio networking standard. J Netw Comput Appl 34:1160–1167

A Deployment of RFID for Manufacturing and Logistic

Patcharaporn Choeysuwan and Somsak Choomchuay

Abstract Widespread adoption of Radio Frequency Identification System (RFID) has been widely used to develop and improve the Supply chain management. The Logistic control system has been focused in this work. In details, the system emphasizes at packaging, inventory and warehouse utilization that actually holds a good impact to final packing process and shipping process. The RFID hardware's specification are UHF frequency at 860–930 MHz on ISO/IEC 18000-6C Class 1 Gen 2 which well known to utilize in supply chain management. Data stored in the tag are the compressed version of all pieces inside the box. No indexing is further needed as required by the conventional barcode system. For the analysis of business deployment, an economic analysis tool is employed. The Net Present Value (NPV) of 5 year-period has been carried out for the purpose of medium-term to long-term investment.

Keywords Radio frequency identification system · Supply chain · Logistic Net present value

1 Introduction

RFID technology has grown rapidly for authentication applications on automatic identification and widely considered to represent the next generation beyond ubiquitous 1-D and 2-D barcode. One of the key emerging technologies is the

---

P. Choeysuwan (✉) · S. Choomchuay
Department of Electronic Engineering, King Mongkut's Institute of Technology
Ladkrabang, Bangkok, Thailand
e-mail: patcharapornc@yahoo.com

S. Choomchuay
e-mail: kchsomsa@kmitl.ac.th

J. J. (Jong Hyuk) Park et al. (eds.), *Multimedia and Ubiquitous Engineering*,
Lecture Notes in Electrical Engineering 240, DOI: 10.1007/978-94-007-6738-6_115,
© Springer Science+Business Media Dordrecht(Outside the USA) 2013

opportunities and challenges related to the deployment of Electronic Product Code (EPC) and RFID as the unique identify object of each manufacturer. The relative impact on EPC/RFID attractiveness will vary based on each product's unique characteristic and each company's specific requirement.

EPC/RFID tag as well as barcode system are generally used for the same purpose as the product identification. To consider a tiny product or pallet packed in a small box with identification number either in a form of a barcode or an EPC tag. Many of these pallets are again packed in a rather bigger box. That box is of course identified unique number. Now, the real time awareness without going back to recheck the database can be achieved. Likewise manual checking can still be possible with some cost savings. These increase overall performance of supply chain.

Nowadays, RFID is rapidly becoming a cost-effective technology. One of the benefit potential is to reduce the cost of the system deployment that can be done in three major keys: hardware, software and services. Tag costs are one of the key considerations in RFID deployment. Tags are available in various design forms and memory size. Basically based on the application or where the tags are used.

In this work we concentrate on the hardware side that the user's memory size of the tag is a constraint. The main contributions are (1) the method to manage the big information to be able to store in a limited memory portion, and (2) the study whether such an effort can be deployed economically. In Sect. 2, RFID system and components are given in brief. In Sect. 3, production line manufacturing of the final packing process and shipping process are elaborated. Then in Sect. 4, the details of RLE-like and lossless compression techniques are given. In Sect. 5, the deployment of RFID system that the compressed information had been accommodated to the tag is detailed out. In Sect. 6, the calculation of cost and investment consideration by using economic tool of NPV is studied. Finally, this work is concluded in Sect. 7.

2 RFID Systems

2.1 Tags

An RFID system uses wireless radio communication technology to uniquely identify tagged objects. RID tag usually holds an amount of memory for the both the system and the users. For example, the EPC Class 1 Gen 2 hold 4 memory banks; Reserve bank, EPC bank, Tag ID bank, and user memory bank. The user-defined data storage size is made vary from hundreds of byte to few kilobytes. The basic RFID system contains three major types of tags.

- *Passive tags*: There is no power source inside of the design. The power is provided by radio frequency wave sending from the reader to activate the tag. The most common RFID tags being used today are passive UHF RFID tags. The

ISO18000-6C defines the communication interfaces for UHF bands that work on 860–960 MHz frequency ranges. The EPC Class-1 Gen 2 standard is well known for UHF tags that also operate as much longer range. We also use this type of the tag in this experiment.

- *Active tag*: The internal power source is installed inside. The signal strength is higher than the passive one and can be read from the longer distance. The useful is applied to transportation truck system.
- *Semi-passive tags*: The design is a combined strength of each passive and active tag. This type of tags are commonly used in the car park system, smart shelve department store, and etc.

2.2 Reader/Interrogator

The components can transmit/receive the information from tags using the radio frequency waves via antenna for communication. The information can be read/written to tags based on the circuits and its associated protocol to protect the anticollision.

2.3 Middleware

The applications connect to hardware as reader to collect the unique number of each tag for processing the data by real-time acknowledgement and keep in back-end database for purpose analysis.

3 Production Line Manufacturing

Shown below in Fig. 1 is one of the final packing process; starting from the receiving of the parts-bundled plastic pack from the cleanroom. Then in the next process is the shipping process. This is shown in Fig. 2. One may notice some redundant procedures, in particular the barcode scanning. This regard is the major concern of our implementation that RFID can be utilized effectively.

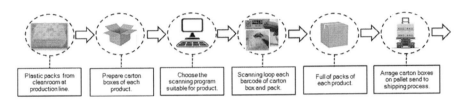

Fig. 1 Final packing process

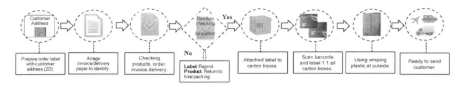

Fig. 2 Shipping process

4 Lossless Data Compression

4.1 Run Length-Like Coding

Upon the investigation of pattern characteristic of product serial that holds similar behavior of character value such as the same batch of product or the same data of built. Instead of compress the whole original serial shipment data [1], we can compress only those variable portions that hold only 5–8 character digits. Digit 5–12th is considered to lie in the variable concentrated portion. Digit 5–7th are found frequently changed while digit 8–9th are found more dynamic. and digit 10–12th are found rarely changed. For the pre-coding to be more efficient all input identifiers are ascending sorted and then compressed with the RLL-like algorithms. Let's consider Fig. 3 where 15 identifiers are given as an example.

- X1: *7,5,B,5,Y,5,6,5,V,5,7,5,8,3,C,1,E,1,L,2,M,1,N,2,1,1,2,3,E,1*
- X2: *276,5,2BV,5,2Y7,5,8,3,C,1,E,1,L,2,M,1,N,2,1,1,2,3,E,1*
- X3: *3768,3,376C,1,376E,1,3BVL,2,3BVM,1,3BVN,2,3Y71,1,3Y72,3,3Y7E,1*
- Y1: *1,10,3,5,0,5,4,5,2,5,1,5,9,5,5,5*
- Y2: *1,10,3,5,201,5,249,5,225,5*
- Y3: *3101,5,3149,5,3325,5*
- Z1: digit 8th has less duplicated symbol. The code is *QQV2C3UPFGY0016*
- Z2: digit 9th also has less duplicated symbol. The code is *RVS2NAXHY6BRW07*.

Fig. 3 Arrangement of each character on 15 identifiers sampled parts

As a result of grouping, 9 patterns can be re-arranged. These are (X1 + Y1 + Z1 + Z2), (X1 + Y2 + Z1 + Z2), (X1 + Y3 + Z1 + Z2), (X2 + Y1 + Z1 + Z2), (X2 + Y2 + Z1 + Z2), (X2 + Y3 + Z1 + Z2), (X3 + Y1 + Z1 + Z2), (X3 + Y2 + Z1 + Z2), and (X3 + Y3 + Z1 + Z2). It is found that the group contains X1 or X2 shows similar compression ratio and better than that contains X3. With this RLL-like technique, the data of 12,144 bits can be reduced to 1,992 bits or 84 % after compression.

4.2 Huffman Coding and Arithmetic Coding

Two most common statistical compression methods are Huffman and Arithmetic coding. Traditional Huffman utilizes a static table to represent all the characters with their frequencies and then generates a probabilities code table to Huffman tree. In a Huffman tree where each node is the sum of its children have weighted sum of the leaves. For Adaptive Huffman coding, the tree and corresponding encoding scheme change accordingly base on technique of algorithm FGK. Arithmetic coding represents frequently characters using low bit and infrequently characters using high bit. The adaptive arithmetic model keeps the symbols, their counts frequencies of occurrence and their cumulative frequencies. The frequencies could be changed each time it is encoded and update the cumulative frequencies. We use the RLE-like compressed output to be the input of either Huffman or Arithmetic compression. As a result the data size can be further reduced by 40 %.

5 Deployment of the RFID System

The significant reduction of raw data provided by the double-stage compression detailed in the previous sections has enabled us to pack the necessary data input the limited RFID user memory. The deployment to an application is quite convinced. The final packing process and shipping process can be combined as shown below in Fig. 4. The tag is write-application by the store section where the typical barcode is attached to the carton box. Data can be revised in next process as read-application to all RFID tags information on pallet during wrapping that must be done before shipment. By doing so, we can save operation times by 2.61 min/unit.

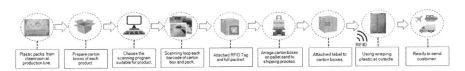

Fig. 4 Flow chart of new process combines final packing process and shipping process

That means if we estimated the production of 100 K/day, we can save 47 min per 8 h rate working. Furthermore the labor force can be reduced from 10 to 7 persons as well as some other assets. Hence the hard cost can be able to quantify to financial statement or comparison. The soft cost such as customer satisfaction, productivity, increasing staff performance, etc. can be included in the key performance indicators (KPIs).

6 Cost and Investment Consideration

As the RFID deployment affects new operation processes and company's financial statements, the cost analysis should be studied carefully. Economic analysis tool should be employed to estimate the value. The Net Present Value (NPV) [2] is the mechanism to understand the cash flow series, where V_t is the cash flow series at t time period, n is the number of analysis time period, and D is the discount rate at that time (i.e. time value of money).

$$NPV = \sum_{t=0}^{n} \frac{V_t}{(1+D)^t} \tag{1}$$

In this work we consider 5-year of period for purpose of medium-term to long-term investment classified by fix cost/one-time cost and annual cost. In connection with the deployment of RFID-based system, all direct cost and effected cost are listed and/or estimated. As a result the total cost is 346,300.00 and estimated benefits and saving cost is 207,212.75 per year when the value assigned to D is 0.12 (12 %). NPV calculation for RFID implementation cost is detailed below.

$$\begin{aligned} NPV_1 &= \frac{-346,300}{(1+0.12)^0} + \frac{-17,980}{(1+0.12)^1} + \frac{-17,980}{(1+0.12)^2} + \frac{-17,980}{(1+0.12)^3} + \frac{-17,980}{(1+0.12)^4} \\ &+ \frac{-17,980}{(1+0.12)^5} = -411,113.88 \end{aligned} \tag{2}$$

And NPV calculations for estimated benefits and saving cost,

$$\begin{aligned} NPV_B &= 0 + \frac{207,212.75}{(1+0.12)^1} + \frac{414,425.50}{(1+0.12)^2} + \frac{621,638.25}{(1+0.12)^3} + \frac{828,851.00}{(1+0.12)^4} \\ &+ \frac{1,036,063.75}{(1+0.12)^5} = 2,072,498.87 \end{aligned} \tag{3}$$

The comparison between RFID implementation cost and estimated benefits and saving cost can be: loss of investment 43.12 % in the first year, in the 2nd year gain a small investment at 8.41 %, the 3rd year at 55.32 %, the 4th year at 98.19 % and the 5th year at 137.52 %. The value investment is double in the next year in this estimation. The long-term trend increases average gain investment 43 % per year.

7 Conclusion

The RFID deployment has been considered under medium-term to long-term project investment. The develop operation process have affected the financial statement, process reengineering, lean manufacturing and section labor (training performance, working hours, etc.). In the first year of investment cannot gain investment due to big component of an RFID installation, but in the next year can achieve the investment and increase by double in next year until time of period. The NPV calculation is the mechanism cash flow series that can be one tool for the investor decision on business.

Acknowledgments This work is supported by Industry/University Cooperative of Data Storage Technology and Applications Research Center (I/UCRC), King Mongkut's Institute of Technology Ladkrabang and National Electronic and Computer Technology Center (NECTEC), National Science and Technology Development Agency (NSTDA) under scholarship HDD-01-52-11 M. We also would like to thank Compart Precision (Thailand) company for their continue support to this research.

References

1. Choeysuwan P, Choomchuay C (2010) A technique for label text compression applied to RFID passive tag. In: The 33rd electrical engineering conference, pp 1001–1004
2. Ozelkan EC, Sireli Y, Munoz MP, Mahadevan S (2006) A decision model to analyze costs and benefits of RFID for superior supply chain performance. In: Technology management for the global future, pp 610–617

Real Time Video Implementation on FPGA

Pham Minh Luan Nguyen and Sang Bock Cho

Abstract Nowadays, real time video becomes popular in a lot of multimedia equipment, video cameras, tablets, camcorders. The more hardware improved the more application is used. Requesting a faster and cost-effective systems there are triggers a shift to Field Programmable Gate Arrays (FPGAs), where the inherent parallelism results in better performance. The implementation is based on efficient utilization of embedded multipliers and look up table (LUT) of target device to improve speed but also saves the general purpose resources of the target device. This paper proposes new hardware architecture for capture NTSC/PAL video stream. The whole system is implemented on a single low cost FPAG chip, capable of real time procession at frequency 60 MHz. In addition, to increase real-time performance, hardware architecture with streamlined data flow are developed.

Keywords FPGA · Video processing · Image processing · Real time processing

1 Introduction

In recent years, automated video surveillance system is developing and applying massive. That is enable when the progress in technology scaling more robust computationally intensive algorithms. The advantage of surveillance automation over traditional television based on system lies in the fact that it is a self contained

---

P. M. L. Nguyen (✉) · S. B. Cho
School of Electrical Engineering, University of Ulsan, Ulsan, Korea
e-mail: npmluan@gmail.com

S. B. Cho
e-mail: sbcho@ulsan.ac.kr

J. J. (Jong Hyuk) Park et al. (eds.), *Multimedia and Ubiquitous Engineering*,
Lecture Notes in Electrical Engineering 240, DOI: 10.1007/978-94-007-6738-6_116,
© Springer Science+Business Media Dordrecht(Outside the USA) 2013

system capable of automatic video processing. That is a request for a real-time video system. The implementation system on FPGA gets more advantages than the other hardware. The FPGA and SOC products improved fast. The traditional hardware implementation of image processing uses Digital Signal Processors (DSPs) or Application Specific Integrated Circuits (ASICs). An advanced system can process video stream real time, process image from video frames in time. As a logic capacity of FPGAs increases, they are being increasing used to implement large arithmetic-intensive applications. Since data-path circuits are designed to process multiple-bit-wide data, FPGAs implementing these circuits often have to transport a large amount of multiple-bit-wide signals form one computing element (such as a logic block, a DSP block, or a multi addressable memory cell) to another. With the advent of FPGAs having greater processing capability, it has been regarded as a useful means for implementing algorithms that massive parallelism is required.

In [9], the authors designed system based on Virtex-4 XC4VLX200-10 from Xilinx and two VCC-8350CL cameras there is more complicated in procession data. There also increase computationally in program. In [2], the system also implemented on Xilinx, Virtex-II pro vp30 FPGA. The camera in [2] is Kodak Kac-9648. In this case, the system has to reduce memory usage but they get more noise in results.

In this paper, we present hardware architecture capable of real time video processing with a screen resolution 30 frames per second. The paper is organized as follows. Section 2 discusses the proposed implementation hardware and results. Finally, conclusions are covered in Sect. 3.

2 Implementation FPGA

2.1 The Observation System

We use the system included: camera ST-400CD, kit Altera DE2_115, LCD Monitor (Flatron Wide). There are some detailed descriptions of camera ST-400CD: Standard Camera Sonny 1/3" Super HAD Color CCD Digital Signal Processing, Video Auto Iris Lens.

The board we use in this paper is Altera Kit DE_115 Development. It was implemented with a Altera FPGA chip, EP4CE115F29C7 (Cyclone IV E), that has 11,4480 LEs, 432 M9 K memory blocks, 3,888 embedded memory (Kbits), 4 PLLs, 528 maximum user I/Os, 230 maximum differential channels. The following hardware is provided on the DE2-115 board: Two 64 MB SDRAM, 2 MB SRAM, 8 MB, flash memory, VGA DAC with VGA out connector. The main improvement of our work include: the development of hardware and software components for a flexible, powerful and low-cost video processing engine, and the use of techniques such as run-time reconfiguration. The implementation has been placed

Real Time Video Implementation on FPGA

Fig. 1 Real-time video system

and routed using Quartus II v.11.1. The operating frequency of the design on DE2-115 is 50 MHz. Video stream is captured by camera. Video signal transmits to FPGA. Processing video stream on board FPGA, the output signal sends to VGA port. LCD Monitor gets video signal and capture it out screen. There is low cost and high performance system (Fig. 1).

Real-time image processing requires high computation power. For example, the NTSC video standard requires 30 frames second, with approximately 0.25 mega pixel per second. PAL video standard has a similar processing load with 25 frames per second, but the frame size is larger. The amount of processing required per pixel depends on the image processing algorithm. In our proposed hardware architecture, we design data flow from captured signal by camera to VGA output signal. In Fig. 2, there is data processing stream from TV decoder chip to VGA chip.

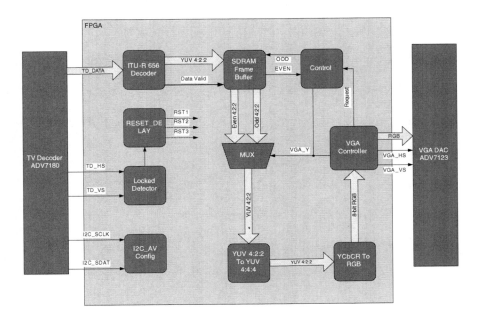

Fig. 2 Data flow in and out FPGA

2.2 Block Processing

- I2C Block (I2C_AV Config): I2C block uses to connect FPGA and TV Decoder chip. Two signals I2C_SClk, I2C_SDATA connect to SCLK pin and SDA pin of Decoder chip. Data signal transmit in 3 bytes per frame.
- Locked dectector block: This block check input conditions from VS and HS to know how the chip worked. When chip works, the TD_Stable is high to make Reset_Delay block work.
- ITU-R656 Block: ITU-R BT 656 describes a simple digital video protocol for streaming uncompressed PAL or NTSC Standard Definition TV (525 or 625 lines) signals. The protocol builds upon the 4:2:2 digital video encoding parameters defined in ITU-R Recommendation BT.601, which provides interlaced video data, streaming each filed separately, and uses the YCbCr Color space and a 13.5 MHz sampling frequency for pixels. The standard can be implemented to transmit either 8-bit values (the standard in consumer electronics) or 10-bit values (sometimes used in studio environments).
- SDRAM Frame buffer, MUX, VGA Controller and Control: We used the SDRAM Frame Buffer and a field selection multiplexer (MUX) which is controlled the VGA controller to perform the de-interlacing operation. Internally, the VGA Controller generates data request and odd/even selection signals to the SDRAM Frame Buffer and filed selection multiplexer (MUX).
- YUV 4:2:2 To YUV 4:4:4. The YUV422 to YUV444 block converts the selected YcrCb 4:2:2 (YUV 4:2:2) video data to the YcrCb 4:4:4 (YUV 4:4:4) video data format (Figs. 3, 4, 5).

2.3 Results

From our architecture, we do some experiments to get results. There are some data about our implementation hardware on Cyclone IV E in Table 1.

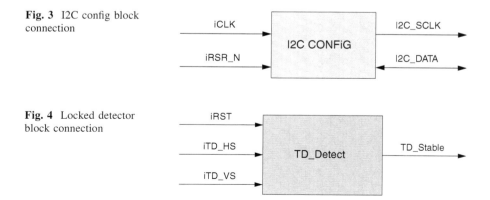

Fig. 3 I2C config block connection

Fig. 4 Locked detector block connection

Real Time Video Implementation on FPGA 947

Fig. 5 ITU-R656 decoder block connection

Table 1 Usage percents on Cyclone IV E

	Origin	Usage	Usage percents (%)
Logic elements	114,800	1,695	1
Memory bits	3,981,312	45,028	1
Embedded multiplier 9-bit elements	432	18	4
Phase-locked loops (PLLs)	4	1	25

Fig. 6 Before download program on FPGA

Fig. 7 After download program on FPGA

The time download program from PC to board about 20 s. The video displays on LCD response real time when we change the screen. In Fig. 6, when we connect LCD and DE2-115 board, the screen on LCD displays the image about DE2-115. That is original program on DE2-115. In Figs. 6 and 7, when we load the program to FPGA, the camera will capture the environment image, and video will display on LCD.

3 Conclusions

In this paper, we proposed new implementation hardware to get real video from camera by FPGA implementation. We get real video stream from camera that is very important detail. We apply some image processing application on real time in future.

We use DE2_115 development kit for our study. This board gives more advance condition with video port, VGA port especial consumer chip Cyclone IV E.

References

1. Abutaleb MM, Hamdy A, Saad EM (2008) FPGA-based real time video object segmentation with optimization schemes. Int J Circ Syst Signal Process 2:78–86
2. Jiang H, Owall V, Ardo H (2006) Real-time video segmentation with VGA resolution and memory bandwidth reduction. In: Proceeding of the IEEE international conference on video and signal based surveillance, Sydney
3. Lapalme FX, Amer A, Wang C (2006) FPGA architecture for real-time video noise estimation. In: IEEE international conference on image processing. Atlanta, pp 3257–3260
4. Chen PP, Ye A (2011) The effect of multi-bit correlation on the design of field-programmable gate array routing resources. IEEE Trans VLSI Syst 19:283–294
5. Cho J, Jin S, Kwon KH, Jeon JW (2010) A real-time histogram equalization system with automatic gain control using FPGA TIIS 633–654
6. DE2-115 user manual (2011)
7. Altera (2011) Introduction to the Altera SOPC builder using verilog designs
8. Altera (2011) Using the SDRAM on Altera's DE2-115 board with verilog designs
9. Jin S, Cho J, Pham XD, Lee KM, Park SK, Jeon JW (2010) FPGA design and implementation of a real-time stereo vision system. IEEE Trans Circ Syst Video Technol 20:15–26

Recovery Algorithm for Compressive Image Sensing with Adaptive Hard Thresholding

Viet Anh Nguyen and Byeungwoo Jeon

Abstract Iterative hard thresholding (IHT) algorithm is one of the representative compressive sensing (CS) reconstruction algorithms. For applying to images, however, it has a problem of lacking in addressing human visual system (HVS) characteristics—its hard thresholding process treats all of coefficients in transform domain equally. To overcome the problem, this paper addresses an adaptive hard thresholding method accounting for the HVS characteristics. For this purpose, a suitable threshold level is adaptively selected for each coefficient in transform domain by utilizing the standard weighting matrix table used in JPEG together with the threshold value which is estimated over the noisy version of image. Experimental results show that the performance of the block compressive sensing with smooth projected Landweber (BCS-SPL) with the proposed adaptive hard thresholding algorithm remarkably outperforms that of the conventional BCS-SPL algorithm.

Keywords Compressive image sensing · Adaptive hard thresholding

1 Introduction

In conventional digital image acquisition and compression system, the encoding process is very time-consuming since all of transform coefficients had to be calculated even though most of them were discarded in quantization process. Obviously, the classical transform-coding procedure demands much of computational power and

---

V. A. Nguyen (✉) · B. Jeon
School of Electrical and Computer Engineering Sungkyunkwan University,
300 Chunchun-dong, Jangan-gu, Suwon, Korea
e-mail: vietanh@skku.edu

B. Jeon
e-mail: bjeon@skku.edu

J. J. (Jong Hyuk) Park et al. (eds.), *Multimedia and Ubiquitous Engineering*,
Lecture Notes in Electrical Engineering 240, DOI: 10.1007/978-94-007-6738-6_117,
© Springer Science+Business Media Dordrecht(Outside the USA) 2013

memory storage. Recently, a novel sampling paradigm called compressive sensing (CS), which directly acquires compressible signal at a sub-Nyquist rate [1], has been being developed to overcome this problem. Due to the simplicity of signal acquisition, CS has drawn great attention. As a result, many efficient reconstruction algorithms have been proposed. Iterative hard thresholding (IHT) algorithm [2] is one of the promising compressive sensing reconstruction algorithms. IHT algorithm not only is very easy to implement in practical application and extremely fast but also has a strong performance guarantees in term of recovery error as shown in [3].

In applying to images, IHT algorithm has some problems. In fact, the hard thresholding process treats all coefficients of image in transform domain equally; any coefficients whose magnitude is lower than a threshold level is considered as noise and will be replaced by zero. However, the low frequency coefficients of image in transform domain are very important because of two reasons: energy in most of the natural images is mostly concentrated at the low frequency bands; the HVS is more sensitive to the loss of low frequency components than that of high frequency components. Moreover, the threshold value (τ) in the hard thresholding process is inaccurate since it is estimated from a noisy version of image. Therefore, applying the threshold level to all coefficients in transform domain may wrongly discard some important low frequency coefficients. As a result, the quality of reconstructed image can be degraded. Note that the iterative hard thresholding algorithm does not account for the HVS characteristics—the coefficients which are discarded in the hard thresholding process may contain visually significant information.

JPEG uses a standard weighting matrix table [4] for quantization which is designed based on human visual system (HVS) characteristics for compressing images without much visible artifacts. The quantization step size at frequency location (u, v), denoted by $Q_{u,v}$, is obtained by multiplying a common factor (which is irrespective of frequency location) to a weight corresponding to the frequency position (u, v) in the matrix table. The weight is chosen as the perceptual threshold [5]. A higher quality factor leads to better image quality. In the quantization process, a coefficient whose magnitude is smaller than half of $Q_{u,v}$ becomes zero [4]; it indirectly suggests that such a small value can be visually less significant. Motivated by this observation, we adaptively select a suitable threshold among $Q_{u,v}/2$ and τ which is estimated using a noisy version of image for applying to each coefficient to avoid wrongly discarding visually significant information in the hard thresholding process. In this way, the hard thresholding method can be made to account for the HVS characteristics.

In this paper, in order to evaluate the performance of the proposed method, we apply an adaptive hard thresholding method to the block compressive sensing with smoothed projected Landweber (BCS-SPL) algorithm—a prominent application of iterative hard thresholding algorithm to images [6, 7]. The objective quality (PSNR) of reconstructed image of BCS-SPL with the proposed adaptive hard thresholding method is compared to that of the conventional BCS-SPL algorithm.

The rest of the paper is organized as follows. Section 2 reviews the fundamental of compressive sensing and the structure of BCS-SPL algorithm. Furthermore, Sect. 3 presents the proposed method. Simulation results are illustrated in Sect. 4. Finally, Sect. 5 draws our conclusion.

2 Background

2.1 Compressive Sensing

Compressive sensing (CS) theory is built upon the work of Candès et al. [8], and of Donoho [9], who show that a finite-dimensional signal having a sparse (contains a lot of zero entries) or compressible representation in a selected domain (e.g., DCT, DWT) can be reconstructed from a small set of linear, non-adaptive measurements. Stated another ways, CS directly acquires signal in a compressed form by projecting it into a sensing matrix. In its study, designing a sensing matrix and designing a reconstruction algorithm are two main problems. The sensing matrix needs to satisfy some requirement conditions [1] (e.g., RIP condition, incoherent condition) to warrant a unique solution. The recovery algorithm should be fast and have a good performance guarantees in term of recovery error [3].

In being applied to images, a challenge for compressive sensing is to reduce the computational complexity in reconstruction process. The commonly used sensing matrix—i.i.d. Gaussian matrix is an unstructured random matrix, and its high complexity of matrix multiplication makes the reconstruction process slow, especially for a large-sized image. Furthermore, a huge memory is necessary to store the large-sized random matrix. In this context, due to the block-based processing, the BCS-SPL (Block Based CS with Smoothed Projected Landweber) algorithm seems to be suitable for applying to a large-sized image.

2.2 Block Based CS with Smoothed Projected Landweber Structure (BCS-SPL)

Figure 1a illustrates the structure of the BCS-SPL algorithm, which consists of block compressive sensing (BCS) at encoder side and the smoothed projected Landweber (SPL) reconstruction algorithm at decoder side. In BCS, an image is divided into small blocks. Then, each block is sampled by projecting it into a compact size of a random measurement matrix (e.g., i.i.d. Gaussian matrix). The measurement vectors of each block are transmitted to decoder. Due to the block-based processing, BCS reduces not only memory for storing the measurement matrix but also computational complexity in reconstruction process.

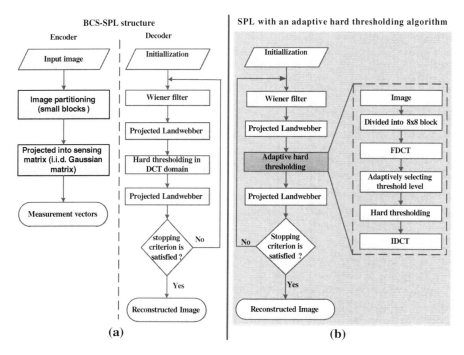

Fig. 1 **a** BCS-SPL structure; **b** SPL algorithm with an adaptive hard thresholding method

In the SPL reconstruction algorithm, an image is iteratively reconstructed. In each iteration, a procedure which incorporates Wiener filter and variant projected Landweber (i.e., namely hard thresholding and projected Landweber framework (PL) [7]) is performed using a noisy version of image. The wiener filter is applied in spatial domain of image signal for removing noise and blocking artifacts as well. The approximation of reconstructed image is calculated by projecting it into projected Landweber (PL) framework. The hard thresholding process imposes sparsity of image signal in transform domain (e.g., DCT) by eliminating any coefficient whose magnitude is less than a given threshold level. The level is calculated using the noisy version of image based on a universal threshold method [10]. In addition, if the stopping criterion is satisfied, SPL algorithm is terminated (see [7] for more details).

3 Proposed Method

As mentioned above, in order to improve the quality of reconstructed image, it is important to adaptively choose an appropriate threshold level for applying to each frequency band of image according to the HVS characteristics. The SPL algorithm with an adaptive hard thresholding method is illustrated in Fig. 1b.

Firstly, the given image is divided into 8×8 blocks each of which is put to the forward discrete cosine transform (FDCT). Before the hard thresholding process, we select a threshold value for each DCT coefficient between $Q_{u,v}/2$ and τ. The standard weighting table is derived from the psycho-visual experiments [5]; the weight values in the table are chosen as perceptual thresholds. If quantization step size $Q_{u,v}$ is larger than the weight values in the standard weighting matrix, visible artifact will occur [4]. Therefore, we regard the weighting values as the quantization step size values and the resulting maximum quantization error is equal to $Q_{u,v}/2$. Thus, the coefficients whose magnitudes are larger than $Q_{u,v}/2$ (i.e., the half of their corresponding quantization step size values) are considered as visually significant information and should be preserved even though their magnitudes are lower than τ. Stated another way, for each coefficient, if $Q_{u,v}/2 \leq \tau$, the threshold level will be set as $Q_{u,v}/2$. Moreover, in the hard thresholding process, τ is used to differentiate signals and noises; the coefficients whose magnitudes are larger than τ are considered as signals and should be preserved. That is, for each coefficient, if $Q_{u,v}/2 > \tau$, the threshold level will be set as τ. By adaptively applying the threshold level to each coefficient, we prevent the hard thresholding process from wrongly eliminating the visually significant information.

In following section, we evaluate the performance of our proposed method by comparing it with the conventional BCS-SPL algorithm.

4 Experimental Result

In our simulation, four 512×512 gray-level benchmark images: Lena, Barbara, Mandrill, and Peppers, are used as input. Block size is set as 8×8. The objective quality (PSNR) is used to compare the performance of the proposed method with that of the conventional BCS-SPL algorithm [6].

Figure 2 shows that the PSNR performance of the proposed method remarkably outperforms that of the conventional BCS-SPL algorithm. Specially, at sub-rate 0.1, the PSNR of the proposed method is much better than that of the conventional BCS-SPL algorithm—PSNR gains of about 4.6, 4, 3.6, and 3 dB in case of Lena, Peppers, Mandrill, and Barbara, respectively. At higher sub-rates (from 0.2 to 0.9), the PSNR performance of the proposed method is also higher than that of the conventional BCS-SPL algorithm—average PSNR gains under sub-rates 0.2–0.9 are about 0.5, 0.5, and 0.8 dB in case of Lena, Peppers, Mandrill, respectively. Actually, at a sub-rate of 0.1, tested image contains a lot of noise, so the estimated threshold value from the noisy version of image is very large. Therefore, a lot of significant coefficients in the low frequency region are lost in the hard thresholding process. Moreover, except Barbara image, in the other images (Lena, Mandrill, and Peppers), most energy is concentrated at low frequency bands; therefore, by adaptively applying a suitable threshold level in each frequency band of transform

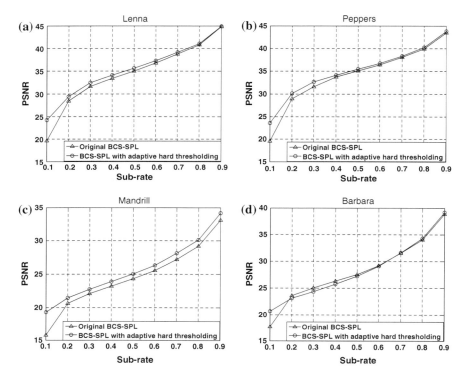

Fig. 2 Performance comparison between the proposed method and the conventional BCS-SPL [6] on test image: **a** Lena; **b** Peppers; **c** Mandrill; and **d** Barbara

domain, we prevent a hard thresholding process from losing the visually significant information of image.

In case of Barbara image (an image with much detail), energy is mostly concentrated at high frequency region. After the filtering process, the high frequency components are distorted since the Wiener filter is used as a low pass filter [7]. Therefore, energy distribution of the image in DCT domain is changed (the energy in low frequency region is more than that in high frequency region), which leads the adaptive hard thresholding algorithm wrongly selects a suitable threshold level for applying to each coefficient. As a result, from sub-rates of 0.2–0.5, the averaged PSNR of the proposed method is less than that of the conventional algorithm (about 0.5 dB). At sub-rates from 0.6 to 0.9, the energy distribution of the image may not be affected much by the filtering process since not much noise is contained in the image. Therefore, the adaptive hard thresholding method may successfully select the appropriate threshold level for applying to each coefficient. As the result, the performance of our proposed method is slightly better than that of the conventional BCS-SPL algorithm.

5 Conclusion

In this paper, we propose an adaptive hard thresholding method accounting for the HVS characteristics. By adaptively applying the suitable threshold level by referring to HVS in each frequency band, the proposed method not only pursues the sparsity of signal but also preserves the visually significant information. Simulation results showed the proposed adaptive hard thresholding algorithm enhanced the BCS-SPL algorithm.

Acknowledgments This work was supported by the National Research Foundation of Korea (NRF) grant funded by the Korea government (MEST) (No. 2011-001-7578).

References

1. Baraniuk RG (2007) Compressive sensing. IEEE Signal Process Mag 24:118–121
2. Blumensath T, Davies ME (2008) Iterative thresholding for sparse approximation. J Fourier Anal Appl 14:629–654
3. Eldar YC, Kutyniok G (2012) Compressive sensing theory and applications. Cambridge University Press, Cambridge
4. Wallace GK (1992) The JPEG still picture compression standard. IEEE Trans Consumer Electron 38:18–34
5. Lohscheller H (1984) A subjectively adapted image communication system. IEEE Trans Commun 32:1316–1322
6. Fowler JE, Mun S, Tramel EW (2012) Block-based compressed sensing of images and video. Found Trends Signal Process 4:297–416
7. Gan L (2007) Block compressed sensing of natural images. In: Proceedings of international conference on digital signal processing. Cardiff, UK, pp 403–406
8. Candès EJ, Romberg J, Tao T (2006) Robust uncertainty principles: exact signal reconstruction from highly incomplete frequency information. IEEE Trans Inf Theory 52:489–509
9. Donoho DL (2006) Compressive sensing. IEEE Trans Inf Theory 52:1289–1306
10. Donoho DL (1995) De-noising by soft-thresholding. IEEE Trans Inf Theory 41:613–627

Estimation Value for Three Dimension Reconstruction

Tae-Eun Kim

Abstract This paper deals with a fundamental problem for 3D model acquisition after camera calibration [1]. We present an approach to estimate a robust fundamental matrix for camera calibration [2, 3]. Single axis motion can be described in terms of its fixed entities, those geometric objects in space or in the image that remain invariant throughout the sequence. In particular, corresponding epipolar lines between two images intersect at the image of the rotation axis. This constraint is then used to remove the outliers and provides new algorithms for the computing the fundamental matrix. In the simulation results, our method can be used to compute the fundamental matrix for camera calibration more efficiently.

Keywords 3D Reconstruction

1 Proposed Calibration Approach

In the past few years, the growing demand of realistic three-dimensional object models for graphic rendering, creation of nonconventional digital libraries, and population of virtual environments has renewed the interest in the reconstruction of the geometry of 3D objects from one or more camera images. One of the simple and robust methods to acquire 3D models from image sequences is using turntable motion. Therefore, it has been widely used by computer vision and graphics researchers. Turntable motion refers to the situation where the relative motion between a scene and a camera can be described as a rotation about a fixed axis. The motion is a practical case of the more general planar motion as all rotations are restricted to be around the same axis. The fundamental task for acquiring 3D

T.-E. Kim (✉)
Department of Multimedia, Namseoul University, Cheonan, Korea
e-mail: tekim5@empas.com

J. J. (Jong Hyuk) Park et al. (eds.), *Multimedia and Ubiquitous Engineering*,
Lecture Notes in Electrical Engineering 240, DOI: 10.1007/978-94-007-6738-6_118,
© Springer Science+Business Media Dordrecht(Outside the USA) 2013

models from single axis motion is to recover the camera parameters and the relative pose of the cameras. The estimation of the camera positions, or simply the rotation angles relative to a static camera, is the most important and difficult part of the modeling process [4, 5]. Traditionally, rotation angles are obtained by careful calibration. Fizgibbon extended the single axis approach to recover unknown rotation angles from uncalibrated image sequences based on a projective geometry approach. However, fundamental matrices and/or trifocal tensors have to be computed for each pair of images or each triplet of images. In the new algorithms, we try to improve the accuracy by estimating the robust fundamental matrices. Figure 1 shows the flow chart of the 3D reconstruction. From the epipolar geometry of circular motion, we can remove the outliers and estimate the robust fundamental matrix for camera calibration.

The fundamental matrix corresponding to a pair of cameras related by a rotation around a fixed axis has a very special parameterization, as shown in Fig. 2, which can be expressed explicitly in terms of fixed image features under circular motion (image of rotation axis, pole, and horizon, jointly holding 5 dof) Consider the pair of cameras P_1 and P_2, given by (Fig. 3)

$$P_1 = K[I|t] \tag{1}$$

$$P_2 = K[R_y(\theta)|t] \tag{2}$$

where

$$t = [0\ 0\ 1]^T \text{ and}$$

$$R_y(\theta) = \begin{bmatrix} \cos\theta & 0 & \sin\theta \\ 0 & 1 & 0 \\ -\sin\theta & 0 & \cos\theta \end{bmatrix}, K = \begin{bmatrix} f_x & s & c_x \\ 0 & f_y & c_y \\ 0 & 0 & 1 \end{bmatrix} \tag{3}$$

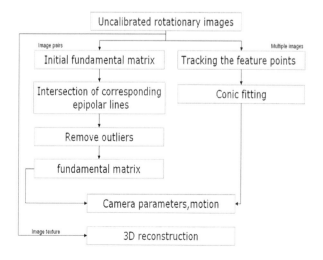

Fig. 1 3D Reconstruction algorithm

Estimation Value for Three Dimension Reconstruction

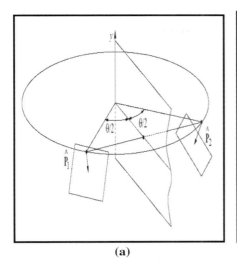
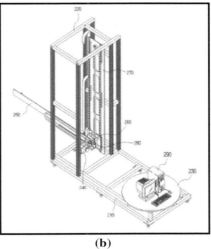

Fig. 2 **a** Geometry of turntable system, **b** Image acquisition system

Fig. 3 The epipolar geometry in circular motion. All corresponding epipolar lines must intersect at m (projection of the rotation axis)

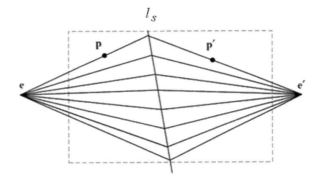

Given a static camera and a generic object rotating on a turntable, single axis motion provides a sequence of different images of the object. Single axis motion can be described in terms of its fixed entities, those geometric objects in space or in the image that remain invariant throughout the sequence. In particular, corresponding epipolar lines between two images intersect at the image of the rotation axis. The epipolar constraint is then used to remove the outliers [6].

If the two views are taken by an uncalibrated moving camera, only the epipolar geometry between the two views was estimated. The epipolar geometry can be nicely coded by a 3 × 3 rank 2 matrix F, called fundamental matrix. Solving fundamental matrix F is the algebraic representation of the epipolar constraint for the uncalibrated cameras. The epipolar constraint is described as follows: For each point m in the 1st image plane, its corresponding point m' lies on its epipolar line l'm and similar for any point m' in the 2nd image plane. This relation can be given as

$$l'_m = Fm \tag{4}$$

$$l_m = F^T m' \tag{5}$$

Since m lies on l'_m m' lies on l_m, following relations are obtained :

$$m'^T Fm = 0 \tag{6}$$

$$m^T F^T m' = 0 \tag{7}$$

where

$$m_i = [ui, \ vi, 1] \tag{8}$$

$$m'_i = \left[ui', \ vi', 1 \right] \tag{9}$$

2 Experimental Results

8-point algorithm. If n corresponding points (at least 8) are given, a set of linear equations is obtained as:

$$Af = \begin{bmatrix} u'_1 u_1 & u'_1 v_1 & u'_1 & v'_1 u_1 & v'_1 v_1 & v'_1 & u_1 & v_1 & 1 \\ \vdots & \vdots & \vdots & \vdots & \vdots & \vdots & \vdots & \vdots & \vdots \\ \vdots & \vdots & \vdots & \vdots & \vdots & \vdots & \vdots & \vdots & \vdots \\ u'_n u_n & u'_n v_n & u'_n & v'_n u_n & v'_n v_n & v'_n & u_n & v_n & 1 \end{bmatrix} \begin{bmatrix} f_{11} \\ f_{12} \\ f_{13} \\ f_{21} \\ f_{22} \\ f_{23} \\ f_{31} \\ f_{32} \\ f_{33} \end{bmatrix} = 0 \tag{10}$$

A robust solution of this equation is the eigenvector corresponding to the smallest singular value of A, that is, the column of V in the singular value decomposition(SVD) of A = UDVT. In order to obtain a unique solution, the rank of A matrix must be equal to 8. Therefore, the closest singular F' matrix to F matrix can be obtained as:

$$F' = Udiag(r,s,0)V^T$$
$$\text{where } D = diag(r,s,t) \text{ where } r \geq s \geq t. \tag{11}$$

8-point algorithm. Hartly proposed a normalized 8-point algorithm to improve the performance. This normalization was performed by translating the center of corresponding points to the origin of the image reference frame and then scaling the corresponding points, so that the average distance from the origin becomes equal to. Finally, after the calculation of matrix using the 8-point algorithm, it is converted to F matrix of corresponding points before normalization as:

$$F = T_2^T \hat{F} T_1 \tag{12}$$

where T1 and T2 are transformation(normalization) matrices for the first and second images, respectively.

The corresponding epipolar lines between two images intersect at the image of the rotation axis. The epipolar constraint is then used to remove the outliers and estimate the fundamental matrix. Corresponding epipolar lines between two views and the epipolar lines meet at the rotation axis.

The algorithm to estimate the robust fundamental matrix can be summarized as follows:

1. Extract and match points across all images.
2. Compute the initial fundamental matrix using the normalized 8-point algorithm.
3. Compute the intersection of two corresponding epipolar lines.
4. Calculate the geometric error between rotation axis and intersection from the previous computations.
5. Estimate the fundamental matrix by minimizing the cost function over all correspondences.

Algorithm

1. Compute the initital F matrix using the normalized 8- point algorithm
2. Robustly estimate the F matrix

   a. Compute the intersection of two corresponding epipolar lines

$$m_i = 1 \times l', m_i(u_i, v_i)$$

   b. Calculate the geometric error between l_s and m_i

$$d(l_s, m_i) = \frac{au_i + bv_i + c}{\sqrt{a^2 + b^2}}, \quad l_s(a,b,c) \text{ image of rotation axis}$$

   c. Minimize the cost function over all correspondences

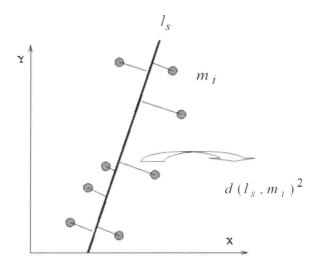

Fig. 4 Intersection of corresponding epipolar and lines geometric distance of rotation axis

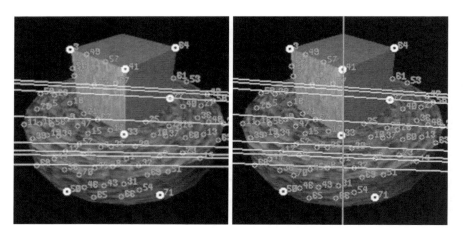

Fig. 5 **a** Initial epipolar lines and **b** Rotation axis of synthetic images

Estimation Value for Three Dimension Reconstruction 963

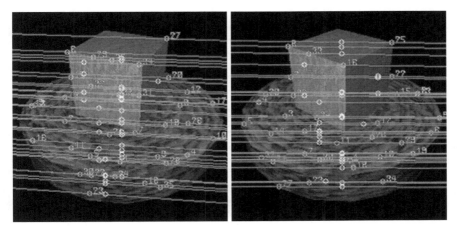

Fig. 6 White circles represent the intersection of two corresponding epipolar lines

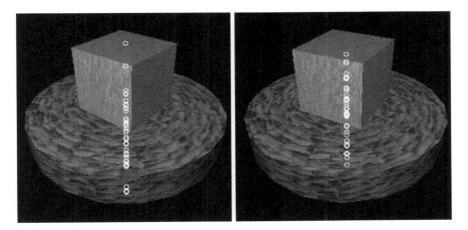

Fig. 7 Inliers of the intersection of epipolar lines where outliers are removed

3 Conclusion

In this paper, we have presented a simple and practical approach for computing the fundamental matrix to estimate the camera parameters. In this system, we need only uncalibrated images of a turntable sequence for input. The strong epipolar constraint can be used for estimating the robust fundamental matrix. The experiments on real and synthetic images demonstrate the usability of proposed algorithm (Figs. 4, 5, 6, 7, 8, 9, 10, 11).

Acknowledgments Funding of this paper was provided by Namseoul University.

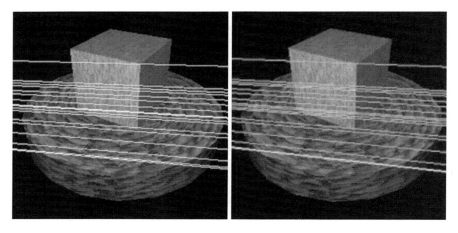

Fig. 8 b Recalculated epipolar lines

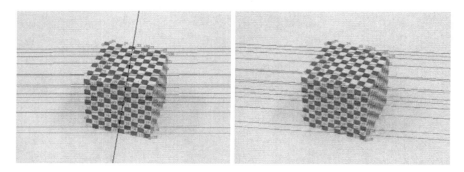

Fig. 9 Recalculated epipolar lines of the real image pair

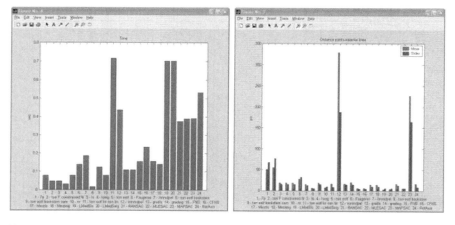

Fig. 10 Fundamental matrix error

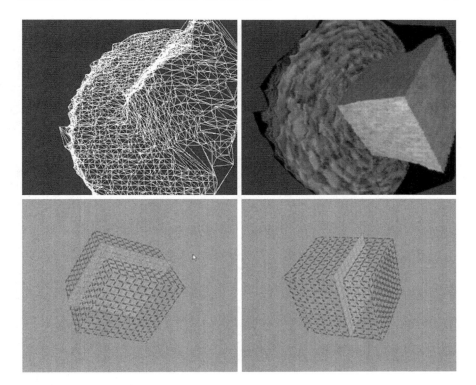

Fig. 11 Results of 3D reconstruction

References

1. Jang G, Tusi HT, Quan L, Zisserman A (2003) Single axis geometry by fitting conics. IEEE Trans Pattern Anal Mach Intell 25(10):1343–1348
2. Zhang Z (1998) Determining the epipolar geometry and its uncertainty: a review. Int J Comput Vis 27(2):161–195
3. Serra J (1982) Image analysis and mathematical morphology, vol 1. Academic Press, New York
4. Canny J (1986) A computation approach to edge detection. IEEE Trans PAMI 8(6):679–698
5. Rao K (1993) Extracting salient contours for target recognition: algorithm and performance evaluation. Opt Eng 32(11):2690–2697
6. Hartly R, Zisserman A (2000) Multiple view geometry in computer vision. Oxford university press, Oxford

Gesture Recognition Algorithm using Morphological Analysis

Tae-Eun Kim

Abstract Recently, research into computer, vision-based methods to recognize gestures are widely conducted as means to communicate the volition of humans to a computer. The most important problem of gesture recognition is a reduction in the simplification treatment time for algorithms by means of real-time treatment. In order to resolve this problem, this research applies mathematical morphology, which is based on geometric set theory. The orientations for the primitive shape elements of hand signal shapes acquired from the application of morphological shape analysis include important information on hand signals. Utilizing such a characteristic, this research is aimed at suggesting a feature vector-based, morphological gesture recognition algorithm from a straight line which connects central dots of major primitive shape elements and minor primitive shape elements. It will also demonstrate the usefulness of the algorithm by means of experimentation.

Keywords Human motion · Computer vision

1 Introduction

In recent days, research is widely conducted to use gestures as a means to communicate between humans and computers. If an interface which understands gestures used during everyday dialogue is developed, subtle expressions and gestures for communicating ideas, emotions or sentiments can be widely used as natural inputs for gesture recognition.

T.-E. Kim (✉)
Department of Multimedia, Namseoul University, Cheonan, Korea
e-mail: tekim5@empas.com

J. J. (Jong Hyuk) Park et al. (eds.), *Multimedia and Ubiquitous Engineering*,
Lecture Notes in Electrical Engineering 240, DOI: 10.1007/978-94-007-6738-6_119,
© Springer Science+Business Media Dordrecht(Outside the USA) 2013

Broadly, gesture recognition techniques may be divided into an appliance attachment method which attaches many types of sensors, like gloves, to the body of humans and a vision-based image treatment method, which treats images acquired from a video camera. A method for attaching a sensor has some problems in that movements are limited, motions are unnatural, and handling is very inconvenient as discomfort and psychological burden is felt when it is worn. In this situation, currently, research is being actively conducted into fields which desire to use vision-based image treatment to recognize gestures in a non-touch manner, except for in special application fields.

The strength of a vision-based image treatment method is that a sensor need not be attached, but encounters problems as well. For example, the resolution limit of video data may be low, making hardware materialization and real-time treatment impossible. This is attributed to the fact that the recognition algorithm for hidden Markov model (HMM), neural network model, and others, which are mainly applied in form-based approach methods (and, also, major objects of research into vision-based image treatment) is complex and cannot be treated real-time. However, the core of gesture recognition is related to the real-time control of hardware as it concerns the interface between humans and machines. Therefore, big obstacles in putting it to practical use are that the algorithm is complex, and that it is difficult to treat the recognition in real-time and convert it into hardware [1–3].

So, this research suggests a morphological hand signal recognition algorithm applied with mathematical morphology in which hardware materialization is easy and high-speed operation is possible. In mathematical morphology, based on logical operations between pixels, various types of useful image treatment technologies, composed of morphological logic operations, are developed. Shape regions extracted from hand signal images are dismantled into primitive shape elements. This is because humans' visual recognition is a basic step for disintegrating complex shapes of objects contained in 2-dimensional images into simple primitive shape elements and expressing them in a stratum-wise manner. Based on this, a method is suggested to use the positional relationship of primitive shape elements and an experiment is conducted to demonstrate the usefulness of the suggested theory and confirm that hand signals can be recognized in search for video contents.

2 Disintegration of Morphological Shapes

Morphological operation. Morphological image treatment is a nonlinear treatment method used to interpret geometric characteristics of images; images in mathematical morphology, based on set theory, are a set of dots that can conduct set operations, such as translation, union and intersection. When A, B, C and K are described as open sets defined in Z^2, which is a 2-dimensional Euclidean space, and O is expressed as the starting point of Z^2, such operations as are widely used in the morphological image treatment field are shown in the following manner [4–6].

$$\text{Dilation } (\backslash) : \mathbf{A} \oplus \mathbf{K} \tag{1}$$

$$\text{Erosion } (\backslash) : \mathbf{A} \ominus \mathbf{K} \tag{2}$$

$$\text{Open } (\backslash) : \mathbf{A} \circ \mathbf{K} = (\mathbf{A} \ominus \mathbf{K}) \oplus \mathbf{K} \tag{3}$$

$$\text{Close } (\backslash) : \mathbf{A} \bullet \mathbf{K} = (\mathbf{A} \oplus \mathbf{K}) \ominus \mathbf{K} \tag{4}$$

B, which is mentioned here as a structuring element, is an image pattern used to convert images. An opening operation dilates the results of an erosion operation, and a closing operation erodes the results of a dilation operation. An opening operation has the nature of a filter that softens sharp corners of an object and eliminates smaller objects that are not contained in the structuring element. Meanwhile, a closing operation tends to fill up small holes on gorge-shaped objects. From these facts, it may be deduced that an opening operation and a closing operation can be used as a filter to remove positive noise components and negative noise components, respectively.

Disintegration of the shape. Disintegrating complex shapes of objects in 2-dimensional images into the elements of a simple primitive shape and stratum-wise expressions is a treatment which corresponds to a basic step of humans' visual recognition. This research utilizes primitive shape elements that are acquired through disintegrating gesture shapes into morphological shapes, so as to express gesture shapes.

The following is a morphological expression of shape disintegration which is intended to use morphological operations disintegrate 2-dimensional shape $\mathbf{X} \in \mathbf{Z}^2$ into many sets or $\{\mathbf{X}_i\}$, which are primitive shape elements.

$$\mathbf{X} = \bigcup_{i=1}^{n} X_i, \quad X_i \in G(\mathbf{Z}^2) \tag{5}$$

Here, $G(\mathbf{Z}^2)$ is a 2-dimensional open set defined in Z, which is a 2-dimensional Euclidean space. When a primitive shape element for generating $\mathbf{X}_i$ is expressed as $\mathbf{Y}_i$ and a structuring element corresponding to $\mathbf{Y}_i$ is expressed as B, the morphological shape disintegration algorithm is expressed in the following manner. Also, the simplest example of primitive shape element $\mathbf{Y}_i$ is $n_i B$, which is the scalar multiple of structuring element B selected as the original plate or a square having a unit area.

$$X_i = X_{n_i B} = \left(X \ominus n_i B^S\right) \oplus n_i B \tag{6}$$

Here, B is a structuring element, n_i is the size of a structuring element, and B^s is a reflection of B against the starting point.

Formula (6) implies that a primitive shape element can be acquired through a dilation operation conducted many times because an erosion operation is conducted of the results of an erosion operation of shape X against B until the shape is reduced to dots or lines. Such a treatment is repeatedly conducted against $X - X_i$.

The following expresses the treatment process mentioned above, in the form of regression [7, 8].

$$X_i = \left(X - X'_{i-1}\right) n_i B, X'_i = \bigcup_{j=1}^{i} X_j, X'_0 = \phi \tag{7}$$

$$\text{Stopping condition:} \left(X - X'_K\right) \ominus B^S = \phi$$

Here, the stopping condition is a condition for disintegrating all and every region of the shape. And, k is the total number of disintegrated primitive shape elements. Each primitive shape element, disintegrated through applying the shape disintegration algorithm is a region generated as, is moved in parallel to track expressed as dot or line. Track of the maximum structuring element inscribed in a primitive shape element may be expressed in the following manner:

$$L_i = \left(X - \bigcup_{0 \le j \le i-1} \left(L_j \oplus n_i B\right)\right) \ominus n_i B^s \tag{8}$$

The relationship between the primitive shape element and track can be understood through combining the above formula (7) with formula (8).

$$X_i = L_i \oplus n_i B \tag{9}$$

3 Recognition of the Shape

Expression of the shape. In the process of morphological shape disintegration, primitive shape elements are extracted starting from larger ones to the smaller ones. Again, these primitive shape elements are stratum-wise expressed in the following manner:

1. The largest primitive shape elements extracted first during shape disintegration set X_i as their main element.
2. $X_i(i = 2, 3, \ldots, n)$, which are extracted during shape disintegration, are set as $1, 2, 3, \ldots, n$th minor elements, according to their order.
3. As is seen in Fig. 1, primitive shape elements positioned between the higher two minor elements are removed from shape expression since they are not required for the recognition of hand signal orientation. The primitive shape elements positioned between the two elements in such a process are primitive shape elements—the distance from which to major elements is shorter compared to the immediately higher elements.
4. The said process is conducted for $i = 2$ through n. Here, n is the frequency of shape disintegration, conducted until such a feature vector that recognizes hand signals can be acquired.

Fig. 1 Primitive shape elements positioned between higher two minor elements

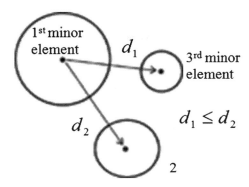

The next step is extracting the central dot of primitive shape elements. The central dot is acquired through repeatedly conducting an erosion operation until primitive shape elements get smaller than structuring elements.

Extraction and recognition feature vectors. A feature vector is referred to as the angle of such a line that connects the central dots of major primitive shape elements positioned in a 2-dimensional space via shape disintegration and shape expression processes, with the central dots of minor primitive shape elements.

$$\mathbf{x} = \{\theta_1, \theta_2, \theta_3, \ldots, \theta_{n-1}\} \quad (10)$$

Here, n is the number of primitive shape elements acquired in the process of shape disintegration and θ_{n-1} is central dots (x,y) of a major primitive shape element. If the central dot of $n-1$st major primitive shape element is (x_{n-1}, y_{n-1}), its value is calculated in the following manner:

$$\theta_{n-1} = \tan^{-1}(|y - y_{n-1}|/|x - x_{n-1}|) \quad (11)$$

The mean value of feature vectors, which are composed of such angles that connect the central dots of major primitive shape elements and minor primitive shape elements, can be calculated in the following way:

$$x = (\theta_1 + \theta_2 + \theta_3 + \cdots + \theta_{n-1})/(n-1) \quad (12)$$

The values of feature vectors calculated through applying Formulas (10) and (11) are based on the thumb and the orientation which have important meanings for hand signals.

Recognition of hand signals. As is seen in Fig. 2, if the mean value of the feature vectors is $224° \sim 43°$, $45° \sim 134°$ and $135° \sim 225°$, feature vectors are recognized as region 2, 4 and 3, respectively, and minor primitive shape elements of region 1 are positioned around major primitive shape elements, then it is recognized as a case where only major primitive shape elements exist and where minor primitive shape elements are removed. This is under the assumption that only major primitive shape elements exist.

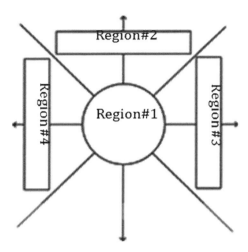

Fig. 2 Regions where feature vectors are recognized

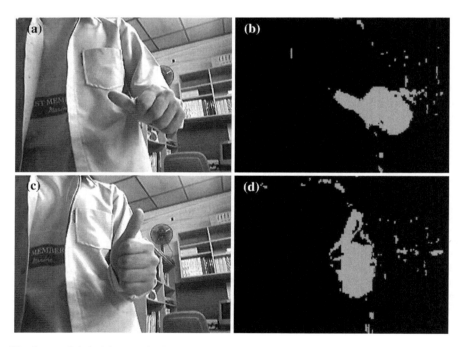

Fig. 3 **a**, **c** Original images, **b**, **d** Results of the detection of skin region

Gesture Recognition Algorithm using Morphological Analysis

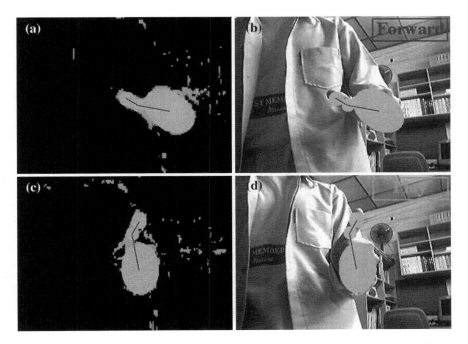

Fig. 4 **a, b** Results of shape disintegration, **c, d** Results of the recognition of hand signals

4 Results of the Experiment

Figure 3 shows images about the detection of a man's skin color after an acquired image(320x240 RGB) is converted into a YcbCr model ($77 \leq C_b \leq 127$, $133 \leq C_r \leq 173$).

In Fig. 4a, c show the results of the display of the higher 3 elements by using shape disintegration; and b, d show the results of the recognition of the higher 3 elements by use of feature vectors.

5 Conclusion

The method, newly suggested in this research, is to recognize hand signals using such feature vectors that connect the central dots of primitive shape elements. These elements' are as are the largest with the central dots of the other primitive shape elements after selecting the central dots of primitive shape elements, which are extracted using morphological shape disintegration. It's expected that the said method will be widely applied to searches for video data and to interface designs related to the running of other electronic systems.

Acknowledgments Funding of this paper was provided by Namseoul University.

References

1. Pavlovic VI, Sharma R, Huang TS (1997) Visual interpretation of hand gestures for human-computer interaction: a review. IEEE Trans. PAMI 19(7):677–695
2. Ahmad T, Taylor CJ, Lanitis A, Cootes TF (1997) Tracking and recognising hand gestures, using statistical shape models. Image Vis Comput 15:345–352, Elsevier
3. Wilson AD, Bobick AF (1999) Parametric hidden markov models for gesture recognition. IEEE Trans PAMI 21(9):884–900
4. Serra J (1986) Introduction to mathematical morphology. Comput Vis Graph Image Process 35(3):283–305
5. Serra J (1982) Image analysis and mathematical morphology. Academic Press, New York
6. Maragos P (1989) A representation theory for morphological image and signal processing. IEEE Trans Pattern Anal Mach Intell 11(6):586–599
7. Pitas I, Venesanopoulos AN (1990) Morphological shape decomposition. IEEE Trans Pattern Anal Mach Intell 12(1):38–45
8. Pitas I, Venetsanopoulos AN (1992) Morphological shape representation, Pattern Recog 25(6):555–565

Omnidirectional Object Recognition Based Mobile Robot Localization

Sungho Kim and In So Kweon

Abstract This paper presents a novel paradigm of a global localization method motivated. The proposed localization paradigm consists of three parts: panoramic image acquisition, multiple object recognition, and grid-based localization. Multiple object recognition information from panoramic images is utilized in the localization part. High level object information is useful not only for global localization but also robot-object interaction. The metric global localization (position, viewing direction) is conducted based on the bearing information of recognized objects from just one panoramic image. The experimental results validate the feasibility of the novel localization paradigm.

Keywords Object recognition · Localization · Omnidirectional camera

1 Introduction

A robot should have the ability to determine its global location in order to successfully handle self-initialization and kidnapping problem. Several approaches were proposed to handle such problem. Park et al. proposed a hybrid map of object and spatial layouts using a stereo camera to localize globally [1]. Angeli et al. proposed a topological visual SLAM (simultaneous localization and mapping) for determining global localization [2]. Visual words were used to handle global

---

S. Kim (✉)
LED-IT Fusion Technology Research Center and Department of Electronic Engineering,
Yeungnam University, 214-1 Dae-dong Gyeongsan-si, Gyeongsangbuk-do, Korea
e-mail: sunghokim@ynu.ac.kr

I. S. Kweon
Department of Electrical Engineering and Computer Science, Korea Advanced Institute of
Science and Technology, 373-1 Guseong-dong Yuseong-gu, Daejeon, Korea
e-mail: iskweon@kaist.ac.kr

location, and odometry information was combined to give metric information. Ramisa et al. also proposed a topological localization method using affine invariant features [3]. Although these approaches can provide global location information, they used additional information, such as stereo and odometry, for global metric localization. An additional requirement is the fast global localization capability using just one image frame. Most approaches can achieve topological localization by recognizing objects or scenes from an image [4]. Metric localization is possible if there is a depth cue (stereo camera) or motion cue (structure from motion) [1, 4]. The last requirement is the capability of robot-object interaction for visual servoing. Robots should have object label and position information.

There are several paradigms for mobile robot localization. Initially, artificial landmark-based approaches were proposed [5, 6]. After then, the SLAM paradigm became a popular approach since it can build a map and localize itself simultaneously by using the extended Kalman filter and an invariant feature such as SIFT [7, 8]. Particle filter-based statistical estimation was also useful in SLAM approach. These paradigms were partially successful since they could estimate relatively accurate location information by matching low level features, such as corner points or invariant features, in multi-frames. However, the location estimation error can be large if they use only one frame. In addition, those approaches can not provide high level information for robot-object interaction.

Then, how can human visual systems (HVSs) localize themselves? HVSs can localize themselves and interact with environment robustly. Do HVSs recognize their locations by point matching as SLAM? Most people will say "No". We surveyed the localization mechanisms of the HVS to get the answer or clue. Although accurate mechanisms are not disclosed, it is evident that object recognition and localization are strongly related according to experimental studies such as lesion of visual cortex (ventral stream and dorsal stream) [9, 10]. This observation means that object recognition and localization are strongly correlated and that they facilitate each other.

Motivated by such biological research results, we propose a novel localization paradigm using only high level object recognition information from one image. The paradigm consists of three parts: omnidirectional panoramic image acquisition, multiple object recognition and grid-based localization. Multiple object recognition is performed from a panoramic image, and mobile robot localization is then conducted using bearing information of objects. This paradigm can estimate both spatial position and viewing direction using only one image. Section 2 overviews the proposed localization system and explains the multiple object recognition method. In addition Sect. 2 represents the mobile robot localization algorithm using object information. Section 3 experimentally validates the feasibility of the proposed paradigm, and Sect. 4 concludes the paper.

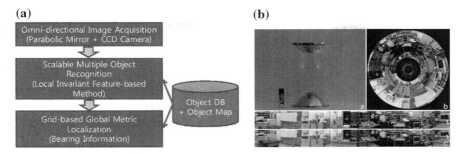

Fig. 1 **a** The proposed novel paradigm of localization using high level object information, **b** Omnidirectional stereo camera system

2 Object Recognition Based Localization

As shown in Fig. 1a, the proposed localization system consists of image acquisition, object recognition and global metric localization. The proposed localization system consists of an off-line database construction module and an on-line localization module. The object database and object-based map are constructed off-line. The object DB module contains learned local feature-based object models representing a 3D object as a set of views. Since it is based on a robust invariant feature, the learned models can handle geometrically, photometrically distorted objects in a general environment. The object-based map is built manually by accurately measuring object locations. On-line localization is conducted through the panoramic image acquisition module via an omnidirectional camera, multiple object recognition module and a bearing angle-based localization module. The large field of view is required for the object-based localization from one image. Although there can be several methods for getting an omnidirectional image, we adopt the parabolic mirror-based panoramic camera. After image acquisition occurs, we extract multiple object information (object label and position in image) by applying the local invariant feature-based method. We use object databases (DBs) that are learned to handle large numbers of objects. After such object recognition occurs, we can have the bearing (angle) information of each object. The final robot localization (spatial position and viewing direction) is estimated by intersecting the bearing information.

The proposed localization method utilizes the omnidirectional camera developed by Jang [11]. Figure 1b shows the omnidirectional camera system. It is composed of 2 parabolic mirrors and an IEEE 1394 camera (1600 × 1200 image resolution). We can acquire omnidirectional stereo images via the camera system. Currently, we use the upper rectified images for the purpose of recognizing objects in these images, which can give higher image resolution than lower images.

In the multiple object recognition module, we can recognize learned objects stored in the object database. Each recognized object can provide an object label and a bearing angle measurement. Since the resolution of a rectified image is

1800 × 161, the bearing measurement resolution of the top-line is 0.2 deg/pixel. Now, we introduce a powerful and efficient 3D object representation, learning and recognition method. Any 3D objects can be represented by a set of multiple views. Each view consists of local features. We apply the sharing concept to the features and views of scalable object representation [12].

How can we fully utilize the shared feature-based view clustering method in object recognition? Basically, we use the well-known hypothesis and verification framework. However, we modify it to recognize multiple objects via the proposed object representation scheme. We can get all possible matching pairs by NN (nearest neighbor) search in feature library. From these, hypotheses are generated by generalized Hough transform in CFCM (common frame constellation model) ID, scale (11 bins), orientation (8 bins) space [8], and grouped by object ID. Then, we determine whether to accept or reject the hypothesized object based on the bin size with an optimal threshold [13]. Finally, we select the optimal hypotheses that can be best matched to the object features in a scene.

In the localization module, the recognized object labels are used to achieve data association of objects in a map, and the intersection of bearing measurements is used to accomplish robot localization. Through object recognition, we can estimate the position of recognized objects in an image. Especially, the column position provides the bearing measurement (ϕ_Z^i) of ith object in panoramic images as shown in Fig. 2a. In this work, we regard the 1st column of an image as 0 radian. An object center is estimated by the similarity transform of a corresponding CFCM. Given a set of object labels and bearing measurements, the robot localization is defined as coordinate transformation from reference coordinates to robot coordinates in 2D space.

Let $\{A\}_Z$ be a set of bearing measurements by a mobile robot through multiple object recognition, $\{A\}_R$ be a set of model bearing measurements after coordinate transformation. Then the robot localization problem is to estimate $T = (x, y, \phi)$ which is the coordinate transformation function from reference coordinates to robot coordinates as shown in Fig. 2b. Shimshoni proposed a direct estimation method based on linear constraints [14]. We applied this method but the estimation

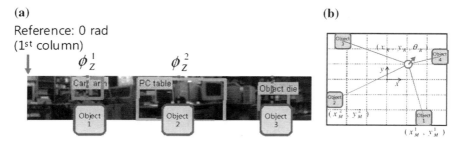

Fig. 2 Bearing measurement information of recognized objects in panoramic images: **a** Visual interpretation, **b** Localization problem is regarded as coordinate transformation from reference coordinates to robot coordinates

results are very unstable due to bearing measurement noise and a few number of measurements (usually 3–6). Fox et al. proposed a Monte Carlo localization method that approximates a posterior by a set of samples [15]. We also applied the latter method, but it takes time to converge. Instead, we use the grid-based localization method. If we divide the coordinate transformation space into moderate resolution (in current implementation $\delta_x = \delta_y = 10$ cm, $\delta_\phi = \pi/180$ rad), then robot location is estimated by Eq. (1). N denotes the number of recognized objects. If we specify the symbols, then the localization problem is the minimization problem of three dimensions as Eq. (2). ϕ_R^i denotes the angle of model object i after transformation with $T = (x_R, y_R, \theta_R)$ as shown in Eq. (3). We can get the optimal robot location with orientation information by minimizing Eq. (2).

$$\hat{T} = \min_T \left[\sum_{i=1}^{N} \left(A_Z^{(i)} - A_R^{(i)}(T) \right) \right] \tag{1}$$

$$\left(\hat{x}_R, \hat{y}_R, \hat{\theta}_R \right) = \min_{(x_R, y_R, \theta_R)} \left[\sum_{i=1}^{N} \left(\phi_Z^i - \phi_R^i(x_R, y_R, \theta_R) \right) \right] \tag{2}$$

$$\phi_R^i(x_R, y_R, \theta_R) = \tan^{-1} \left(\frac{y_M^i - y_R}{x_M^i - x_R} \right) + \theta_R \tag{3}$$

3 Experimental Results

We apply the object recognition-based localization method to a complex laboratory environment. There are bookshelf, PC table, air cleaner, wash stand, printer and so on. Note that the image quality of an individual object is very low. We use every two views for object modeling. The total number of objects is 9 with multiple views. According to the results of object learning, part clustering reduces the size by 44.2 %, while view clustering reduces the size by 39.8 %.

Figure 3a shows localization examples of a mobile robot, KASIRI IV, which can move accurately according to the planned path. In each result, the top image shows the recognized objects with object centers that are equal to the bearing measurements. In the bottom image, the red arrow represents the location (position with direction) of the mobile robot, and data association is linked by the dotted blue line. Note that multiple objects are recognized and used for robot localization. Figure 3b summarizes the overall localization performance. The red dotted line represents the true path of the mobile robot and the blue square represents the estimated robot location using our algorithm. The average location error is $(x, y = 14.5$ cm, 18.5 cm), which is relatively large compared to those of the range sensor-based approaches or interesting point-based approaches (usually within 5 cm) in a 10×10 m environment. However, our proposed system can

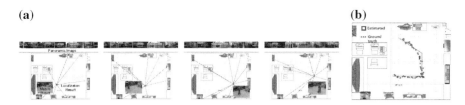

Fig. 3 The overall localization performance of the test sequence: **a** Examples of robot localization using the proposed method, **b** Final localization results

provide high level information of an object that is useful for robot-environment interaction. Note that human visual systems (HVSs) can recognize relative locations with very low metric accuracy but can well interact in an environment with object information.

4 Conclusions and Discussions

In this paper, we proposed a new robot localization method using the object recognition method. Instead of fragile low level features, we regard objects as natural landmarks for localization. For this system, we introduce the multiple object recognition method based on a learned object model and grid-based localization using bearing measurements. The experimental results validate the feasibility of the proposed system. There are several research directions. Currently, we do not use the tracking of objects. If we utilize the temporal continuity, then we can get smother localization. In addition, the map is generated manually. We have to investigate automatic object-map generation. If we combine it with topological localization, then the working space can be increased.

Acknowledgments This research was supported by Basic Science Research Program through the NRF funded by the Ministry of Education, Science and Technology (No. 2012-0003252) and by National Strategic R&D Program for Industrial Technology, Korea. It was also supported by the 2012 Yeungnam University Research Grants.

References

1. Park S, Kim S, Park M, Park SK (2009) Vision-based global localization for mobile robots with hybrid maps of objects and spatial layouts. Inf Sci 179(24):4174–4198
2. Angeli A, Doncieux S, Meyer JA, Filliat D et al (2009) Visual topological slam and global localization. In: ICRA'09: Proceedings of the 2009 IEEE international conference on robotics and automation, pp 2029–2034
3. Ramisa A Tapus A, de Mántaras RL et al (2008) Mobile robot localization using panoramic vision and combinations of feature region detectors. In: Proceedings of the ICRA, pp 538–543

4. Murillo AC, Guerrero JJ, Sagues C (2007) Topological and metric robot localization through computer vision techniques. In: Proceedings of the ICRA Workshop-from features to actions: unifying perspectives in computational robot vision, pp 79–85
5. Scharstein D, Briggs AJ (2001) Real-time recognition of self-similar landmarks. Image Vis Comput. 19(11):763–772
6. Jang G, Kim S, Lee W, Kweon I et al (2002) Color landmark based selflocalization for indoor mobile. In: Proceedings of the ICRA, pp 1032–1042
7. Durrant-Whyte H, Bailey T (2006) Simultaneous localisation and mapping (slam): part i the essential algorithms. IEEE Robot Autom Magazine 13:99–110
8. Lowe D (2004) Distinctive image features from scale-invariant keypoints. Int J Comput Vision 60(2):91–110
9. Himmelbach M, Karnath H (2005) Dorsal and ventral stream interaction: Contributions from optic ataxia. J. Cogn Neurosci 17(4):632–640
10. Blangero A et al (2008) Dorsal and ventral stream interaction: Evidence from optic ataxia. Brain Cogn 67:2
11. Jang G, Kim S, Kweon I (2006) Single-camera panoramic stereo system with single-viewpoint optics. Opt Lett 31(1):41–43
12. Kim S, Kweon IS (2008) Scalable representation for 3D object recognition using feature sharing and view clustering. Pattern Recogn 41(2):754–773
13. Murphy-Chutorian E, Triesch J (2005) Shared features for scalable appearance-based object recognition. In: Proceedings of the seventh IEEE workshops on application of computer vision, vol 1 pp 16–21
14. Shimshoni I (2002) On mobile robot localization from landmark bearings. IEEE Trans Rob 18(6):971–976
15. Fox D, Burgard W, Dellaert F et al (1999) Monte carlo localization: Efficient position estimation for mobile robots. In: Proceedings of the national conference on artificial intelligence

Gender Classification Using Faces and Gaits

Hong Quan Dang, Intaek Kim and YoungSung Soh

Abstract Gender classification is one of the challenging problems in computer vision. Many interactive applications need to exactly recognize human genders. In this paper, we are carrying out some experiments to classify the human gender in conditions of low captured video resolution. We use Local Binary Pattern, Gray Level Co-occurrence Matrix to extract the features from faces and Gait Energy Motion, Gait Energy Image for gaits. We propose to combine face and gait features with the combination classifier to enhance gender classification performance.

Keywords Gender Classification · Faces · Gaits · Local binary pattern · Gait Energy Motion · Gait Energy Image

1 Introduction

Nowadays, gender classification plays an important role in many practical applications in medical, social and security fields. And many mentioned applications critically depend on the correct gender classification. Based on gender, surveillance system can improve their performance of tracking and store managers can estimate the difference of male or female interest to increase profits. Automatic gender classification can be based on the human characteristics such as voice, faces and gaits.

Early studies about gender classification from faces began in 1990s with works of Golomb et al. [1] using multiple layer neural networks. Since then, many feature extraction algorithms have been developed to enhance the performance. Discrete cosine transform algorithm [2] is used to reduce the dimensionality of the data set.

H. Q. Dang (✉) · I. Kim · Y. Soh
Department of Information and Communication Engineering,
Myongji University, Yongin, South Korea
e-mail: kit@mju.ac.kr

J. J. (Jong Hyuk) Park et al. (eds.), *Multimedia and Ubiquitous Engineering*,
Lecture Notes in Electrical Engineering 240, DOI: 10.1007/978-94-007-6738-6_121,
© Springer Science+Business Media Dordrecht(Outside the USA) 2013

Some methods reduce the redundant information of high density such as local binary pattern (LBP) and principal component analysis [3].

Meanwhile, gender classification by gait has received much attention in computer vision community due to its advantage of classification without attracting attention and distance limitation. Gait data can be captured at far distance without demanding physical human information. Therefore, there is a great potential for recognition applications based on cameras with low-resolution videos which are affected by the camera devices or indoor and outdoor environment. In gait, we have two kinds of features to extract. One is model-based features and the other is model-free features. Model-based features explore static and dynamic body parameters while model-free features just use binary silhouettes, without model of moving person requirement [4].

In this paper, we present the classification of gender with features extracted from face and human gait. We use LBP method and Grey Level Concurrence Matrix (GLCM) to extract face features. Gait Energy Motion (GEM) and Gait Energy Image (GEI) are used to extract gait features. Then we further propose to classify gender by combining them. Support Vector Machine (SVM) algorithm is used for classification with linear kernel and majority voting for a combination classifier. The frame size in the dataset is 320-by-240 pixel and the frame rate is 25 fps. Section 2 explains the feature extraction from face and gait, respectively. In Sect. 3, we experimented with several methods and the combinations of the methods depicted in previous sections. The conclusion is made in the last section.

2 Feature Extraction

2.1 Feature Extraction from Faces

Texture is an important characteristic to identify the regions of interest in an image. And the LBP operator is a gray-scale texture measure which summarizes the local spatial structure and gray scale contrast of an image. The original LBP operator, a nonparametric 3×3 kernel, labels the pixels of an image by thresholding the 3×3 neighborhood of each pixel with the center value and considers the results as a binary number. The advantage properties of the LBP are highly discriminative and invariance to the monotonic gray level changes under the rotation effect and the influence of illumination. In the case of rotated image, the sampling neighborhoods will be moved correspondingly along the perimeter of the circle around the center pixel with same direction of rotation of the image. The value of LBP label gets changed because the obtained binary number is shifted left or right. Keeping the LBP label constant at all rotation angles, the LBP value corresponds to the smallest shifted binary number. The rotation invariance approach gets 36 LBP labels in all 256 possibilities and it is presented as a histogram with 36 bins [5]. Another extension to the original operator is defined as uniform patterns, which can be used

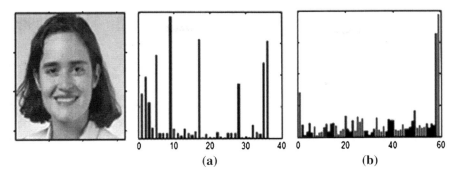

Fig. 1 a Rotation invariant LBP histogram. **b** Uniform LBP histogram

to reduce the length of the feature vector and implement a simple rotation invariant descriptor. A local binary pattern is called uniform if the binary pattern contains at most 2 bitwise transitions from 0 to 1 or vice versa. There is a separate label for each uniform pattern and all the non-uniform patterns are labeled with a single label. With 3 × 3 neighborhoods, there are total of 256 patterns, 58 of them are uniform, which yields in 59 different labels [5]. Rotation invariant LBP histogram and uniform LBP histogram are shown in Fig. 1.

Grey Level Co-occurrence Matrix (GLCM) characterizes the texture by considering the spatial relationship of pixels over a sub-region of an image [6]. GLCM calculates how often different combinations of the reference pixel and neighbor pixel occur in the image. The combination defines the relationship between pixels in four directions such as horizontal, vertical, left and right diagonal. The number of occurrences a pixel with intensity i is adjacent to a pixel with intensity j, would be counted and stored in matrix with dimensions corresponding to the number of intensity values of an image. Because co-occurrence matrices are typically large and sparse, some statistic features can be extracted such as contrast, correlation and energy homogeneity to get more usefulness of feature. Contrast measures the local variations in the GLCM. Correlation measures the joint probability occurrence of the specified pixel pairs. Energy provides the sum of the squared elements in the GLCM. Homogeneity measures the closeness of the distribution of diagonal elements in the GLCM.

2.2 Feature Extraction from Gait

Dynamic features of human gait are used mostly in many approaches of individual recognition as well as gender classification. Two dynamic features of Gait Energy Image (GEI) and Gait Energy Motion (GEM) have proved its effectiveness for representing the characteristics of human gait. These features are from regular human walking, which is the repetitive motion of body with a differently stable frequency for each person. In complicated environments, human movement is firstly extracted from the image by using background subtraction techniques using Gaussian mixture model or kernel density estimation. Some further image processing is applied to get silhouette images with normalization and horizontal alignment. Given the processed binary images at time t in a sequence, the gait energy image is defined in Eq. 1:

$$F(i,j) = \frac{1}{N}\sum_{t=1}^{N} I_t(i,j) \qquad (1)$$

where $I(i, j)$ is the intensity of a pixel and N is the number of frames in the sequence. In [7], a modification of GEI is explored by taking the motion between each frame of the sequence as spatial data for human gait, which is called Gait Energy Motion (GEM). It is defined as:

$$F(i, j) = \frac{1}{N}\sum_{t=2}^{N} |I_t(i, j) - I_{(t-1)}(i, j)| \qquad (2)$$

where N is the number of frames in a sequence of walking person, $I(i, j)$ is the intensity of an image at time t and time $(t - 1)$. The difference between GEI and GEM is illustrated in Fig. 2.

3 Experiments

We carried out the experiments with the CASIA Gait Database of Dataset B collected by Institute of Automation, Chinese Academy of Science [8]. In our experiments, we used 29 female and 29 male gait dataset for training and testing. From 11 views of cameras, we just chose frontal view to get face information and silhouette sequences provided in Dataset B to get GEM. The images of volunteers

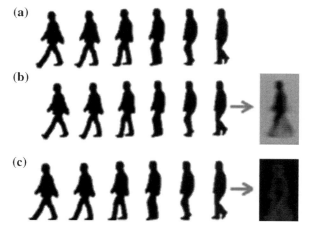

Fig. 2 a Silhouette images. b GEI c GEM

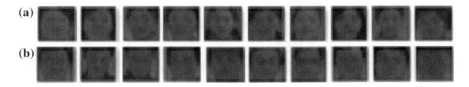

Fig. 3 Face dataset: **a** Female. **b** Male

were captured in 320 × 240 resolutions with 25 fps of frame rate. We used the last frame of each video to detect the face regions of each subject. Therefore, detected face are different in appearance of expression, head pose variations, hair and glasses wearing. Faces are normalized as 28 × 28 pixels images as shown in Fig. 3.

Human gait energy motions are calculated from the silhouette sequences in different of status walking: normal walking, carrying a bag and walking with a coat. 49 frames for each person are used to calculate GEM and GEI. With 29 female images and 29 male images, 80 % are used for training and 20 % for testing. As there are training images and test images, we selected images randomly, so that no overlapping could exist between them. We calculated average Correct Classification Rate (CCR). We used Support Vector Machine (SVM) as a classification method with linear kernel and used 10 cross fold validation as a training method. The results of testing with each feature of face and gait are shown in the Table 1a. Then we fused the label outputs using majority voting combining classifier [9]. The principal component analysis approach is used to reduce the feature dimensions for the combination. In our case, the average CCR of gender classification with face features gets lower performance because the noise has impacts to the extracted features. Better performance of 93.04 % is archived with the proposed combination of GEM and LBP as shown in the Table 1b.

Table 1 **a** Gender classification with each feature of face and gait. **b** Gender classification with the combination classifier

Method	Correct classification rate
GLCM	80.43 %
LBP	86.62 %
GEI	88.31 %
GEM	89.15 %

Method	Correct classification rate
GLCM + LBP	87.23 %
GEM + GEI	90.49 %
GEI + LBP	92.36 %
GEM + LBP	93.04 %

4 Conclusion

In this paper, we proposed the combination of face and gait for gender classification. Combining classifier appears as a natural step forward when a large amount of knowledge of single classifier models has been explored. Combining methods obtain better predictive performance than could be obtained from any of the constituent models. By combining face and gait using combination classifier, we increased the correct rate of gender classification for the low video resolution.

Acknowledgments This work (Grants No. C0005448) was supported by Business for Cooperative R&D between industry, Academy, and Research Institute funded by Korea Small and Medium Business Administration in 2012.

References

1. Golomb B, Lawrence D, Sejnowski T (1990) Sexnet: a neural network identifies sex from human faces. In: Advance in neural information processing systems, California, vol 3, pp 572–577
2. Nazir M, Ishtaiq M, Batool A, Jaffar A, Mirza AM (2010) Feature selection for efficient gender classification. In: Proceedings of the WSEAS international conference, Wisconsin, pp 70–75
3. Fang Y, Wang Z (2010) Improving LDP features for gender classification. In: IEEE international conference on wavelet analysis and pattern recognition, pp 1203–1208
4. Wang J, She M, Nahavandi S, Kouzani A (2010) A review of vision-based gait recognition methods for human identification. In: international conference on digital image computing: techniques and applications, pp 320–327
5. Ojala T, Maenpaa T (2002) Multiresolution gray_scale and rotation invariant texture classification with local binary pattern. IEEE Trans Pattern Anal Mach Intell 24:971–887 (Washington)
6. Clausi DA (2002) An analysis of co-occurrence texture statistics as a function of grey level quantization. J Remote Sens 28:45–62 (Canada)
7. Arai K, Asmara RA (2012) Human gait gender classification in spatial and temporal reasoning. Int J Adv Res Artif Intell 1:1–6
8. CASIA Gait Database. http://www.cbrs.ia.ac.cn/English/index.asp
9. Lam L, Suen CY (1997) Application of majority voting to pattern recognition: an analysis of its behavior and performance. IEEE Trans Syst Man Cybern 27(5):553–568

Implementation of Improved Census Transform Stereo Matching on a Multicore Processor

Jae Chang Kwak, Tae Ryong Park, Yong Seo Koo and Kwang Yeob Lee

Abstract Traditionally, sub-pixel interpolation in stereo-vision systems has been used for the block-matching algorithm. In this paper, Census transform algorithm which has been on area-based matching algorithm is improved and it's compared with existing census transform algorithm. Two algorithms are compared using Tsukuba stereo images provided by Middlebury web site. As a result, disparity map error rate is decreased from 16.3 to 11.8 %.

Keywords Census transform · Stereo matching · Area-based matching · Multi-core processing

1 Introduction

The stereo vision is a method to extract image depth information using two different images that have been captured by right and left view points. Among overall procedures of stereo vision, a step to find matching points is called stereo

---

J. C. Kwak (✉)
Department of Computer Science, Seo Kyeong University, Seoul, Korea
e-mail: jckwak@skuniv.ac.kr

T. R. Park · K. Y. Lee
Department of Computer Engineering, Seo Kyeong University, Seoul, Korea
e-mail: trpark@skuniv.ac.kr

K. Y. Lee
e-mail: kylee@skuniv.ac.kr

Y. S. Koo
Department of Electronic Engineering, Dan Kook University, Yongin, Korea
e-mail: yskoo@dankook.ac.kr

J. J. (Jong Hyuk) Park et al. (eds.), *Multimedia and Ubiquitous Engineering*,
Lecture Notes in Electrical Engineering 240, DOI: 10.1007/978-94-007-6738-6_122,
© Springer Science+Business Media Dordrecht(Outside the USA) 2013

matching. The stereo matching is a core of stereo vision system. The Census Transform (CT) stereo matching algorithm is a method finding matching points from two images having different viewpoints using structural information of pixels in regions. CT algorithm has less computational complexity [1].

In this paper, the CT algorithm is improved and the improved algorithm has been compared with original CT algorithm for its accuracy of stereo matching. Also, the efficiency in parallel processing system using ARM 11 MP-Core conditions is proved through parallel processing of multi-core. In embedded environment, the effort to reduce computations is important as well as the accuracy of algorithm.

2 The Census Transform Algorithm

The CT algorithm [2] transforms images comparing intensity of pixels with their neighbors. The CT algorithm compares center pixels with their neighbor pixels and presents the result with bit string. For the comparison, the transformation to bit string can be expressed as Eq. (1).

$$P_{xy} = \begin{cases} 0 & \text{if } P_{center} \geq P_{xy} \\ 1 & \text{if } P_{center} < P_{xy} \end{cases} \quad (1)$$

In Eq. (1), the P_{xy} means pixel intensity within sub-windows, and P_{center} means an intensity index of the center pixel. Each pixel in the sub-window is expressed with 0 and 1, comparing their intensity with the center pixel. Using the extracted P_{xy}, pixels are transformed to a bit string. Based on the pixel that has minimum Hamming Distance value, the disparity is extracted and the stereo matching is performed.

The MCT algorithm [3] is an improved version of CT algorithm. The MCT algorithm is an algorithm to reduce errors that can be caused by changes of intensity in source images. The MCT algorithm performs transformation based on the average values of pixels in sub window, while CT algorithm performs based on center pixels. The Fig. 1 shows difference between CT and MCT algorithms.

$$P_{xy} = \begin{cases} 0 & \text{if } P_{avg} \geq P_{xy} \\ 1 & \text{if } P_{avg} < P_{xy} \end{cases} \quad (2)$$

In Fig. 2, each values are determines to be 0 based on 5 in CT algorithm, and they are to be 0 or 1 based on average value, 2.3 in MCT algorithm. Thus, it has benefits to reduce errors that can happen according to changes of the intensity.

Fig. 1 Difference of CT and MCT algorithms

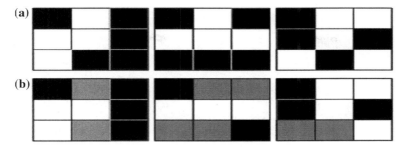

Fig. 2 Kernel index of CT **a** and proposed CT **b** algorithms

3 The Proposed Census Transform Algorithm

The proposed algorithm uses increased number of kernel index cases for more accurate stereo matching. As shown in Eqs. (1) and (2), previous algorithms express neighbors of center pixels as 0 or 1, so that there are 256 kernel indexes from 0 to 255. On the other hands, the proposed algorithm expresses more number of cases for kernel index, as Eqs. (3) and (4) show.

$$P_{xy} = \begin{cases} 2 & \text{if } P_{center} \geq P_{xy} - c \\ 1 & \text{if } P_{center} < P_{xy} + c \\ 0 & \text{otherwise} \end{cases} \quad (3)$$

$$P_{xy} = \begin{cases} 2 & \text{if } P_{avg} \geq P_{xy} - c \\ 1 & \text{if } P_{avg} < P_{xy} + c \\ 0 & \text{otherwise} \end{cases} \quad (4)$$

c in Eqs. (3) and (4) is a value for subdivision of ranges, so it sets the range of index pixel. Comparing to previous CT algorithms, the number of kernel index cases have been increased from $2^8 = 256$ to $3^8 = 6561$ in the proposed algorithm. For comparing pixel intensity, previous CT algorithms classify neighbors to small and big based on the intensity of center pixel as shown in Fig. 2a, however the proposed algorithm classifies neighbors to similar, small, and big for the higher precision as shown in Fig. 2b.

4 Multi-Core Processing

ARM 11 MP-Core system has four 320 MHz ARM11 processors and 32 Kb L1 command cache, 32 Kb L1 data cache, 1 Mb L2 share cache, and interrupt distributor. The consistency of L1 cache is managed by SCU (Snoop Control Unit). The ARM 11 MP-Core supports OpenMP [4] so that users can convert sequential code to parallel code only with directive insertion. The proposed algorithm is

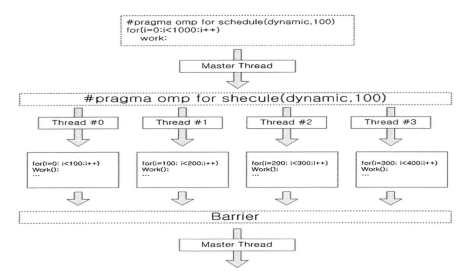

Fig. 3 Dynamic scheduling using OpenMP

paralleled using multi-core condition of OpenMP and the performance speed is compared.

The parallel process divides an input image with the unit of x axes and distributes to multi core as shown in Fig. 3. Once a particular core finishes its allocated iteration, it returns to get another one from the iterations that are left. In this way, the core waiting time has been minimized using dynamic scheduling.

5 Experiment

For experiments, images and stereo pairs, provided by Stereo vision Research Page of Middlebury, are used. Using provided source images and ground truth of each images, errors in disparity map, block section, and discontinuous points are extracted. The disparity maps created by CT algorithm and the proposed algorithm have been compared. Sizes and search ranges for each image in experiments are shown on the Table 1. Figure 4 shows disparity maps. The left images are disparity maps extracted by previous Census Transforms Algorithm, and the right images are disparity maps extracted by the proposed Census Transform algorithm. In

Table 1 Parameters of stereo images

Image	Image size	Search range	Scale
Tsukuba	384 × 288	16	16
Venus	434 × 383	20	8
Cones	450 × 375	60	4

Implementation of Improved Census Transform Stereo

Fig. 4 The result of Stereo matching. **a** Tsukuba. **b** Venus. **c** Cones

disparity map, error rates have been calculated using the method provided by Middlebury College excluding block section. The equation for error rate calculation is expressed as Eq. (5).

$$B = \frac{1}{N}\sum_{(x,y)}(|d_c(x, y) - d_T(x, y)| > \delta_d) \qquad (5)$$

$d_c(x, y)$ is a displacement value at x, y of stereo matching and $d_T(x, y)$ is a provided actual displacement value. The threshold δ_d is set to be 1. If the absolute value of a gap between two displacement values at x, y in two disparity map is bigger than threshold, the pixel will be regarded as a bad one, and total number of bad pixels is divided by number of whole pixels, N to calculate the error rate in percentage. When Tsukuba is used as a source image, the error rate in the disparity maps of CT and PCT (Proposed CT) algorithms is presented on the Table 1. Also, error rates for MCT and PMCT (Proposed MCT) algorithms are compared on the Tables 2 and 3.

In Tsukuba image, the Census Transform algorithm has 39.5 % error rate in case of using 5 × 5 sub windows, while the proposed algorithm showed 26.9 % error rate that has been improved 9 %. Moreover, the proposed algorithm shows better performance than the precious algorithms with step bigger sub-windows. The proposed algorithm has been experimented using ARM 11 MP-Core, and the

Table 2 Comparison of error rate for CT and PCT

Size	CT			PCT		
	Non occ'	All	Disc'	Non occ'	All	Disc'
5 × 5	34.7	35.9	35.1	25.5	26.9	27.7
7 × 7	26.7	28	30.5	18.9	20.3	24.3
9 × 9	21.2	22.5	28.3	14.8	16.2	23
11 × 11	17.8	19	27.4	12.4	13.7	22.3
13 × 13	15.1	16.3	26.5	10.6	11.8	22.4

Table 3 Comparison of error rate for MCT and PMCT

Size	MCT			PMCT		
	Non occ'	All	Disc'	Non occ'	All	Disc'
5 × 5	36.7	38.1	39.2	29.9	31.3	33
7 × 7	26.8	28.4	34.6	22.6	24.3	30.6
9 × 9	21.2	22.5	28.3	18.7	20.4	32.1
11 × 11	17.8	19	27.4	16.3	18.1	33.8
13 × 13	17.1	18.8	37.9	14.9	16.7	35.4

execution speed has been improved using parallel processing. When multi-core is applied for the proposed algorithm, the performance is shown in Fig. 5 for the parallel processing result using PCT with Tsukuba. The proposed algorithm with quad-core processor showes 3.69 times faster than the case of using single core processor.

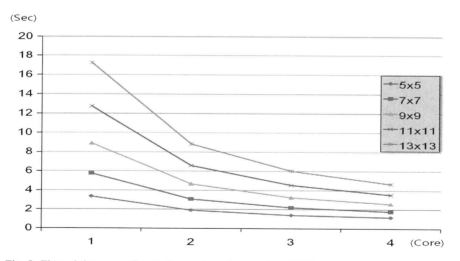

Fig. 5 Elapsed time according to the number of cores about PCT

6 Conclusion

In general, the development of stereo matching algorithm is focused on the improvement of accuracy rather than the execution speed. As the market of mobile devices is becoming prosperous, however, proper algorithm for embedded system is also important. In this paper, the improved version of CT algorithm is proposed for the stereo matching. The proposed algorithm is applied to multi-core processor and is compared with previous CT algorithm. Since the previous CT algorithm uses less kernel index cases, it has higher error rate. In case of the proposed CT algorithm, kernel index cases have been more diversified, so that the error rate has been decreased. In addition, the execution speed has been also improved through the parallel processing. There are two goals for further progress. First, constant values of the proposed algorithm should be extracted in variable ways according to the size of sub-windows. Second, hardware systems should be designed using the proposed CT algorithm, so that it performs faster stereo matching at embedded conditions.

Acknowledgments This work was sponsored by Industrial Strategic Technology Development Program funded by the Ministry of Knowledge Economy (10039188, SoC platform development for smart vehicle info-tainment system) and Industrial Strategic Technology Development Program funded by the Ministry of Knowledge Economy (10039145, the development of system semiconductor technology for IT fusion revolution).

References

1. Humenberger M, Zinner C, Kubinger W (2009) Performance evaluation of a census-based stereo matching algorithm on embedded and multi-core hardware. In: Proceedings of the international symposium on image and signal processing and analysis, vol 6
2. Zabih R, Woodfill J (1994) Non-parametric local transforms for computing visual correspondence. In: Proceedings of the European conference on computer vision, pp 151–158
3. Fröba B, Ernst A (2004) Face detection with the modified census transform. In: Proceeding of IEEE Conference on automatic face and gesture recognition, pp 91–96
4. ARM11MPCore http://www.arm.com/products/CPUs/ARM11MPCoreMultiprocessor.html
5. Schastein D, Szeliski R (2002) Ataxonomy and evaluation of dense two-frame stereo correspondence algorithm. Int J Comput Vis 47(1):7–42

A Filter Selection Method in Hard Thresholding Recovery for Compressed Image Sensing

Phuong Minh Pham, Khanh Quoc Dinh and Byeungwoo Jeon

Abstract Compressed sensing has been widely researched since the beginning of 2000s. Although there are several well-known signal recovery algorithms, its reconstruction noise cannot be avoided completely, thus requiring good filters to remove the noise in the reconstructing process. Since each different filter has its own advantages and disadvantages depending on specific reconstruction algorithm, the reconstruction performance can be varied according to the choice of filter. This paper proposes an inner filter selection method according to the sampling rate and the property of image to be sensed.

Keywords Compressed sensing · Wiener filter · Median filter

1 Introduction

In Nyquist-Shannon theorem, two times of signal bandwidth (Nyquist rate) is the slowest rate at which sampling of any band-limited signal should be done to guarantee perfect reconstruction [1]. However, in many low-cost practical applications of image or video, such a high Nyquist rate might be an expensive choice. David Donoho, Emmanuel Candes and Terence Tao introduced the compressed sensing [2, 3] which allows a signal to be sampled at a sub-Nyquist rate while still

---

P. M. Pham (✉) · K. Q. Dinh · B. Jeon
School of Electrical and Computer Engineering, Sungkyunkwan University, Seoul, Korea
e-mail: phamphuong@skku.edu

K. Q. Dinh
e-mail: diqkhanh@gmail.com

B. Jeon
e-mail: bjeon@skku.edu

J. J. (Jong Hyuk) Park et al. (eds.), *Multimedia and Ubiquitous Engineering*,
Lecture Notes in Electrical Engineering 240, DOI: 10.1007/978-94-007-6738-6_123,
© Springer Science+Business Media Dordrecht(Outside the USA) 2013

attaining near-optimal reconstruction [4]. It relies on the two basic assumptions: signals are sparse and samples are linear functional.

Several factors affect the overall quality of reconstructed images—sparsity of signal, sub-rate, incoherence between sparsifying transform—measurement matrix, and smoothing algorithm (at the decoder's side), etc. [2].

Some researchers [5] have designed adaptive smoothing algorithms, but they mainly focused on an iterative method which holds the estimation error of signal below a specified threshold. Both theoretical and practical studies proved that the selection of a sub-rate would directly impact reconstructed result. Moreover, some authors investigated the effects of different inner filters in the smoothing process [6].

In this paper, we investigate a filter selection method in the hard thresholding reconstruction algorithm for compressively sensed image. At each iteration, based on characteristic of image and sub-rate of transmitted signal, a better filter is chosen to obtain a higher PSNR value.

Our paper is organized as follows. Section 2 introduces some basic knowledge about the compressed sensing and the Block Based Compressed Sensing and Smoothed Projected Landweber reconstruction algorithm [5]. Section 3 represents the proposed method. Section 4 shows our experimental results. Finally, in Sect. 5 we conclude our work.

2 Background

2.1 Compressed Sensing Overview

Compressed sensing (CS) is a mathematical algorithm which reconstructs a signal with length N from M measurements where $M \ll N$. Suppose that x is a real-valued signal represented by an Nx1 vector, y is a sampled vector of length M:

$$y = \Phi x \tag{1}$$

where Φ is called a measurement matrix of size MxN, and the ratio M/N is called the sub-rate. Even if M is much smaller than N, if x is sparse, it is known to be able to reconstruct exactly or approximately x from y [2]. x is called K-sparse if it has at most K non-zeros coefficients. Since natural signals may not be sparse, we may need to represent x in a transform domain as:

$$x = \Psi s \tag{2}$$

where Ψ is a transform matrix with N columns $[\psi_1 | \psi_2 | \ldots | \psi_N]$, each column $\{\psi_i\}$ ($i = 1 \sim$ N) is a basis vector of length N. If s satisfies Eq. (2) and has at most K non-zero coefficients ($K \leq M \ll N$), we call x as K-sparse in a transform domain represented by Ψ. Ψ is named as sparsifying transform or sparsity basis.

The main issue in CS is how to reconstruct x from y. A large number of reconstruction algorithms were introduced for CS [5, 7–10]. Taking both the complexity and stability of reconstruction process into account, BCS-SPL (Block Based Compressed Sensing with Smoothed Projected Landweber) [5], which is formed by successively projecting and iterative hard thresholding, provides reduced computational complexity and possibly offers additional optimization criteria.

2.2 Block Based Compressed Sensing with Smoothed Projected Landweber

In essence, BCS-SPL combines Block Based Compressed Sensing (BCS) at the encoder's side with Smoothed Projected Lanweber (SPL) reconstruction algorithm at the decoder's side. At the encoder side, a natural image (x) is divided into blocks of size BxB. The jth block is denoted by x_j. Then the measurement is done for each block using a measurement matrix Φ_B of size $M_B \times B^2$ where $M_B = B^2*$sub-rate. The measurement, denoted by y_j, is computed as: $y_j = \Phi_B x_j$.

At the decoder side, the measurement of each vector is processed by using the SPL algorithm as below: For each block j,

Step 1: Compute initial reconstructed vector:

Compute $x_j^{(0)} = \Phi_B^T y_j$ and set iteration number i $= 1$

Step 2: Compute the reconstructed vector at iteration i:

Step 2.1: Set: $x' = x^{(i-1)}$

Step 2.2: Filter x' to impose smoothness and remove noise in spatial domain: $x' \leftarrow$ **wiener2**

Step 2.3: Apply hard thresholding: $x^{(i)} \leftarrow$ *hard* **thresholding** (x')

Step 2.4: Check stopping condition:

$$\tau = \left| D^{(i)} - D^{(i-1)} \right|$$

where $D^{(i)}$ is a Mean Squared Error (MSE) and calculated by Eq. (3):

$$D^{(i)} = 1/\sqrt{N} \left\| x^{(i)} - x^{(i-1)} \right\|_2 \tag{3}$$

If $\tau < 10^{-4}$ go to **Step 3**, *else* go to *Step 2.1* with i $=$ i $+ 1$.

Step 3: End $x^{(i)}$ is a reconstructed image for the block j.

Note that in this procedure, the compressed-sensed signal is supposed to be exactly available at the decoder side.

3 Proposed Method

As mentioned in Sect. 2.2, a Wiener filter is incorporated into the SPL algorithm to make a reconstructed image smoother but it sometime over-smoothes the image [11], and noise still exists in reconstructed image as in Fig. 1.

In spatial domain, a median filter is a very popular non–linear filter which can preserve edges, remove impulse noise (also known as salt and pepper noise), and avoid excessive smoothing [12]. Hence, in case image is prone to be oversmoothed, a median filter can be better to use than the Wiener filter. On the contrary, for an image with much texture, a reconstructed image using Wiener filter has higher quality than that using the median. In this paper, we design and implement a method which can flexibly choose the median filter or the Wiener filter in each Projected Landweber framework's iteration. The proposed method is illustrated as in Fig. 2.

The measurement vector is processed via successive functional blocks:

- Linear Initialization: As Step 1 in SPL algorithm, an initial vector is computed by multiplying measurement vector with the transposed matrix Φ_B: $x_j^{(0)} = \Phi_B^T y_j$.
- Projected Landweber: iterative hard thresholding operator for each block as Step 2.3 in SPL algorithm: $x^{(i)} =$ **hard thresholding** (x').
- Filter Selection.

Correlation measure can give the similarity between two signals. After the Projected Landweber, we have $x^{(i)}$ by hard thresholding x' (see step 2.3 in SPL algorithm): some coefficients of x' that are below a threshold will be replaced by "0". Let A_j be the correlation of the jth blocks in x' and $x^{(i)}$. Obviously, the higher A_j is the more similar they are. Denote that R = average (A_j)/maximum (A_j) (j = 1 $\sim$ number of blocks). Due to this definition of R, we can observe the followings: If R is high, the values A_j are approximate, the number of coefficients

Fig. 1 A part of reconstructed image with sub-rate = 0.1. **a** Barbara. **b** Lena. **c** Cameraman

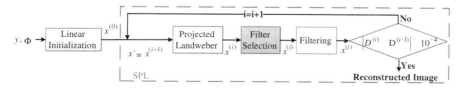

Fig. 2 Filter selection in an hard thresholding recovery of compressively-sensed image

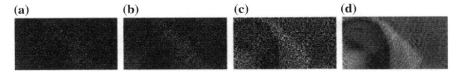

Fig. 3 A part of Barbara image. **a** and **c** are initial images with sub-rate 0.1 and 0.5, respectively, **b** and **d** are images after the first iteration of hard thresholding

which are less than the threshold in every block are close. In this case, $x^{(i)}$ is seen to be smooth. In contrast, if there is a remarkable difference in the number of zeroed points from all partitioned blocks, there is significant difference between A_j values, and R becomes a small value. In this case, $x^{(i)}$ is seen to be textured.

An initial vector is calculated by the equation: $x_j^{(0)} = \Phi_B^T y_j = \Phi_B^T \Phi_B x_j$. At low sub-rate ($M \ll N$), noise is supposed to be at high level ($\Phi_B^T \Phi_B$ is small), number of coefficients which are less than the threshold in blocks is large. After each iteration of hard thresholding, all blocks in image are changed equaly. Therefore, image is smoother. When M progresses towards N (high sub-rate), noise becomes harder to perceive, and image converges fast to the original. Some blocks with many coefficients nearly "0" will have remarkable change after hard thresholding, some blocks with few coefficients less than a hard threshold will not change much after hard thresholding. The difference between changes of blocks is impressive. In this case, image is more textured (Fig. 3).

Based on the experiments and observations related to comparing R with sub-rate, we decide using a median filter if R > sub-rate, and using the Wiener filter otherwise.

- Filtering: Use a *filter* which is chosen in Filter Selection for $x^{(i)}$:

$$x^{(i)} \leftarrow filter\left(x^{(i)}\right)$$

4 Experimental Results

In this paper, we evaluate the performance of our proposed method in comparison with original BCS-SPL at sub-rate from 0.1 to 0.5 with images of size 512×512. We use sparsifying transform as Discrete Wavelet Transform and block size 32×32 to implement. Tested images are Barbara, Lena, Cameraman, and Girl.

Table 1 collects PSNR values of various reconstructed images when using the proposed method and when using BCS-SPL. In most cases, the proposed method introduces much higher PSNR than BCS-SPL. When the original image has texture as Barbara or Lena, the proposed method shows higher PSNR than BCS-SPL (0.5 dB at sub-rate = 0.1 and 0.2 dB at sub-rate = 0.5). Bold-faced in Table 2 shows significant improvements of the proposed method. In case image is smooth

Table 1 PSNR of image with block size 32 × 32 at sub-rate from 0.1 to 0.5

Images	Sub-rate	0.1	0.2	0.3	0.4	0.5
Barbara	Proposed method	22.9	24.0	25.2	26.7	28.2
	BCS–SPL Wiener filter	22.4	23.7	25.1	26.5	28.0
Lena	Proposed method	28.0	31.1	33.1	34.8	36.3
	BCS–SPL Wiener filter	27.5	30.7	32.9	34.6	36.1
Clown	Proposed method	**27.0**	**29.7**	**31.7**	**33.6**	**35.2**
	BCS–SPL Wiener filter	25.2	29.3	31.6	33.4	35.1
Cameraman	Proposed method	**27.5**	**31.2**	**33.5**	**36.3**	**38.6**
	BCS–SPL Wiener filter	25.7	29.9	32.8	35.4	37.8
Girl	Proposed method	29.8	32.6	34.4	36.2	37.9
	BCS–SPL Wiener filter	29.2	32.0	34.2	36.0	37.7

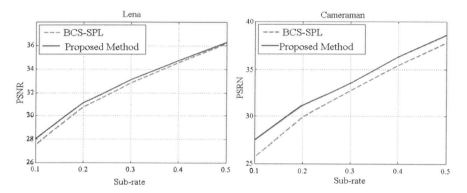

Fig. 4 PSNR of lena image (*left*) and cameraman image (*right*)

(Girl), our idea produces objective quality higher than BCS-SPL (0.6 dB at sub-rate = 0.1 and 0.2 dB at sub-rate = 0.5).

Figure 4 shows results of coding Lena and Cameraman image. Our proposed method (solid lines) offers a higher PSNR than BCS-SPL (dotted lines).

5 Conclusion

This paper proposed a filter selection method in the hard thresholding recovery for compressed sensing image coding. It is shown that the proposed adaptive method inherits the advantage of both median and Wiener filter. In each iteration of hard thresholding, based on property of image and sub-rate, a better filter will be chosen. Consequently, it is more effective on reconstructed image's quality than using only Wiener filter in BCS-SPL algorithm.

Acknowledgments This work was supported by the National Research Foundation of Korea (NRF) grant funded by the Korea government (MEST) (No. 2011-001-7578).

References

1. Cover TM, Thomas JA (2006) Elements of information theory, 2nd edn. Wiley, New York
2. Donoho DL (2006) Compressed sensing. IEEE Trans Inform Theory 52(4): 1289–1306
3. Candes EJ, Romberg J, Tao T (2006) Robust uncertainty principles: exact signal reconstruction from highly incomplete frequency information. IEEE Trans Inform Theory 52(2): 489–509
4. Baraniuk RG (2007) Compressive sensing [lecture notes]. IEEE Signal Process Mag 24(4):118–121
5. Fowler JE, Mun S, Tramel EW (2012) Block–Based compressed sensing of image and video. Found Trends Signal Process 4(4):297–416
6. Kumar S, Kumar P, Gupta M, Nagawat AK (2010) Performance comparison of median and Wiener filter in image de-noising. Int J Comput Appl (0975–8887) 12(4): 27–31
7. Candes E, Romberg J (2005) ℓ1-magic: recovery of sparse signals via convex programming. Technical report, California Institute of Technology
8. Figueiredo MAT, Nowak RD, Wright SJ (2007) Gradient Projection for sparse reconstruction: application to compressed sensing and other inverse problems. IEEE J Sel Topics Signal Process 1(4):586–597
9. Chen SS, Donoho DL, Saunders MA (1998) Atomic decomposition by Basis Pursuit. SIAM J Sci Comput 20(1):33–61
10. Ji S, Xue Y, Carin L (2008) Bayesian compressive sensing. IEEE Trans Signal Process 56(6):2346–2356
11. Khireddine AK, Benmahammed, Puech W (2007) Digital image restoration by Wiener filter in 2D case. Adv Eng Softw 38(7):513–516
12. Church JC, Yixin C, Rice SV (2008) A spatial median filter for noise removal in digital images. In: IEEE Southeastcon, Alabama, pp 618–623

Facial Expression Recognition Using Extended Local Binary Patterns of 3D Curvature

Soon-Yong Chun, Chan-Su Lee and Sang-Heon Lee

Abstract This paper presents extended local binary patterns (LBP) for facial expression analysis from 3D depth map images. Recognition of facial expressions is important to understand human emotion and develop affective human computer interaction. LBP and its extensions are frequently used for texture classification and face identification and detection. In the 3D surface analysis, curvature is very important characteristics. This paper presents an extension of LBP for modeling curvature from 3D depth map images. The extended curvature LBP (CLBP) is used for facial expression recognition. Experimental results using Bosphorus facial expression database show better performance by 3D curvature and the combination of 3D curvature and 2D images than by conventional 2D or 2D + 3D approaches.

1 Introduction

Facial expressions are one of the key components for understanding human emotional states. The recognition of human emotion is very important for affective computing, human robot interaction, and smart devices such as smart TV, smart phone, and smart lighting system. Conventionally, 2D image-based face detection, and feature extraction are used for facial expression recognition. Recently consumer 3D depth cameras like Kinect® are available to be used in personal computer. 3D depth information can be captured robust to illumination change and

---

S.-Y. Chun (✉) · C.-S. Lee
Department of Electronic Engineering, Yeungnam University, 214-1 Dae-dong,
Gyeongsan-si, Gyeongsangbook-do 712-749, Korea

S.-H. Lee
Daegu Gyeongbuk Institute of Science and Technology, 50-1 Sang-ri,
Hyeongpung-myeon, Dalseong-gue, Daegu 711-873, Korea

J. J. (Jong Hyuk) Park et al. (eds.), *Multimedia and Ubiquitous Engineering*,
Lecture Notes in Electrical Engineering 240, DOI: 10.1007/978-94-007-6738-6_124,
© Springer Science+Business Media Dordrecht(Outside the USA) 2013

view change. Not only 3D depth information as well as 2D images is available from Kinect®. This paper presents a new feature for 3D depth map image and applies the feature for facial expression recognition from the combination of 2D and 3D depth map features.

Local binary patterns are introduced by Ojala et al. [1] for local shape analysis robust to illumination change. It is originally used for local texture analysis and applied for many other applications such as face identification, face detection, facial expression recognition [2]. A global description of texture from local descriptor can be achieved by dividing whole textures into local regions using regular grid and extracting LBP histograms from each sub regions independently [3, 4]. Patterns of oriented edge magnitudes are also used for face recognition from 2D texture images [5].

For the facial expression recognition from 2D, local binary patterns are also used [6]. Division of sub-region and weighting for dissimilarity measurement shows improvement of the facial expression recognition performance. Support vector machine, linear discriminant analysis, and linear programming are used for the recognition of facial expression from the LBP features and PCA subspaces.

In 3D depth map, depth value in z axis can be represented by gray value in x, y coordinate. Feature extraction in 2D gray image can be applied similarly to the gray value from 3D depth map. However, depth value changes smoothly and does not show clearly motion characteristics directly. Geometrically localized features and surface curvature features are used for better discrimination of facial expression from 3D [7]. 3D patch shape distance of curve shape [8] and histogram of curvature type [9] are also used for 3D facial expression recognition. Many of these features are related to the curvature of the 3D facial surfaces.

To improve the performance of the depth data using LBP, the difference of the depth data can also be coded in addition to the sign compared with the center [10], which shows improved performance in facial expression recognition from depth map image. However, the coding is complicated and does not use the characteristics of the curvature. This paper presents a new LBP extension which is directly connected to the curvature of the 3D facial surface and useful for facial expression recognition.

2 An Extension of LBP for 3D Surface Curvature

2.1 Introduction to Local Binary Patterns

Local binary patterns (LBP) are introduced for 2D image texture analysis. The basic principle of LBP operator was based on the assumptions that a texture has a pattern and its intensity, that is, its strength. As the texture pattern is more important and needs to be encoded invariant to the intensity variations, relative strength of neighbor points are described using binary patterns compared with

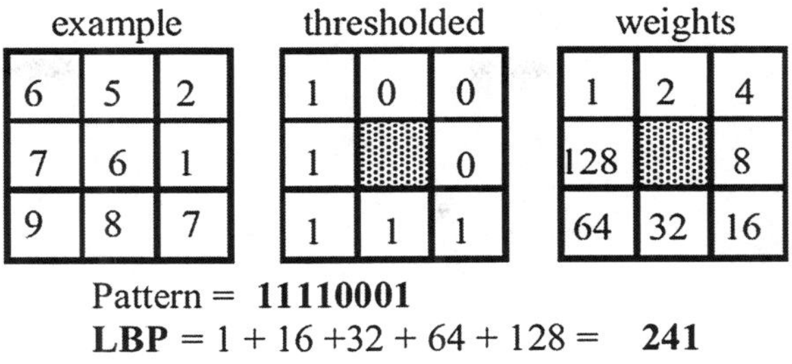

Fig. 1 A simple example of LBP

example		
6	5	2
7	6	1
9	8	7

thresholded		
1	0	0
1		0
1	1	1

weights		
1	2	4
128		8
64	32	16

Pattern = **11110001**
LBP = 1 + 16 +32 + 64 + 128 = **241**

central points. Figure 1 shows a simple example of LBP patterns for 3×3 rectangles. Threshold values are computed by comparing with the middle value 6. Weights are used to convert a threshold binary number to a decimal number.

The basic LBP operator was extended into multiscale using variations of radius of the sampling points and rotation invariance using circularly rotated code mapping into its minimum value [1]. Many other LBP variants are proposed in preprocessing, neighborhood topology, threshold and encoding, multiscale analysis, handling rotation, handling color, and so on [2].

2.2 Computation of Curvature and Curvature Local Binary Patterns

Geometrically, the curvature of straight line is defined to be zero. The curvature of a circle of radius R is defined to be the reciprocal of the radius R:

$$\kappa = \frac{1}{R} \tag{1}$$

For the plane curve given explicitly as y = f(x), the curvature can be computed by Eq. (2). The equation can further be simplified when the slope is small compared with unity by Eq. (3).

$$\kappa = \frac{|y''|}{(1 + y'^2)^{3/2}} \tag{2}$$

$$\kappa \approx \left| \frac{d^2 y}{dx^2} \right| \tag{3}$$

Primitive geometric features such as ridge, peak, saddle, convex hill can be used for facial expression recognition [9]. In order to estimate the geometry features of facial surface, fitting a smooth polynomial patch onto the local surface patch is required. When local patch height z(x, y) is approximated by polynomial surface as follows:

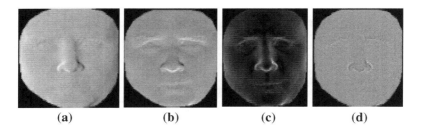

Fig. 2 An example of depth map derivative and its magnitude. **a** dx, **b** dy, **c** drv and **d** 2nd drv

$$z(\bar{x}, \bar{y}) = \left(\frac{1}{2}A\bar{x}^2 + B\bar{x}\bar{y} + \frac{1}{2}C\bar{y}^2 + D\bar{x}^3 + E\bar{x}^2\bar{y} + F\bar{x}\,\bar{y}^2 + G\bar{y}^3\right), \quad (4)$$

where $\bar{x}$, $\bar{y}$, are local coordinate value. Weingarten matrix for the surface fitting becomes as follows:

$$W = \begin{bmatrix} A & B \\ B & C \end{bmatrix} \quad (5)$$

After the eigenvalue decomposition, the principal directions can be estimated.

In this paper, we approximate the curvature of 3D depth image from Eq. (3). When the magnitude of the derivative of x and y axis is given by Eq. (4), the LBP for the magnitude can approximate the second derivative because the LBP also compare the value of center and its neighborhood and have a role as a derivative. We call this approximation of 3D depth curvature as curvature local binary pattern (CLBP).

Computation of CLBP is easy and fast. First, the depth image derivative is computed for x axis and y axis as in Fig. 2a and b. Then, its magnitude is computed by Eq. (6) as in Fig. 2c. Conventional LBP or its variations are applied to the magnitude of the derivatives. Figure 2d shows the second derivative of the depth map image, which shows noisy characteristics of depth image data.

$$dz = \sqrt{dx^2 + dy^2} \quad (6)$$

3 Facial Expression Recognition Using Curvature LBP

This section explains how to recognize facial expressions using Curvature LBP from 3D depth image database. Bosphorus database [10] is used to evaluate the performance.

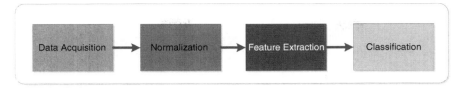

Fig. 3 Overview of facial expression recognition system from 2D and 3D

3.1 System Configuration

There are four steps to recognize facial expressions: data acquisition, normalization, feature extraction, and classification. 2D and 3D facial expression data and their landmark point data was collected from the Bosphorus database [10]. A mean landmark shape is estimated using Procrustes algorithm. Each 2D image and 3D depth data are normalized by geometric transformation to fit landmark points to the mean landmark shape. 2D appearance image, 3D depth gray image, 3D curvature image and their combinations are used. For the classification of facial expressions, Chi square distance-based NN classification, and support vector machine (SVM) are used. Figure 3 shows the whole steps for facial expression recognition.

3.2 Database for Facial Expression Recognition

In this paper, Bosphorus facial expression database [10] is used. This database provides high resolution 2D images and low resolution 3D map, and their landmark points, 3D pose variations, facial action unit (AU), and facial expression types. 41 subjects whose data contains all six different facial expressions are used (Fig. 4).

3.3 Feature Extraction for Facial Expression Recognition

The 2D facial images and 3D depth images are normalized based on landmark points and cropped to remove background images and boundaries, which are not

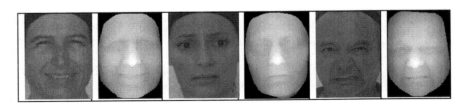

Fig. 4 Sample examples of facial expression in 2D and 3D from Bosphorous DB

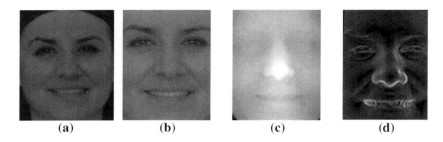

Fig. 5 Cropped 2D, 3D, and 3D curvature images used for feature extraction. **a** Original image, **b** Cropped 2D image, **c** Cropped 3D depth image and **d** Cropped 3D curvature image

relevant to facial expression recognitions. A subject's original image, cropped 2D and 3D depth images are shown in Fig. 5a, b, and c. The cropped 3D curvature image in Fig. 5d represents the magnitude of the image derivatives in the x axis and y axis.

Not only conventional uniformity 2 LBP (LBP U2), but also uniformity 4 LBP (LBP U4), Local derivative pattern (LDP) [11], center-symmetric local binary patterns (CS-LBP) [12] are used to extract local pattern histogram from 2D, 3D, and 3D curvature images to compare performance in different type of LBPs.

3.4 Facial Expression Recognition

Extracted binary pattern features are histogram distributions according to applied local pattern descriptors and input source types. Chi square statistics of distance measurement in Eq. (7) and nearest neighborhood classifications are applied for facial expression recognition from different histogram distributions.

$$x^2(S,M) = \sum_{i,j} \frac{(S_{i,j} - M_{i,j})^2}{S_{i,j} + M_{i,j}} \qquad (7)$$

In addition, SVM classifiers are learned after converting histogram distribution into vector representation. As shown in the experimental results, SVM performs better than nearest neighborhood search based on Chi square distance.

4 Experimental Results

We tested the performance of the proposed Curvature LBP and existing methods (LBP, CS LBP, LDP) for facial expression recognition using Bosphorus facial expression database. Leave one-subject out test method, which use 40 subject for

Facial Expression Recognition Using Extended Local Binary Patterns

Table 1 Facial expression recognition average results with 2D, 3D, 3D Curvature using Chi square method

	2D	3D	3D Curvature	2D + 3D	2D + 3D Curvature
LBP (u2)	60.98	54.88	65.85	62.20	65.85
LBP (u4)	57.72	59.73	63.01	61.38	**66.26**
LDP	58.54	58.13	55.69	58.94	63.41
CS LBP	60.98	52.85	60.98	61.38	60.98

training and one subject for testing, is used for the evaluation of the subject-independent facial expression recognition performance in different features and classifiers.

4.1 Facial Expression Recognition Performance in Different Feature

For the facial expression recognition, we used Chi square distance with LBP (u2), LBP (u4), CS LBP, LDP, CLBP. We tested the performance of each feature extraction methods using leave-one subject-out test method. Table 1 shows the estimated performance in the combination of 2D, 3D, and 3D curvature faces with four different feature extraction methods. In the experiment, images were resized into 110×150 pixels with 6×7 sub-regions. According to experimental result, 2D + 3D Curvature LBP (u4) shows the best performance in Chi square distance evaluation.

4.2 Facial Expression Recognition Performance by SVM

For better classification of facial expression, we used SVM with C-SVC type with RBF (Radial Basis Function) kernels. We used 120 images for training, others for evaluation among 246 images. We tested the performance of each feature extraction method. Table 2 shows the estimated facial expression recognition performance by SVM. According to experimental result, 2D + 3D CLBP (u2) shows the best performance in overall experiments.

Table 2 Facial expression recognition results with 2D, 3D, 3D Curvature using SVM

	2D	3D	3D curvature	2D + 3D	2D + 3D curvature
LBP (u2)	76.19	65.08	64.29	69.05	76.98
LBP (u4)	67.46	65.87	63.49	69.84	75.40
LDP	70.63	59.52	58.73	24.29	74.60
CS LBP	74.60	65.08	62.70	71.43	76.19

5 Conclusions

This paper presented a new curvature binary pattern analysis to extract curvature information from 3D depth map images. The curvature binary from 3D depth image with the combination of 2D LBP shows the best performance in the facial expression recognitions from the Bosphorus facial expression database.

Acknowledgments This work was supported by the DGIST R&D Program of the Ministry of Education, Science and Technology of Korea (13-IT-03).

References

1. Ojala T, Pietikenen M, Harwood D (1996) A comparative study of texture measures with classification based on featured distribution. Pattern Recognit 29
2. Pietikainen M, Hadid A, Zhao G, Ahonen T (2011) In computer vision using local binary patterns. Springer (2011)
3. Ahonen T, Hadid A, Pietikainen M (2004) Face recognition with local binary patterns. In: Proceedings of ECCV, pp 469–481
4. Ahonen T, Hadid A, Pietikainen M (2006) Face description with local binary patterns: application to face recognition. IEEE Trans PAMI 28
5. Vu NS, Caplier A (2011) Enhanced patterns of oriented edge magnitudes for face recognition and image matching. IEEE Trans Image Process 21:135–1365
6. Shan G, Gong S, McOwan PW (2009) Facial expression recognition based on local binary patterns: a comprehensive study. Image Vis Comput 27(6):803–816
7. Sha T, Song M, Bu J, Chen C, Tao D (2011) Feature level analysis for 3D facial expression recognition. Neurocomputing 74:2135–2141
8. Maalej A, Amor BB, Daoudi M, Srivastava A, Berretti S (2011) Shape analysis of local facial patches for 3D facial expression recognition. Pattern Recognit 44:1581–1589
9. Wang J, Yin L, Wei X, Sun Y (2006) 3D facial expression recognition based on primitive surface feature distribution. In: IEEE Conference on computer vision and pattern recognition, pp 1399–1406
10. Savran A, Alyuz N, Dibeklioglu H, Gokberk O, Sankur B, Akarun I (2008) Bosphorus database for 3D face analysis. In: Proceedings of the Workshop on BIOID
11. Zhang B, Gao Y (2010) Local derivative pattern versus local binary pattern: face recognition with high-order local pattern descriptor. IEEE Trans Image Process 19:533–544
12. Kalra P, Peleg S (2006) Description of interest regions with center-symmetric local binary patterns. In: Proceedings of the ICVGIP, LNCS 4338, pp 58–69

Overview of Three and Four-Dimensional GIS Data Models

Tuan Anh Nguyen Gia, Phuoc Vinh Tran and Duy Huynh Khac

Abstract The paper focuses on brief presenting again the 3D, 4D GIS data models that have been proposed in the past. The paper uses the tables to compare models based on various criteria in the applications of 3D, 4D GIS. In 3D data models, there were some special models because they can represent the spatial objects by multiple levels of detail. The paper also compares the differences of these models in tables. These tables are the foundation for the developers to build the GIS applications.

Keywords 3D · 4D data model · GIS

1 Introduction

3D GIS are a system that be able represent, manage, and manipulate, analysis information links with 3D phenomena. 3D GIS data model is the key to 3D GIS [1] and is a big topic in the five major topics of 3D GIS: WebGIS, data presentation, spatial analysis, data model and data collection [2]. The selecting a data model to represent 3D objects 3D GIS for a specific application will determine methods to store, access. There are many models of the authors have suggested [1, 3–8] for spatial and [9–17] for spatial-temporal. The purpose of the paper provides an

---

T. A. N. Gia (✉)
University of Science, Ho Chi Minh City, Vietnam
e-mail: anhngt2003@yahoo.com

P. V. Tran · D. H. Khac
University of IT, Ho Chi Minh City, Vietnam
e-mail: Phuoc.gis@gmail.com

D. H. Khac
e-mail: huynhkhacduy@gmail.com

J. J. (Jong Hyuk) Park et al. (eds.), *Multimedia and Ubiquitous Engineering*, Lecture Notes in Electrical Engineering 240, DOI: 10.1007/978-94-007-6738-6_125, © Springer Science+Business Media Dordrecht(Outside the USA) 2013

overview of 3D, 4D data models have been proposed, compared the models on important criteria by tables. These tables will support to recognize the development trend of the 3D, 4D GIS data models in the future. The attempts to classify the models will be the foundation for the researches related to 3D GIS and 4D GIS. Previously several authors have made this issue for spatial model and spatial-temporal. However, the works still missed some of models appearing recently and some criteria.

The paper proposes four tables to compare and classify the spatial models: 3D objects are represented by its boundaries; voxel elements; a combination of the 3D basic block and by a combination of the above methods. The paper also proposed two tables to compare the spatial-temporal 3D, 4D models. The 3D, 4D models include either 2D spatial + 1D temporal or 3D spatial + 1D temporal. This paper structure includes three sections. Section 1 introduces about the 3D, 4D GIS data models. Section 2 has five comparison tables of models on many different criteria. Section 3 describes and compares the 3D, 3.5D, 4D spatial-temporal data modes.

2 Compare Models

There are many data models, which were proposed in many past years. These modes are classified by four the approaches. They include B-REP, CGS, Voxel and hybrid method. B-REP approach has the models: 3D-FDS (Format Data Structure), TEN (Tetrahedral Network), OO Model (Object Oriented Model), SSM (Simplified Spatial Model), SOMAS (Solid Object Management System), UDM (Urban data Model), OO 3D (Object Oriented 3D), CITYGML model, LUDM model. Voxel has two modes: 3D Array and Octree and hybrid approach has also two models: V-3D, B_REP + CSG.

2.1 Comparing the Models on the Following Criteria: Surface Representation Method, Objects Inside Representation

See Table 1.

2.2 Comparing the Models on the Following Criteria: Primitive Elements, Geometry Elements and Application of the Model

See Table 2.

Table 1 Comparing models for 2.1

Style of model	Authors	Model	Surface representation	Objects inside representation
BREP	Molenaar 1990	3DFDS	Non triangular	No
	Pilouk 1996	TEN	Triangular	Yes
	Zlatanova 2000	SSM	Non triangular	No
	Delalosa 1999	OO	Triangular	Yes
	Pfund 2001	SOMAS	Non triangular	No
	Coors 2003	UDM	Triangular	No
	Shi et al. 2003	00 3D	Triangular	Yes
	Groger et al. 2007	City GML	Triangular	No
	Anh N.G Tuan 2011	LUDM	Non triangular	No
Voxel	NULL	3D Array	Non triangular	Yes
	Meagher 1984	Octree	Non triangular	Yes
CSG	Samet 1990	CSG	Non triangular	Yes
Combi nation	Xinhua et al. 2000	V-3D	Non triangular	Yes
	Chokri et al. 2009	B_REP CSG	Non triangular	No

2.3 Comparing the Models on the Following Criteria: Spatial Structure, Direction, Measurement and Topology

See Table 3.

2.4 Comparing Models on Query Criteria: Attribute, Location and Topology

See Table 4.

2.5 Comparing Models for Representation: Lod, Curve Surface, Semantic, History

See Table 5.

3 The Spatial-Temporal 3D, 3.5D and 4D Models

There were several spatial-temporal models proposed in the past. The Snapshot model was proposed in 1984 by Dangermon, Space Time Composite by Chrisman 1988, SpatioTime Cube by Szego 1987. Base-State was proposed by Armenakis

Table 2 Comparing the models for 2.2

Style of model	Model, primitive elements	Geometry elements	Main ideal	Data size	App
BREP	3DFDS. Point, Line, surface, body	Node, arc, face	2D GIS model	Large	3D UM
	TEN. Point, line, surface, body	Node, Arc triangle, tetra	Tetrahedron	Large	GA
	OO. 0-simplex, 1-2-3 simplex	Node, arc, face	Oriented-Object, n-simplex	Large	3D UM
	SSM. Point, line, surface, body	Node, face	Oriented-Object, topology	Small	3D WEB Urban
	SOMAS. Point, line, polygon, solid	Vertex, edge face, solid	Oriented-Object	Large	3D UM
	UDM. Point, line, surface, body	Node, face	Triangular	Small	3D UM
	OO3D. Point, line, surface, volume	Triangle, segment, node	Oriented-object, triangular	Large	3D UM
	CityGML. Point, curve, surface, solid	Polygon, linestring	Define standards of object	Large	3D UM
	LUDM. Point, line, surface, solid, LOD	Node, face	Represent LOD for solid	Large	3D UM
Voxel	3D Array. Elements/	None	Partition object by array	Very large	GA
	Octree. Voxel	None	Partition object by voxel	Very large	GA
CSG	CSG. Basic 3D block	NULL	Combine basic 3D block	Small	CAD CAM
Combination	V3D. Point, line, surface, body, raster	Node, edge, Face	Combine BREP and raster	Large	3D UM
	B-REP + CSG. Point, Line, Surface, Body, Basic 3D block.	Linestring, Face	Combine BREP and CSG	Large	3D UM

3D *UM* 3D Urban management; *GA* Geological application

Overview of Three and Four-Dimensional GIS Data Models

Table 3 Comparing the models for 2.3

Model	Spatial structure	Direction	Measurement	Topology
3DFDS	V	Yes	No	Yes
TEN	V	No	Yes	No
SSM	V	No	No	Yes
OO	V	Yes	No	No
SOMAS	V	Yes	No	No
UDM	V	Yes	Yes	No
OO 3D	V	No	No	No
CityGML	V	No	No	No
LUDM	V	No	Yes	No
3DArray	R	No	Yes	No
Octree	R	No	Yes	No
CSG	V	No	Yes	No
B_REP + CSG	V	No	Yes	No
V-3D	VR	No	No	No

Table 4 Comparing the models for 2.4

Model	Attribute query	Position query	Topology relationship query
3DFDS	No	Yes	Yes
TEN	No	Yes	No
SSM	No	Yes	Yes
OO	No	Yes	No
SOMAS	No	Yes	No
UDM	No	Yes	No
OO 3D	No	Yes	No
CityGML	Yes	Yes	No
LUDM	No	Yes	No
3D Array	No	No	No
Octree	No	No	No
CSG	No	Yes	No
B_REP + CSG	Yes	Yes	No
V-3D	No	Yes	No

1992, Object- Oriented by Worboys. Three-Domain was proposed by Yuan 1994, Event-Oriented by Peuget 1995, History graph by Renolen 1996, STER by Tryfona 1997. MADS was proposed by Spaccapietra 1999, STUML by Tryfona 2000. Moving objects was proposed by R.H.Guting 2000, by Balovnev 2002. OO was proposed by Shouheil Khaddaj, Event Based by Shuo Wang 2005, Geotoolkit + Geodeform was in 2002 and TUDM in 2012. Tables 6 and 7 compare the models for the criteria: the number of dimension, change history of objects, the time types and their application.

Table 5 Comparison the models for 2.5

Model	LOD	Curve surface	Semantic	History
3DFDS	1	No	No	No
TEN	1	Yes	No	No
SSM	1	No	No	No
OO	1	Yes	No	No
SOMAS	1	No	No	No
UDM	1	Yes	No	No
OO 3D	1	Yes	No	No
CityGML	5	Yes	Yes	No
LUDM	n	Yes	No	No
3D Array	1	Yes	No	No
Octree	1	Yes	No	No
CSG	1	Yes	No	No
B_REP + CSG	1	Yes	No	No
V-3D	1	No	No	No

Table 6 Comparison between the spatial- temporal models

Model	The number of dimension	Object
Snapshot	3	Face
Space time composite	3	Face
Spatio time cube	3	Face
Base-state	3	Face
Object-oriented	3	Face
Three-domain	3	Face
Event-oriented	3	Face
History graph	3	Face
STER	3	Face
MADS	3	Face
STUML	3	Face
Moving objects	3	Point, face
OO	3	Face
Event based	3	Face
Geotoolkit + Geodeform	4	Surface
TUDM	4	Body, surface, line, Point

Table 7 Comparison between TUDM and Geotoolkit + Geodeform

Model	Time	Application
Geotoolkit + Geodeform	Discrete, continuous	GM
TUDM	Discrete	UM

UM Urban management; *GM* Geology management

4 Conclusion

The main purpose of the paper is to present the development history, the main features of the 3D, and 4D data model. The models divided two groups: spatial model and spatial-temporal model. Each model has different advantages and limitations. The advantages of this model may be difficult for the other models. Choosing model depends on to develop a specific 3D, 4D GIS application. This paper classifies models for the nature of data structure on each model. The paper has created seven comparison tables to models on universal criteria in the areas of GIS, which was based on their characteristics. The tables again help the researchers have an overview of 3D, 4D GIS data model in past and recent years. It is the basis theory to envision in the next work of research and the important foundation for researchers to build models of 4D GIS later.

References

1. Alias AR (2008) Spatial data modeling for 3D GIS. Springer, Berlin
2. Stoter J, Zlatanova S (2003) 3D GIS, where are we standing. In: Spatial, temporal and multi-dimensional data modeling and analysis
3. Tuan Anh NG, Vinh PT, Vu TP, Sy AT, Dang VP (2011) Representing multiple levels for objects in three-dimensional GIS model, iiWAS2011. ACM Press, Ho Chi Minh City, pp. 591–595 ISBN 978-1-4503-0784-0
4. Billen R, Zlatanova S (2003) 3D spatial relationships model: a useful concept for 3D cadastre. Comput Environ Urban Syst 27(4):411–425
5. Chokri K, Mathieu, K (2009) A simplified geometric and topological modeling of 3D building enriched by semantic data: combination of SURFACE-based and SOLID-based representations. In: ASPRS 2009 annual conference, Baltimore, Maryland
6. Coors V (2003) 3D-GIS in networking environments. Comput Environ Urban Syst 27:345–357
7. OGC (2007) City geography markup language (Citygml) encoding standard. Open Geospatial Consortium inc
8. Wang X, Gruen A (2000) A hybrid GIS for 3D city models. In: IAPRS, vol 23, Amsterdam
9. Anh N, VinhPT, Duy HK (2012) A study on four-dimensional GIS spatio-temporal data model. In: KSE 2012 the fourth international conference on knowledge and systems engineering, published in IEEE Xplore, Da nang Vietnam
10. Breunig M, Balovnev O, Cremers AB, Shumilov S (2002) Spatial and temporal database support for geologists—an example from the lower Rhine basin. Neth J Geosci 81(2):251–256
11. Pelekis N, Theodoulidis B, Kopanakis I, Theodoridis Y (2005) Literature review of spatio-temporal database models. Knowl Eng Rev 19:235–274
12. Raza A (2001) Object-oriented temporal GIS for urban applications. PhD thesis, University of Twente. ITC Dissertation 79. ISBN 90-3651-540-8
13. Thapa RB, Murayama Y (2009) Examining spatiotemporal urbanization patterns in Kathmandu Valley, Nepal: remote sensing and spatial metrics approaches. Remote Sens J. ISSN 2072-4292
14. Güting RH (2000) A foundation for representing and querying moving objects. Geoinformatica ACM Trans Databases Syst 25:1–42

15. Khaddaj S, Adamu A, Morad M (2005) Construction of an integrated object oriented system for temporal GIS. Am J Appl Sci 2:1584–1594
16. Wang S, Nakayama K, Kobayashi Y, Maekawa M (2005) An event-based spatiotemporal approach. ECTI Trans Comput Inf Theory 1:15–23
17. Zhang N (2006) Spatio-temporal cadastral data model: geo-information management perspective in China. Master thesis, International Institute for Geo-information science and earth observation enschede, The Netherlands

Modeling and Simulation of an Intelligent Traffic Light System Using Multiagent Technology

Tuyen T. T. Truong and Cuong H. Phan

Abstract In this paper, we describe an approach of modeling and simulation based on multi-agent theory to build the model of traffic system. Our model presents a traffic system including traffic light system at a '+' junction. This paper controls duration of red-light/green-light to release traffic congestion on roads. In addition, we establish six Vietnamese traffic rules in our model. The simulation model represents real traffic system in GAMA platform with GAML language. All systems are visualization in GUI to show the participant of vehicles and traffic light on roads.

Keywords Modeling · Multiagent-based simulation · Traffic light system · GAMA platform · GAML

1 Introduction

Nowadays, in Vietnam, many large cities have to face traffic jam such as Hanoi, HoChiMinh as well as CanTho city. Maybe the main reason of this problem is that the number of vehicles increases too fast. Besides, a traffic light system which is unchangeable duration of red/green-light—called normal traffic light system- are not useful if the density of vehicle on two roads is different. As a result, it is necessary to have an intelligent traffic light system which can change red/green duration of traffic light based on real time conditions. There were many researches that related to this topic. They used extension neural network, fuzzy logic and

---

T. T. T. Truong (✉) · C. H. Phan
College of Information Technology, CanTho University, CanTho, Vietnam
e-mail: ttttuyen@cit.ctu.edu.vn

C. H. Phan
e-mail: phcuong@cit.ctu.edu.vn

J. J. (Jong Hyuk) Park et al. (eds.), *Multimedia and Ubiquitous Engineering*, Lecture Notes in Electrical Engineering 240, DOI: 10.1007/978-94-007-6738-6_126, © Springer Science+Business Media Dordrecht(Outside the USA) 2013

multi-agent to control traffic light system [1–4]. These articles had just mentioned how to control the time of traffic light system at '+' junctions. They did not concentrate on the behaviours of vehicles. Besides, in Vietnam, almost traffic light systems are set up with unchangeable-duration of red-light/green-light. So that it is easy to occur traffic congestion. This problem is not only waste of time, money and fuel, but also directly affect the ecological environment and people as well. The intelligent traffic light systems can automatically adjust duration of the traffic light based on density of vehicle that existed on road. In fact, to establish an intelligent traffic light system will cost much higher in comparison with a normal traffic light system [3]. And, the efficiency of this system depends on the characteristics of particular crossroad. Besides, to install the initial parameters for the system is important too. Thus, before deploying an intelligent traffic light system at cross-road, we need to scrutinize. The previous researches about intelligent traffic light system did not toward a simulation software based on multi-agent simulation system. It is necessary to have a software to visualize and adjust duration of traffic light system to release traffic congestion. Moreover, in Vietnam, some traffic rules are quite difference in comparison with in other countries. So that, this paper describes the model of the traffic system in "+" junction which includes many types of vehicles and traffic rules in Vietnam only. And then, we use multi-agent technology to simulate the behaviour of vehicles on roads and control duration of traffic light based on the number of vehicles on roads with GAML language in GAMA platform.

This paper includes five sections. The Sect. 1 is an introduction which briefly shows the motivation of this paper and some related works. Models will be presented in the Sect. 2. The Sect. 3 is simulation section. In this section, we describe how to simulate the models. Another Sect. 4 is experimentation section, which shows some scenarios and evaluates the simulation results. The Sect. 5 is conclusion.

2 The Model of Intelligent Traffic Light System

There are many types of vehicles such as car, truck, moto, bike and etc. In this paper, we divide them into three types: car, truck and moto. The attributes and behaviour of these types are nearly the same as shown in Fig. 1. In some areas in Vietnam, motobike is allowed to turn right while the red-light is turning on. So that, we also set up parameter for this rule (true (by default) if allow to turn right while red-light is turning on otherwise false). The attributes and behaviours of vehicles, traffic light, and road are pointed out in Fig. 1.

The moving direction of vehicle named h and its value is in [0..359]. In general, while moving, vehicle has to avoid collision with other neighbour vehicles. Figure 2 shows the safety areas in left, right and front of vehicle. The safety areas are divided by two lines: f1 and f2 whereas (x, y) is current location of vehicle. The radius of this circle is depended on current speed of vehicle.

Modeling and Simulation of an Intelligent Traffic

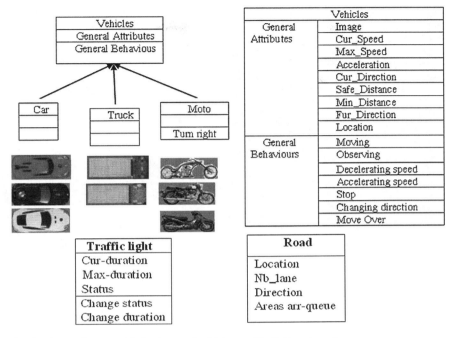

Fig. 1 The attributes and behaviours of vehicles, traffic light and road

Fig. 2 The safety areas of vehicles

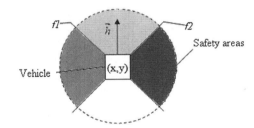

Traffic light is also built as an agent as mentioned in Fig. 1. Its attributes are: status (value in [green, red, yellow]), the cur-duration (how many seconds are kept before changing another light's status), the max-duration (the maximum duration of green-light status). The cur-duration is computed based on the density of vehicles in A, B, C, D arrival queue of previous step (in Fig. 3) by following rules:

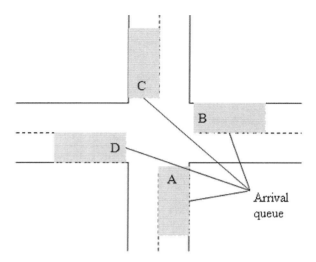

Fig. 3 The arrival queue at crossroad

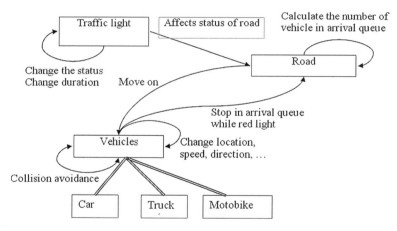

Fig. 4 The intelligent traffic light system model

$$\text{Cur-duration} = \begin{cases} \text{Minimum-time if the rate of density of vehicle in these arrival} \\ \text{queue is nearly 0.} \\ \\ \text{(Density of vehicle on B, D in previous light time cycle)(Density of} \\ \text{vehicle on A, C in previous light time cycle)}^*\text{Avg\_duration.} \\ \\ \text{Maximum-duration if the rate of density in these arrival queue is too large} \\ \text{(if cur-duration} > \text{Maximum-duration).} \end{cases}$$

The relationship between agents is presented in Fig. 4.

Our traffic light system model is built to adjust the traffic rules in Vietnam. We establish six rules:

- Traffic light system must match the red-light and green-light duration to ensure that at crossroad, in specific time only vehicles on one road are allowed to go and the other must stop. For example: if vehicles on B, D roads are allowed to go, vehicles on A, C roads must stop.
- All vehicles have to move on right lanes.
- Vehicles must comply traffic light signals at crossroad. Vehicles stop in front of pedestrian bar at the crossroad while red-light is turning on. Especially, motobike can turn right while the red-light is turning on (as optional rule based on initial parameter of model).
- The speed of vehicles must lower than max-speed. Vehicles can change the speed to avoid a collision with the other vehicles or reach the '+' junction.
- All vehicles must comply safe distance with neighbour vehicles.
- The red-light duration is equal to sum of the green-light duration and yellow-light duration at the crossroad.

3 Intelligent Traffic Light System Simulation

With the characteristics and rules mentioned above, we set up the simulation environment in a grid (GRID) as shown in Fig. 5. It represents '+' junction of two roads and each road has three lanes. The A, B, C, D letters are the name of the arrival queue of roads at crossroad. Vehicles will appear from the rear of the roads and will move with six established rules as mentioned in part II. When vehicles move to the end of road, they will disappear.

Simulation model is established with the input parameters as shown in Fig. 6. The parameters of the model are:

Fig. 5 The grid environment

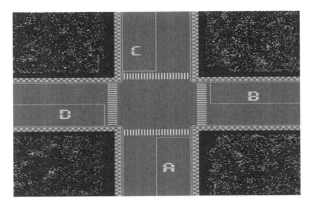

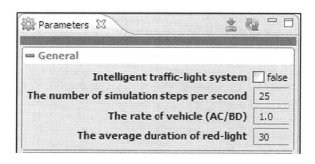

Fig. 6 The parameters of simulation model

- The intelligent traffic light system (true/false): allow enable/disable intelligent traffic light system.
- The number of simulation steps per second.
- The rate of vehicle (AC/BD): The rate of the number of vehicle on the AC road per on BD road.
- Avg-duration: The average duration of red-light.

4 Experimentation

4.1 Scenario 1: Normal Traffic Light System

With density of vehicles on two roads are equal as initial parameters in Fig. 6, the simulation results after 7500 steps (~ 5 min) is not crowded (Fig. 7). If vehicle's density of two roads is equal, traffic jam has not occurred yet as shown in Fig. 7. After that, we change density of road BD up to 3 times (the rate of vehicle (AC/BD) = 0.33) and after step 9000th, there are many vehicles on BD road (Fig. 8). And then, at step 13500th (~ 9 min), traffic jam occurs at the crossroad as well as in BD road (Fig. 9).

4.2 Scenario 2: Intelligent Traffic Light System

When we change the input parameters as shown in Fig. 10 and after 21367 simulation steps (~ 14 min), traffic congestion has been still not occurred. The duration of red/green-light also changes depending on density of vehicles that exist on roads. Figure 10 shows the intelligent traffic system model which maintained for a long time (more than 14 min). Green-light duration can reach 44 s to release the traffic jams on the road.

Modeling and Simulation of an Intelligent Traffic 1027

Fig. 7 The simulation results after 7500 steps with the vehicle on AC and BD are equal

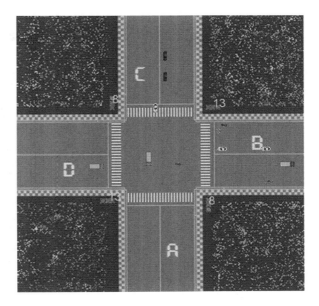

Fig. 8 The simulation results after 9000 steps with the vehicles on AC and BD are different

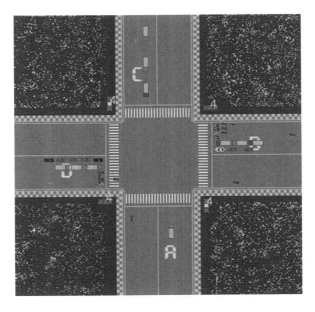

Fig. 9 The simulation results after 13500 steps with the vehicles on AC and BD are different

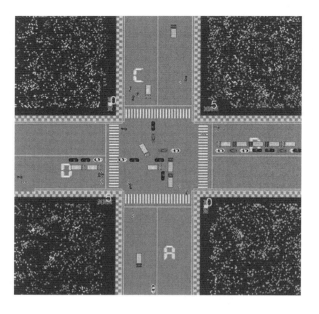

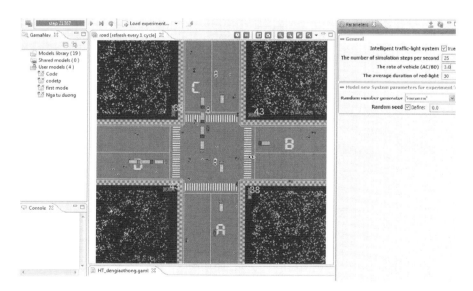

Fig. 10 The results of intelligent traffic light system after 21367 steps

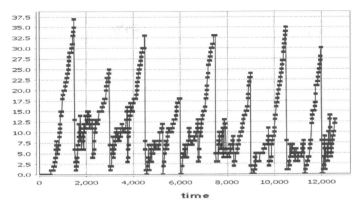

Fig. 11 The number of existed vehicles on normal traffic light system

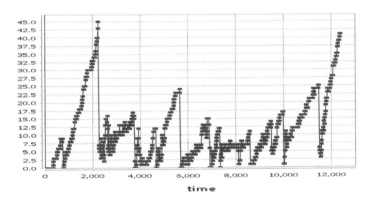

Fig. 12 The number of existed vehicles on intelligent traffic light system

4.3 The Comparison Simulation Results of the Two Traffic Lights System

Figures 11 and 12 illustrate the number of vehicles that are being existed on traffic system for the normal traffic light system and the intelligent traffic light system respectively. The results show that the intelligent traffic light system, the number of vehicles decreases faster and keeps close to the horizontal axis in comparison with the normal traffic light system. Through two simulation scenario results, we can see the effectiveness of intelligent traffic light system is better than normal traffic light system. That means the number of waiting vehicles at crossroad is reduced as a result, drivers can save money, time as well as keep good-atmosphere for residents who live near crossroad.

5 Conclusion

This paper presents a multiagent approach to simulate an intelligent traffic light system as well as a normal traffic light system. This article also builds simulation model in GAMA platform. All processes of participation of vehicles, traffic light system and traffic rules in Vietnam are visualized in our simulation model. Furthermore simulation results can help to prove the need for an intelligent traffic light system at '+' junction to save time, money and help protect the green environment. Our system can simulate to be able to choose a suitable duration of red-light/green-light in specific crossroad before establishing real traffic light system.

References

1. Krajzewicz D, Brockfeld E, Jürgen M, Julia R, Rössel C, Tuchscheerer W, Wagner P, Wösler R (2005) Simulation of modern traffic lights control systems using the open source traffic simulation SUMO. In: Proceedings of the 3rd industrial simulation conference 2005, pp 299–302
2. Chao K-H, Lee R-H, Wang M-H (2008) An intelligent traffic light control based on extension neural network. In: Proceedings of the KES 2008, Part I, LNAI 5177, Springer, pp 17–24
3. Chi NN (2011) A method for smart traffic light control system design. In: Proceedings of the national conference on control and automation VCCA-2011, pp 639–644
4. Chinyere OU, Francisca OO, Amano OE (2011) Design and simulation of an intelligent traffic control system. Int J Adv Eng Tech 1(5):47–57
5. Drogoul A (2003) Multi-agent based simulation: where are the agents? Springer, 2581/2003, pp 43–49
6. Drogoul A (2008) Lecture notes simulation. Cantho University
7. Wooldridge M (2002) An introduction to multiagent systems. John Wiley & Sons Ltd, Chichester
8. http://code.google.com/p/gama-platform/ (2013)

A Numerical Approach to Solve Point Kinetic Equations Using Taylor-Lie Series and the Adomian Decomposition Method

Hag-Tae Kim, Ganduulga, Dong Pyo Hong and Kil To Chong

Abstract The point kinetic equations in nuclear dynamics, various analytical methods have been used. In this paper, a numerical approach of point kinetic equations using an inherently large sampling interval and multiple inputs is developed and analyzed. To implement this method, Taylor-Lie Series under the Zero Order Hold (ZOH) is used to approximate the neutron density and precursor concentrations at each corresponding time step. Afterwards, an additional technique, the Adomian Decomposition Method, is used based on its merit of algorithmic and computational advantages in carrying out the discretization.

Keywords Point kinetic equations · Numerical solution · Taylor-Lie series · Zero order hold (ZOH) approximation · Adomian decomposition method

1 Introduction

The point kinetics model is flawless such as problems mentioned in obtaining essential parameters that explain the reactor within the method. As point kinetic equations are systems which feature stiff nonlinear ordinary differential equations,

---

H.-T. Kim
Korea Atomic Energy Research Institute, Daejeon, Republic of Korea

Ganduulga · K. T. Chong
Department of Electronics Engineering, Jeonbuk National University,
Jeonju, Republic of Korea

D. P. Hong
Department of Mechanical Engineering, Jeonbuk National University,
Jeonju, Republic of Korea

K. T. Chong (✉)
Advanced Research Center for Electronics and Information, Jeonbuk National University,
Jeonju, Republic of Korea
e-mail: kitchong@chonbuk.ac.kr

J. J. (Jong Hyuk) Park et al. (eds.), *Multimedia and Ubiquitous Engineering*,
Lecture Notes in Electrical Engineering 240, DOI: 10.1007/978-94-007-6738-6_127,
© Springer Science+Business Media Dordrecht(Outside the USA) 2013

by solving the equations, estimations of various system variables of the reactor core and the transient behavior of the reactor's power can be obtained [1]. In addition to solving these equations, the neutron density and delayed neutron precursor concentrations are solved in a tightly coupled reactor as a function of time. Many of the cases, however, use a model reactor with a minimum of six delayed precursor groups to solve the point kinetic equations which further results in a system consisting of seven or more coupled differential equations.

In the field of reactor kinetics, much research has focused primarily on finding the numerical solutions of point kinetic equations. Sanchez [2] proposed an A-Stable Runge–Kutta Method to combat the problem of stiffness. Kinard and Allen [3] devised a piecewise constant approximation (PCA) method to overcome the previously mentioned. Specifically, this method deals with point kinetic equations using the Zero Order Hold (ZOH) assumption for linearity within time-discretization intervals, allowing an exact solution by using exponential matrices to be obtained. The PWS method introduced by Sathiyasheela [4] was dependent on the truncated Taylor series or those of the exponential matrices under the Zero Order Hold (ZOH) assumption. Since then, several other meaningful attempts have been made such as the Taylor series method proposed by Nahla to solve point kinetic equations [1] with a small sampling interval needed for accurate results.

In this paper, a numerical solution based on the Zero Order Hold (ZOH) assumption and Adomian Decomposition Method is proposed in anticipation of providing a remedy to the problems mentioned above. More specifically, the proposed method makes use of the Taylor-Lie series of the neutron density and delayed precursor functions at each time step, then approximates the exact values using the Adomian Decomposition Method.

2 Discretization of the Point Kinetics Equation

The point kinetic equations with m delayed groups are [10, 11]:

$$\frac{dn(t)}{dt} = \frac{\rho(t) - \beta}{\Lambda} n(t) + \sum_{i=1}^{m} \lambda_i C_i(t) + F(t) \tag{1}$$

$$\frac{dC_i(t)}{dt} = \frac{\beta_i}{\Lambda} n(t) - \lambda_i C_i(t) \quad i = 1, 2, \ldots, m \tag{2}$$

where $n(t)$ is the time-dependent neutron density, $\rho(t)$ the time-dependent reactivity function, β_i the ith delayed fraction, $\beta = \sum_{i=1}^{m} \beta_i$ the total delayed fraction, $F(t)$ the time-dependent neutron source function, $C_i(t)$ the ith precursor density, Λ the neutron generation time, and λ_i the ith group decay constant.

The point kinetics equation can then be expressed with the state-space representation of the form:

A Numerical Approach to Solve Point Kinetic Equations

$$x_1 = n(t), \quad \dot{x}_1 = \frac{dn(t)}{dt}$$

$$x_2 = C_1(t), \quad \dot{x}_2 = \frac{dC_1(t)}{dt}$$

$$x_i = C_{i-1}(t), \quad \dot{x}_i = \frac{dC_{i-1}(t)}{dt} \tag{3}$$

$$u_1(t) = \rho(t), u_2(t) = F(t), u_3(t)\ldots u_i(t) = 0 \tag{4}$$

Therefore, the system can then be rewritten as:

$$\dot{x}_1 = -\frac{\beta}{\Lambda}x_1 + \lambda_i x_2 + \frac{1}{\Lambda}x_1 * u_1(t) + u_2(t)$$

$$\dot{x}_2 = -\frac{\beta_1}{\Lambda}x_1 + \lambda_1 x_2$$

$$\dot{x}_i = -\frac{\beta_{i-1}}{\Lambda}x_1 + \lambda_{i-1}x_i \tag{5}$$

It is also assumed that systems (1) and (2) are driven by an input piecewise constant for the sampling interval, therefore, the zero order hold (ZOH) assumption holds true.

The solutions of Eqs. (1) and (2) are then expanded into a uniformly convergent Taylor series resulting in coefficients that hold true under the ZOH assumption while within the sampling interval. The solutions of Eqs. (1) and (2) are then easily identified using successive partial derivatives of the right hand side of each equation.

$$x_i(k+1) = x_i(k) + \sum_{l=1}^{\infty} \frac{T^l}{l!}\frac{d^l x_i}{dt^l}\bigg|_{t_k} = x_i(k) + \sum_{l=1}^{\infty} A_i^{[l]}(x(k), u(k))\frac{T^l}{l!} \tag{6}$$

where $x_1(k), x_2(k), \ldots, x_i(k)$ are the values of the state vector $x_1, x_2, \ldots, x_i$ at time $t = t_k = kT$ and $A_1^{[l]}(x, u), A_2^{[l]}(x, u), \ldots, A_i^{[l]}(x, u)$ produced by the following recursive procedure [12]:

$$A_1^1(x_1, u) = f_1(x) + u_1 g_1(x) + \cdots + u_m g_m(x)$$

$$A_2^1(x_2, u) = f_2(x) + u_1 g_1(x) + \cdots + u_m g_m(x)$$

$$A_i^1(x_i, u) = f_i(x) + u_1 g_1(x) + \cdots + u_m g_m(x)$$

$$A_i^{l+1}(x_i, u) = \frac{\partial A_i^l(x_i, u)}{\partial x_1}A_1^1 + \frac{\partial A_i^l(x_i, u)}{\partial x_i}A_i^1 + \frac{\partial A_1^{[l]}(x, u)}{\partial u_1}\frac{du_1}{dt} + \cdots + \frac{\partial A_1^{[l]}}{\partial u_m}\frac{du_m}{dt} \tag{7}$$

3 Adomian Decomposition Approximations

Developed by G. Adomian, the Adomian Decomposition Method (ADM) possesses unique algorithmic and computational advantages, particularly when used to calculate approximations to numerical solutions of nonlinear differential or partial differential equations [5–9].

To approximate, the Adomian Decomposition Method must first be used to discrete the results utilizing the Taylor-Lie series. Within the discrete results of the point kinetic equations obtained, the ith-dimensional system should be considered given that the decomposition method requires that:

$$x_i(t) = \sum_{n=0}^{\infty} x_i^n(t) \tag{8}$$

When given the nonlinearities $f_1, f_2, \ldots, f_i$ under the zero order hold (ZOH) assumption, the corresponding Adomian polynomials are calculated such that:

$$f_i(x_1, x_2, u(k)) = \sum_{n=0}^{\infty} A_{fi}^n(x_1, x_2, u(k)) \tag{9}$$

where $A_{f1}^n, A_{f2}^n, \ldots, A_{fi}^n (n = 0, 1, 2 \ldots)$ are the corresponding Adomian polynomials of the nonlinearities $f_1, f_2, \ldots, f_i$, respectively. Since we have [5–7],

$$N_1 = \int_{t_0}^{t} f_1(x_1(s), x_2(s), \bar{u}) ds,$$
$$N_2 = \int_{t_0}^{t} f_2(x_1(s), x_2(s), \bar{u}) ds,$$
$$\ldots,$$
$$N_i = \int_{t_0}^{t} f_i(x_1(s), x_2(s), \bar{u}) ds,$$

the following can be acquired:

$$N_i = \int_{t_0}^{t} \sum_{n=0}^{\infty} A_{fi}^n(x_1, x_2, u(k)) ds = \sum_{n=0}^{\infty} \int_{t_0}^{t} A_{fi}^n(x_1, x_2, u(k)) ds \tag{10}$$

Also, the following can be attained [5–7]:

$$N_i = \sum_{n=0}^{\infty} \int_{t_0}^{t} A_{fi}^n(x_1, x_2, u(k)) ds = \sum_{n=0}^{\infty} A_i^n \tag{11}$$

Therefore, Eq. (11) can be used to calculate $A_1^n, A_2^n, \ldots, A_i^n$:

$$A_i^n = \int_{t_0}^{t} A_{fi}^n(x_1, x_2, u(k)) ds \tag{12}$$

Using $x^0 = g$, $g_1 = x_1(t_0) + \int_{t_0}^{t} b_1 \bar{u} ds = x_1(t_0) + b_1(t - t_0)\bar{u}$, $g_2 = x_2(t_0) + \int_{t_0}^{t} b_2 \bar{u} ds = x_2(t_0) + b_2(t - t_0)\bar{u}$, $\ldots$, $g_i = x_i(t_0) + \int_{t_0}^{t} b_i \bar{u} ds = x_i(t_0) + b_i(t - t_0)\bar{u}$, the zero-order terms can be calculated as [5–7]:

$$x_i^0(t) = x_i(k) + b_i(t - t_k)u(k) \tag{13}$$

Finally, using

$$x^1 = L(x^0) + A^0(x^0), \ L_{1,1} = \int_{t_0}^t a_{11}x_1(s)ds, \ L_{1,2} = \int_{t_0}^t a_{12}x_2(s)ds,$$

$$L_{2,1} = \int_{t_0}^t a_{21}x_1(s)ds, \ L_{2,2} = \int_{t_0}^t a_{22}x_2(s)ds, \ A_1^n = \int_{t_0}^t A_{f1}^n(x_1, x_2, u(k))ds,$$

$$A_2^n = \int_{t_0}^t A_{f2}^n(x_1, x_2, u(k))ds, \dots, \ A_i^n = \int_{t_0}^t A_{fi}^n(x_1, x_2, u(k))ds,$$

the following can be calculated:

$$x_1^1(t) = L_{1,1} + L_{1,2} + A_1^0$$

$$= a_{11} \int_{t_k}^t (x_1(k) + b_1(s - t_k)u(k))ds + a_{12} \int_{t_k}^t (x_2(k) + b_2(s - t_k)u(k))ds$$

$$+ \int_{t_k}^t A_{f1}^0(x_1(k) + b_1(s - t_k)u(k), x_2(k) + b_2(s - t_k)u(k))ds \qquad (14)$$

$$x_2^1(t) = L_{2,1} + L_{2,2} + A_2^0$$

$$= a_{21} \int_{t_k}^t (x_1(k) + b_1(s - t_k)u(k))ds + a_{22} \int_{t_k}^t (x_2(k) + b_2(s - t_k)u(k))ds$$

$$+ \int_{t_k}^t A_{f2}^0(x_1(k) + b_1(s - t_k)u(k), x_2(k) + b_2(s - t_k)u(k))ds$$

or:

$$x_1^1(t) = a_{11}\left(x_1(k)(t - t_k) + \frac{1}{2!}b_1(t - t_k)^2 u(k)\right)$$

$$+ a_{12}\left(x_2(k)(t - t_k) + \frac{1}{2!}b_2(t - t_k)^2 u(k)\right)$$

$$+ \int_{t_k}^t A_{f1}^0(x_1(k) + b_1(s - t_k)u(k), x_2(k) + b_2(s - t_k)u(k))ds \qquad (15)$$

$$x_2^1(t) = a_{21}\left(x_1(k)(t - t_k) + \frac{1}{2!}b_1(t - t_k)^2 u(k)\right)$$

$$+ a_{22}\left(x_2(k)(t - t_k) + \frac{1}{2!}b_2(t - t_k)^2 u(k)\right)$$

$$+ \int_{t_k}^t A_{f2}^0(x_1(k) + b_1(s - t_k)u(k), x_2(k) + b_2(s - t_k)u(k))ds$$

The above process presents higher-order terms in a recursive method using Adomian series [5–7]. This means that in all these expressions, time t enters as $t - t_k$ owing to the autonomous nature of the differential equations under the zero order hold (ZOH) assumption. Therefore

$$x_1^n(t) = x_1^n(x_1(k), x_2(k), u(k), t - t_k), \ x_2^n(t) = x_1^n(x_1(k), x_2(k), u(k), t - t_k), \ \ldots,$$
$$x_i^n(t) = x_1^n(x_1(k), x_2(k), u(k), t - t_k), \ \text{and}$$

$$x_i(t) = \sum_{n=0}^{\infty} x_i^n(x_1(k), x_2(k), u(k), t - t_k) \tag{16}$$

Finally, by letting $t = t_{k+1}$, the above associative properties lead to the following exact sample-data representation:

$$x_i(k+1) = \Phi_i(x_1(k), x_2(k), u(k)) = \sum_{n=0}^{\infty} x_i^n(x_1(k), x_2(k), u(k), T) \tag{17}$$

Approximate sample-data representations of order N are attributed to the finite truncation of N orders:

$$x_i(k+1) = \Phi_i^N(x_1(k), x_2(k), u(k)) = \sum_{n=0}^{N} x_2^n(x_1(k), x_2(k), u(k), T) \tag{18}$$

In the previous analysis, the Adomian Decomposition Method was applied as an effective solution to nonlinear systems to the discrete results using Taylor-Lie series to approximate exact values.

4 Computational results

In order to verify the effectiveness, the proposed methods' results are compared with exact values. Secondly, the procedure is implemented using various initial conditions and input data such as single step, single ramp, multiple inputs of reactivity, and source function cases. Finally, the results of the simulation are presented and analyzed below:

The single step and single ramp reactivity are considered with all simulations executed using MAPLE software. The initial conditions used within the simulations are identical for simplicity and assumes a source-free equilibrium:

$$\dot{x}_1(0) = 1, \dot{x}_2(0) = \tfrac{\beta_i}{\lambda_i \Lambda} \quad i = 1 \ldots 6$$

For single step reactivity, the parameters are defined as:

$$\Lambda = 0.00002$$

$$\beta = 0.007$$

$$\lambda_i = (0.0127, 0.0317, 0.115, 0.311, 1.4, 3.87)$$

$$\beta_i = (0.000266, 0.001491, 0.001316, 0.002849, 0.000896, 0.000182)$$

In addition, the one prompt subcritical $\rho(t) = u(t) = 0.003$ is considered.

First, Fig. 1 shows the results of the proposed method (ADM) compared with exact values. Using a sampling time of 0.01 as well as a single step for input data,

A Numerical Approach to Solve Point Kinetic Equations

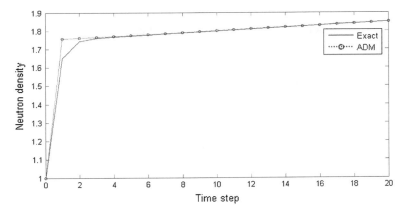

Fig. 1 The comparison of the exact values and the ADM results with single step input

Table 1 The results using sampling time 0.01 with single step input

Time step	Exact	ADM	Error
1	1.649936078	1.75599774	0.106061662
4	1.766607476	1.771017817	0.004410341
7	1.783618313	1.786037893	0.00241958
10	1.799177796	1.801057968	0.001880172
13	1.814993972	1.816078039	0.001084067
16	1.830780794	1.831098109	0.000317315
19	1.84605792	1.846118175	6.0255E–05

it is evident that apart from time step 3 and before, the method shows no noticeable differences as time passes.

Table 1 outlines the errors and specified results of the ADM in comparison with exact values. Using the data, it is palpable that ADM is in line with exact values. Additionally, the RMSE of the ADM is an estimated 0.108177528.

5 Conclusion

In the previous paper, a new numerical solution for point kinetic equations, based on the zero order hold (ZOH) assumption and the Adomian Decomposition Method (ADM) in nuclear reactor dynamics was proposed. Moreover, the proposed method uses a Taylor-Lie series of the neutron density and delayed precursor functions at each corresponding time interval and then approximates the exact values using the ADM. Regarding the performance of the proposed numerical solution, detailed simulations using various initial conditions and input data such as single step, single ramp, and multiple inputs were chosen. These simulations only verified the increased performance and accuracy. Furthermore, compared to analytical methods, the proposed method is far less complex and does not require small time intervals. For these reasons, solving point kinetic equations is simplified. In the future, this method will be developed further using the first order hold (FOH) assumption, a more complicated approach, to potentially increase accuracy and performance.

Acknowledgments This work was supported by the National Research Foundation of Korea (NRF) grant funded by the Korea government (MEST) (No. 2012-038978) and (No. 2012-0002434).

References

1. Nahla AA (2011) Taylor's series method for solving the nonlinear point kinetics equations. Nucl Eng Des
2. Sanchez J (1989) On the numerical solution of the point kinetics equations by generalized Runge–Kutta methods. Nucl Sci Eng 103:94–99
3. Kinard M, Allen EJ (2003) Efficient numerical solution of the point kinetics equations in nuclear reactor dynamics. Ann Nucl Energy 31:1039–1051
4. Sathiyasheela T (2008) Power series solution method for solving point kinetics equations with lumped model temperature and feedback. Ann Nucl Energy 36:246–250
5. Adomian G, Rach R, Meyers R (1991) Numerical algorithms and decomposition. Comput Math Appl 22:57–61
6. Adomian G (1991) A review of the decomposition method and some recent results on nonlinear equations. Comput Math Appl 21:101–127
7. Adomian G (1993) Solving frontier problems of physics: The decomposition method. Springer, London
8. Cherruault Y, Adomian G (1993) Decomposition methods. In: A new proof of convergence. Math Comput Model 18:103–106
9. Deeba E, Yoon JM (2002) A decomposition method for solving nonlinear systems of compartment models. J Math Anal Appl 266:227–236
10. Hetrick DL (1971) Dynamics of nuclear reactors. The University of Chicago Press, Chicago
11. Petersen CZ, Dulla S, Vilhena MTMB, Ravetto P (2011) An analytical solution of the point kinetics equations with time-variable reactivity by the decomposition method. Prog Nucl Energy 53:1091–1094
12. Kazantzis N, Chong KT, Park JH, Parlos AG (2005) Control-relevant discretization of nonlinear systems with time-delay using Taylor-Lie series. J Dyn Syst Meas Contr 127:153–159

Regional CRL Distribution Based on the LBS for Vehicular Networks

HyunGon Kim, MinSoo Kim, SeokWon Jung and JaeHyun Seo

Abstract To protect the members of the vehicular networks from malicious users and malfunctioning equipments, certificate revocation list (CRL) should be distributed as quickly and efficiently as possible without over-burdening the network. The common theme among existing methods in the literature to reduce distribution time is to reduce the size of the CRL, since smaller files can be distributed more quickly. Our proposal has been concerned with the problem of how to reduce the size of CRL effectively. We propose a regional CRL distribution method that introduces partitioned CRLs corresponding to certificate authority (CA) administrative regions. A regional CRL includes only neighbouring vehicle's revoked certificates and distributed to vehicles within one CA region. Consequently, since there is no need to process full CRLs by all vehicles, the method can reduce computational overhead, long authentication delay, message signature and verification delay, and processing complexity imposed by full CRL distribution methods.

Keywords Regional CRL · VANET · PKI · CRL

This research was supported by Basic Science Research Program through the National Research Foundation of Korea (NRF) funded by the Ministry of Education, Science and Technology (NRF-2011-0027-006).

H. Kim (✉) · M. Kim · S. Jung · J. Seo
Department of Information Security, Mokpo National University,
560 Muanno CheongGye-Myeon, Muan-Gun, Jeonnam 534-729, Korea
e-mail: hyungon@mokpo.ac.kr

M. Kim
e-mail: phoenix@mokpo.ac.kr

S. Jung
e-mail: jsw@mokpo.ac.kr

J. Seo
e-mail: jhseo@mokpo.ac.kr

J. J. (Jong Hyuk) Park et al. (eds.), *Multimedia and Ubiquitous Engineering*,
Lecture Notes in Electrical Engineering 240, DOI: 10.1007/978-94-007-6738-6_128,
© Springer Science+Business Media Dordrecht(Outside the USA) 2013

1 Introduction

Vehicular network has been one of the emerging research areas and promising way to facilitate road safety, traffic management, and infotainment dissemination of drivers and passengers. However, without the integration of strong and practical security and privacy enhancing mechanisms, vehicular communication system can be disrupted or disabled, even by relatively unsophisticated attackers [1].

Security and privacy are essential components for the successful deployment of vehicle networks. Those components need to be carefully assessed and addressed in the design of the vehicular communication system, especially because of the life-critical nature of the vehicular network operation. The IEEE 1609.2 standard [2] defines security services for vehicular networks. It defines secure message formats and techniques for processing these secure messages using the public key infrastructure (PKI). In traditional PKI architecture, the most commonly adopted certification revocation scheme is to use CRL that is a list of revoked certificates stored in repositories prepared by CAs. In vehicular networks, the CA adds the identification of the revoked certificate(s) to a CRL. The CA then publishes the updated CRL to all vehicular network participants, and instructing them not to trust the revoked certificate. Timely access to revocation information is important for the robustness of its operation: message faulty, compromised, or otherwise illegitimate, and overall potentially dangerous, vehicles can be ignored.

CRL is straightforward and widespread used. Since the validity period of certificates is long and the number of users is immense, CRLs can grow extremely larger. A great amount of data needs to be transmitted especially, over the air. Also, according to the Dedicated Short Range Communication (DSRC) [3], each on-board unit (OBU) has to broadcast a message every 300 ms indicating its current position, speed, and the road conditions. In such scenario, each OBU may receive a large number of messages every 300 ms, and it has to check the current CRL for all the received certificates, which may incur long authentication delay depending on the size of CRL.

The common theme among existing methods in the literature to reduce distribution time is to reduce the size of the CRL, since smaller files can be distributed more quickly. To address these challenges, our proposal has been concerned with the problem of how to reduce the size of CRL effectively. We propose a regional CRL distribution method to reduce the size of CRL. The basic idea is that vehicles within a CA administrative region operate only one regional CRL, which include neighboring vehicle's revoked certificates.

The reminder of the paper is organized as follows: Section 2 presents problem statements of CRL for vehicular networks and related works; Sect. 3 introduces a proposed regional CRL distribution method, an algorithm used to make regional CRLs at location application server, and network architecture for obtaining precise vehicles location; and finally Sect. 4 summarizes results and end with some conclusions.

2 Problem Statements and Related Works

To know global revocation information for all CAs and all vehicles in vehicular networks, full CRLs have to be distributed but, it would be a costly operation. Full CRLs could be large size, which is directly proportional to the revocation rate, the number of nodes in the system, and, the number of certificate used by each vehicle. This requires computational overhead, long authentication delay, message signature and verification delay, and processing complexity imposed by full CRL distribution methods. In addition, in a situation where the rate of revocation is very high, full CRLs will be large and will change often.

The size of CRL in vehicular networks is expected to be large for the following reasons: firstly, to preserve the privacy of the drivers, i.e., to abstain the leakage of the real identities and locations information of the drivers from any external eavesdropper [4, 5], each OBU should be preloaded with a set of anonymous digital certificates, where it has to periodically change its anonymous certificate to mislead attackers [6]. Consequently, an OBU revocation results in revoking all the certificates carried by that OBU leading to a large increase in the CRL size; secondly, the vehicular network scale is very large. CRLs for all the received certificates are stored in OBU of vehicle. Upon receiving signed or encrypted messages from neighboring road-side unit (RSU) or vehicles, each OBU has to check the current CRL for all the received certificates, which may incur long processing delay depending on the size of CRL. The ability to check a CRL for a large number of certificates in a timely manner forms an inevitable challenge to vehicle networks.

The problem of certificate revocation in vehicular networks has hardly attracted any attention. The [7] aims at achieving scalable and efficient mechanism for the distribution of large CRLs across wide regions by utilizing a very low bandwidth at each RSU. CRLs are encoded into numerous self-verifiable pieces, so vehicles only get from the RSUs those pieces of the CRLs. The [8] made many experiments on the size of CRL and how to distribute the CRL in the vehicular networks. The result said when the size of CRL is high the delay time for receiving it will be high. Another idea proposed in [9] said that CRL will store entries for less than a year old. This idea used to decrease the size of CRL, but still suffer from huge size.

3 Design of Regional CRL Distribution Method

The method involves segmenting a full CRL into regional CRLs according to a number of CA administrative regions; the regional CRLs will be significantly smaller than the full CRL. Each regional CRL includes only neighboring vehicle's revoked certificates and it would be distributed to vehicles within one CA administrative region. Consequently, all vehicles in vehicular networks may receive and process only relevant CRLs associated with their neighboring vehicles (Fig. 1).

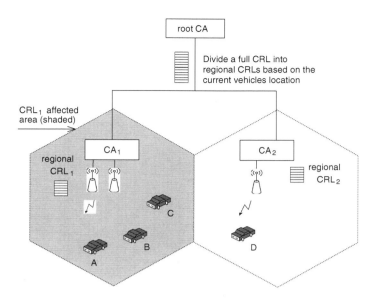

Fig. 1 Regional CRL distribution for reducing the size of CRL

3.1 Regional CRL Distribution Procedure

The proposed regional CRL distribution method is shown in Fig. 2. For segmentation, the root CA divides a full CRL into regional CRLs corresponding to the CA administrative regions. The regional CRL_1 is only distributed to vehicles within the CA_1 administrative region. Thus, the regional CRL_1 affects only CA_1 administrative region. Those vehicles would receive the regional CRL_1 and only need to process it.

However, since a regional CRL includes neighboring vehicle's revoked certificates within a CA administrative region, the root CA has to know all vehicles location when it bundles all the revoked certificates corresponding to the given CA administrative region into single regional CRL. For this, vehicle's mobility should be carefully considered since vehicles can move from one CA administrative region to another dynamically. Thus, how to efficiently know vehicles location corresponding to CA administrative regions represents a major challenge for the proposed method. Our approach envisioned to achieve this is via the location-based services (LBS) defined by open mobile alliance (OMA) [10]. The LBS can be a vehicular application that provides information and functionality to location application servers based on vehicles geographical location. LBS for the purpose of regulatory compliance and/or commercial services are already commonly supported in today's deployed 2G and 3G wireless networks.

Figure 2 presents network architecture and regional CRL distribution procedure. We assume that each vehicle is capable of a LBS client called, location services client (LCS) [10]. Serving mobile location center (SMLC) and gateway

Regional CRL Distribution Based on the LBS for Vehicular Networks 1043

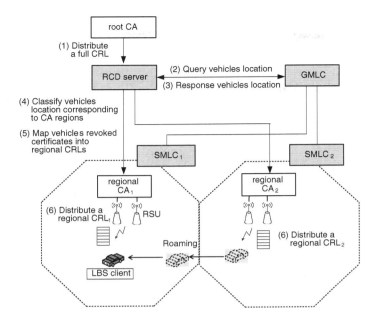

Fig. 2 Regional CRL distribution method

mobile location center (GMLC) are entities of vehicular networks and they provide functionality required to support LBS. The SMLC is responsible to calculate the location of vehicles and the GMLC is an entity obtaining geographic location information of all vehicles by interrogation with SMLCs. The regional CRL distribution server called RCD server, acts as a location application server [10]. To obtain all vehicles location from the GMLC, the mobile location protocol (MLP) [11] can be used that is an application-level protocol for getting the position of vehicles independent of underlying network technology.

The regional CRL distribution procedure involves the following steps: (1) the root CA sends a full CRL to RCD server, (2) the RCD server queries vehicles location to GMLC, (3) it obtains vehicles location from GMLC, (4) it classifies a full CRL into regional CRLs corresponding to CA administrative regions, (5) it maps vehicle's revoked certificates based on the current location into the regional CRLs, (6) each regional CA distributes the regional CRL to vehicles within its region.

3.2 Algorithm for Making Regional CRLs

As a location application server, the RCD server performs main functionalities of the proposed method. Figure 3 presents an algorithm used to make regional CRLs at the RCD server involving the following steps: (1) receive vehicles location obtained in

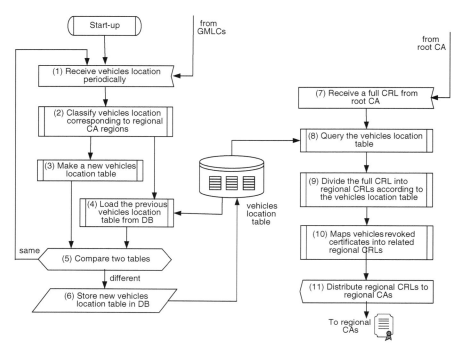

Fig. 3 Algorithm used to make regional CRLs at RCD server

the form of x, y coordinates from GMLCs periodically, (2) classify vehicles location corresponding to CA administrative regions, (3) make a new vehicles location table, for example $\{CA_{1area} = \{Vehicle_1(x_1, y_1), Vehicle_2(x_2, y_2), \ldots, \}$, (4) load the previous vehicles location table from database, (5) compare the previous table and the new table. If two tables are the same then, repeat the first step. This means that vehicles do not move between CA administrative regions for given time period, (6) If different then, store the new table in database, (7) receive a full CRL from CA periodically, (8) when receive a full CRL, query the stored vehicles location table, (9) divide the full CRL into regional CRLs according to the vehicles location table, (10) maps vehicle's revoked certificates into the related regional CRLs, (11) distribute them to the designated regional CAs and also the regional CAs distributes them to all vehicles within their regions.

3.3 Precise Vehicles Location

To obtain precise vehicles location, we can utilize combined access networks capable of LBS as shown in Fig. 4. The RCD server gathers vehicles location from different access networks such as WAVE network, 2G/3G network, 4G LTE

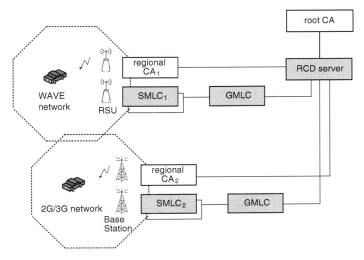

Fig. 4 Network architecture for obtaining precise vehicles location

network, WiBro, and WiFi etc. [2]. Combined LBS can allow the RDS server to know precise vehicles location and to make regional CRLs effectively.

4 Conclusions

We have presented a regional CRL distribution method that introduces regional CRLs partitioned by certificate authority administrative regions. The proposed method has been concerned with the problem of how to reduce the size of CRL effectively. The method involves segmenting a full CRL into regional CRLs according to a number of CA administrative regions; the regional CRLs will be significantly smaller than the full CRL. Each regional CRL includes only neighboring vehicle's revoked certificates and it would be distributed to vehicles within one CA administrative region. The advantage of the method is that it can reduce overhead, long authentication delay, message signature and verification delay, and processing complexity imposed by full CRL distribution methods since all vehicles in vehicular networks may receive and process only relevant small CRLs associated with their neighboring vehicles. The disadvantage is that location application server has to manage vehicles location corresponding to CA administrative regions in a real time manner. As a part of further work, we have been planning to evaluate performance analysis comparing full CRL distribution and regional CRL distribution method.

References

1. Lin X, Lu R, Zhang C, Zhu H, Ho PH, Shen X (2008) Security in vehicular ad hoc networks. IEEE Commun Mag 46(4):88–95
2. IEEE Std 1609.2: Trial-Use standard for wireless access in vehicular environments—Security services for applications and management message. IEEE Standard 1609.2 (2006)
3. Wasef A, Shen X (2009) MAAC: Message authentication acceleration protocol for vehicular ad hoc networks. In: Proceedings of the IEEE Globecom 2009, pp 1–6
4. Lin X, Sun X , Ho P-H , Shen X (2007) GSIS: A secure and privacy preserving protocol for vehicular communications. IEEE Trans Veh Technol 56: 3442–3456
5. Papadimitratos P, Kung A, Hubaux JP, Kargl F (2006) Privacy and identity management for vehicular communication systems: a position paper. In: Proceedings of workshop on standards for privacy in user-centric identity management
6. Laberteaux KP, Haas JJ, Hu YC (2008) Security certificate revocation list distribution for VANET. In: 15th ACM international workshop on VehiculAr InterNETworing (VANET), pp 88–89
7. Papadimitratos P, Mezzour G, Hubaux JP (2008) Certificate revocation list distribution in vehicular communication systems. In: 15th ACM international workshop on Vehicular InterNETworing (VANET), pp 1–10
8. Kamat P, Baliga A, Trappe W (2006) An identity-based security framework for VANETs. In: Proceedings of the third ACM workshop on vehicular networks
9. Raya M, Jungels D, Papadimitratos P, Aad I, Hubaux JP (2006) Certificate revocation in vehicular networks. In: Laboratory for computer communications and applications (LCA) School of Computer and Communication Sciences
10. OMA standard (2011) Secure user plan location. V3.0
11. OMA standard (2011) Mobile location protocol. V3.1

Study of Reinforcement Learning Based Dynamic Traffic Control Mechanism

Zheng Zhang, Seung Jun Baek, Duck Jin Lee and Kil To Chong

Abstract A traffic signal control mechanism is proposed to improve the dynamic response performance of a traffic flow control system in an urban area. The necessary sensor networks are installed in the roads and on the roadside upon which reinforcement learning is adopted as the core algorithm for this mechanism. A traffic policy can be planned online according to the updated situations on the roads based on all the information from the vehicles and the roads. The optimum intersection signals can be learned automatically online. An intersection control system is studied as an example of the mechanism using Q-learning based algorithm and simulation results showed that the proposed mechanism can improve traffic efficiently more than a traditional signaling system.

Keywords Intelligent transportation system · Cooperative vehicle-highway systems · Reinforcement learning · Traffic control mechanism · Intersection signal control

---

Z. Zhang
Department of Mechanical Engineering, Xian Jiaotong University, Xian,
Peoples Republic of China

S. J. Baek · K. T. Chong
Department of Electronics Engineering, Jeonbuk National University, Jeonju,
Republic of Korea

D. J. Lee
Department of Mechanical Engineering, Jeonbuk National University, Jeonju,
Republic of Korea

K. T. Chong (✉)
Advanced Research Center for Electronics and Information, Jeonbuk National University,
Jeonju, Republic of Korea
e-mail: kitchong@chonbuk.ac.kr

J. J. (Jong Hyuk) Park et al. (eds.), *Multimedia and Ubiquitous Engineering*,
Lecture Notes in Electrical Engineering 240, DOI: 10.1007/978-94-007-6738-6_129,
© Springer Science+Business Media Dordrecht(Outside the USA) 2013

1 Introduction

Intelligent Transportation Systems (ITS) utilizes synergistic technologies and systems engineering concepts to develop and improve transportation systems of all kinds [1]. Machine intelligence on the road has been a popular research area with the advent of modern technologies especially artificial intelligence, wireless communication and advanced novel sensors.

Current traffic signal control system design is based on historic traffic flow data which cannot adapt itself to the rapidly varying situations at a crossroad. In some extreme situations, there are no vehicles during a green light and lots of vehicles waiting at a red one.

Many researchers have proposed schemes to solve the afore-mentioned problems like Choy et al. [2] who introduced hybrid agent architecture for real-time signal control. He suggested in his paper a dynamic database for storing all recommendations of the controller agents for each evaluation period. Liu et al. [3] proposed a calculating method of intersection delay under signal control while Bao et al. [4] studied an adaptive traffic signal timing scheme for an isolated intersection. However all these papers solve the problem according to the history flow data but not the current information [5, 6].

This paper makes the following contributions in particular:

(a) A novel traffic flow control mechanism is proposed based on the cooperation of the vehicle, road and traffic management systems. A roadside wireless communication network supports a dynamic traffic flow control method.
(b) Reinforcement learning is introduced as the core algorithm to dynamically plan traffic flow in order to improve efficiency. A Q-learning based intersection traffic signal control system is studied as an example of the proposed mechanism.

2 Study of Intersection Signal Control

In this section, a Q learning algorithm will be used to create a real time cooperation policy for an isolated intersection control under the proposed Traffic Control Mechanism. The algorithm and the simulation are both described in detail. The result shows the advantage of the proposed method.

2.1 Q-Learning Algorithm

Q learning, a type of reinforcement learning, can develop optimal control strategies from delayed rewards, even when an agent has no prior knowledge of the effects of its actions on the environment [7].

The agent's learning task can be described as follows. We require that the agent learn a policy π that maximizes $V^\pi(s)$ for all states s. We will call such a policy an optimal policy and denote it by π^*

$$\pi^* \equiv \arg\max_\pi V^\pi(s), (\forall s) \tag{1}$$

To simplify notation, we will refer to the value function $V^{\pi^*}(s)$ of such an optimal policy as $V^*(s)$. $V^*(s)$ gives the maximum discounted cumulative reward that the agent can obtain starting from state s; that is, the discounted cumulative reward obtained by following the optimal policy beginning at state s.

However, it is difficult to learn the function $\pi^* : S \rightarrow A$ directly, because the available training data does not provide training examples of the form $<s, a>$. Instead, the only training information available to the learner is the sequence of immediate rewards $r(s_i, a_i)$ for $i = 0, 1, 2, \ldots$. As we shall see, given this kind of training information it is easier to learn a numerical evaluation function defined over states and actions, then implement the optimal policy in terms of this evaluation function.

What evaluation function should the agent attempt to learn? One obvious choice is V^*. The agent should prefer state s_1 over state s_2 whenever $V^*(s_1) > V^*(s_2)$, because the cumulative future reward will be greater from s_1. The agent's policy must choose among actions, not among states. However, it can use V^* in certain settings to choose among actions as well. The optimal action in state s is the action a that maximizes the sum of the immediate reward $r(s, a)$ plus the value V^* of the immediate successor state, discounted by γ.

$$\pi^*(s) = \arg\max_a [r(s, a) + \gamma V^*(\delta(s, a))] \tag{2}$$

where $\delta(s, a)$ denotes the state resulting from applying action a to state s.

Thus, the agent can acquire the optimal policy by learning V^*, provided it has perfect knowledge of the immediate reward function r and the state transition function δ. When the agent knows the functions r and δ used by the environment to respond to its actions, it can then use Eq. (2) to calculate the optimal action for any state s.

Unfortunately, learning V^* is a useful way to learn the optimal policy only when the agent has perfect knowledge of δ and r.

Let us define the evaluation function $Q(s, a)$ so that its value is the maximum discounted cumulative reward that can be achieved starting from state s and applying action a as the first action. In other words, the value of Q is the reward received immediately upon executing action a from state s, plus the value (discounted by γ) of following the optimal policy thereafter.

$$Q(s, a) \equiv r(s, a) + \gamma V^*(\delta(s, a)) \tag{3}$$

Note that $Q(s, a)$ is exactly the quantity that is maximized in Eq. (3) in order to choose the optimal action a in state s. Therefore, we can rewrite Eq. (3) in terms of $Q(s, a)$ as

$$\pi^*(s) = \arg\max_a Q(s, a) \tag{4}$$

Why is this rewrite important? Because it shows that if the agent learns the Q function instead of the V^* function, it will be able to select optimal actions even when it has no knowledge of the functions r and δ. As Eq. (4) makes clear, it need only consider each available action a in its current state s and choose the action that maximizes $Q(s, a)$. This is exactly the most important advantages of Q learning, and also is the reason why we choose Q learning in this paper.

How should the Q learning algorithm be implemented? The key problem is finding a reliable way to estimate training values for Q, given only a sequence of immediate rewards r spread out over time. This can be accomplished through iterative approximation. To see how, notice the close relationship between Q and V^*, $V^*(s) = \max_{a'} Q(s, a')$, which allows rewriting Eq. (3) as follows:

$$Q(s, a) = r(s, a) + \gamma \max_{a'} Q(\delta(s, a), a') \tag{5}$$

Equation (5) provides the basis for algorithms that iteratively approximate Q. In the algorithm, $\overline{Q}$ will be the learner's estimate, or hypothesis of the actual Q function. $\overline{Q}$ will be represented by a large table with a separate entry for each state-action pair. The table can be initially filled with random values (though it is easier to understand the algorithm if one assumes initial values of zero). The agent repeatedly observes its current state s, choose some action a, executes this action, then observes the resulting reward $r = r(s, a)$ and the new state $s' = \delta(s, a)$. It then updates the table entry for $\overline{Q}(s, a)$ following each such transition, according to the rule:

$$\overline{Q}(s, a) \leftarrow r(s, a) + \gamma \max_{a'} \overline{Q}(s', a') \tag{6}$$

Note that the above training rule uses the agent's current $\overline{Q}$ values for the new state s' to refine its estimate of $\overline{Q}(s, a)$ for the previous state s.

The iterative training rule (6) will be replaced by

$$\overline{Q}(s, a) \leftarrow g(s, a) + \gamma \min_{a'} \overline{Q}(s', a'). \tag{7}$$

It means that the learning target is to minimize the Q function by minimizing the total cost when acting based on the optimum action sequences. This is exactly the algorithm used in this paper.

2.2 Model of the Intersection Signal System

A traffic system consists of various components, among which the traffic intersection is one of the most important [8]. Our method is applied to a traffic intersection that consists of two intersecting roads, each with several lanes and a set of synchronized traffic lights that manage the flow of vehicles, as shown in Fig. 1.

Fig. 1 Isolated intersection

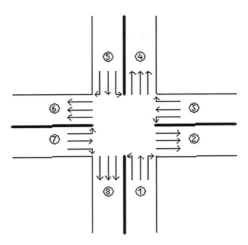

In this intersection, the rule of traffic management is right-hand based, which is used in China and South Korea. The vehicles in lanes ①, ③, ⑤ and ⑦, are approaching the intersection. Vehicles in ②, ④, ⑥ and ⑧, are leaving the intersection. For each of the approaching lanes, there are three directions for vehicles to choose: turn left, turn right and go straight, as shown in Fig. 1.

We will not consider the turn right direction because it does not impact other directions. In order to make this problem easy to model, we will not consider the pedestrian crossing the road. It will be very easy to add an additional rule for a pedestrian under our proposed mechanism.

Therefore, this problem can be modeled as 8 queues for different paths, as shown in Table 1.

We assume that there are a random number of vehicles spreading on different queues at the beginning of a signal period. This is the initial state of the environment. The final state must be that all the vehicles in the initial state have crossed the intersection. The intersection signal control system is modeled as a leader agent to manage the actions of all vehicle agents around the intersection. Since the action libraries of vehicle agents include actions from A1 to A8, the leader agent can choose any one action or their reasonable combination to reach the final state.

If two of the actions from A1 to A8 are nonintervention, they are possible action combinations. We call these different combinations a signal phase. All possible combinations are shown in Table 2.

Therefore, the problem can be described as how to find the optimum sequence of action combinations to reach the final state. This is the main function of the intersection signal control agent.

For each of the discrete states from the initial state to the final state, the optimum policy will be independent of the previous state. The successor state will be deterministic after one action combination is done. Therefore, this problem can be modeled as a deterministic Markov decision process.

Table 1 Basic action definition of different queues

Queue	Basic action symbol	Path
Que1	A1	① → ④
Que2	A2	① → ⑥
Que3	B1	⑤ → ⑧
Que4	B2	⑤ → ②
Que5	C1	③ → ⑥
Que6	C2	③ → ⑧
Que7	D1	⑦ → ②
Que8	D2	⑦ → ④

Table 2 Action combination symbol

Phase	Action combination symbol	Component
Ph1	Ac1	A1 + A2
Ph2	Ac2	A1 + B1
Ph3	Ac3	B1 + B2
Ph4	Ac4	C1 + C2
Ph5	Ac5	C1 + D1
Ph6	Ac6	D1 + D2

2.3 *Parameters of Learning Process*

(1) *Cost function*

We suppose that the vehicle number is n at state s. After the selected action a completed, the current vehicle number will be n_1. The cost of this action depends on the waiting time t, and the remainder of vehicles n_1.

$$g(s, a) = n_1 \times (t + t_{transition}). \qquad (8)$$

where $t_{transition}$ equals one of the three numbers $\{0, 1.5, 3\}$ shown in Table 3. The average time for each vehicle passing the crossroad is supposed to be 3 s.

Table 3 $t_{transition}$ of different phase transition

Phase transition type	Comment	$t_{transition}(s)$
No transition	Current phase is the same as the previous one	0
Half transition	$Ac1 \Leftrightarrow Ac2; Ac2 \Leftrightarrow Ac3;$ $Ac4 \Leftrightarrow Ac5; Ac5 \Leftrightarrow Ac6;$	1.5
Full transition	Phase transfer except half transition	3

(2) *Discount factor*

In the simulation we set the discount factor, $\gamma = 0.8$.

2.4 Simulation and Results

We wrote some MATLAB code to complete the simulation with the following configuration.

CPU: Intel Pentium 4 Processor 2.40 GHz,
Memory: 1047792 KB,
Operation System: Microsoft Windows XP Professional (SP3).

In order to show the advantage of our proposed mechanism, the traditional signal mechanism was introduced to create a comparative study. In the traditional mechanism, the signal phase transition is in a fixed sequence as shown by Ph1, Ph2, Ph3, Ph4, Ph5 and Ph6. However, our proposed method can determine the optimum phase sequence automatically based on the updated situation.

In the following, we will show the comparative result for three different periods T and different phase time interval t_{phase}.

In the above-mentioned tables, Ps is the simulation period series, NIV is the total number of vehicles at the initial state, Random Queues the number of vehicle queues that are randomly created, TIQ is the time interval from the initial state to the final state for a Q learning method, TWQ is the total waiting time for the Q learning method, TIT is the time interval from the initial state to the final state for the traditional method, $T_{IT} = 6 \times t_{phase}$,
TWT is the total waiting time for the traditional method,

$$P_{EI} = \frac{T_{IT} - T_{IQ}}{T_{IT}} \times 100\ \% \tag{9}$$

Equation (9) determines the percent improvement in the traffic efficiency,

$$P_{WD} = \frac{T_{WT} - T_{WQ}}{T_{WT}} \times 100\ \% \tag{10}$$

Equation (10) shows the percent decrease in total waiting time.

OA is the optimum phase sequence from Q learning, TL is the running time of the Q learning program on the above mentioned computer.

2.5 Analysis of the Results

From Table 4, we find that all the running times of the Q learning program TL in every period are less than one second. This is short enough for the application of the intersection signal control system.

Table 4 Simulation result when $t_{phase} = 60$ s

P_a	N_{IV}	Random queues	T_{IQ} (s)	T_{WQ} (s)	T_{WT} (s)	P_{EI} (%)	P_{WD} (%)	O_A	T_L (s)
1	180	{20 20 40 20 20 20 20 20}	312	18900	24000	13.33	21.25	{4 5 6 1 2 3}	0.8438
2	175	{20 20 19 19 38 19 20 20}	303	17916	23280	15.83	23.04	{1 2 3 6 5 4}	0.9375
3	176	{20 20 18 18 40 20 20 20}	306	18048	23640	15.00	23.65	{1 2 3 6 5 4}	0.4375
4	202	{17 17 36 18 36 18 40 20}	345	28479	30600	4.17	6.93	{1 2 3 6 5 4}	0.8438
5	183	{38 19 16 16 17 17 40 20}	345	21939	26880	4.17	18.38	{3 2 1 4 5 6}	0.8906
6	189	{19 19 36 18 20 20 38 19}	351	23418	28380	2.50	17.48	{1 2 3 4 5 6}	0.9063
7	157	{14 14 16 16 38 19 20 20}	276	14331	22740	23.33	36.98	{3 2 1 6 5 4}	0.8750
8	174	{38 19 26 13 20 20 19 19}	282	18852	22020	21.67	14.39	{4 5 6 1 2 3}	0.9063
9	123	{30 15 14 14 13 13 12 12}	219	8916	14280	39.17	37.35	{4 5 6 3 2 1}	0.8906
10	133	{15 15 17 17 30 15 12 12}	234	10413	16440	35.00	36.66	{3 2 1 6 5 4}	0.9219
11	130	{11 11 34 17 22 11 12 12}	249	11007	16140	30.83	31.80	{6 5 4 1 2 3}	0.8594
12	152	{22 11 15 15 16 16 38 19}	285	14904	24480	20.83	39.12	{3 2 1 4 5 6}	0.8750
13	171	{36 18 32 16 24 12 22 11}	273	19292	19500	24.17	1.07	{4 5 6 1 2 3}	0.9531
14	183	{26 13 36 18 26 13 34 17}	300	22265	25560	16.67	12.89	{6 5 4 3 2 1}	0.9375
15	120	{32 16 20 10 20 10 6 6}	216	10221	11760	40.00	13.09	{6 5 4 1 2 3}	0.8750
16	128	{19 19 20 10 15 15 15 15}	219	9768	15900	39.17	38.57	{4 5 6 1 2 3}	0.8906
17	112	{14 14 9 9 28 14 16 8}	189	7905	14220	47.50	44.41	{1 2 3 4 5 6}	0.8906
18	100	{16 8 8 8 34 17 6 3}	195	6480	11760	45.83	44.90	{3 2 1 6 5 4}	0.9375
19	128	{16 16 32 16 16 8 16 8}	228	9960	15120	36.67	34.13	{4 5 6 1 2 3}	0.8906

(continued)

Table 4 (continued)

P_a	N_{IV}	Random queues	T_{IQ} (s)	T_{WQ} (s)	T_{WT} (s)	P_{EI} (%)	P_{WD} (%)	O_A	T_L (s)
20	128	{15 15 30 15 147 16 16}	237	10431	16620	34.17	37.24	{6 5 4 1 2 3}	0.8906
21	108	{16 8 36 18 14 14 1 1}	189	6906	10860	47.50	36.41	{4 5 6 3 2 1}	0.8750
22	81	{5 5 24 12 18 9 4 4}	165	4365	10140	54.17	56.95	{6 5 4 1 2 3}	0.4844
23	109	{14 14 18 9 30 15 6 3}	207	7740	11400	42.50	32.11	{1 2 3 6 5 4}	0.4375
24	144	{9 9 16 16 40 20 17 17}	258	11772	21660	28.33	45.65	{3 2 1 6 5 4}	0.5156
25	77	{5 5 30 15 8 4 5 5}	156	3501	9060	56.67	61.36	{6 5 4 1 2 3}	0.4219

At the same time, the percent traffic efficiency improvement PEI, is located in [4.17 % 47.5 %], the percent total waiting time decrease PWD is located in [1.07 % 56.95 %]. The average percents of PEI are 32.2 % and the average percents of PWD are 37.5 %.

3 Conclusion

A new traffic control based mechanism based on a combination of machine learning and multiagent modeling methods is proposed for future intelligent transportation systems. The control systems, the vehicles, and some necessary roadside sensors are all modeled as intelligent agents in the proposed systems, therefore the ITS system will be a multiagent system. It is possible to improve the traffic control efficiency by some artificial intelligence algorithm.

The control method for an isolated intersection was studied specifically. The intersection signal was first modeled according to the proposed mechanism then a new algorithm based on reinforcement learning, especially Q-learning, was proposed and studied in detail. A simulation for such an intersection system was finally carried out and a comparative study with the traditional intersectional signal method was done.

Simulation results showed that the proposed intersection control mechanism can improve traffic efficiency by more than 30 % over the traditional method and simultaneously bring the drivers some benefit by decreasing the waiting time by more than 30 %. This proves that the proposed traffic control mechanism is applicable in the near future.

Acknowledgments This work was supported by the National Research Foundation of Korea (NRF) grant funded by the Korea government (MEST) (No. 2012-038978) and (No. 2012-0002434).

References

1. http://www.ewh.ieee.org/tc/its/
2. Choy MC, Srinivasan D, Cheu RL (2003) Cooperative, hybrid agent architecture for real-time traffic signal control. IEEE Trans Syst Man Cybern Part A Syst Hum 33(5):597–607
3. Liu G, Zhai R, Pei Y (2007) A calculating method of intersection delay under signal control. In: Proceedings of the 2007 IEEE intelligent transportation systems conference, Seattle, pp 1114–1119
4. Bao W, Chen Q, Xu X (2006) An adaptive traffic signal timing scheme for bus priority at isolated intersection. In: Proceedings of the 6th world congress on intelligent control and automation, Dalian, pp 8712–8716
5. Srinivasan D, Choy MC (2006) Cooperative multi-agent system for coordinated traffic signal control. IEE Proc Intell Transp Syst 153(1):41–50
6. Lee JH, Lee-Kwang H (1999) Distributed and cooperative fuzzy controllers for traffic intersections group. IEEE Trans Syst Man Cybern C Appl Rev 29:263–271
7. Mitchell TM (1997) Machine learning. McGraw-Hill, New York. ISBN: 0070428077
8. D'Ambrogio A et al (2008) Simulation model building of traffic intersections. Simul Model Pract Theory

Understanding and Extending AUTOSAR BSW for Custom Functionality Implementation

Taeho Kim, Ji Chan Maeng, Hyunmin Yoon and Minsoo Ryu

Abstract AUTOSAR (Automotive Open System Architecture) is a de factor standard for automotive software development. It addresses crucial topics such as software architecture, application interfaces and development methodology, thereby providing a basic infrastructure for software development. However, the current AUTOSAR standard is too complex to learn and has significant dependence upon tool chains. As a result, it is very difficult to implement custom functionality in BSW (Basic Software) without special support from tool vendors. In this paper, we present how custom functionality can be implemented within AUTOSAR BSW obviating the need for tool vendor's support. We first examine the internal structure and function of AUTOSAR software stack with an emphasis on the interfaces and execution of BSW modules. We then describe how a new BSW functionality can be incorporated into AUTOSAR BSW. Our approach is illustrated through a simple BSW module implementation with EB tresos Auto-Core and Infineon TriCore TC1797.

Keywords AUTOSAR · Automotive · Software architecture · BSW (basic software) · Custom functionality

---

T. Kim · J. C. Maeng · H. Yoon
Department of EECS, Hanyang University, Ansan, Korea
e-mail: thkim@rtcc.hanyang.ac.kr

J. C. Maeng
e-mail: jcmaeng@rtcc.hanyang.ac.kr

H. Yoon
e-mail: hmyoon@rtcc.hanyang.ac.kr

M. Ryu (✉)
Department of CSE, Hanyang University, Ansan, Korea
e-mail: msryu@hanyang.ac.kr

J. J. (Jong Hyuk) Park et al. (eds.), *Multimedia and Ubiquitous Engineering*,
Lecture Notes in Electrical Engineering 240, DOI: 10.1007/978-94-007-6738-6_130,
© Springer Science+Business Media Dordrecht(Outside the USA) 2013

1 Introduction

AUTOSAR (Automotive Open System Architecture) is receiving wide attention as the best means of automotive software development. Since electronics and software are becoming increasingly important in vehicle technology, the vehicle industry faces new challenges for software development such as standardization of system-level functionality, component-based software development and integration, model-driven development, software reuse and maintainability. To overcome these challenges, AUTOSAR has been designed to address crucial topics such as software architecture, application interfaces and development methodology, thereby providing a basic infrastructure for automotive software development. Currently, many OEMs and suppliers are accepting AUTOSAR as a de facto standard for automotive software development.

AUTOSAR has many features from a software engineering point of view. One of them is to provide complete abstraction between applications and system software so that developers can focus on the design and implementation of application components. Such abstraction is made possible by AUTOSAR tools such as Vector's DaVinci and Elektrobit's EB tresos. These tools can be used to configure system-level functionality and automatically generate source code for Run-Time Environment (RTE) and Basic Software (BSW).

Although the abstraction supported by AUTOSAR is useful for application development, it also severely limits the accessibility of system software to developers. During automotive system development, engineers often need direct access to system software for many reasons such as system-level tracing and debugging, system-wide performance profiling and optimization. For instance, consider that a developer needs to analyze and optimize system-level performance. To do so, some custom functionality should be implemented within AUTOSAR BSW. However, implementing custom functionality in BSW is not easy without special support from the tool vendor since the current AUTOSAR standard is too complex to learn and has significant dependence upon tool chains.

In this paper, we describe how custom functionality can be implemented within AUTOSAR BSW. We first introduce the internal structure and behavior of AUTOSAR software stack with an emphasis on the interfaces and execution of BSW modules. We then present how a new BSW functionality can be incorporated into AUTOSAR BSW. Our approach is illustrated through a simple BSW module implementation with EB tresos AutoCore and Infineon TriCore TC1797. We implemented a BSW module that is able to blink LED while handling CAN (controller area network) interface interrupts. We could successfully incorporate this BSW module into EB tresos AutoCore without any technical support from the tool vendor.

2 Implementing Custom Functionality Within AUTOSAR BSW

The AUTOSAR software stack consists three layers, application, Run-Time Environment (RTE) and Basic Software (BSW). The application layer can be modeled as a set of interconnected software components (SW-C) that encapsulate application-specific functionality. Software components have well-defined interfaces through which components can communicate in various ways such as sender-receiver or client-server fashions. Software components also have runnables that can be invoked by an OS task or other software component. The notion of runnable is very similar to the Runnable interface in Java, which is intended to be executed by a separate thread of control.

The RTE layer plays a role of middleware between application and BSW layers. Its main responsibility is to provide communication services to application software, communications between software components and communications between software components and basic software modules.

The BSW layer is responsible for managing hardware resources and providing common services for application software. BSW is further divided into four parts, Microcontroller Abstraction Layer (MCAL), ECU Abstraction Layer (EAL), Services Layer and Complex Device Drivers (CDD). The Microcontroller Abstraction Layer is the lowest software layer in BSW and provides device drivers for microcontrollers and internal peripherals. The ECU Abstraction Layer provides device drivers for external devices. The Services Layer offers many functions including OS functionality, network management, memory management and diagnostic services. The Complex Drivers Layer provides special purpose functionality that is not specified within the AUTOSAR standard (Fig. 1).

In order to understand and extend BSW functionality, it is important to understand AUTOSAR OS and BSW scheduler in the Services Layer. The main

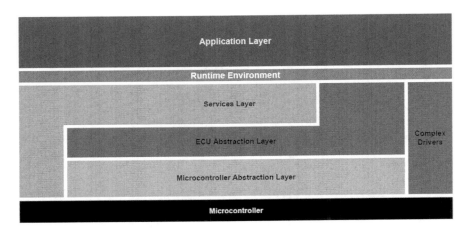

Fig. 1 Software layers in AUTOSAR

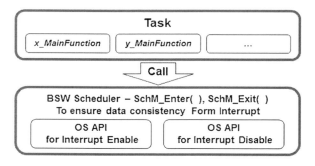

Fig. 2 Invocation of BSW scheduler APIs to protect the critical section

role of AUTOSAR OS is the management and execution of tasks and ISRs. AUTOSAR OS is responsible for generating, executing and scheduling tasks at runtime. It is also responsible for invoking interrupt handlers.

BSW Scheduler, another important component in the Services Layer, provides two services. First, it executes main processing functions of BSWs in the form of tasks. Second, it provides mutual exclusion APIs needed for protecting critical sections in BSW modules. Specifically, BSW Scheduler provides "SchM_Enter()" and "SchM_Exit()" APIs that are used to disable and enable interrupts for protecting critical sections accessible by other interrupts (Fig. 2).

Since BSW modules are mostly used for hardware-related functionality, we need to know how their main processing functions and interrupts handlers are specified and executed when we create a custom BSW module. First, when we create a new BSW module, we must follow the AUTOSAR naming convention. Every main processing function in BSW module must be named as shown in the following code.

```
<Module name>_MainFunction_<Extension name>( );
```

"<Extension name>" is used to distinguish between main processing functions in a BSW module. Interrupt handlers should also be named as shown in the following code. Generally, "<Service name>" is used to express a service of an interrupt handler.

```
<Module name>_<Service name>( );
```

Second, interrupt handlers should be specified using AUTOSAR ISR macros. Each ISR can be associated with a specific interrupt type and appropriate interrupt handling code.

```
ISR ( /* interrupt type */ ) {
     /* Driver Module User Code to handle ISR */
}
```

Understanding and Extending AUTOSAR BSW

Third, we may use SchM_Enter(Instance, Exclusive Area) and SchM_Exit(Instance, Exclusive Area) to avoid any possible data races in main processing functions.

```
SchM_Enter_<Module name>(Instance, ExclusiveArea);
SchM_Exit_<Module name>(Instance, ExclusiveArea);
```

In above code, "Instance" is a main processing function in a BSW module "<Module name>". In BSW modules, there are several types of critical sections such as initialization, writing and reading shared data. The second parameter, "Exclusive Area", is used to specify the type of critical sections and invoke an appropriate API.

3 Case Study

We implemented a simple custom BSW module within EB tresos AutoCore on Infineon TriCore TC1797 hardware. It is able to blink LEDs while handling controller area network (CAN) interface interrupts. We could successfully incorporate this BSW module into EB tresos AutoCore without any technical support from AUTOSAR tool vendors (Fig. 3).

Specifically, we first implemented eight main processing functions for LED control as shown in the following code.

Fig. 3 LED controller and CAN controller on Infineon TriCore TC1797

```
LedControl_MainFunction_1();
LedControl_MainFunction_2();
...
LedControl_MainFunction_7();
LedControl_MainFunction_8();
```

We then mapped the main processing functions onto a single task whose activation period is one second. Thus, all LED sequentially blinks every one second.

```
TASK (Ex_LedControl_Sequence_1second) {
    LedControl_MainFunction_1();
    LedControl_MainFunction_2();
    ...
    LedControl_MainFunction_7();
    LedControl_MainFunction_8();
}
```

We also implemented CAN message reception functionality in our BSW module by using an interrupt handler. We associated an ISR with the CAN_RECEIVE interrupt. Whenever a CAN_RECEIVE interrupt occurs, the ISR is invoked to blink LED.

```
ISR (CAN_RECEIVE) {
    Can_IsrReceiveHandler();
}
```

4 Conclusion

In this paper, we described how custom functionality can be implemented within AUTOSAR BSW. We first introduced the internal structure and behavior of AUTOSAR software stack with an emphasis on the interfaces and execution of BSW modules. We then presented how a new BSW functionality can be incorporated into AUTOSAR BSW. Our approach was illustrated through a simple BSW module implementation with EB tresos AutoCore and Infineon TriCore TC1797 hardware.

Acknowledgment This work was supported partly by Mid-career Researcher Program through National Research Foundation (NRF) grant NRF-2011-0015997 funded by the Ministry of Education, Science and Technology (MEST), partly by the IT R&D Program of MKE/KEIT [10035708, "The Development of Cyber-Physical Systems (CPS) Core Technologies for High Confidential Autonomic Control Software"], partly by Seoul Creative Human Development

Program (HM120006), and partly by the The Ministry of Knowledge Economy (MKE), Korea, under the Convergence Information Technology Research Center (CITRC) support program (NIPA-2013-H0401-13-1009) supervised by the National IT Industry Promotion Agency (NIPA).

References

1. Daehyun K, Gwang-Min P, Seonghun L, Wooyoung J (2008) AUTOSAR migration from existing automotive software, In: International conference on control, automation and systems, pp 558–562
2. Wang D, Zheng J, Zhao G, Bo H, Liu S (2010) Survey of the AUTOSAR complex drivers in the field of automotive electronics. In: Intelligent computation technology and automation, pp 662–664
3. Diekhoff D (2010) AUTOSAR basic software for complex control units. In: Design automation and test in europe conference and exhibition, pp 263–266

A Hybrid Intelligent Control Method in Application of Battery Management System

T. T. Ngoc Nguyen and Franklin Bien

Abstract This paper presents a hybrid adaptive neuro-fuzzy algorithm in application of battery management system. The proposed system employed the Cuk converter as equalizing circuit, and utilized a hybrid adaptive neuro-fuzzy as control method for the equalizing current. The proposed system has ability for tracking dynamic reactions on battery packs, due to taking advantages of adaptability and learning ability of adaptive neuro-fuzzy algorithm. The current output generated from learning process drives Pulse-Width-Modulation (PWM) signals. This current output is observed and collected for next coming learning process. The feedback line is provided for current output observation. The results demonstrate the proposed scheme has the ability to learn previous stages. Therefore, the proposed system has adaptability to deal with changing of working conditions.

Keywords Fuzzy logic · Adaptive neuro-fuzzy system · dc-dc converter · Battery equalization

1 Introduction

Nowadays, lithium-ion batteries are more and more replacing the traditional acid batteries. However, the lithium-ion battery cell provides very low voltage around 4–4.5 V that is not enough to supply in electronic vehicles. Therefore, the requirement of connecting lithium-ion battery cells is on demand. Normally, in

---

T. T. Ngoc Nguyen (✉) · F. Bien
School of Electrical and Computer Engineering, Ulsan National Institute
of Science and Technology, Ulsan, South Korea
e-mail: ngocntt@unist.ac.kr

F. Bien
e-mail: bien@unist.ac.kr

J. J. (Jong Hyuk) Park et al. (eds.), *Multimedia and Ubiquitous Engineering*,
Lecture Notes in Electrical Engineering 240, DOI: 10.1007/978-94-007-6738-6_131,
© Springer Science+Business Media Dordrecht(Outside the USA) 2013

very high voltage systems, the hundreds of battery cells are connected together that are called serially-connected battery systems. The intelligent controllers are required for these systems due to imbalance between cells when they are connected in series. The differences in working conditions between cells cause severe problems such as high risk of explosion, depth discharging or over-charging. Therefore, balancing voltages between the battery cells is essential for cell protection and prolongation of the battery life. A number of research activities focusing on control method to equalizing battery voltages have been published. In recent times, integrated individual cell equalizer (ICE) is under development regarding equalization scheme [1–3]. The equalization control restricts battery cell's operation range to prevent over-discharging and over-charging, which cause subsequence damage to the active materials in the battery such as electrodes. During charging state, the battery voltage should be limited to this value. Otherwise, the internal pressure and temperature would produce a high risk of explosion.

On the other hand, the copper in the electrolyte are easy to be dissolved during over-discharging that shortens battery life. Normally, the low voltage threshold of a lithium-ion battery cell is in the range of 2.6–2.8 V. Obtaining maximum usable capacity from battery cells is one of purposes of battery controller, in order to have high energy efficiency. In the battery system using individual cell equalizers (ICE), the controller control equalizing current between two neighboring cells in an ICE. The imbalance problems caused by changes in the internal impedance and cell capacity are exacerbated while the battery is working. Meanwhile, most of previous control methods based on imbalanced state-of-charge (SOC) between cells, and gradients of ambient temperature of the battery pack. These characteristics are non-linear and difficult to model exactly due to objective and subjective parameters, such as temperature, workload, etc. As a result, these conventional methods are short of adaptability to the dynamic system.

The proposed scheme of equalization scheme employed adaptive neuro-fuzzy algorithm. The cell voltages are controlled by the driving pulse-width modulation (PWM) signals, based on the equalization algorithm scheme in the battery unit controller. This adaptive neuro-fuzzy algorithm provides off-line training, system tracking and estimation of the proposed BMS scheme. The advantage of the proposed system is inheriting adaptability of back propagation neuro network for training and estimating that are suitable with dynamic system like serially connected battery cells.

2 The Proposed Hybrid Adaptive Neuro-Fuzzy Control Method

Figure 1 show the proposed system, the framework of the proposed battery equalization system is based on individual cell equalizer (ICE) that balances each pair of the neighboring cells. In the proposed system, a complete cell voltage

A Hybrid Intelligent Control Method

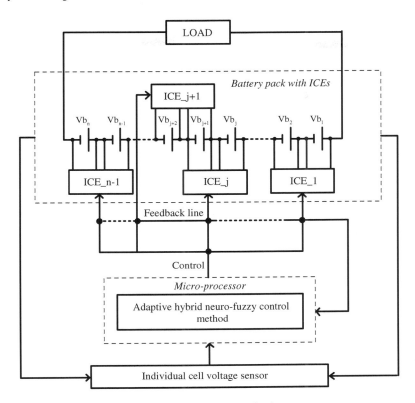

Fig. 1 The proposed hybrid adaptive neuro-fuzzy control scheme

equalizer is performed using a bi-directional dc-dc converter modified from the Cuk converter. In conventional schemes, it is difficult to handle the battery cell model for describing the equalizing characteristic of the lithium-ion battery strings due to electrochemical reactions and ambient temperature. Fuzzy logic control has advantage in modeling the non-linear characteristic of equalizing current during charging and discharging. However, the limitation of fuzzy system is that it cannot automatically acquire the rules to make output decisions. In addition, it is challenging to model the membership functions in small change of voltage offsets with conventional fuzzy-based control methods that cause degradation accuracy of the system employing such control methods. In this work, combining neuron network in fuzzy logic control is used to tune membership functions of fuzzy system, which enables adaptability and learning ability. In the proposed system, a feedback line is provided to allow updating and tracking the changes of battery strings.

The equalizing circuit works on a cell-to-cell basis, therefore only one ICE_1 is analyzed for the proposed equalization as illustrated in Fig. 2. The voltage battery of cell 1 and cell 2 are represented by Vb_1 and Vb_2, respectively. Control signal from PWM turns on or off MOSFET Q1 or Q2 with control frequency of f_s or

Fig. 2 Individual cell equalization circuit

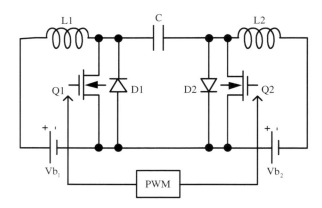

period of $T_s = 1/f_s$. Assuming $Vb_1 > Vb_2$, battery cell 1 will be discharged and battery cell 2 will be charged. At the initial state, the voltage of the capacitor C equals to $(Vb_1 + Vb_2)$. During the first period (T1) while Q1 is turned on, the energy stored in the capacitor transfers to cell 2 and the inductor L1 stores energy transferred from Vb_1 of cell 1. Likewise, the inductor L2 also stores energy. In the remaining period (T2) of the PWM control signal, Q1 is turned off, and D2 is forced to turn on. Hence, the capacitor is charged by energy stored in L1, and the energy stored in L2 continues to charge cell 2. The equivalent circuit for the equalization scheme is shown in Fig. 3. The energy transformation is continuous until the circuit reaches stable state. The quantitative analysis for the equivalent circuit for in Fig. 3 can be expressed with the following equations:

When Q1 is turned on: $(0 \leq t < T_1)$

$$\begin{cases} Vb_1 = L_1 \dfrac{di_{L1}}{dt} \\ Vb_2 = -L_2 \dfrac{di_{L2}}{dt} + \dfrac{1}{C}\int i_{L2}dt \end{cases} \quad (1)$$

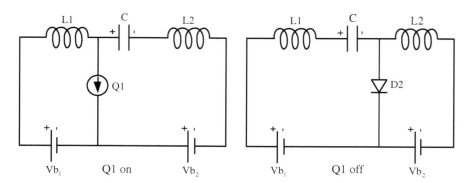

Fig. 3 Cell equalization circuit analysis

A Hybrid Intelligent Control Method

When Q1 is turned off: $(T_1 < t \leq T_s)$

$$\begin{cases} Vb_1 = L_1 \dfrac{di_{L1}}{dt} + \dfrac{1}{C} \displaystyle\int i_{L1} dt \\[3mm] Vb_2 = -L_2 \dfrac{di_{L2}}{dt} \end{cases} \tag{2}$$

Combining with the initial conditions and limitations, the average of inductor currents through L1 and L2 is computed as followed:

$$\begin{cases} I_{L1} = \dfrac{1}{2} T_s \left(\dfrac{Vb_1}{L_1} T_1^2 + \dfrac{V_C - Vb_1}{L_1} T_2^2 \right) \\[3mm] I_{L2} = \dfrac{1}{2} T_s \left(\dfrac{Vb_2}{L_2} T_2^2 + \dfrac{V_C - Vb_2}{L_2} T_1^2 \right) \end{cases} \quad (T_2 = T_s - T_1) \tag{3}$$

For simplicity, the value of L1 and L2 are chosen equally so that equal currents flow through these inductors. The output current I_{turn} from the adaptive neuron-fuzzy controller is used to switch the period of the PWM driving signal of the proposed scheme.

In the proposed control method, the battery cell voltage Vb and derivation of neighboring cells De are the inputs of the controller. The architecture of hybrid adaptive neuro-fuzzy algorithm consists of 5 layers.

- Layer 1 The first layer generates the membership functions of inputs {Vb, De}.

$$A_i(V_b) = \left[1 + \left(\dfrac{V_b - a_{i1}}{b_{i1}} \right)^2 \right]^{-1} \tag{4}$$

$$B_i(D_e) = \left[1 + \left(\dfrac{D_e - a_{i2}}{b_{i2}} \right)^2 \right]^{-1}$$

where $\{a_i, b_i\}$ is the *parameter set* while A, B are linguistic labels.

- Layer 2: each node computes the firing strength of the associated rules. The outputs of these neurons are labeled by T because of choosing *T-Norm* for modeling, and are calculated as following:

$$\alpha_1 = A_1(V_b) \times B_1(D_e) = A_1(V_b) \wedge B_1(D_e) \tag{5}$$

$$\alpha_2 = A_2(V_b) \times B_2(D_e) = A_2(V_b) \wedge B_2(D_e)$$

The node in this layer are called rule nodes

- Layer 3: every node in this layer is labeled by N to indicate the normalization of firing levels.

$$\beta_1 = \frac{\alpha_1}{\alpha_1 + \alpha_2}, \beta_2 = \frac{\alpha_2}{\alpha_1 + \alpha_2} \tag{6}$$

- Layer 4: the outputs of these nodes at this layer are the product of the normalized firing level and the individual rule output of associated rule.

$$\beta_i s_i = \beta_i (p_i V_b + q_i D_e) \tag{7}$$

where $\{p_i, q_i\}$ are the set of *consequent parameters* associated the i-rule.

- Layer 5: the single output of this layer computes the overall system output as the sum of all the incoming signals. The overall output is current output to drive PWM signals.

$$I_{turn} = \sum_i \beta_i s_i \tag{8}$$

The above layered structure is viewed as neural network. This structure can adapt its antecedent and consequent parameters to improve performance of the system by observing outputs. By employing these advantages of adaptive neuron-fuzzy architecture, the proposed battery equalization system has ability of learning and adaptation to deal with dynamic changes.

A hybrid learning rule based on least-square estimator (LSE) and gradient-descent method (GDM) is widely used [7]. A partition p of i-input is defined by giving an offset value a_i^p and distance b_i^p. Therefore, the total number of antecedent parameters to be trained with NP_i antecedents per input is $2NP_i$. Concerning the consequent parameters, the number of crisp consequents to be adapted is equal to the number of rules, where the maximum is $m = \prod_{i=1}^{n} NP_i$ in which all possible rules are considered. The training data pair $\{(V_b^k, D_e^k; I_{turn}^{k'}), k \in (1, 2, \ldots, K)\}$ is used in this learning process. After the initial antecedent parameters and consequence parameters are identified, GDM will optimize the bell-shaped membership functions by computing the error function:

$$E = \frac{1}{2K} \sum_{k=1}^{K} (I_{turn}^k - I_{turn}^{k'})^2 \tag{9}$$

where I_{turn}^k is the actual output of the network for the k-training data, and $I_{turn}^{k'}$ is the desired output.

3 Experimental Results

The following part represents the experimental results of the proposed system. The initial training data set is composed of 1,000 pairs of input set $\{V_b, D_e\}$ and current output I_{turn}. The membership functions for each input are determined after the training process as can be seen from Fig. 4. As illustrated in Fig. 4, after training

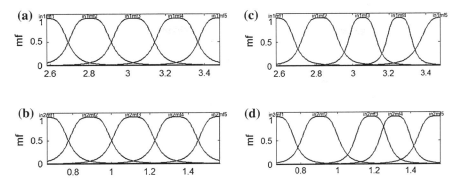

Fig. 4 Membership functions of input voltages before and after learning process with initial training datasets. **a** *Vb* (*V*) in initialization. **b** *De* (*V*) in initialization. **c** *Vb* (*V*) after learning. **d** *De* (*V*) after learning

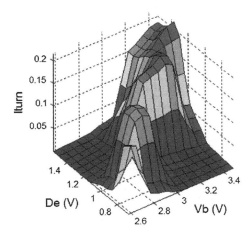

Fig. 5 Control current output *Iturn* model from learning process according to input voltages *De* and *Vb*

process, the membership functions of voltage inputs are changed by updating adaptive parameters. The output current I_{turn} is model based on adaptive membership functions of inputs in Fig. 5. In the experiments, the battery cell is the RHV-320064 model that has nominal voltage of 3.2 V. The circuit parameters are L1 = L2 = 200 μH, C = 450 μF. The switching frequency for initialization is selected at 50 kHz, with duty cycle of 50 %.

4 Conclusion

The proposed system employing a hybrid adaptive neuro-fuzzy algorithm is demonstrated successfully. An intelligent learning and adaptable algorithm of hybrid adaptive neuro-fuzzy is employed for tuning driving current to control

PWM. Based on the proposed control method, the battery system has ability of learning previous stages and enabled adaptability to deal with dynamic reactions in serially connected battery cells. The experimental result demonstrated the advantages of the proposed system.

Acknowledgment This work was supported by the development program of local science park funded by the ULSAN Metropolitan City and the MEST (Ministry of Education, Science and Technology).

References

1. Park HS, Kim CE, Kim CH, Moon GW, Lee JH (2009) A modularized charge equalizer for an HEV lithium-ion battery string. IEEE Trans Ind Electron 56(2):1464–1476
2. Lindemark B (1991) Individual cell voltage equalizers (ICE) for reliable battery performance. In: Proceedings of the 13th Annual International Telecommunications Energy Conference, Kyotopp, pp 196–201
3. Moore SW, Schneider PJ (2001) A review of cell equalization methods for lithium-ion and lithium polymer battery systems. In: Proceedings of the SAE 2001 world congress, Detroit
4. Lee YS, Cheng GT (2006) Quasi-resonant zero-current-switching bidirectional converter for battery equalization applications. IEEE Trans Power Electron 21(5):1213–1224
5. Lee YS, Jao CW (2003) Fuzzy controlled lithium-ion battery equalization with state-of-charge estimator. In: IEEE International Conference on Systems Man Cybernetics, vol 5, pp 4431–4438
6. Campo L, Echanobe J, Bosque G, Tarela JM (2008) Efficient Hardware/Software implementation of an adaptive Neuro-Fuzzy system. IEEE Trans Fuzzy Syst 16(3):761–778
7. Jang J (1993) ANFIS: adaptive-network-based Fuzzy inference system. IEEE Trans Syst Man Cybern 23(3):665–685

Interpretation and Modeling of Change Patterns of Concentration Based on EEG Signals

JungEun Lim, Soon-Yong Chun and BoHyeok Seo

Abstract It is very important to understand the brain's biological cognition data processing mechanism for human cognitive ability and concentration enhancement. Based on biological data processing area and information flow, concentration indicators were defined to interpret the brain data processing mechanism in concentration by engineering, and cognitive concentration model based on this was proposed. The cognitive concentration model is the change of concentration patterns shown by EEG signal. The value of cognitive concentration model was verified with the EEG signals acquired from Subjects solving mathematical questions with different difficulties.

Keywords Brain information processing · Attention · Concentration · EEG · Cognitive concentration model

1 Introduction

Body organ engaging in 'thought', like cognition of things and thinking, is the brain. It is known that this brain constitutes the nervous system in which cells called neurons are entangled complicatedly [1]. Studies of brain science on human cognition and thinking are actively carried out. Among studies using this, various

---

J. Lim · B. Seo
School of Electrical Engineering and Computer Science, The Graduate School,
Kyungpook National University, Daegu, Korea
e-mail: euny1122@knu.ac.kr

B. Seo
e-mail: bhsuh@knu.ac.kr

S.-Y. Chun (✉)
Dongyang University, Yeongju, Korea
e-mail: control@dyu.ac.kr

J. J. (Jong Hyuk) Park et al. (eds.), *Multimedia and Ubiquitous Engineering*,
Lecture Notes in Electrical Engineering 240, DOI: 10.1007/978-94-007-6738-6_132,
© Springer Science+Business Media Dordrecht(Outside the USA) 2013

studies were attempted to increase human cognition processing ability related to memory, attention, and emotion, etc. [2–5]. Concentration is defined as examination of some stimulus or keeping attention to a stimulus selected. Attention is defined as concentration or focus of consciousness kept in mind vividly after selecting one among objects or thoughts [6]. In other words, attention can be said to be an ability to maintain attention to selected stimulus in a limited time. Thus, in a prior step to increase human cognition processing abilities, this paper defined general cognitive abilities, the maximum attention, and the duration of concentration. We set up a cognitive concentration model in order to interpret and model the change patterns of concentration. Also, it carried out a test to prove these using hard, normal and easy level questions in the field of mathematics.

2 Biological Cognition Mechanism

When you selectively focus on certain information, the frontal lobe, thalamus, and amygdala, etc. are involved [7–9]. The frontal lobe actions allow us to concentrate and process an assignment using attention, and among them, thanks to prefrontal cortex. If the actions of prefrontal cortex are not active, people cannot concentrate one thing for a long time paying attention and mind is easily dispersed. As a result of the brain image photography, the left brain prefrontal cortex of a person who is solving a hard question by concentrating attention was strongly activated, and this was activated when he or she processed information concentrating on anything other than mathematical question [8]. In a child with a problem in attention, the prefrontal cortex of the frontal lobe acts weakly while limbic system acts very actively, the limbic system decreases ability to pay attention and control emotions and increases instinctive and primary emotional roles. The amygdala, one of the limbic system can receive various data through thalamus before the cerebral cortex, and as emotional information flows directly into the amygdala, before the cerebral cortex analyzes and judges information, already one gets favor or displeasure for the stimulus [9].

In addition, in the process of filtering unnecessary stimuli other than important information, reticular formation and thalamus act, and the reticular formation is located at the center of the brain stem, controls awakening level, so the cerebral cortex can pay attention or process information by the actions of the reticular formation. Norepinephrine, a neurotransmitter for keeping functions of the reticular formation rapidly increases when an organism gets excited or nervous and when it is abnormally low, the awakening function of the reticular formation is deteriorated.

Thalamus located in inner center of the frontal brain takes sense information entering in each part of the body, selects important things only and concentrates and conveys command from the cerebral cortex to muscles. Approximately 100 million stimuli per second comes to the central nerves and among them, about 100 important stimuli are selected for the cerebral cortex to make a right decisions [8].

The reticular formation, thalamus and neurons in the multiple sensory systems of the brain serve as filters for numerous stimuli coming into the sensory receptors, and allow men to pay attention to appropriate stimuli. If the roles of the amygdala in the frontal lobe get stronger or the functions of the reticular formation decline by norepinephrine, with the problem in the thalamus's ability of selecting information, there may be a problem in attention ability. Like this, attention consists of roles of various areas of the brain and neurotransmitters. Among existing studies on selective attention, Koch and Ullman in 1985 tried to show selective concentration using the concept of Saliency Map, and selecting Winner-Take-All through the most remarkable part and then make the part to concentrate. In the 1990s, Wolf and Cave insisted on Guided Search model of parallel process in pre-attention and sequential process in attention. This is gradually developed on the basis of visual attention model currently. However, it was concentrated on theories not developed to application system. Later in 1998, Itti made a four-step model combining the two methods [10]. Human cognition process can be viewed as a combination of bottom-up process of receiving external stimuli and top-down process of creating information inside the brain by active attention [11], and the systems of selective attention based on this biological cognition mechanism are proven as new cognition systems that combine human attention abilities with learning theory.

3 Propose of Cognitive Concentration Model

For an engineered model of selective attention, when there are data entered, each datum is processed after passing through a filter based on previously learned knowledge, which is estimated in a certain form, and the final conclusion is made according to the reliability of conclusions. This study aims to graft the concept of time in the engineer model of selected attention, and the inputting EEG power spectrum should be used for changing and providing model about the attention.

Figure 1 is the implementation of back-propagation algorithm for the single neuron as cognitive model. Once input, patterns are presented and attention is paid to one datum among them, the cost function E is defined Eq. (1). The distance between the desired value d_i of output neuron and the actual output (attended output) value y is minimized.

$$E = \frac{1}{2} \sum_i (d_i - y_i)^2 \tag{1}$$

The steepest descent gradient rule as cognitive weighting factor for each input datum x_i is expressed as the Eq. (2).

$$\frac{dw_{ij}}{dt} = -\mu \frac{\partial E_i}{\partial w_{ij}} \tag{2}$$

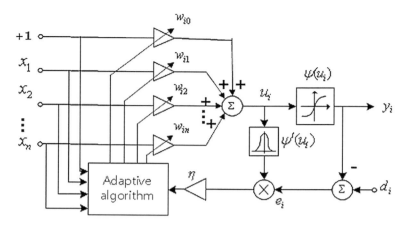

Fig. 1 Implementation of the single neuron as cognitive model

where μ is a positive learning parameter determining the speed of convergence to targeted input datum x_i.

In this study, the cognitive ability in the daily life set as μ, and it can be low or high according to age or experience, and can be different by people.

Pattern y with the greatest reliability to the kind of the entire pattern estimated as the outcome of cognition. The parameter τ is the immersion constant and the cognitive concentration model is defined as the Eq. (3). $y(t)$ means the concentration degree, and the bigger it is, the higher the concentration degree. But, it doesn't mean the relative value for ability of concentration. When a certain assignment is given, the maximum degree of concentration is defined as $c_{\max}$ the maximum immersion, and as of the point of time when an event takes place, the duration of concentration was defined as T. The input was used after measuring and processing EEG datum.

$$y(t) = C + (C - \mu)e^{-\tau t} \qquad (3)$$

where c is immersion degree.

4 Experiment on the Change Patterns of Concentration

An experiment to verify the usefulness of the proposed cognitive concentration model was conducted as follows: the prefrontal lobes, Fp1 and Fp2 have important functions for cognition, thinking, and creativity, which play central roles of the brain functions related to learning behaviors [12]. Thus, Fp1 and Fp2 areas were measuring points of which the signals were measured using Biopac MP150 and EEG 100C. To interpret the change patterns of concentration, hard, normal and

Interpretation and Modeling of Change Patterns

easy mathematical questions were issued and the brain wave was measured while performing them step by step [13].

The collected EEG was classified to several band related to attention. The bands of the brain wave with high correlation to concentration are known as Beta wave, SMR wave, and high Beta wave. For the collected time domain data, using moving window analysis with Window size 2[s] and Overlap Ratio 40[%], data in each section were converted to frequency domain data and characteristics were examined. Power spectrum is found when time domain data was translated to frequency domain data. Discrete Fourier Transform (DFT) is as following Eq. (4):

$$H(f_n) = \sum_{k=0}^{N-1} h_k e^{-j2\pi n/N} = H_n \tag{4}$$

This equation is organized by Inverse Fast Fourier Transform (IFFT) as Eq. (5) and square the absolute value on both sides of this equation and then take $\sum_{k=0}^{N-1}$ to sum to draw out Eq. (6).

$$h_k = \frac{1}{N} \sum_{n=0}^{N-1} H_n e^{-2\pi kn/N} \tag{5}$$

$$TotalPower \equiv \sum_{k=0}^{N-1} |h_k|^2 = \frac{1}{N} \sum_{n=0}^{N-1} |H_n|^2 \tag{6}$$

This means the total power value of the signal is the same in time domain or frequency domain, which is called Parseval theorem. One side power spectrum that satisfies this theorem is as Eq. (7).

$$P(f_0) = P(0) = \frac{1}{N^2} |H_0|^2 \tag{7}$$

Using this, the brain wave by each was found and when the subjects perform hard, normal and easy assignments, changes of the brain wave by time were compared.

As elements with a big difference during concentration, beta wave and theta wave were selected to use in the verification of the cognitive concentration model.

5 EEG Interpretation and Cognitive Concentration Model Verification

To verify the defined model using the brain wave, the change patterns of concentration were interpreted with the difference between the selected theta and beta waves by comparing each power spectrum of the brain wave. Figure 2a shows change of $y(t)$ by each difficulty. As the assignment got more difficult, the performing time and difficulty of the assignment got higher along with the value of $y(t)$. However, it decreased again as time passed. This can be interpreted that after a certain time, concentration fell. Figure 2b shows the value to find the basic brain

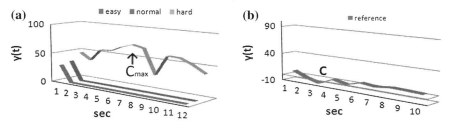

Fig. 2 a Concentration by difficulty; b general cognitive ability

wave before performing assignments, which was defined as the mean after 10 time of execution. It had the value of $|y(t)| \leq 10|$. Approximately it had a value in $25 \leq y(t) \leq 35$ for easy questions, $35 \leq y(t) \leq 45$ for normal ones and $30 \leq y(t) \leq 70$ for hard ones. This is consistent with the result that they could solve assignments with a lower difficulty without great attention, and the higher the difficulty was, the higher the attention should get. Duration of immersion T appears around 10 s. This differs depending on the degree of individual learning.

6 Conclusions

This study defined general cognitive ability, maximum attention, and duration of concentration. As a prior step to increase human cognition processing ability, we defined cognitive concentration model to interpret and modeling change pattern of concentration, the ability to keep attention to stimulus selected for a limited time. In addition, using hard, normal and easy questions in mathematical area, the brain wave experiment was conducted to prove this. It measured the brain wave, converted it to frequency domain, examined characteristics by time and verified the usefulness of the cognitive concentration model.

As a result of this paper, using human cognitive concentration model to enhance attention will be a foothold to system development that can get a better result in emotional stability and enhancement of academic record through enhancing attentiveness.

References

1. Crick F (1996) The astonishing hypothesis: the scientific search for the soul. Contemp Psychol 41(5):427–428
2. Robbins J, Wired for miracles (neurofeedback therapy). Psychology Today, May 1st
3. Anna W (1995) High performance mind. Tarcher Putnam, New York
4. Nak CL (1992) Correlates of EEG hemispheric integration. PhD Indiana University
5. Jung-Eun L, Bo-Hyeok S, Soon-Yong C (2012) Study on EEG feature extraction under LED color exposure to enhance the concentration. Adv Eng Forum 2–3:261–265

6. Sterman MB (1977) Sensorimotor EEG operant conditioning and experimental and clinical effects. Pavlov J Biol Sci 12(2):65–92
7. You-Me K (2001) The Journal of Elementary Education studies 8(1):1–32
8. Man-sang F (2007) Creating an intelligent brain. Jisiksanupsa, Seoul
9. Joo-yun C (1998) Educational applications of cognitive sciences discoveries about learning/memory. J Elem Educ 12(2):5–27
10. Itti L, Koch C, Niebur E (1998) Model of saliency-based visual attention for rapid scene analysis. IEEE Trans Pattern Anal Mach Intell 20(11):1254–1259
11. Posner M, Raichle M (1994) Images of mind. Scientific American Library, New York
12. Lawson AE (1997) The role of the prefrontal lobes in scientific reasoning. J Korea Assoc Sci Educ 17(4):525–540
13. Jung-Eun L, Un-ho J, Bo-hyeok S, Soon-yong C (2012) Measurements of color effects on the brain activity for eye-attention, ICROS

Design of Autonomic Nerve Measuring System Using Pulse Signal

Un-Ho Ji and Soon-Yong Chun

Abstract As for studies on autonomic nerve reactions, many researchers have published study results for long time, but in the standpoint of reproducibility and usefulness, no satisfactory outcomes have been obtained so far. As a new way to measure the change in autonomic nerve system, the method of using pulse signal the oriental medicine diagnosis measure is proposed. Specifically the autonomic nerve reaction measuring device to verify the usefulness of the proposed method is designed directly, and new method to process the measured pulse signals is proposed.

Keywords Autonomic nerve · Radial pulse · Pressure pulse wave · Fast fourier transform

1 Introduction

Methods to measure the change of human body autonomic nerve include electrocardiogram, heartbeat variability, skin resistance, electromyogram, electroencephlogram and electro-oculogram etc., and such methods predict the change of autonomic nerve by analyzing the utilization of sympathetic nerve and parasympathetic nerves [1–7]. Such methods can compare the utilization degrees of sympathetic nerve and parasympathetic nerve by identifying the characteristics like increase in the number of heartbeats and decrease in variance of heartbeats that occur due to activation of sympathetic nerve. Existing measurement methods

---

U.-H. Ji · S.-Y. Chun (✉)
Dongyang University, Yeongju-si, Gyeongbuk, Korea
e-mail: control@dyu.ac.kr

U.-H. Ji
e-mail: jiunho@hotmail.com

J. J. (Jong Hyuk) Park et al. (eds.), *Multimedia and Ubiquitous Engineering*, Lecture Notes in Electrical Engineering 240, DOI: 10.1007/978-94-007-6738-6_133, © Springer Science+Business Media Dordrecht(Outside the USA) 2013

are useful to analyze the utilized degree of synthetic nerve and parasympathetic nerve comparatively, and express the utilized degrees of synthetic nerve and parasympathetic nerve by particular disease or internal/external stimulations. Also, as for the comparison of activity degrees of synthetic nerve and parasympathetic nerve which are determined through repetitive measurements, the activity degrees measured before and after measurement were found to be constant at two levels "increase" and "decrease". However, it is judged that the methods require complementation for use as measurement methods to provide reliable data for doctors who use instruments clinically in quantitative aspect. That is because the bio-electrical characteristics of human body are liable to change, but are kept constant by the activities of autonomic nerve system that works to keep the constancy of human body [8]. Therefore, in order to diagnose the medical condition of disease by measuring the change in autonomic nerve system, there needs be a method to specifically judge the long term change in the condition of human body consisting the inside based on the change in the reaction of autonomic nerve system rather than just the level of activity like in existing method. It is judged that to that end, measuring and analyzing the change in pulse that occurs as change in human body autonomic nerve would be useful. Pulse is characteristic that it changes by the speed and flow rate of blood flow the characteristics of blood vessel linked to all elements of human body, and the change in the characteristics of blood vessel occurs as the blood circulates human body elements. The method that can be utilized the most effectively with the use of the correlation of the characteristics of pulse change and human body is pulse method a method effectively used in oriental medicine. Pulse signals which are considered as important measure for diagnosis of lesions in oriental medicine, signals that include the characteristics of blood vessel flow rate and pressure that changes according to elements conditions as the blood circulates human body, are judged as capable to identify the characteristics of various changes in human body occurring by the change in autonomic nerves.

In this study, method to judge the change in human body elements based on the change in human body autonomic nerve reactions by measuring pulse signals was proposed, and after designing of system for measurement of pulse signal, pulse signals measured through experiment were analyzed.

2 Measuring Autonomic Nerve Reaction

2.1 Actions to Measure Autonomic Nerve Reactions

Methods used the most as device to measure autonomic nerve reaction include the method of measuring the variance of heartbeat [1–3]. In general, HRV can be calculated from the change in ECG signal R–R interval, and the power spectrum can be divided into three parts depending on frequency areas: [LF(0.01–0.08 Hz),

MF(0.08–0.15 Hz), and HF(0.15–0.5 Hz)]. While LF reflects the activity of sympathetic nerve system mostly and those of parasympathetic nerve system a little, HF reflects the activities of parasympathetic nerve system exclusively, so LF/HF has been used as indicator to measure the balance of sympathetic nerve system and parasympathetic nerve system. It is known that MF represents mixed activities of sympathetic nerve system and parasympathetic nerve system, but it represents the activities of parasympathetic nerve system more. Bio electrical testing method is a method that recognizes human body as a circuit with electrical characteristics and analyzes the physiological/pathological characteristics by detecting skin resistance value and conductivity when static current or static voltage is flown based on the fact that sin electrical activity is closely related to sweat gland and autonomic nerve activity [9–13].

Such measurement of change in autonomic nerve system have been subjects in many studies by being recognizes as important device to predict diseases that can occur in human body and diagnose health condition, but there has been found no device to measure, predict and analyze specific changes in health condition and judgment of particular disease symptom. To judge change in health condition and judge particular disease symptom after analyzing the result of measurement of change in autonomic nerve system, specific actions for measurement and analysis are judged required, and to that end, in this study, the method of measuring and analyzing pulse signal used as important diagnosis action in oriental medicine is proposed.

2.2 Designing of Autonomic Nerve Reaction Measurement System Using Pulse Signal

Figure 1a shows the overall composition of measurement system designed and produced as action to measure pulse signal in this study. Figure 1b shows the pulse signal measurement system for measurement of autonomic nerve reaction in the method proposed in this study.

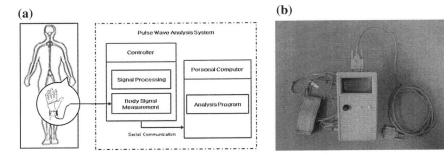

Fig. 1 a Block diagram of measurement system. b Pulse measurement system

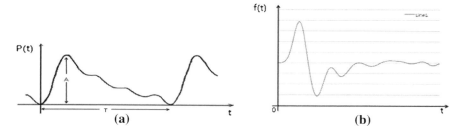

Fig. 2 a Human body pulse. **b** Differential pulse

Pulse signals, signals of waves measured in hipbone artery of wrist, measure the capacity pulse of hipbone artery utilizing piezoelectric element. Pulse is wave that occurs by the change in the blood vessel amount of blood and speed of blood flow by constriction and relaxation of heart, and one cycle of pulse occurs through a series of process like beginning of constriction of left ventricle, culmination of construction, expansion of aorta wall, reduction of blood vessel and elastic wave of heart valve and heart muscle etc. Also, pulse signals include the characteristics that they pass capillary vessels of body organs. Figure 2a shows the flow of pulse signal.

Here, 'T' indicates the cycle of pulse and 'A' the maximum point of constriction. In particular, the maximum point of constriction in pulse, which constitutes the largest interval and an important point in detecting change in heartbeat, can be used in analyzing the activity of autonomic nerve system. Figure 2b shows the differential pulse put out through the system designed in this study. Differential pulse can be used to identify the characteristics of change in blood vessel capacity from beginning of heart constriction to culmination of constriction, and effectively represents the characteristics of change of volume occurring due to the expansion of aorta wall, reduction of blood efflux and resilience of heart valve and heart muscle after reduction of ventricles constriction. Analysis of such elements is used as indicator for evaluation of blood vessel age [14]. The procedure of processing the pulse signals in the system designed in this study can be summed up as in Fig. 3.

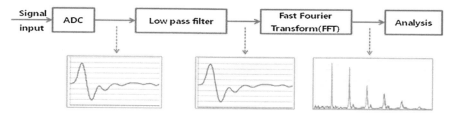

Fig. 3 The procedure of processing pulses

Design of Autonomic Nerve Measuring System 1085

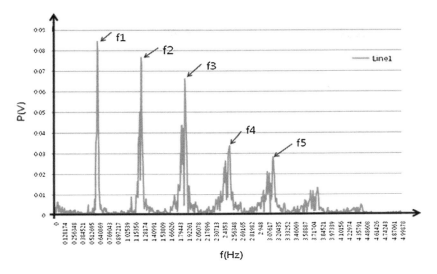

Fig. 4 The FFT of differential pulse

To remove the noise elements that can be included in the procedure of processing signals in measurement system, 2nd low bandwidth filter was used, and to create indicator for autonomic nerve reaction, FFT results of differential pulse were used. Figure 4 shows the data obtained through FFT transformation, and as shown in Fig. 4, the power of particular frequency bandwidth appear high in FFT

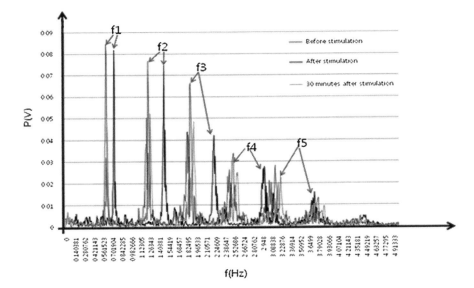

Fig. 5 The characteristics of pulse signals before and after outside stimulation

Table 1 The characteristics indicator of change in pulse signal

	f1	$\pm\varepsilon1$	F2	$\pm\varepsilon2$	F3	$\pm\varepsilon3$	F4	$\pm\varepsilon4$	F5	$\pm\varepsilon5$
Before stimulation	0.6225	0.0305	1.2512	0.0732	1.8859	0.1037	2.5268	0.1037	3.1494	0.1647
After stimulation	0.7446	0.0305	1.4892	0.0366	2.2399	0.0427	2.9907	0.0610	3.7353	0.0549
30 min after stimulation	0.6408	0.0366	1.2878	0.0610	1.9348	0.1098	2.5756	0.1159	3.2348	0.1342

Design of Autonomic Nerve Measuring System 1087

Fig. 6 Central frequency locations of f1 through f5 before and after stimulation and after 30 min rest

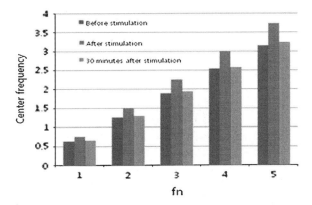

Fig. 7 Standard deviations before and after stimulation and after 30 min rest

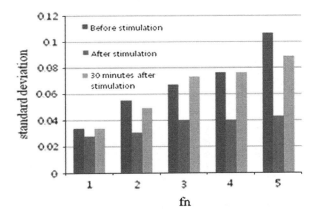

signal. The central frequencies in this part are defined respectively as f1, f2, f3, f4 and f5, and the change width of the central frequencies occurring by repetitive measurement as standard deviation $\pm \varepsilon$.

2.3 *Measurement Experiment of Autonomic Nerve Reaction*

To check the potential for measurement of autonomic nerve reactions through the pulse signal measurement system designed in this study, experiment was conducted. As for experiment method, the characteristics appearing usually and the characteristics of change in autonomic nerves for particular outside stimulation were analyzed comparatively.

Measurement of pulse signal: In the experiment, as for outside stimulation element to raise autonomic nerve change, the experiment participant was set to run about 400 m distance with full force, and with the characteristics of the usual pulse signals of experiment participant measured 10 times repeatedly before experiment, the averages were taken and compared with experiment data. Figure 5 shows the comparison results.

Results of pulse signal analysis: Table 1 shows the results of analyzing the characteristics of pulse before and after outside stimulation from the results of Fig. 5.

Figure 6 shows the comparison of the central frequency location of fn that occurs before and after stimulation and 30 min rest in graph. It is found that the central frequencies of f1 through f5 increase before and after stimulation, and after 30 min rest, the central frequency moves toward usual frequency range again.

Figure 7 shows that once stimulation is applied due to change in standard deviation occurring before and after stimulation and 30 min rest, standard deviation decreases.

3 Conclusion

In this study, a paper of experiment study on the device to measure autonomic nerve system reaction, experiment device to measure autonomic nerve reaction was designed and produced. Through the experiment of using the produced measurement system, the usefulness of autonomic nerve system reaction measurement was found through experiment.

Unlike in existing method that indicated the activity degree of sympathetic nerve and parasympathetic nerve generally occurring depending on the change in autonomic nerve system, in this study a device that measures and analyzes the change in autonomic nerve system was proposed with the use of pulse signals used as important diagnosis measure in oriental medicine.

As for analysis of pulse signal, the data measured from sensor are amplified and filtered through hardware signal processing part, and the processed results were made digital through ADC. The digitalized values were analyzed on frequency components through FFT, and $f1$–$f5$, $\pm\varepsilon1$–$\pm\varepsilon5$ the quantitative evaluation indicators were specified from the analysis results.

For experiment to measure autonomic nerve reaction, change in human body was induced through exercise as outside stimulation element, and through analysis of pulse signals measured here, the changes were measured and analyzed by being divided into before stimulation, after stimulation and after rest.

As a result of analysis, as shown in Figs. 6 and 7, evaluation indicator changed due to the change in autonomic nerve system.

Based on this study results, future studies must be done on the characteristics of autonomic nerve system that reacts according the inside/outside stimulation or condition change of human body for development of medical advances.

References

1. Lee Y-H (2004) Designing of oriental medicine diagnosis system by bio-electrical reaction. Korea Ocean IT Soc J 420–429 (Book 8, 2nd edn)
2. Park C-W (2004) A study on the effects of far-infrared ray heat on human autonomic nerve function. Korea Med Eng Soc J 25(6): 623–628 (87th edn)
3. Lee J-H (2000) Designing of the time frequency analysis system of heartbeat change signal for evaluation of autonomic nerve system operation. Yonsei University, Seoul
4. Choi E-M (2006) A study on the skin resistance variance of ovarian insufficiency using bio-electrical autonomic reaction measurer. Korea Orient Med Gynecol Soc J 19(3): 247–256
5. Oh D-H (2003) Mental physiology of post-trauma stress disability. Ment Health Study 22:24–37
6. Han T-R (1994) The effects of heat sympathetic reaction on nerve conduction and autonomic nerve function. Korea Rehabil Med J 18(1): 28–34
7. McCraty R, Atkinson M, Tiller WA, Rein G, Watkins AD (1995) The Effects of Emotions on Short-Term Power Spectrum Analysis of Heartbeat Variability. Am J Cardiol 76: 1089–1093
8. Lee B-C (1995) Analysis of the characteristics of bio non-linear dynamic system using chaos theory. Yonsei University, Seoul
9. Park Y-J (2001) A study on skin resistance variance. Korea Orient Med Diagn J 5(2): 365–376
10. Boucsein W (1992) Electrodermal Activity. Plenum press, New York, pp 1–42
11. Nam D-H (2001) The effects of Korean adults males and females with healthy deep breath capability on skin electrical autonomic reaction. Korea Orient Med Diagn J 5(1): 139–152
12. Park C-W (1990) Autonomic nerve pharmacology. Seoul National University Press, Seoul, pp 63–80
13. Chun S-Y, Ji U-H (2004) Measurements of the current change on acupuncture spots at the meal time. IEEE, CBMS 2004, p 59
14. Takazawa K, Tanaka N, Fujita M, Matsuoka O, Saiki T, Aikawa M, Tamura S, Ibukiyama C (1998) Assessment of vasoactive agents and vascular aging by the second dericative of photoplethysmogram waveform. Am Heart Assoc Hypertens 32(2) 365–370

Semiconductor Monitoring System for Etching Process

Sang-Chul Kim

Abstract In this paper, we developed the semiconductor monitoring system for the etching process. Process monitoring techniques has an important role to give an equivalent quality and productivity to produce semiconductor. The proposed monitoring system is mainly focused on the dry etching process using plasma and it provides the detailed observation, analysis and feedback to managers.

Keywords Etching · Plasma · Optical emission spectroscopy (OES) · End-point detecting · Real-time monitoring · Dynamic linked library (DLL) interface

1 Introduction

The semiconductor industry began to develop in earnest from the mid-1950s, interlinking with space development and weaponeering. At the present time it has entered the microelectronic age, with the need for subminiature, ultra-light, high-reliability electronic components. Here at, many countries are keenly competing with each other for it [1]. In particular, there has been an increasing interest in endpoint sensing to prevent defects in the manufacturing process. A defect in the semiconductor etching process causes great economic loss. The real-time observation of the semiconductor etching process, the application of systematic process control and the maximization of automatic process control (APC) efficiency are required for the prevention of defects. In this regard, this study is on a monitoring system to implement the foregoing. Section 2 is to deal with involved technologies and usable platforms. Section 3 is to explain the system structure, and Sect. 4 is to show the test results. Lastly, chapter Sect. 5 is to make a conclusion and make mention of future plans.

S.-C. Kim (✉)
School of Computer Science, Kookmin University, Seoul, Korea
e-mail: sckim7@kookmin.ac.kr

J. J. (Jong Hyuk) Park et al. (eds.), *Multimedia and Ubiquitous Engineering*, Lecture Notes in Electrical Engineering 240, DOI: 10.1007/978-94-007-6738-6_134, © Springer Science+Business Media Dordrecht(Outside the USA) 2013

2 Related Works

A semiconductor etching process is to etch the wafer, made with processed silicon, with a circuit, which is divided into wet etching and dry etching. In the case of wet etching; a chemical solution is applied to the membrane of the wafer. It was widely used early in the semiconductor industry. However, it causes undercut as shown in Fig. 1a because chemical reactions occur vertically as well as horizontally.

The biggest problem is that it is difficult to form the vertical core layer because the pattern line width becomes narrower. That is why plasma-based dry etching has been widely used in recent times [1]. The dry etching is based on a chemical reaction between elements in plasma and the surface of the wafer, and the reaction is accelerated as active species in plasma crash against the surface. Since the process can be controlled in atomic unit, it is possible to etch the wafer in any shape as shown as in Fig. 1b [2].

The existing etching process, putting electrodes into plasma, mostly used the Langmuir probe to measure electron density, electron temperature and plasma potential, a interferometer to make material waves into single waves and thus to observe the interference fringes of neutral atoms and a mass spectrometer to gauge the mass of the ion made from specimen particles in a vacuum.

The optical emission spectroscopy (OES) sensor as shown in Fig. 2 makes a real-time analysis of the spectrum of gas caused by the plasma reaction in a semiconductor etching process, and the processed data to interface devices [2]. The OES sensor has the merit of being installed outside process equipment in which the plasma reaction occurs. Thus, it is not affected by internal variables such as voltage, frequency, electromagnetic field and gas pressure.

3 Proposed Monitoring System

The system, shown in Fig. 3, is largely divided into three parts. The OES sensor that transmits data after observing and analyzing the semiconductor-etching process, Dynamic-linked library (DLL) Interface that processes data, transmitted from the sensor, and transmits them to the monitoring system, and the monitoring system that reprocesses the transmitted data in accordance with the logic made by the administrator and checks whether the process proceeds as planned.

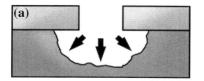 

Fig. 1 a Model of wet-etching. b Model of dry-etching

Semiconductor Monitoring System for Etching Process

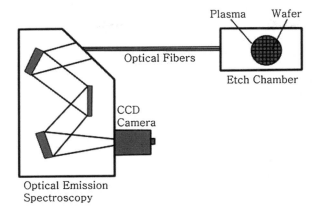

Fig. 2 Optical emission spectroscopy (OES) sensor

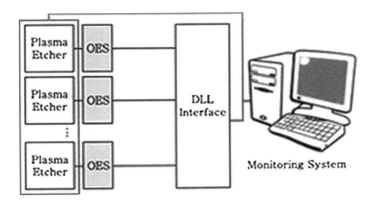

Fig. 3 System architecture

The OES sensor is installed outside plasma etch equipment. The inside can be observed through an optical fiber cable connected from transparent lens to an OES sensor. The observed image falls on the charge-coupled device (CCD) array, which is transmitted from USB interface to DLL Interface [2]. In this study, the monitoring system is equipped with the OES sensor that can detect 200–900 nm wavelength, which is equivalent to ultraviolet, visible and infrared lights. The OES sensor divides the 200–900 nm wavelength into 3648 sections and measures the strengths. Specifically, the CCD Array is formed with a length of 3648.

The OES sensor is controlled by DLL Interface. The OES sensor provides DLL Interface with information in the pre-work stage. The sensor information is composed of sensor ID, reserved name, wavelength and wavelength table. In case many sensors are interlocked with each other, respective sensors can be distinguished through the transmitted information. Figure 4 defines the structure that saves sensor information in DLL Interface.

```
typedef struct OesUnitInformation {
    s32 id;
    char model_name[256];
    char serial[256];
    s32 wave_length;
    double wave_length_table[4096];
} OesUnitInformation;
```

Fig. 4 Structure of OES sensor information

The main class of monitoring system is App class. It can approach to all the objects in the system, but at the same time, allows all of them access to itself, which can be implemented as the points of objects in the system are listed in App class. Figure 5 shows the structures of objects in the monitoring system. In the monitoring system, all the information related to OES sensors is organized in the object of the sensor module (more hereof later).

The user interface of monitoring system is largely divided into three parts, *configuration dialog*, *chart view* and *recipe view*. *Configuration dialog* is implemented. *Chart view* is composed of a full spectrum chart and a time trend chart as shown in Fig. 6. The full spectrum chart includes the data source list that tells of the sources of respective wavelengths. The time trend chart includes the equation list premade by the administrator. If an item is chosen in the data source list, the corresponding wavelength range is highlighted in the full spectrum chart. In case an item is selected in the equation list, it is shown in the time trend chart in chronological order after data was processed within the range predetermined by the administrator. *Recipe view* is to show the actual state of process; to be specific; it shows where the process is on the scenario made by the administrator.

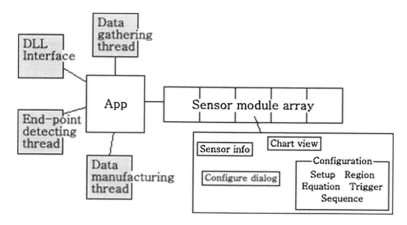

Fig. 5 Map of structured objects

Semiconductor Monitoring System for Etching Process

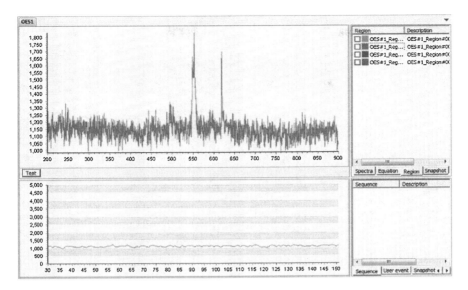

Fig. 6 Chart view

The ultimate purpose of the semiconductor etching process monitoring system is to detect the endpoint. The endpoint is detected as per the scenario made by the administrator. The administrator can get various forms of data that were usably processed, by using the setups of *configuration*. The scenario, based on the data, is a recipe for the corresponding process. A scenario is made by the administrator in *sequence* in *configuration*, and the logic that the scenario is applied to the monitoring system.

The monitoring system judges whether the manufactured data meet what was stipulated in the present sequence item or the described technique, whenever they are updated. The index is *chart data* that reflects the present process. With the completion of endpoint detection (EPD), the EPD signal is transmitted to process equipment to stop the process.

4 Experimental Results

The OES hardware, used in the present system, makes it possible to transmit data at intervals of at least 10 ms. But the interface, implemented in the monitoring system, was so designed that it could receive data at intervals of at least 100 ms, and thus memory can be shared without conflicts when data are processed in the system. Accordingly, the test should be conducted at a sampling per 10 counts, as mentioned above, and at intervals of 100 ms, the maximum performance of interface. Since it is realistically difficult to conduct the test in a real process, it is

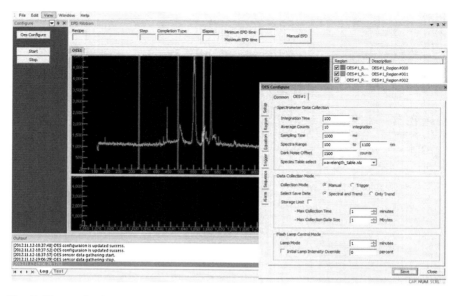

Fig. 7 The monitoring system after test

necessary to check whether data are normally gathered, whether the processing results, produced by configuration, are accurately reflected in *chart view* and how long the test is on the log. Figure 7 shows that the data have been successfully gathered and processed as per given time and environment.

The test is focused on checking how often data samplings are conducted and whether it meets the expected value. The source code is to put on record when data begin to be gathered, when data gathering is finished, and how many data samplings are conducted while data are gathering, of which records are saved in the log. In result, data samplings were conducted 1712 times for 28.5 min, which means that data samplings were conducted 60 times per min. It met the data sampling cycle determined in advance. The results showed that the data processing method could be controlled in real time as to the processing environment, and that data reliability could be heightened through normalization even in the shortest process cycle. As a result, it was proved to be highly effective to heighten control efficiency in different processing environments.

5 Conclusion

The monitoring system, developed in this study, has superiority over others. First, the optical sensor can accurately observe the inside of process equipment from outside it. Second, logic could be made once to N times by use of *equation* and thus data could be processed with almost infinite numbers. Lastly, the automated check system prevents damage, taking primary measures in unforeseen circumstances. The semiconductor etching process monitoring system observes the visible state of process, analyzing sensor data, and at the same time conducts real-time logic that diagnoses the present process, and thus detects the endpoint. In this study, the monitoring system was focused on the semiconductor etching process, but in reality it is expected to be applied to more fields.

Acknowledgments This research was supported by Business for Cooperative R&D between Industry, Academy, and Research Institute funded Korea Small and Medium Business Administration in 2011 (Grants No. 0045590) and the Global Leading Technology Program of the Office of Strategic R&D Planning (OSP) funded by the Ministry of Knowledge Economy, Republic of Korea (grant number: 10042421).

References

1. Kang Ho Y An Introduction of Etch Process, hynix semiconductor
2. Roawen C OES-based sensing for plasma processing in IC manufacturing. Doctor of Philosophy in Engineering, Electrical Engineering and Computer Sciences, University of California at Berkeley

Enhancing the Robustness of Fault Isolation Estimator for Fault Diagnosis in Robotic Systems

Ngoc-Bach Hoang and Hee-Jun Kang

Abstract Fault diagnosis and fault tolerant control are increasingly importance in robotic systems. A number of researchers have proposed the generalized observer scheme for fault isolation when a fault happened. One of the key issues in this scheme is based on the sensitive of the residual with the corresponding adaptive threshold. In this paper, we present a new method to derive the adaptive threshold in order to enhance the robustness of fault isolation estimator and reduce the fault isolation time. Mathematical proof and computer simulation are performed to show the effectiveness of the proposed method.

Keywords Fault detection · Fault isolation · Nonlinear observer · Robotics

1 Introduction

Nowadays, robotic systems are widely used in complex engineering applications which demand very high performance, productivity and safe operation. Robotic fault can potentially result both in the loss of productivity and in unsafe operation of the system. Moreover, difficult and often dangerous environments limit the ability of humans to perform any supervisory and corrective tasks. Hence, automated fault detection, isolation and accommodation play a key role in the operation of modern robotic systems.

A number of researchers have worked on the problem of designing automated fault diagnosis schemes for robotic systems using analytical redundancy methods.

N.-B. Hoang
Graduate School of Electrical Engineering, University of Ulsan, Ulsan 680-749, Korea
e-mail: hoangngocbach@gmail.com

H.-J. Kang (✉)
School of Electrical Engineering, University of Ulsan, Ulsan 680-749, Korea
e-mail: hjkang@ulsan.ac.kr

J. J. (Jong Hyuk) Park et al. (eds.), *Multimedia and Ubiquitous Engineering*,
Lecture Notes in Electrical Engineering 240, DOI: 10.1007/978-94-007-6738-6_135,
© Springer Science+Business Media Dordrecht(Outside the USA) 2013

A neural network based learning methodology is described in [1]. In [5, 6], a robust fault diagnosis scheme for abrupt and incipient faults in nonlinear uncertain dynamic systems is proposed. The extension of [5] to the case of robotic systems is presented in [7]. The architecture in [5, 6] consists of a bank of nonlinear adaptive estimators, one of which is used for the detection and the approximation of a fault, whereas the rest are used for online fault isolation. First, a nonlinear observer is designed based on the robotic model. The fault detection is carried out by comparing the observer states with their signatures. Then, multiple state observers are constructed based on possible fault function set. The fault isolation is implemented by checking each residual generated by observer state estimation and the corresponding threshold. One of the most importance criteria in fault diagnosis is the fault isolation time, which refers to the time taken by the fault isolation scheme to identify a fault that has occurred. In order to increase the capability of successful isolation and decrease the fault isolation time, we propose a more robust threshold in this paper. Both mathematical prove and simulation results are presented to show the effectiveness of our proposed threshold.

This paper is organized as follows: In Sect. 2, the robot dynamics in the presence of faults is presented. In Sect. 3, the design of fault detection and isolation scheme, including the derivation of new robust adaptive threshold, is described. In Sect. 4", the effectiveness of the proposed robust threshold is demonstrated via computer simulations. Section 5 has some concluding remarks.

2 Fault in Robot Manipulator

In the presence of a fault, the robot dynamics can be represented by [7]

$$\ddot{q} = M^{-1}(q)[\tau - V_m(q, \dot{q}) - F(\dot{q}) - G(q) - \tau_d] + \beta(t - T)\phi(q, \dot{q}, \tau) \quad (1)$$

The fault profile $\beta(t - T)$ is a diagonal matrix of the form

$$\beta(t - T) = diag(\beta_1(t - T), \beta_2(t - T), \ldots, \beta_n(t - T)) \quad (2)$$

Each time profile can be represented by

$$\beta_i(t - T) = \begin{cases} 0 & t < T \\ 1 - e^{-\alpha_i(t-T)} & t \geq T \end{cases} \quad (3)$$

where $\alpha_i(i = 1, 2\ldots, n)$ are unknown constants that represent the development of fault.

The following assumptions will be used throughout this paper:

(1) The system states remain bounded after the occurrence of the fault: $q(t), \dot{q}(t) \in L_\infty$.

Enhancing the Robustness of Fault Isolation Estimator 1101

(2) The friction satisfies $F(\dot{q}) = F_v\dot{q} + F_d$ with $\|F_d\| \le k_B\|\dot{q}\| + k_F$, $(k_B, k_F > 0)$, where F_v and F_d are the coefficient matrix of viscous friction and the dynamic friction term, respectively.

(3) The load disturbance is bounded by $\|\tau_d\| \le \tau_B$, where τ_B is a known constant.

3 Fault Diagnosis Scheme

3.1 Fault Detection

Let $x = \dot{q}^T(t)$. The dynamic model of Eq. (1) can be rewritten as

$$\dot{x} = M^{-1}(q)[\tau - V_m(q, \dot{q}) - F_v\dot{q} - F_d - G(q) - \tau_d] + \beta(t - T)\phi(q, \dot{q}, \tau) \quad (4)$$

The following estimated model is considered

$$\dot{\hat{x}} = M^{-1}(q)[\tau - V_m(q, \dot{q}) - F_v\dot{q} - G(q)] + \Lambda(\hat{x} - x) \quad (5)$$

where $\hat{x} \in R^n$, $\Lambda = diag(-\lambda_1, -\lambda_2, .., -\lambda_n)$ are the estimation vector of x and a stable matrix, respectively. From Eqs. (4) and (5), we can derive the error dynamic equation

$$\dot{\varepsilon} = \Lambda(\hat{x} - x) - M^{-1}(F_d + \tau_d) + \beta(t - T)\psi(q, \dot{q}, \tau) \quad (6)$$

By taking the norm of Eq. (6), we obtain the upper bound of $\|\varepsilon\|$

$$\|\varepsilon\| \le \alpha e^{-\upsilon t}\|\varepsilon_0(0)\| + \int_0^t \alpha e^{-\upsilon(t-\xi)}(k_B\|\dot{q}\| + k_F + \tau_B)d\xi = {}^D\varepsilon \quad (7)$$

Fault detection decision scheme: The decision about the occurrence of the fault is made when the residuals $\|\varepsilon\|$ exceeds its corresponding threshold ${}^D\varepsilon$ [7].

3.2 Fault Isolation with Robust Threshold Derivation

After a fault is detected, the following isolation estimators are activated:

$$\dot{\hat{x}}^s = \Lambda^s(\hat{x}^s - x) + M^{-1}(q)[\tau - V_m(q, \dot{q}) - F_v\dot{q} - G(q)] + \hat{\phi}^s(q, \dot{q}, \tau, \hat{\theta}^s) \quad (8)$$

$$\hat{\phi}^s(q, \dot{q}, \tau) = [(\hat{\theta}_1^s)^T g_1^s(q, \dot{q}, \tau), \ldots, (\hat{\theta}_n^s)^T g_n^s(q, \dot{q}, \tau)]^T \quad (9)$$

where $\hat{x}^s$ and $\hat{\theta}_i^s$ are the estimated states and parameters x and θ_i^s, respectively. $\Lambda_s = diag(-\lambda_1, .., -\lambda_n)$ $(\lambda_1, .., \lambda_n \ge 0)$ is a stable matrix. The online updating law for $\hat{\theta}_i^s$ is $\dot{\hat{\theta}}_i^s = P_{\Theta_i^s}\{\Gamma_i^s g_i^s(q,\dot{q},\tau)\varepsilon_i^s\}$ where Γ_i^s, $g_i^s(q,\dot{q},\tau)$, $\varepsilon_i^s = (x_i^s - \hat{x}_i^s)$ $(i = 1,..,n)$ are the positive definite learning rate, the corresponding smooth vector field, the error state, respectively.

From Eq. (8), the error dynamics is given by

$$\dot{\varepsilon} = \Lambda^s \varepsilon - M^{-1}(F_d + \tau_d) + \beta(t - T)\phi^s(q,\dot{q},\tau) - \hat{\phi}^s(q,\dot{q},\tau,\hat{\theta}^s) \qquad (10)$$

Thus, each element of state estimation error is given by:

$$\dot{\varepsilon}_i^s = -\lambda_i \varepsilon_i - \sum_{j=1}^{n} m v_{ij}(f_{dj} + \tau_{dj}) + (1 - e^{-\alpha_i(t-T)})(\theta_i^s)^T g_i^s(q,\dot{q},\tau) - (\hat{\theta}_i^s)^T g_i^s(q,\dot{q},\tau)$$

$$\qquad (11)$$

$$\varepsilon_i^s = \varepsilon_i^s(T_d)\exp(-\lambda_i(t - T_d)) + \int_{T_d}^{t} \exp(-\lambda_i(\xi - T_d)) \sum_{j=1}^{n} m v_{ij}(f_{dj} + \tau_{dj})d\xi$$

$$+ \int_{T_d}^{t} \exp(-\lambda_1(\xi - T_d))((1 - e^{-\alpha_i(\xi-T)})(\theta_i^s)^T g_i^s(q,\dot{q},\tau) - (\hat{\theta}_i^s)^T g_i^s(q,\dot{q},\tau))d\xi$$

$$\qquad (12)$$

In this paper, a new threshold functions are derived as below:
The following function is defined to use later for deriving the threshold as:

$$^1h_i = \begin{cases} g_i^s(q,\dot{q},\tau) & (g_i^s(q,\dot{q},\tau) \geq 0) \\ 0 & (g_i^s(q,\dot{q},\tau) < 0) \end{cases}; {}^2h_i = \begin{cases} 0 & (g_i^s(q,\dot{q},\tau) \geq 0) \\ g_i^s(q,\dot{q},\tau) & (g_i^s(q,\dot{q},\tau) < 0) \end{cases} \qquad (13)$$

Because $[\theta_i^s]$ is assumed to belong to a known compact set $\Theta_i^s \subset R^{q_i^s}$, there exist two values ${}^m\theta_i, {}^M\theta_i$ the minimum and maximum of $\Theta_i^s \subset R^{q_i^s}$. So we get the inequality:

$$^m\theta_i \leq \theta_i^s \leq {}^M\theta_i \leq 0 \quad or \quad 0 \leq {}^m\theta_i \leq \theta_i^s \leq {}^M\theta_i \qquad (14)$$

Moreover, we assume that there exists a lower bound of fault evolution rate which satisfies $\alpha_i \geq \bar{\alpha}_i$, so that the following inequality can be established

$$(1 - e^{-\bar{\alpha}_i(t-T_d)}) \leq (1 - e^{-\alpha_i(t-T_d)}) \leq 1 \qquad (15)$$

If $0 \leq {}^m\theta_i \leq {}^M\theta_i$, from Eq. (14) and (15), we have

$$m_i = (1 - e^{-\bar{\alpha}_i(t-T_d)})^m\theta_i \leq (1 - e^{-\alpha_i(t-T_d)})\theta_i^s \leq {}^M\theta_i = M_i \qquad (16)$$

Or if ${}^m\theta_i \leq {}^M\theta_i \leq 0$ then

$$m_i = {}^m\theta_i \leq (1 - e^{-\alpha_i(t-T_d)})\theta_i^s \leq (1 - e^{-\bar{\alpha}_i(t-T_d)})^M\theta_i = M_i \qquad (17)$$

Now, let's combine Eqs. (13) (16) and (17) and put them in the matrix form

$$(m_i)^{T.1}h_i + (M_i)^{T.2}h_i \le (1 - e^{-\alpha_i(t-T_d)})(\theta_i^s)^T g_i^s \le (M_i)^{T.1}h_i + (m_i)^{T.2}h_i \quad (18)$$

Thus, the following threshold (upper bound and lower bound) can be chosen for isolation decision

$$^U\varepsilon_i = \exp(-\lambda_i(t - T_d))e_i(T_d) + \int_{T_d}^t \exp(-\lambda_i(\xi - T_d)) \sum_{j=1}^n |mv_{ij}| (f_{Bj}|\dot{q}_j| + f_{Fj} + \tau_{Bj})d\xi$$

$$+ \int_{T_d}^t \exp(-\lambda_i(\xi - T_d))((M_i)^{T.1}h_i + (m_i)^{T.2}h_i - (\hat{\theta}_i^s)^T g_i^s(q, \dot{q}, \tau))d\xi$$

$$(19)$$

$$^L\varepsilon_i = \exp(-\lambda_i(t - T_d))e_i(T_d) - \int_{T_d}^t \exp(-\lambda_i(\xi - T_d)) \sum_{j=1}^n |mv_{ij}| (f_{Bj}|\dot{q}_j| + f_{Fj} + \tau_{Bj})d\xi$$

$$+ \int_{T_d}^t \exp(-\lambda_i(\xi - T_d))((m_i^T).^1h_i + (M_i)^{T.2}h_i - (\hat{\theta}_i^s)^T g_i^s(q, \dot{q}, \tau))d\xi$$

$$(20)$$

The threshold given by Eqs. (19) and (20) can be implemented with a linear filter with the transfer function $1/(s + \lambda_i)$ and with the input

$$^UI_i = -(\hat{\theta}_i^s)^T g_i^s(q, \dot{q}, \tau) + ((M_i)^{T.1}h_i + (m_i)^{T.2}h_i) + \sum_{j=1}^n |mv_{ij}| (f_{Bj}|\dot{q}_j| + f_{Fj} + \tau_{Bj})$$

$$(21)$$

$$^LI_i = -(\hat{\theta}_i^s)^T g_i^s(q, \dot{q}, \tau) + ((m_i)^{T.1}h_i + (M_i)^{T.2}h_i) - \sum_{j=1}^n |mv_{ij}| (f_{Bj}|\dot{q}_j| + f_{Fj} + \tau_{Bj})$$

$$(22)$$

4 Simulation Results

A two-link planar robotic system now is used to illustrate for the performance of our proposed threshold. We consider 2 types of possible faults being given by

$$F = \left\{ \begin{bmatrix} 10 \sin(3q_1).\dot{q}_2 \\ 15q_1 \end{bmatrix}, \begin{bmatrix} 20 \sin(3q_1).\dot{q}_2 \\ 20 \sin(3q_1).\dot{q}_1 \end{bmatrix} \right\}$$

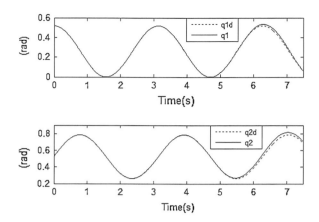

Fig. 1 Position normal control with fault occurrence

The fault #1 is assumed to be triggered at T = 5(s). Figure 1 shows the position control when fault #1 occurs at $T = 5(s)$ without a fault accommodation scheme. By using fault detection scheme, a fault is detected at $T_d = 5.1(s)$.

After the fault is detected, two observers are activated. The outputs of two observers are showed in Fig. 2. Figure 2a and b belong to observer 1, Fig. 2c and d belong to observer 2. In Fig. 2, old upper bound and old lower bound stand for the derived threshold in [5–7], new upper bound and new lower bound stand for our proposed threshold. The fault #1 is successful isolated at $T_{isol} = 5.67(s)$. In order to show the effective of our derived threshold, we enlarge Fig. 2c and d in Fig. 2e,

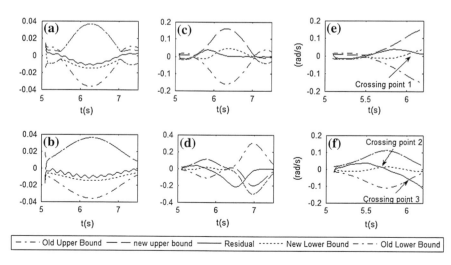

Fig. 2 Fault isolation. **a, b** Output of fault 1 observer. **c, d** Output of fault 2 observer. **e, f** Inlarge of **c, d**

f. In Fig. 2e, f if our threshold is applied, the fault is successfully isolated at $T_2 = 5.67(s)$ by checking crossing point 2, earlier than $T_3 = 6.05(s)$ by checking crossing point 3. This results show that our derived thresholds are better than which are proposed in [5–7].

5 Conclusions

In this paper, we present a new method to derive the adaptive threshold for fault isolation in robotic systems. With our derived threshold, the capability of successful isolation is improved and the fault isolation time is reduced. Hence, the fault tolerant controller is activated sooner to reduce the effect of fault to the robotic systems. The detailed simulation for two-link manipulator has been given to show the effectiveness of the proposed method.

Acknowledgments This work was supported by the Ministry of Knowledge Economy under the Human Resources Development Program for Convergence Robot specialists and under the Robot Industry Core Technology Project.

References

1. Vemuri AT, Polycarpou MM (1997) Neural-network-based robust fault diagnosis in robotic systems. IEEE Trans Neural Netw 8:1410–1420
2. Wang H, Daley S (1996) Actuator fault diagnosis: an adaptive observer-based technique. IEEE Trans Autom Control 41:1073–1078
3. Gang T, Xiaoli M, Joshi SM (2000) Adaptive state feedback control of systems with actuator failures. In: Proceedings of the American control conference, vol 4, pp 2669–2673
4. Visinsky ML, Cavallaro JR, Walker ID (1994) Expert system framework for fault detection and fault tolerance in robotics. Comput Electr Eng 20:421–435
5. Xiaodong Z, Parisini T, Polycarpou MM (2004) Adaptive fault-tolerant control of nonlinear uncertain systems: an information-based diagnostic approach. IEEE Trans Autom Control 49:1259–1274
6. Xiaodong Z, Polycarpou MM, Parisini T (2002) A robust detection and isolation scheme for abrupt and incipient faults in nonlinear systems. IEEE Trans Autom Control 47:576–593
7. Huang SN, Kok Kiang T (2008) Fault detection, isolation, and accommodation control in robotic systems. IEEE Trans Autom Sci Eng 5:480–489

Software-Based Fault Detection and Recovery for Cyber-Physical Systems

Jooyi Lee, Ji Chan Maeng, Byeonghun Song, Hyunmin Yoon, Taeho Kim, Won-Tae Kim and Minsoo Ryu

Abstract Cyber-physical systems demand higher levels of reliability for several reasons. First, unlike traditional computer-based systems, cyber-physical systems are more vulnerable to various faults since they operate under harsh working conditions. For instance, sensors and actuator may not always obey their specification due to wear-out or radiation. Second, even a minor fault in cyber-physical systems may lead to serious consequences since they operate under minimal supervision of human operators. In this paper we propose a software framework of fault detection and recovery for cyber-physical systems, called Fault Detection and Recovery for CPS (FDR-CPS). FDR-CPS focuses on specific types of faults related to sensors and actuators, which seem to be the likely cause of critical system failures such as system hangs and crashes. We divide such critical failures into four classes and then present the design and implementation of FDR-CPS that

---

J. Lee (✉) · J. C. Maeng · B. Song · H. Yoon · T. Kim
Department of EECS, Hanyang University, Seoul, Korea
e-mail: jylee@rtcc.hanyang.ac.kr

J. C. Maeng
e-mail: jcmaeng@rtcc.hanyang.ac.kr

B. Song
e-mail: bhsong@rtcc.hanyang.ac.kr

H. Yoon
e-mail: hmyoon@rtcc.hanyang.ac.kr

T. Kim
e-mail: thkim@rtcc.hanyang.ac.kr

W.-T. Kim
Embedded SW Research Division, ETRI, Daejeon, Korea
e-mail: wtkim@etri.re.kr

M. Ryu
Department of CSE, Hanyang University, Seoul, Korea
e-mail: msryu@hanyang.ac.kr

J. J. (Jong Hyuk) Park et al. (eds.), *Multimedia and Ubiquitous Engineering*,
Lecture Notes in Electrical Engineering 240, DOI: 10.1007/978-94-007-6738-6_136,
© Springer Science+Business Media Dordrecht(Outside the USA) 2013

can successfully handle the four classes of critical failures. We also describe a case study with quadrotor to demonstrate how FDR-CPS can be applied in a real world application.

Keywords Cyber-physical system · Reliability · Fault · Detection · Recovery

1 Introduction

A cyber-physical system is defined as a network of interconnected physical and computational elements. In typical cyber-physical systems, physical elements are monitored and controlled through sensors and actuators while computational elements process the information collected by sensors and send appropriate commands to actuators. This style of system organization is very similar to automatic control system structures, and thus enables various levels of autonomy in cyber-physical systems.

Cyber-physical systems demand higher levels of reliability for several reasons. First, unlike traditional computer-based systems, cyber-physical systems are more vulnerable to various faults since they operate under harsh working conditions. For instance, sensors and actuators may not always obey their specification due to wear-out or radiation. Second, even a minor fault in cyber-physical systems may lead to serious consequences since they operate under minimal supervision of human operators.

In this paper, we propose a software framework of fault detection and recovery for cyber-physical systems, called FDR-CPS (Fault Detection and Recovery for CPS). The primary goal of FDR-CPS is to provide resilience against critical faults that can cause system hangs or crashes. To that end, FDR-CPS focuses on specific types of faults that are related to sensors and actuators. Note that device-related faults are the main cause of critical failures like system crashes in many computer-based systems [1]. Since most software is written assuming the reliability of hardware devices, a single device fault may seriously impact the whole system. In this work we divide such critical failures into four classes and then present the design and implementation of FDR-CPS that can successfully handle the four classes of critical failures. We also describe a quadrotor case study to show how FDR-CPS can be applied in practice.

2 Critical Failures in Cyber_Physical Systems

Sensors and actuators are highly error-prone since they are often exposed to harsh physical environments. There exist many sources of faults such as wear-out, EMI and radiation that may cause various types of failures in sensors and actuators. Among them, we identify four classes of critical failures as follows.

- **Indefinite waiting**: Some specific event being waited for does not occur. For instance, a sensor or actuator may not be able to generate interrupts because of some internal failure while the program is waiting for the interrupt. In this case, the system may wait indefinitely. We will show this type of failure can be easily handled by using a timeout mechanism.
- **Infinite loop**: The program's workflow goes into infinite loop. For instance, a stuck-at fault may occur at some I/O channel that is referenced by a conditional loop. For instance, the program cannot escape from an infinite loop if a device's busy flag has a stuck-at fault and does not satisfy the loop's stopping condition. In this case, we can use a counter to detect and finish infinite loop iterations.
- **Erroneous sensor data**: Sensor data are not correct. This type of failure can be further divided into two subtypes. First, sensor data are inaccurate due to some reasons like noises or variations in ambient conditions. Second, sensor data are invalid because they exceed the normal range or telling inaccurate data constantly due to some internal failure.
- **Repeated operation failures**: A sensor or actuator repeatedly fails to complete program's requests. For instance, a sensor or actuator may constantly return error code as a return value in response to program's request. In case a transient fault causes this type of failure, we may recover the device through a reset leveraging the shadow driver technique [5].

3 Fault Detection and Recovery

FDR-CPS consists of three main components, fault detection module, fault recovery module and knowledge base. The fault detection module provides a set of skeleton functions that should be specialized by the programmer. A skeleton function is associated with each type of failure and provides a generic structure needed to diagnose a specific failure. The fault recovery module also provides a set of skeleton functions, each of which provides a generic structure to handle a specific type of failure. The fault detection and recovery modules may consult the knowledge base for failure handling policies and guidelines. The knowledge base contains several policies and guidelines about fault identification, logging, reporting and recovery (Fig. 1).

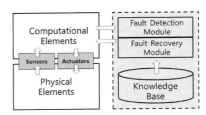

Fig. 1 FDR-CPS architecture

3.1 Direct Code Insertion

In order to apply FDR-CPS, we need to implement fault handling functionality in the target CPS program's code as well as specializing skeleton functions. FDR-CPS offers two options of implementation, direct code insertion (DCI) and function call insertion (FCI). The DCI technique is to insert some source code directly into the target program's source code. This technique is more efficient than FCI since it does not incur function call overhead.

Figure 2 illustrates the application of DCI technique to NS8390 ethernet device driver in Linux kernel 3.5.4. The original device driver had a potential infinite loop, but we inserted code to count the number of iterations at line 6, report and recover from line 8 through line 9 and escape the infinite loop at line 10.

Figure 3 illustrates another application of DCI technique. The example came from DE600 ethernet driver in Linux kernel 3.5.4. This example shows how invalid data can be detected and handled.

Fig. 2 Illustration of DCI for infinite loop handling

```
1    static void el2_get_8390_hdr(....)
2    {
3        ...
4        while ((inb(E33G_STATUS) & ESTAT_DPRDY) == 0)
5        {
6                    if(!boguscount--)
7                    {
8                            pr_notice("...", dev->name);
9                            el2_reset_8390(dev);
10                           goto blocked;
11                   }
12           ...
13       }
```

Fig. 3 Illustration of DCI for invalid data failure handling

```
1    static void de600_rx_intr(...)
2    {
3        size = de600_read_byte(RX_LEN, dev);
4        size += (de600_read_byte(RX_LEN, dev) << 8);
5        size -= 4;
6        ...
7        if ((size < 32) || (size > 1535)) {
8                printk(...);
9                if (size > 10000)
10                       adapter_init(dev);
11               return;
12       }
13       ...
```

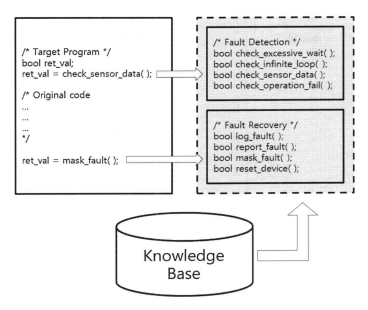

Fig. 4 Illustration of FCI for fault detection and recovery

3.2 Function Call Insertion

Function Call Insertion (FCI) may cause some performance overhead due to function call processing. However, it provides many benefits from a software engineering point of view such as modularity and standardized interfaces. Figure 4 shows the brief overview of FCI technique. With FCI, programmers just need to insert some function calls at appropriate places while minimizing source code modification.

4 Case Study

In order to evaluate our approach, we conducted a case study with a commercial quadrotor, called AR. Drone. The target has ARM9 processor, Linux 2.6.27, various types of sensors including accelerometer, gyrometer and ultrasound altimeter, and electric motors for driving high efficiency propellers. We implemented a prototype of FDR-CPS in the quadrotor's control program, and also constructed a simple knowledge base with several criteria needed for diagnosing some specific faults. One of them is that "if the quadrotor keeps moving up for more than 30 s while the setpoint value does not change, there is a stuck-at fault at ultrasound altimeter."

Initially, the quadrotor was configured to be hovering 50 cm high from the ground. We then injected a stuck-at fault into the ultrasound altimeter so that it tells the aircraft is 30 cm high from the ground. After this, we could observe the target kept moving up to reach the setpoint. FDR-CPS detected this fault by using a time-out mechanism and the knowledge base. FDR-CPS then started its recovery operation and the aircraft immediately returned to the original correct setpoint.

5 Conclusion

In this paper we proposed a software framework of fault detection and recovery for cyber-physical systems, called FDR-CPS (Fault Detection and Recovery for CPS). FDR-CPS focuses on specific faults related to sensors and actuators, which seem to be the likely cause of critical system failures such as system hangs and crashes. We divided such critical failures into four classes and then presented the design and implementation of FDR-CPS that can successfully handle the four classes of critical failures. We also described the quadrotor case study how FDR-CPS can be applied in real world applications.

Acknowledgments This work was supported partly by Mid-career Researcher Program through NRF (National Research Foundation) grant NRF-2011-0015997 funded by the MEST (Ministry of Education, Science and Technology), partly by the IT R&D Program of MKE/KEIT [10035708, "The Development of CPS (Cyber-Physical Systems) Core Technologies for High Confidential Autonomic Control Software"], partly by Seoul Creative Human Development Program (HM120006), and partly by the MKE (The Ministry of Knowledge Economy), Korea, under the CITRC (Convergence Information Technology Research Center) support program (NIPA-2013-H0401-13-1009) supervised by the NIPA (National IT Industry Promotion Agency).

References

1. Kadav A, Renzelmann MJ, Swift MM (2009) Tolerating Hardware Device Failures in Software. In: Proceedings of the ACM SIGOPS 22nd symposium on operating systems principles, pp 59–72
2. Graham S (2002) Writing drivers for reliability, robustness and fault tolerant systems. http://www.microsoft.com/whdc/archive/FTdrv.ms
3. Ploski J, Rohr M, Schwenkenberg P, Hasselbring W (2007) Research issues in software fault categorization. ACM SIGSOFT Softw Eng Notes 32(6): 1–8 (article No. 6)
4. Ball T, Bounimova E, Cook B, Levin V, Lichtenberg J, McGarvey C, Ondrusek B, Rajamani SK, Ustuner A (2006) Thorough static analysis of device drivers. In: Proceedings of the 1st ACM SIGOPS/EuroSys European conference on computer systems, pp 73–85
5. Candea G, Fox A (2003) Crash-only software. In: Proceedings of HotOS IX: The 9th workshop on hot topics in operating systems

Sample Adaptive Offset Parallelism in HEVC

Eun-kyung Ryu, Jung-hak Nam, Seon-oh Lee, Hyun-ho Jo and Dong-gyu Sim

Abstract We propose a parallelization method for SAO, in-loop filter of HEVC. SAO filtering proceeds along CTB lines and there exists data dependency between inside and outside of CTB boundaries. Data dependency makes data-level parallelization hard. In this paper, we equally divided an entire frame into sub regions. With a little amount of memory, proposed method shows 1.9 times of performance enhancement in terms of processing time.

Keywords Sample adaptive offset · SAO · SAO parallelism · HEVC parallelism · Multi-core parallelism · In-loop filter

1 Introduction

Needs for the realistic video service, such as higher resolution and higher quality, are increasing with the developments of multimedia-related hardware and software technologies. In addition, requirements for the advanced video coding standard have arisen on the multimedia industrial market. Dependent on the demands, Joint

---

E. Ryu · J. Nam · S. Lee · H. Jo · D. Sim (✉)
Computer Engineering, Kwangwoon University, Seoul, Republic of Korea
e-mail: dgsim@kw.ac.kr

E. Ryu
e-mail: dms0314@kw.ac.kr

J. Nam
e-mail: qejixfyza@kw.ac.kr

S. Lee
e-mail: seon-oh@kw.ac.kr

H. Jo
e-mail: idjhh@kw.ac.kr

J. J. (Jong Hyuk) Park et al. (eds.), *Multimedia and Ubiquitous Engineering*,
Lecture Notes in Electrical Engineering 240, DOI: 10.1007/978-94-007-6738-6_137,
© Springer Science+Business Media Dordrecht(Outside the USA) 2013

Collaborative Team on Video Coding (JCT-VC) which is composed by ISO/IEC Moving Picture Experts Group (MPEG) and ITU-T Video Coding Experts Group (VCEG) has been standardizing HEVC, whose target coding efficiency is twice better than that of H.264/AVC [1].

To introduce the state-of-art video codec to various multimedia industries as soon as standardization process of HEVC is completed, researches on algorithms lowering computational complexity and methods of optimization for real time processing are indispensable. Even during standardizing process, parallel processing schemes for real-time processing have been contributed and accepted for HEVC. Thinking of trends on multi-core and various mobile devices and high complexity of HEVC, researches on parallelization for HEVC should be done.

Lowering quantization error of reconstructed frames, SAO has dependency in pixel decoding process. To develop SAO with parallel scheme, we should think of a feature of SAO dependency.

In this paper, we propose a parallel core assigning method which resolves CTB-level dependency of SAO. We assign the same number of CTBs into each core.

This paper is organized as follows. In Sect. 2, SAO method is introduced. In Sect. 3, dependency between adjacent CTBs is analyzed, and the proposed method is shown. Experimental results and evaluation are given in Sect. 4. Finally, conclusions are stated in Sect. 5.

2 Sample Adaptive Offset for HEVC

New in-loop filter, SAO has been adopted for HEVC. Different to the other tools of video codec, quantization process causes data loss between original and reconstructed videos. Larger transforms and longer-tap interpolation of HEVC than those of the other video codecs can introduce ringing artifacts due to quantization error of transformed coefficients and loss of high frequency components. Ringing artifacts or quantization errors are compensated with parameters within compressed bitstream during SAO process such as Edge Offset (EO) and Band Offset (BO) (Fig. 1).

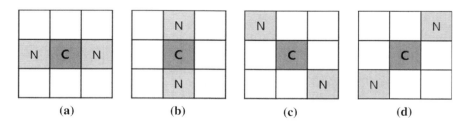

Fig. 1 Four types of edge offset class: **a** 1-D 0-dgree, **b** 1-D 90-dgree, **c** 1-D 135-dgree **d** 1-D 45-dgree

Sample Adaptive Offset Parallelism in HEVC   1115

Table 1 Category conditions

Category	Conditions
1	C < N1 && C < N2
2	(C < N1 && C == N2) || (C == N1 && C < N2)
3	(C > N1 && C == N2) || (C == N1 && C > N2)
4	C > N1 && C > N2
0	None of the above

EO reduces ringing artifacts and quantization errors by making local valleys and concave corners become smoother. Four classes are used for EO and the classification is based on edge direction derived from comparison between current and neighboring pixel values on the encoder side. Side information for EO indicates which one of four classes is applied and instructs offset values to be added.

For a given EO class, each sample inside the CTB is classified into one of five categories. The current sample value, labeled as 'C', is compared with its two neighbors along the selected 1D pattern. The classification rules for each sample are summarized in Table 1. Categories 1 and 4 are associated with a local valley and a local peak along the selected 1D pattern, respectively. Categories 2 and 3 are associated with concave and convex corners along the selected 1D pattern, respectively [2]. If the current CTB does not belong to EO categories 1–4, then it are category 0, and SAO is not applied. Category determining process needs neighboring samples which are not processed by SAO and this referencing structure derives a data dependency problem for multi-core parallelization.

BO reduces quantization errors by adding offset values to all samples of the selected bands. The starting band position and offsets of four consecutive bands determined on the encoder side is signaled to decoder. One offset is added to all samples of the same band. In case of BO, neighboring samples are not required. So, we need not consider data dependency between adjacent CTBs for BO parallelization [2] (Fig. 2).

First of all, data dependency derived from coding mode using neighboring samples or context should be thought for the data-level parallelization of video codec. To parallelize SAO, it is important to know that the sample classification is based on comparison between current samples and neighboring samples whose values are not filtered by SAO.

Figure 3 shows data dependency between adjacent CTBs for EO of SAO. Neighboring samples for EO used for reference should be not processed by SAO. In case of raster order SAO process, boundary samples of CTBs should be stored

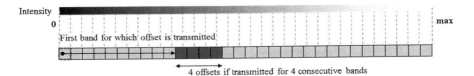

Fig. 2 Bands

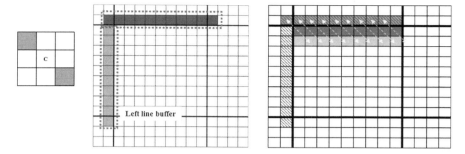

Fig. 3 Data dependency between adjacent CTBs for EO of SAO

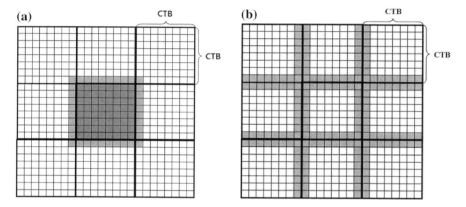

Fig. 4 Four directional data dependency for parallel processing. **a** Data dependency of a CTB. **b** Line buffer in a frame to remove the data dependency

until SAO filtering for all of the adjacent CTBs is completed. HM software using raster order for SAO utilizes additional memory buffer to store left and upper boundary samples.

3 Proposed Method

Different to a case of raster order SAO process which has left and upper directional data dependency, there exists four directional data dependency for parallel processing of SAO. As shown in Fig. 4a, decoder should store four directional neighboring samples for all of CTBs and it allows parallel processing.

Conventional data-level parallelizing methods for SAO use temporal memory buffer storing an entire frame [3]. Amount of memory for an entire frame causes overhead on the decoder side. For the reason, we used line buffer memory which temporally stores two pixel lines composed of inside and outside of CTB boundaries in both vertical and horizontal directions, as shown in Fig. 4b. The line buffer needs 1/16 amount of memory than an entire frame buffer, which allows parallel SAO filtering on the decoder side. In this paper, we selected a method dividing a frame into same size of areas among the conventional data-level parallelizing methods.

4 Experiment Results

4.1 Experiment Conditions

To evaluate the performance of the proposed method, three test sequences from common test conditions of HEVC were used. Test sequences were encoded using all intra, low delay (LD) and random access (RA) modes. Values of quantization parameters used to encode are 22, 27, 32 and 37 Table 2.

Table 2 Experiment conditions

Operating system	Microsoft Windows 7 (32 bit)
CPU	Intel Core^{TM}2 Quad CPU
RAM	4 GB
Profile	Main
Reference software	HM 9.0
OpenMP version	OpenMP 2.0
Compiler	MS Visual Studio 2008 Release mode
#Iteration	3
Measurement	Decoding time (ms)
QP	22, 27, 32, 37

Table 3 Time saving in all intra mode

Sequence	QP	Reference	Proposed	Speed up
BasketballDrive	22	2544.9	1170.6	2.2 x
	27	2783.0	1228.4	2.3 x
	32	2182.5	1050.1	2.1 x
	37	1360.6	760.0	1.8 x
KristenAndSara	22	1501.8	639.8	2.3 x
	27	1266.5	549.1	2.3 x
	32	798.7	414.0	1.9 x
	37	409.3	262.3	1.6 x
NebutaFestiva	22	1632.9	877.0	1.9 x
	27	2685.2	1226.3	2.2 x
	32	3482.1	1448.5	2.4 x
	37	2660.1	1323.3	2.0 x
Average				2.1 x

Table 4 Time saving in low delay mode

Sequence	QP	Reference	Proposed	Speed up
BasketballDrive	22	2282.1	1035.9	2.2 x
	27	1274.9	730.8	1.7 x
	32	459.8	317.4	1.4 x
	37	217.2	182	1.2 x
KristenAndSara	22	393.3	214.7	1.8 x
	27	178.6	114.4	1.6 x
	32	59.4	51.5	1.2 x
	37	42.8	41.5	1.0 x
NebutaFestiva	22	1431.8	732.8	2.0 x
	27	1042.4	477.4	2.2 x
	32	1183.2	526.5	2.2 x
	37	839.9	448.6	1.9 x
Average				1.7 x

Table 5 Time saving in random access mode

Seqeunce	QP	Reference	Proposed	Speed up
BasketballDrive	22	2118.3	964	2.2 x
	27	835.2	493.3	1.7 x
	32	341	195	1.7 x
	37	200	132.9	1.5 x
KristenAndSara	22	340.1	165.8	2.1 x
	27	149.4	77.5	1.9 x
	32	88.8	52.8	1.7 x
	37	43	31	1.4 x
NebutaFestiva	22	1631.4	858.9	1.9 x
	27	2545	1007.9	2.5 x
	32	2974.6	1070.7	2.8 x
	37	1036.4	462.8	2.2 x
Average				1.9 x

4.2 *Experimental Results*

Tables 3, 4 and 5 show performance of HM 9.0 and proposed method in terms of processing time of only SAO module for all intra, low delay and random access coding mode. Generally, lower QP values cause more SAO enable CTBs while the higher QP values derive the less number of SAO enable CTBs. So, reduced processing time for higher QP cases shows higher performance enhancement. The proposed method shows 1.9 times of performance enhancement in terms of average processing time.

5 Conclusion

We propose a parallelization method for SAO, in-loop filter of HEVC. SAO filtering for CTBs needs neighboring sample information, which derives data dependencies. For the parallelization, though conventional methods use temporal memory buffer for an entire frame to resolve data dependency problem of SAO, which causes memory overhead problems. In this paper, we used only 1/16 amount of memory than a frame buffer for data-level parallelization. The proposed method of data-level parallel processing for multi-core processors shows 1.9 times than the conventional method of HM 9.0 in terms of processing time.

Acknowledgments This research was partly supported by the Samsung Electronics, and partly supported by the Ministry of Knowledge Economy (MKE), Korea, under the Information Technology Research Center (ITRC) support program (NIPA-2012-H0301-12-1011) supervised by the National IT Industry Promotion Agency (NIPA).

References

1. Wiegand T, Ohm J-R, Sullivan GJ, Han W-J, Joshi R, Tan TK, Ugur K (2010) Special section on the joint call for proposals on high efficiency video coding (HEVC) standardization. IEEE Trans Circuits Syst Video Technol 20(12):1661–1666
2. Fu C-M, Alshina E, Alshin A, Huang Y-W, Chen C-Y, Tsai C-Y, Hsu C-W, Lei S-M, Park JH, Han W-J (2012) Sample adaptive offset in the HEVC standard. IEEE Trans Circuits Syst Video Technol 22(12):1755–1764
3. Chi CC, Alvarez-Mesa M, Juurlink B, Clare G, Henry F, Pateux S, Schierl T (2012) Parallel scalability and efficiency of HEVC parallelization approaches. IEEE Trans Circuits Syst Video Technol 22(12):1827–1838

Comparison Between SVM and Back Propagation Neural Network in Building IDS

Nguyen Dai Hai and Nguyen Linh Giang

Abstract Recently, applying the novel data mining techniques for anomaly detection-an element in Intrusion Detection System has received much research alternation. Support Vector Machine (SVM) and Back Propagation Neural (BPN) network has been applied successfully in many areas with excellent generalization results, such as rule extraction, classification and evaluation. In this paper, we use an approach that is entropy based analysis method to characterize some common types of attack like scanning attack. A model based on SVM with Gaussian RBF kernel is also proposed here for building anomaly detection system. BPN network is considered one of the simplest and most general methods used for supervised training of multilayered neural network. The comparative results show that with attack scenarios that we create and through the differences between the performance measures, we found that SVM gives higher precision and lower error rate than BPN method.

Keywords Back propagation neural network · Denial of service · Entropy · RBF kernel · Support Vector Machine

---

N. D. Hai (✉)
School of Information and Communication Technology, Hanoi University of Science and Technology, Hanoi, Vietnam
e-mail: haidnguyen0909@gmail.com

N. L. Giang
Department of Communication and Computer Networks, Hanoi University of Science and Technology, Hanoi, Vietnam
e-mail: giangnl@soict.hut.edu.vn

J. J. (Jong Hyuk) Park et al. (eds.), *Multimedia and Ubiquitous Engineering*, Lecture Notes in Electrical Engineering 240, DOI: 10.1007/978-94-007-6738-6_138, © Springer Science+Business Media Dordrecht(Outside the USA) 2013

1 Introduction

Intrusion Detection Systems (IDS) have become popular tools for identifying anomalous and malicious activities in computer systems and networks. Anomaly detection is a key element of intrusion detection and other detection system in which perturbations from normal behavior suggest the presence of attacks, defects etc. Anomaly detection is performed by building a model that contains metrics derived from system operation and flagging any observation as intrusive that has a significant deviation from the model.

In this paper, we consider two common types of attack including Bandwidth flood like ICMP flood, TCP flood, and Scanning attack like TCP scan, Null scan. There are two problems we need to resolve, one problem is that the amount of traffic data does not allow real-time analysis of details and other is that the specific characteristics events are not known in advance. So, there is a need for analysis methods that are real-time capable and can handle large amounts of traffic data. We will apply an entropy based approach proposed in [1], that determines and reports entropy contents of traffic parameters such as IP addresses, port. Changes in the entropy content indicate a massive network event.

In order to improve the performance of Anomaly-based IDS, many Artificial Intelligence methods such as neural network have been widely used. These methods are based on an empirical and have some disadvantages such as local optimal solution, low convergence rate, and epically poor generalization when the number of class samples is limited [2]. In 1990s, Support Vector Machine (SVM) was introduced as a new technique for solving a variety of learning, classification and prediction problems. SVM is a set of related supervised learning methods used for classification and regression. SVM originated as an implementation of Vapnik's [3] structural risk minimization principle which minimizes the generalization error. The paper aims at investigating the capabilities of SVM when compared with BPN network model for building anomaly-based Intrusion Detection System.

This paper is organized as follows: Section 2 explains the fundamental idea related to entropy based analysis method and characteristics of two types of attack. Section 3 deals with the foundation of SVM and BPN network, in this section, we have procedures for anomaly-based intrusion detection. A discussion of implementation details and experimental results is given in Sect. 4. Section 5 give concluding remarks and discuss related work.

2 Entropy Based Analysis Method

In this paper, we consider two common types of attack including Bandwidth flood and Scanning Attack. In bandwidth flood, a small number of hosts send large amounts of traffic to a single destination. So, characteristic of this type of attack is that number of packets in some particular protocol such as TCP, UDP, ICMP increase significantly. We can use number of packets in TCP, UDP and ICMP protocols as attributes of input data of classifier.

Scan attacks can be characterized by using entropy based analysis method. Generally speaking entropy is a measure of how random a dataset is. The more random it is, the more entropy it contains. Entropy contents of a sequence of values can be measured by representing the sequence in binary form and then using data compression on that sequence. The size of the compressed object corresponds to the entropy contents of the sequence. If the compression algorithm is perfect in the mathematical sense, the measurement is exact. The change in IP addresses characteristics seen on flow level. When scanning hosts try to connect to a lot of other hosts, the source IP addresses of the scanning hosts will be seen in many flows and since they are relatively few hosts, the source IP address fields will contain less entropy per address seen then normal traffic. On the other hand the target IP addresses seen in flows will be much more random than in normal traffic. A similar thing happens on the port level. If an attacker scans for a specific vulnerability, these scans often have to go to a specific target port. If scanning traffic with this characteristic becomes a significant component of the overall network traffic, the entropy contents of the destination port fields in flows seen in the network will decrease significantly. So IP address and port entropy change can be used to detect scanning attack.

We compared three different lossless compression methods, the well-known bzip2 [4] and gzip [5] compressors as well as the LZO (Lempel–Ziv–Oberhumer) [6] real-time compressor. We did not consider lossy compressors. Bzip2 is slow and compresses very well, gzip is average in all regards and lzo is fast but does not compress well. Direct comparison of the three compressors on network data shows that while the compression ratios are different, the changes in compressibility are very similar. Because of its speed advantage LZO was selected as preferred algorithm for our work.

In our experiments we choose one minute as measurement interval length. We collect packets in each time interval, create string arrays containing IP addresses, ports and then by using LZO compression algorithm we get compression ratio. These ratios are represented as content entropy of IP addresses, ports (Table 1).

Table 1 Compression rate represents entropy

# Compresses all attributes		
my $csip	= **compress** $strsip;	#contain source IP addresses
my $csport	= **compress** $strsport;	#contain source ports
my $cdip	= **compress** $strdip;	#contain destination IP addresses
my $cdport	= **compress** $strdport;	#contain destination ports
# **get ratio** ($strsip,$strsport,$strdip,$strdport);		
$this-> {_ratio_sip} = length($csip)/length($strsip);		
$this-> {_ratio_sport} = length($csport)/length($strsport);		
$this-> {_ratio_dip} = length($cdip)/length($strdip);		
$this-> {_ratio_dport} = length($cdport)/length($strdport);		

3 Classification Methods

We will describe briefly two classification methods that we use in this paper.

3.1 Overview of SVM

Support Vector Machines (SVMs) are a useful technique for data classification. A classification task usually involves separating data into training and testing sets. Each instance in the training set contains one "target value" (i.e. the class label) and several "attributes" (i.e. the feature or observed variables). The goal of SVM is to produce a model (based on the training data) which predicts the target values of the test data given only the test data attributes. Given a training set of instance-label pairs (x_i, y_i), $i = 1,...l$ where $x_i \in R^n$ and $y \in \{1, -1\}^l$, SVM require the solution of the following optimization problem:

$$\min_{w,b,\xi} \frac{1}{2} w^T w + C \sum_{i=1}^{l} \xi_i \text{ Subject to } y_i(w^T \phi(x_i) + b) \geq 1 - \xi_i \text{ and } \xi_i \geq 0$$

The decision function is formulated in terms of these kernels:

$$f(x) = sign\left(\sum_{i=1}^{n} \alpha_i^* y_i(\phi(x)\phi(x_i) - b^*)\right) = sign(\sum_{i=1}^{n} \alpha_i^* y_i(K(x, x_i) - b^*)).$$

In general, the Gaussian RBF kernel is a reasonable first choice. This kernel nonlinearly maps samples into a higher dimensional space so it, unlike the linear kernel, can handle the case when the relation between class labels and attributes in nonlinear.

3.2 Overview of Back Propagation Neural Network

Back Propagation Neural (BPN) is a common method of training artificial neural networks so as to minimize the objective function. Back propagation works by approximating the non-linear relationship between the input and the output by adjusting the weight values internally. A BPN model consists of an input layer, one or more hidden layers, and output layer. There are two parameters including learning rate $(0 < \alpha < 1)$ and momentum $(0 < \eta < 1)$ required to define by user. The theoretical results showed that one hidden layer is sufficient for a BPN to approximate any continuous mapping from the input patterns to the output patterns to an arbitrary degree freedom [7]. The selection nodes of hidden layers primarily affect the classification performance.

4 Empirical Illustration

4.1 Data Set

In this study, the measured attributes are (11 attributes): (1–2) Entropy-compression rate of the source/destination IP address and (3–4) source/destination port, (5) number of packets, (6–7) total/average size of the packets, (8) standard deviation of packet size and (9–11) number of TCP/UDP/ICMP packets. So each instance will be represented by a vector including 11 attributes and the input of each classifier is differential vector of current vector and reference vector which refer to normal state.

4.2 Experiment

In this section, we will test the system's ability of detecting anomaly-based intrusion activities using two methods: SVM and BPN. We will proceed on the four attack scenarios including ICMP flood, TCP flood, UDP flood and port scan. Each attack will change significantly the number of ICMP, TCP, UDP packets and entropy.

4.2.1 Testing Environment

The system was tested on virtual LAN 100 Mps environment using VMware tool, including two Window XP computers and a Ubuntu computer installed the Anomaly IDS. These computers are connected to each other through a virtual switch.

4.2.2 Testing Scenarios

Two Window XP computers implement TCP flood, UDP flood, ICMP flood refer to bandwidth flood attacks using tools like hping3, udpflood.exe, ping respectively or scan port in range 1–300 on Ubuntu computer installed anomaly IDS. Our program will collect and analysis packets in order to detect anomalous in traffic. Fig. 1.

As we can see, the characteristic of the port scan attack is that entropy source port, destination change significantly compared with the normal state.

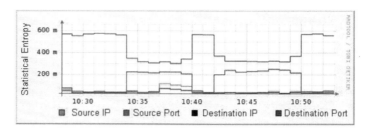

Fig. 1 Statistical entropy when port scanning

4.2.3 Experimental Result

With the attack scenarios as above, we tested IDS's ability of detecting some common attack types using two methods SVM and BNP.

(1) SVM method

We use **Algorithm::SVM** library [8] with 5-fold cross-validation to find the best parameter C and γ [3]. The results confirmed that the classification precision of the SVM with RBF kernel was high as 98 % when γ and C are 0.000015 and 2, respectively. Then we use the best parameter C and γ to train the whole training set, we have 9 support vectors, the outputs are:

Accuracy = 99 % (49/50) (classification), Iteration = 123.

(2) BPN method

We use **Algorithm::NNFlex** library to implement BPN method. In the training model, there are 11 attributes as input nodes, four hidden nodes and one output node standing for normal (+1) or abnormal (−1). >0 as a target value of the normal and <0 as a target value of abnormal. We adopted the range of 0.6–0.9 and 0.1–0.4 to be the decisions of learning rate and momentum. Through several trail-and-error experiments, the structure of 11-4-1 model had the best performance. The outputs are:

Accuracy = 92 % (46/50), Iteration = 411, Time training: 45 s.

We conducted simulating attacks with different scenarios then collecting 200 data samples used for predicting process. The prediction result is listed in Table 2. SVM method shows the best overall prediction accuracy level at 99 % and BPN method, shows the best overall accuracy level at 96 %.

Here we define the attack detection rate and false-positive rate as [9].

Table 2 Detected attack rate and false-positive rate

Methods	Attack Detected	Real	Normal Misclassified	Real	Attack detection rate (%)	False positive rate (%)
SVM	50	50	2	150	100	1.3
BPN	48	50	6	150	96	4

5 Conclusion

This study constructed an Anomaly-based intrusion detection model based on Support Vector Machine and Back Propagation Neural Network. First, we introduce an approach based on entropy to characterize some kinds of scanning attack. Second, we simply define the SVM and BPN methods. Third, a proposed for building an Intrusion Detection System using SVM and BPN is discussed. Finally, experimental results proved the validity of our proposes and we found that SVM method has better accuracy with lower misclassification rate than MLP method based on these results.

References

1. Nychis G, Sekar V, Andersen DG, Kim H, Zhang H (2008) An empirical evaluation of entropy-based traffic anomaly detection. In: Proceedings of the 8th ACM SIGCOMM conference on internet measurement
2. Yuan, SF, Chu FL (2006) Support vector machine based on fault diagnosis for turbo-pump rotor. Mech Syst Signal Process 20:939–952
3. Ben-Hur A, Weston J (2010) A user's guide to support vector machines. Methods Mol Biol 609:223–239
4. The bzip2 and libbzip2 official home page. http://sources.redhat.com/bzip2/
5. The gzip home page. http://www.gzip.org/
6. http://www.oberhumer.com/opensource/lzo/. LZO compression library
7. Randall SS, Dorsey RE (2000) Reliable classification using neural networks: a genetic algorithm and back propagation comparison. Decis Support Syst 30:11–22
8. http://search.cpan.org/~lairdm/Algorithm-SVM-0.13/
9. Liao Y, Vemuri VR (2002) Use of k-nearest neighbor classifier for intrusion detection. Comput Secur 21:439–448
10. Chang CC, Lin CJ (2009) LIBSVM: a library for support vector machines. Software available at http://www.csie.ntu.edu.tw/cjin/libsvm. 18 Nov 2009
11. Fausett L (1994) Fundamentals of neural networks: architectures, algorithms and applications. Prentice-Hall, New Jersey

Anomaly Detection with Multinomial Logistic Regression and Naïve Bayesian

Nguyen Dai Hai and Nguyen Linh Giang

Abstract Intrusion Detection by automated means is gaining widespread interest due to the serious impact of Intrusions on computer system or network. Several techniques have been introduced in an effort to minimize up to some extent the risk associated with Intrusion attack. In this paper, we have used two novel Machine Learning techniques including Multinomial Logistic Regression and Naïve Bayesian in building Anomaly-based Intrusion Detection System (IDS). Also, we create our own dataset based on four attack scenarios including TCP flood, ICMP flood, UDP flood and Scan port. Then, we will test the system's ability of detecting anomaly-based intrusion activities using these two methods. Furthermore we will make the comparison of classification performance between the Multinomial Logistic Regression and Naïve Bayesian.

Keywords DoS · Logistic regression · Naïve Bayesian · Intrusion detection system

1 Introduction

Intrusion Detection is a process of gathering intrusion related knowledge that occurred in the computer networks or systems and analyzing them for detecting future intrusions. Intrusion Detection can be divided into two categories: Anomaly

---

N. D. Hai (✉)
School of Information and Communication Technology, Hanoi University
of Science and Technology, Hanoi, Vietnam
e-mail: haidnguyen0909@gmail.com

N. L. Giang
Department of Communication and Computer Networks, Hanoi University
of Science and Technology, Hanoi, Vietnam
e-mail: giangnl@soict.hut.edu.vn

J. J. (Jong Hyuk) Park et al. (eds.), *Multimedia and Ubiquitous Engineering*,
Lecture Notes in Electrical Engineering 240, DOI: 10.1007/978-94-007-6738-6_139,
© Springer Science+Business Media Dordrecht(Outside the USA) 2013

detection [2] and Misuse detection. The former analyses the information gathered and compares it to a defined baseline of what is seen as "normal" service behaviors, so it has ability to learn how to detect network attacks that are currently unknown. Misuse detection is based on signatures for known attacks, so it is only as good as the database of attack signatures that it uses for comparison. Misuse detection has low false positive rate, but can not detect novel attacks. However, anomaly detection can detect unknown attacks, but has high false positive rate.

The Naïve Bayesian (NB) method is based on the work of Thomas Bayesian. In Bayesian classification, we have a hypothesis that the given data belongs to a particular class. We then calculate the probability for the hypothesis to be true. This is among the most practical approaches for certain types of problems. The approach requires only one scan of the whole data.

A Multinomial Logistic Regression (MLR) model is used for data in which the dependent variable is unordered or polytomous, and independent variables are continuous or categorical predictors. This type of model is therefore measured on a nomial scale and was introduced by McFadden (1974). Unlike a binary logistic model in which a dependent variable has only a binary choice (e.g., presence/absence of a characteristic), the dependent variable in a multinomial logistic model can have more than two choices that are coded categorically, and one of the categories is taken as the reference category.

In this paper, we propose two methods MLR and NB in building anomaly-based IDS and compare the performance of two linear classifier of Naïve Bayesian (NB) and multinomial Logistic Regression (MLR) based on attack scenarios which we created, and search for the characteristics of the data that determine the performance. The comparison between LR and MNB has been studied theoretically by Ng and Jordan (2002).

This paper is organized as follows: Sect. 2 deals with the description of data set for our experiment. Section 3 deals with foundation of methods including naïve Bayesian, multinomial logistic regression, In this section we will consider the problem of applying the two methods in building anomaly-based IDS. In Sect. 4, we give an illustration and experimental results with four attack scenarios. It help in understanding of this procedure, a demonstrative case is given to show the key stages involving the use of the introduced concepts. Section 5 is conclusion.

2 Dataset

Our data set is created by the following activities:

Data collection activity: collection attribute-value of the flow in terms of packet data (IP, port, TCP, UDP, ICMP). Based on these attributes, the program will build Profile (bin level) which contains the characteristic parameters for network traffic in a given time, including: (1–2) Entropy compression rate of the source/destination IP address, (3–4) Entropy compression rate of the source/destination port, (5) number of packets, (6) total size of the packets, (7) average size of packets, (8)standard deviation of packet size, (9) number of TCP packets, (10) number of UDP packets and (11) number of ICMP packets.

Statistical analysis activity: This activity is based on the data have been analyzed from the data collected to build the corresponding bin arrays. The bin is divided into the following levels: hours, days, months correspond to the three classes of data is the current class, reference class and the differential classes:

Cur_bin: represent for each instance "bin" (bin is the smallest time unit, in my program one minute).These instances is continuously created in the processes monitoring network traffic.

Ref_bin: represents the reference model corresponding to one unit of time reference. Reference model is adaptably updated, based on values of Cur_bin in the absence of intrusion detection.

Dif_bin: represents the difference between the current value and the reference value and is the input of classifiers.

3 Methods

3.1 Naïve Bayesian

Naïve Bayesian classifiers assume that the effect of an attribute value on a given class is independent of the values of the other attributes. This assumption is called class conditional independence. Naïve Bayesian classifiers allow the representation of dependencies among subsets of attribute [9]. Through the use of Bayesian networks has proved to be effective in certain situations, the result obtained, are highly dependent on the assumption about the behavior of the target system, and so a deviation in these hypotheses leads to detection errors, attributable to the model considered [10]. The NB classifier work as follows: Let T be a training set of samples, each with their class labels. There are k classes $C_1, C_2, \ldots, C_k$, each sample is represented by an n-dimensional vector $X = \{X_1, X_2, \ldots, X_n\}$.

Given a sample X, The classifier will predict that X belongs to the class having the highest a posteriori probability, conditional on X. That is X is predicted to belong to the class C, if and only if $P(C_i|X) > P(C_j|X)$ for $1 \leq j \leq m, j \neq i$.

By bayes' theorem, we have $P(C_i|X) = \frac{P(X|C_i)P(C_i)}{P(X)}$. As P(X) is the same for all classes and only $P(C_i)$ are not known, then it is commonly assumed that the classes are equally likely, that is, $P(C_1) = P(C_2) = \cdots P(C_m)$ we would therefore maximize $P(X|C_i)$.

In order to reduce computation in evaluating $P(X|C_i)$. The naïve assumption of class conditional independence is made. Mathematically this means that $P(X|C_i) \approx \sum_{k=1}^{n} P(X_k|C_i)$. The probabilities $P(X_k|C_i)$ can easily be estimated from the training set. If X is continuous-valued, then we typically assume that the values have a Gaussian distribution with a mean μ and standard deviation σ. So that

$P(X_k|C_i) = g(X_k, \mu_{ci}, \sigma_{ci})$. We need to compute μ_{ci}, σ_{ci} in training stage. In order to predict the class label of X, $P(X|C_i)P(C_i)$ is evaluated for each class C_i. The classifier predicts that the class label of X is C_i if and only if it is the class that maximizes $P(X|C_i)P(C_i)$.

3.2 Multinomial Logistic Regression

A multinomial logistic regression model is used for data in which the dependent variable is unordered or polytomous, and independent variables are continuous or categorical predictors. This type of model is therefore measured on a nomial scale and was introduced by McFadden (1974). Unlike a binary logistic model in which a dependent variable has only a binary choice (e.g., presence/absence of a characteristic), the dependent variable in a multinomial logistic model can have more than two choices that are coded categorically, and one of the categories is taken as the reference category. This study used "0" (normal) as the reference category. Suppose y_i is the dependent variable with five categories for individual connection i-th, and the probability of being in category s (s = "1" [TCP flood], "2" [ICMP flood], "3" [UDP flood], "4" [Scan Port]) can be denoted $\pi_i^{(s)} = \Pr(y_i = s)$ with the chosen reference category, $\pi_i^{(0)}$. Then, for a simple model with one independent variable x_i, a multinomial logistic regression model with logit link can be represented as:

$$\log\left(\frac{\pi_i^{(s)}}{\pi_i^{(0)}}\right) = \beta_0^{(s)} + \beta_1^{(s)}x_i, s = 1, 2, 3, 4.$$

An alternative way to interpret the effect of an independent variable, x, is to use predicted probabilities $\pi_i^{(s)}$ for different of x:

$$\pi_i^{(s)} = \frac{\exp(\beta_0^{(s)} + \beta_1^{(s)}x_i)}{1 + \sum_{k=1}^{4}\exp(\beta_0^{(k)} + \beta_1^{(k)}x_i)}.$$

Then, the probability of being in the reference category, "0" (normal), can be calculates by subtraction:

$$\pi_i^{(0)} = 1 - \sum_{k=1}^{4}\pi_i^{(k)}$$

4 Experiment and Results

In this section, we summarize our experimental results to detect network intrusion detections using Naïve Bayes and Multinomial Logistic Regression over dataset we created based on four attack scenarios including: TCP flood, ICMP flood, UDP flood and Port Scan.

4.1 Purpose of Study

The objective of this study is to detect some common attack types in computer systems and networks. We furthermore make the comparison of classification performance between the NB and MLR model.

4.2 Dataset

In this study, the measured attributes are (in particular, 11 attributes): entropy compression rate of the source/destination IP address and source/destination port, number of packets, total/average size of the packets, standard deviation of packet size and number of TCP/UDP/ICMP packets, So each instance will be represented by a vector including 11 attributes and the input of each classifier is differential vector of current vector and reference vector which refer to normal state (Table 1).

4.3 Experiment

We will test the system's ability of detecting anomaly-based intrusion activities using two methods: Naïve Bayes and Multinomial Logistic Regression. We will proceed on the four attack scenarios including ICMP flood, TCP flood, UDP flood and port scan. Using with each attack will change significantly the number of ICMP, TCP, UDP packets and entropy source/target.

Table 1 Number of examples in dataset we created

Attack types	Training samples
Normal	110
TCP flood	205
ICMP flood	200
UDP flood	150
Scan port	180

4.3.1 Testing Environment

The system was tested on virtual LAN 100 Mps environment using VMware tool, including two Window XP computers and a Ubuntu computer installed the Anomaly IDS. These computers are connected to each other through a virtual switch.

4.3.2 Testing Scenarios

Two Window XP computers implement TCP flood, UDP flood, ICMP flood refer to bandwidth flood attacks using tools like hping3, udpflood.exe, ping respectively or scan port in range 1–300 on Ubuntu computer installed anomaly IDS. Our program will collect and analysis packets in order to detect anomalous in traffic.

4.3.3 Experimental Results

A "confusion matrix" is sometime used to represent the result of, as shown in Table 2 (Naïve Bayes) and Table 3 (Multinomial Logistic Regression). The advantage of using this matrix is that is not only tells us how many got misclassified but also what misclassification occurred. We define the Accuracy, Detection rate and false-alarm:

$$Accuracy = \frac{TP + TN}{TP + TN + FP + FN} \quad Detection - rate = \frac{TP}{TP + FP}$$
$$False - Alarm = \frac{FP}{FP + TN}$$

FN: False Negative, TN: True Negative, TP: True Positive and FP: False Positive (Table 4).

Table 2 Confusion matrix for naïve bayes

Actual	Predicted normal	Predicted TCP flood	Predicted ICMP flood	Predicted UDP flood	Predicted scan port	Accuracy (%)
Normal	110	0	0	0	0	100
TCP flood	1	201	0	0	3	98
ICMP flood	2	0	197	1	0	98.5
UDP flood	0	0	1	147	2	98
Scan port	1	2	1	4	172	95.6

Table 3 Confusion matrix for multinomial logistic regression

Actual	Predicted normal	Predicted TCP flood	Predicted ICMP flood	Predicted UDP flood	Predicted scan port	Accuracy (%)
Normal	110	0	0	0	0	100
TCP flood	0	204	0	0	1	99.5
ICMP flood	1	0	203	1	0	99
UDP flood	0	0	0	149	1	99.3
Scan port	1	0	0	1	177	98.33

Table 4 Comparison between BN and MLR

	Naïve bayes		Multinomial logistic regression	
	Detection rate	False alarm	Detection rate	False alarm
Normal	100	1	100	1
TCP flood	99	0.33	100	0
ICMP flood	98.99	0.4	100	0
UDP flood	96.7	0.625	98.7	0.66
Port scan	97	0.45	98.8	0.5

5 Conclusion

This study constructed an Anomaly-based Intrusion Detection Model based on Naïve Bayes and Multinomial Logistic Regression algorithm. We also experiment IDS's ability of detection using both these methods in the data sets that we created based on four attack scenarios including ICMP flood, UDP flood, TCP flood and Scan Port. The experimental results show that both two methods give very high accuracy and could be applied in practice. However, this is still only the initial test, and more research is needed, in the future we will continue to improve and experiment in a real network environment.

References

1. Lippmann R, Haines JW, Fried DJ, Korba J, Das K (2000) The 1999 DARPA off-line intrusion detection evaluation. Comput Netw 34:597–595
2. Stillerman M, Marceau C, Stillman M (1999) Intrusion detection for distributed systems. Commun ACM 42(7):62–69
3. Chang CC, Lin CJ (2009) LIBSVM: a library for support vector machines. Software available at http://www.csie.ntu.edu.tw/cjin/libsvm. 18th November 2009
4. Anderson J (1980) Computer security threat monitoring and surveillance. James P. Anderson Co, Washington
5. Yu Y, Hao H (2007) An ensemble approach to intrusion detection based on improved multi-objective genetic algorithm. J Softw 18(6):1369–1378

6. Luo J, Bridges SM (2000) Mining fuzzy association rules and fuzzy frequency episodes for intrusion detection. Int J Intell Syst 15(8):687–703
7. Barbard D, Wu N, Jajodia S (2001) Detecting novel network intrusions using bayes estimator. In: Proceeding of the 1st SIAM international conference on data mining
8. Kuchimanchi G, Phoha V, Balagani K, Gaddam S (2004) Dimension reduction using feature extraction methods for real-time misuse detection systems. In: Fifth annual IEEE proceedings of information assurance workshop, pp 195–202
9. Han J, Kamber M, (2012) Data mining: concepts and techniques. Elsevier, San Francisco
10. Garcia-Teodoro P, Díaz-Verdejo JE, Maciá-Fernández G, Vázquez E (2009) Anomaly-based network intrusion detection: techniques, systems and challenges. Comput Secur 28(1–2):18–28
11. Phoha VV (2002) The springer dictionary of internet security. Springer, New York
12. Vapnik VN (1999) Statistical learning theory. Wiley-Interscience, New York

Implementation of Miniaturized Automotive Media Platform with Vehicle Data Processing

Sang Yub Lee, Sang Hyun Park, Duck Keun Park, Jae Kyu Lee and Hyo Sub Choi

Abstract Among the variety of vehicle technology trend issues, the biggest one is focus on automotive network system. Especially, optical network system is mentioned. The outstanding point in optical network system for vehicular environment is that realization of automotive media platform. This paper is introduced the implementation of media platform which is consist of optical network as called Media Oriented System Transport (MOST) and engraftment of vehicle data processing module collected from On-Board Diagnosis (OBD) via Controller Area Network (CAN) transformed into Extensible Marked Language (XML) schema structure. Also, for the reliability of the development, it is executed the MOST compliance test of designed platform.

Keywords In vehicle network system · MOST · OBD · CAN · XML · Automotive media platform · Compliance Test

---

S. Y. Lee (✉) · S. H. Park · D. K. Park · J. K. Lee · H. S. Choi
Jeonbuk Embedded System Research Centre, Korea Electronics Technology Institute, Dunsan-ri, Bongdong-eup, Wanju-gun, Jeolabuk-do 565-902, Republic of Korea
e-mail: syublee@keti.re.kr

S. H. Park
e-mail: shpark@keti.re.kr

D. K. Park
e-mail: parkdk@keti.re.kr

J. K. Lee
e-mail: jae4850@keti.re.kr

H. S. Choi
e-mail: hschoi@keti.re.kr

J. J. (Jong Hyuk) Park et al. (eds.), *Multimedia and Ubiquitous Engineering*, Lecture Notes in Electrical Engineering 240, DOI: 10.1007/978-94-007-6738-6_140, © Springer Science+Business Media Dordrecht(Outside the USA) 2013

1 Introduction

Most of people want to be experienced in high quality audio streaming service while they drive. According to customer's demands, MOST system is developed to provide an efficient and cost effective fabric to transmit audio data between any devices attached to the harsh environment of automobile. For the simply network setup, it can be obtained the convenient installation, high fuel efficiency from the usage of plastic optical fiber and high reliability without electro-magnetic problems. In realization terms of car audio system, the streaming service gathering automobile information via CAN be enables to tune the volume automatically for their status in automobile. To develop the self-contained audio system, the complex information having vehicular status is needed to be accessed and interlocked.

Section 2 is MOST network system and MOST network service is explained. Vehicle data processing module which transforms into XML data format from vehicular status information to be collected by CAN network is described in Sect. 3. In Sect. 4, developed platform and demonstration of automotive media system linked with vehicle communication module are shown. Particularly, for MOST network system, compliance test is executed whether the designed platform is working properly as a network device or not. This paper is concluded in Sect. 5.

2 MOST Network System

MOST is the de-factor standard for efficient and cost effective networking of automotive multimedia and infotainment system [1, 2]. The current MOST standards released MOST150, 150 means that 150 Mbps network bandwidth with quality of services is available. To satisfy the demands from various automotive applications, MOST network system provides three different message channels: control, synchronous used in streaming service and asynchronous channel only for packet data transmission. Described in Fig. 1, proposed network system is consist of MOST devices with CAN bus line.

2.1 MOST Network Services for External Host Controller

Network service is based on the network interface controller and provides a programming interface for the application which basically consists of the function blocks. In this paper, it comprise designed platform for transferring streaming data in synchronous channel. It contains mechanisms and routines for operating and managing the network and it ensures dynamic behavior of the network. The network service is implemented on the external host controller as shown in Fig. 2.

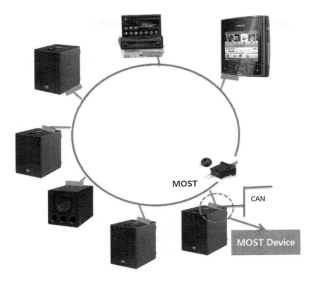

Fig. 1 Automotive media service model

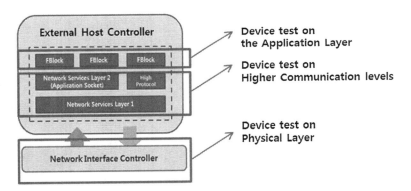

Fig. 2 The development structure applied to platform and compliance test

Being explained the network services; MOST devices have the unique interface which transfers the data between the physical layer network and processor as an external host controller [3]. Working in conjunction with the clock manager, the network port recovers the network clock for time synchronization. And then, it is executed for decoding the received data and delivered to the microprocessor. Transferred data is routed into appropriate memory destination on and off in platform.

2.2 MOST Compliance Test

The integration of network system and modules is particularly challenging due to their complexity. An effective measure is to introduce a certification test process so that the devices and modules are tested before they are integrated into the system. As described in Fig. 2, MOST compliance test process is divided into three courses: device test on physical layer, higher communication levels and application layer.

With regard to the physical layer compliance, the measurement point is to check the signal characteristics for constancy. For the normal behavior, power, error, ring break diagnosis and network management, higher communication level compliance tests are performed. Distinctive features and the dynamic behaviors are tested in the application layer compliance test.

3 Vehicular Status Information

3.1 Vehicle Data Processing Module with XML Schema

As described in this paper, the specific module collecting and analyzing automobile data is called as the vehicle data processing module. In the processing module, vehicle data transmission and channel connection is served in data convergence engine. Existed analysis and process of vehicle information has a limit that it cannot be used in the mobile or telematics terminal, since these devices cannot support CAN communication sockets. Also, as the number of sensors connected with CAN bus are increasing, data management and transmission are faced with high bus load traffic. It causes the system breaking and errors. In order to recover these problems, XML data classification method is proposed. Data Convergence Engine (DCE) provides the process of XML type of collecting data via OBD. As depicted Fig. 3, at the first step, Data Processing Module set configuration parameter from the targeted platform and defines data format and schema to be fit XML format on. And then, from the VSF Library, the variety of the vehicle information such as automobile speed, gear shift and steering etc. are regenerated to XML regards to the type of schema. Through the schema fetcher and reader, data exchange and extraction are performed in Data Processing Module (Fig. 4).

Implementation of Miniaturized Automotive Media Platform

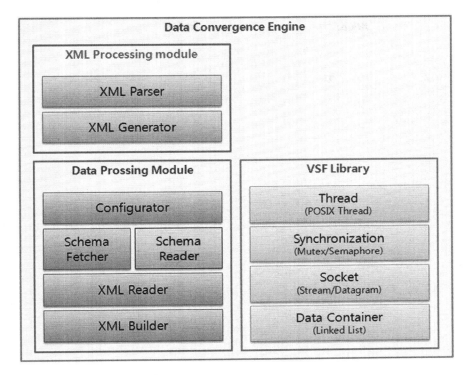

Fig. 3 Data convergence engine

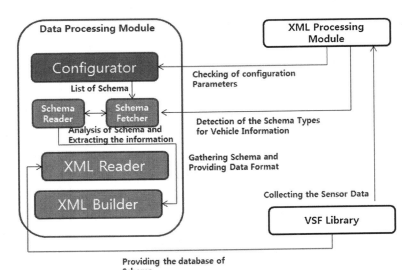

Fig. 4 The flow of data processing module

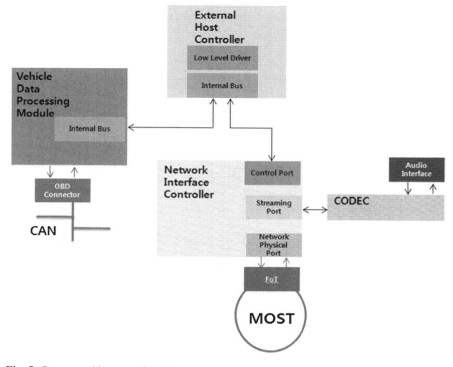

Fig. 5 System architecture of media gateway

4 Implementation of Automotive Media Platform

4.1 System Architecture

The external host controller communicates with network interface controller via I2C bus and connected with OBD via CAN bus [4, 5]. As shown below, the streaming port included in network interface controller can be used for stereo audio exchange between the network and physical audio port. The external host controller may support for managing the streaming audio exchange remotely.

As described in Fig. 5, in order to make the data connection path, Data transferred from the internal bus accesses into embedded memory space. External host controller is used for low graded microprocessor and internal local bus. It is facilitated commercial product.

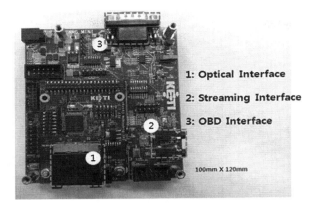

Fig. 6 Streaming service platform

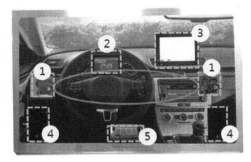

Fig. 7 System demonstration

4.2 Developed Media Platform

As mentioned above that external host controller having network service for network management and schema structure generation for vehicle information processing includes XML converting block. XML converting block conducts exchange of the command line used to produce assembler and C header files from the XML. Being in this method, the XML converting block has to be embedded in external host controller. With this convenient property, for other in-vehicle network devices connecting with heterogeneity communication, it is useful exchanging vehicle status information between data processing terminals which can show indication information in car. In Fig. 6, the interfaces of automotive media platform are shown. And system demonstration is built with designed platform connected with audio speaker and amplifier, vehicle parameter generator which makes the virtual automobile environment via OBD interface in Fig. 7.

Measurement of SP2 output signal of the DUT(5_Tmax_Umin_Pmax_DTmin)

$T_{23^\circ C}/U_{typ}/P_{opt3}$ (max) / Duty Cycle $_{min}$	Setup 1				
Parameter	Min	Max	Unit	Test Value	Result
Optical power					
Optical output power Popt2	-8.5	-1.5	dBm		not tested
Data consistency					
unlock ET Block		-	status		not tested
bit error ET Block		0	error		not tested
unlock PhLSTT		-	status		not tested
bit error PhLSTT		0	error		not tested
Signal Integrity					
Rise time 20% 80% (tr2)	-	0.5	UI	0.246	passed
Fall time 80% 20% (tf2)	-	0.5	UI	0.199	passed
Transferred Jitter (Jtr2) (RMS)		112	ps	10.113	passed
Positive Overshoot		0	Error	0	passed
Undershoot 2UI		0	Error	0	passed
Undershoot 3UI		0	Error	0	passed
Undershoot 4UI		0	Error	0	passed
Undershoot 5UI		0	Error	0	passed
Undershoot 6UI		0	Error	0	passed
Alignment Jitter acc. To Eye MASK	-	0	Error	0	passed
Bit Rate	147.4265	147.4854	Mbits/s	147.455	passed

Fig. 8 The result of physical layer compliance test

```
                  expected                    received              State
                  result                      result
DUT responds correctly to all queries
DUT responds correctly to all queries
DUT ok

The DUT has passed the test
```

Fig. 9 The summary of higher communication level compliance test

4.3 Compliance Test Result

Figure 8 shows the test result for physical layer compliance for MOST150. Based on an analysis of the higher communication level, the compliance test cases were determined as network behavior, system configuration and notification method in Fig. 9.

5 Conclusion

This paper is introduced the development of the automotive media platform based on optical network system. Particularly, optical network system, MOST, to be optimized sound level depending on vehicle environment is satisfied with reducing the weight and ensuring the reliability for free of electro-magnetic problems. With vehicle status information, designed network platform realized and demonstrated that self-contained MOST speaker and woofer can be controlled and tuned their volume level automatically to be served to passengers more conveniently. Especially, outstanding point for implementation is represented XML data interchange from vehicle status information. For being ensured reliability of designed platform, it is performed compliance test of physical and higher communication level.

Acknowledgments This work was supported by the IT R&D program of MKE/KEIT [1004091, the development of Automotive Synchronous Ethernet combined IVN/OVN and Safety Control System for the 1 Gbps class].

References

1. Grzemba A (2011) MOST book from MOST25 to MOST150. MOST cooperation. Franzis, Deggendorf
2. Lee SY, Park SH, Choi HS, Lee CD (2012) MOST network system supporting full-duplexing communication. IEEE ICACT, Korea, pp 1272–1275
3. Lee SY, Kim BC, Choi HS, Lee CD (2012) Development of automotive media streaming device with MOST network system. SMA2012, China
4. Otto S, Rindha R, Jan L (2007) Communication in automotive system principles, limits and new trends for vehicles, airplanes and vessels. IEEE ICTON, pp 1–6
5. Godavarty S, Broyles S, Parten M (2000) Interfacing to the on-board diagnostic system. In: Proceedings of IEEE vehicular technology conference, 52nd-VTC, vol 4, pp 24–28

Design of Software-Based Receiver and Analyzer System for DVB-T2 Broadcast System

M. G. Kang, Y. J. Woo, K. T. Lee, I. K. Kim, J. S. Lee and J. S. Lee

Abstract In this paper, a receiver and an analyzer system of Digital Video Broadcasting_2nd Generation Terrestrial (DVB-T2) were designed using Digital Video Broadcasting_2nd Generation Satellite (DVB-S2), and Digital Video Broadcasting_2nd Generation Cable (DVB-C2). This software-based receiver and analyzer system were implemented by memory sharing techniques for minimization of system overloads.

Keywords DVB-S2 · DVB-T2 · DVB-C2 · Receiver · Analyzer

M. G. Kang (✉) · Y. J. Woo (✉)
Hanshin University, #411 Hanshindae-gil, Osan-si 447-491, Korea
e-mail: Kangmg@hs.ac.kr

Y. J. Woo
e-mail: cosch0610@gmial.com

K. T. Lee
Korea Electronics Technology Institute, 10FL, Electronics Center, #1599,
Sangam-dong, Mapo-gu, Seoul, Korea
e-mail: ktechlee@keti.re.kr

I. K. Kim
Innodigital Co. Ltd., #907, KINS Tower, 25-1 Jeongja-dong, Bundang-Gu, Seongnam-Si
463-782, Korea
e-mail: ikkim@innodigital.net

J. S. Lee
Haesung Optics Co. Ltd., 3B/3L 921 GosaeK-dong, Kwonsun-gu, Suwon-si 441-813, Korea
e-mail: jsyi@hso.co.kr

J. S. Lee
LG CNS Co. Ltd., LG Twin Towers, West 9F, #20, Yoido-dong,
Youngdungpo-gu, Seoul 150-721, Korea
e-mail: jsunglee@lgcns.com

J. J. (Jong Hyuk) Park et al. (eds.), *Multimedia and Ubiquitous Engineering*,
Lecture Notes in Electrical Engineering 240, DOI: 10.1007/978-94-007-6738-6_141,
© Springer Science+Business Media Dordrecht(Outside the USA) 2013

1 Introduction

As Analogue Switch off (ASO) announced an end to the era of analogue TV, broadcasting system throughout the world is swiftly being changed into digital. In the case of European digital broadcast, DVB-S and DVB-C were established in 1994. And DVB-T was established in early 1997 and experimental broadcast took place for the first time in 1998 in UK.

Most European countries are fully covered with digital television and many has switched off analogue TV. According to the DVB organization, DVB 2nd Generation Broadcasting System performance gaining over DVB 1st Broadcasting System is around 30 % at the same transponder bandwidth and emitted signal power.

By utilizing broadcasting frequency that will be newly acquired by ASO, DVB 2nd Generation Broadcasting System is expected to provide multi-channeled High-Definition TeleVision (HDTV) service and data service in various forms.

This paper realizes integrated receiver and analyzer system that can receive DVB-S2/T2/C2 data using File or Ethernet packet and analyzed received frame. Also, this paper introduces technology about DVB 2nd Generation Broadcasting System.

2 S/W Design and Analysis of DVB-T2 Receiver

2.1 DVB 2nd System Overview

DVB-S2 standard is based on DVB-S, and DVB-S2 is envisaged for broadcast services including standard and HDTV, interactive services including in internet access, and data content distribution.

DVB-S2 have two key features that were compared to the DVB Standard are a powerful coding scheme based on a LDPC code.

Next is Variable Coding and Modulation (VCM) and Adaptive Coding and Modulation (ACM) modes, which allow optimizing bandwidth utilization by dynamically changing of transmission parameters [1].

DVB-T2 standard is an expanded form of DVB-T standard, its core being improving digital TV transmission rate, as well as robustness and flexibility of transmission network usage. It also focuses on receiving digital TV contents via mobile devices and receiving broadcasting service smoothly while on transportation such as subway, train, and bus [2].

DVB-C2 standard is broadcasting transmission of digital television over cable. To compare with DVB-C, by using LDPC coding and new modulation techniques, DVB-C2 is greater than DVB-T2, showing 30 % higher spectrum efficiency under the same conditions, and the gaining in downstream channel capacity will be greater than 60 % [3].

2.2 Baseband Frame and Input Modes

By principle, DVB standards are developed to be compatible with each other. Therefore, if an outstanding solution is developed, this solution can also be adopted by other standards. According to such principal, Baseband Frame for data packaging established in DVB-S2 is also used in DVB-T2, DVB-C2.

DVB 2nd Generation Broadcasting System input mode utilizes Generic Encapsulated Stream (GSE), Generic Continuous Stream (GCS), Generic Fixed-length Packetized Stream (GFPS) in addition to MPEG2-TS (Transport Stream).

These 4 types of input modes are transmitted separately or collectively through a transmission method titled Physical Layer Pipes (PLPs) [5].

2.3 Processing Design of DVB 2nd Receiver and Analyzer

DVB-T2 Receiver and analyzer system received input streams, depending on whether user chooses File or Ethernet stream. Input stream is transmitted in DVB Piping format. Layer signal information is parsed from the reconstructed X2MI frame which then is displayed on the application windows, and the system type is extracted on Layer signal information.

Baseband Frame is an extracted base on L1 information. From the extracted baseband frame header information, stream input method is acquired. Decoding process per each input method follows. Basic processing procedure of receiver and analyzer system is shown in Fig. 1.

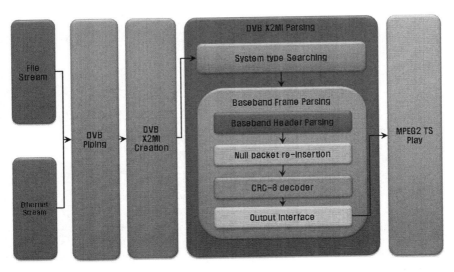

Fig. 1 Process diagram of DVB 2nd generation broadcasting receiver and analyzer system

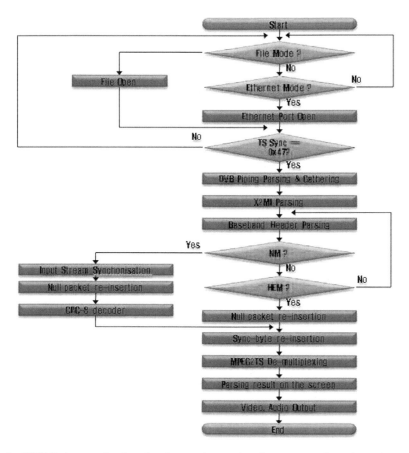

Fig. 2 DVB 2nd generation broadcasting receiver and analyzer system flow chart [6, 7]

To explain the processing procedure from paper in further detail, stream that needs to be processed according to their File or Ethernet packets mode is selected through realized application.

As inputted DVB 2nd stream is transmitted through DVB-Piping method, Input packet is parsed using TS module and X2MI frame restructured. Layer signal acquired from restructured X2MI frame is parsed and used to understand inputted broadcast system type and overall frame structure.

Baseband Frame is extracted and structured from this restructured X2MI frame. Composed Baseband Frame Header is parsed to determine system parameters (e.g., input types (TS or GSE), Mode types (Normal Mode or High Efficiency Mode), and whether additional modules (Input Stream Synchronization (ISSY)), Null Packet Deletion (NPD) is needed). According to acquired Header information, ISSY, NPD, CRC8 decoder and Sync-byte re-insertion module are applied.

Design of Software-Based Receiver and Analyzer System 1151

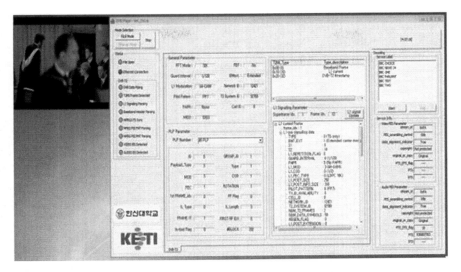

Fig. 3 Performance analysis of receiver and analyzer software result

This TS packet is restructured from Baseband Frame's Data Field. TS module is once again used on these restructure TS packet to compose Program Specific Information (PSI)/Service Information (SI) table. System flow procedure is shown in Fig. 2.

These supposed DVB-T2 receiver, and analyzer system have minimized the system overload by reconstructing TS system used as transmitting method of DVB-S2, DVB-T2, and DVB-C2 into a common module block. Method tiled X2MI, it's made from changed T2MI feature.

This realization result depends on the stream transmission through File or Ethernet packets, and on DVB 2nd Generation broadcast output analysis window on X2MI structure that transmits system type information, baseband frame and information of actual contents in Fig. 3.

This system analyzes transmission information of the system that transmits DVB-S2, DVB-T2 and DVB-C2 and confirms normal function by replaying the transmitted contents.

3 Conclusion

In this paper, the design, implementation of a receiver, and analyzer for DVB 2nd Generation Broadcasting System with window system based software were shown. These receiver and analyzer system were suggested for normal function by analyzing input system and Baseband Frame of received stream that varies depending on each system and playing the contents successfully.

Acknowledgments This work was supported by Hanshin University, and Korea government (MKE; Ministry of Knowledge Economy, #10039717/#10035547, NIPA; IT Industry Promotion Agency, and KOFST; The Korean Federation of Science and Technology Societies, 2012 Science and Technical Support specialists Supporters (H5701-12-1002), Small and Medium Business Administration (#S2064061), and KORIL-RDF: Korea-Israel Industrial R&D Foundation).

References

1. ETSI EN 302 307 (2009) Digital Video Broadcasting (DVB), Second generation framing structure, channel coding and modulation systems for Broadcasting, Interactive Services, News Gathering and other broadband satellite applications (DVB-S2)
2. ETSI EN 302 755 (2009) Digital Video Broadcasting (DVB), Frame structure channel coding and modulation for a second digital terrestrial television broadcasting system (DVB-T2)
3. ETSI EN 302 769 (2009) Digital Video Broadcasting (DVB), Frame structure channel coding and modulation for a second generation digital transmission system for cable system (DVB-C2)
4. Seo JW, Kang MG, Jeon ES, Kim DK (2011) Performance evaluation of a DVB-T2 receiver with iterative de-mapping and decoding in MISO transmission mode. KSII 12(2)
5. ETSI TS 102 773 (2009) Digital Video Broadcasting (DVB), Modulator Interface (T2-MI) for a second generation digital terrestrial television broadcasting system (DVB-T2)
6. Kang M et al (2012) Method of processing DVB-T2/S2/C2 piping-format broadcasting signals with memory sharing,and computer-readable recording medium with broadcasting signal processing program. USA Patent Pending (PCT/KR#2012/004913)
7. Paik JH et al (2011) Design of window based DVB-T2 receiver with file, and ethernet modes. KSII International conference on internet (ICONI)

Age-Group Classification for Family Members Using Multi-Layered Bayesian Classifier with Gaussian Mixture Model

Chuho Yi, Seungdo Jeong, Kyeong-Soo Han and Hankyu Lee

Abstract This paper proposes a TV viewer age-group classification method for family members based on TV watching history. User profiling based on watching history is very complex and difficult to achieve. To overcome these difficulties, we propose a probabilistic approach that models TV watching history with a Gaussian mixture model (GMM) and implements a feature-selection method that identifies useful features for classifying the appropriate age-group class. Then, to improve the accuracy of age-group classification, a multi-layered Bayesian classifier is applied for demographic analysis. Extensive experiments showed that our multi-layered classifier with GMM is valid. The accuracy of classification was improved when certain features were singled out and demographic properties were applied.

Keywords Age-group classification · Gaussian mixture model · Feature selection

---

C. Yi
Research Institute of Electrical and Computer Engineering, Hanyang University,
Seoul, Korea
e-mail: d1uck@hanyang.ac.kr

S. Jeong (✉)
Department of Information and Communication Engineering, Hanyang Cyber University,
Seoul, Korea
e-mail: sdjeong@hycu.ac.kr

K.-S. Han · H. Lee
Electronics and Telecommunications Research Institute, Smart TV Service
Research Team, Daejeon, Korea
e-mail: kshan@etri.re.kr

H. Lee
e-mail: hkl@etri.re.kr

J. J. (Jong Hyuk) Park et al. (eds.), *Multimedia and Ubiquitous Engineering*,
Lecture Notes in Electrical Engineering 240, DOI: 10.1007/978-94-007-6738-6_142,
© Springer Science+Business Media Dordrecht(Outside the USA) 2013

1 Introduction

Due to the advent and popularization of interactive TV and internet TV, there is a need for research into user profiling based on viewer watching history to support targeted advertising [1]. In a previous study, we proposed a method for computing TV viewer preferences for nine types of genre based on viewing history data [2]. To avoid including meaningless behavior such as random/aimless channel changing, we modeled the preference function as a beta distribution. However, extensive experiments showed that simple preference profiling based on an averaged preference for a few genres was not enough to recognize age-groups. There are several reasons for this: TV watching data have a multi-modal characteristic, no feature by itself can be used to identify an age-group, and one person's watching data are typically corrupted by other members of his/her family (watching data are almost never limited to one person but rather include the behavior of multiple viewers).

Here, we propose four methods to overcome these difficulties. First, the types of preferences are more specifically divided, as they are divided into 70 types of genre by the hour and day of the week. This results in 645 features. Then, to reflect the multi-modal distribution of the data, the distribution of features according to viewer age-group is modeled using a Bayesian Gaussian mixture model (GMM). Second, we propose a multilayered classifier to improve the performance of the system through the analysis of demographic properties. Third, because whole features complicate the classification of viewer age-groups, we propose a feature-selection method based on weighting calculated with a training set. Finally, *a priori* knowledge based on the demographic composition of the training data is used in the classification process. In experiments, TV viewer age-groups were inferred using our trained, multi-layered classifier. We evaluate and discuss the results of the classification below.

2 The Proposed Method

2.1 Basic Bayesian Model

Preference values, which are used as observation data in the present study, are computed by dividing each viewer's watching time by the total viewing time. Preference features consist of 70 types of program genre, eight time periods in a day, and 7 days of the week. Hence, a total of 645 profiling selected features are used in the training and classification steps.

Every extracted feature is assumed to be independent and thus is individually processed in the training step. If observation F has N number of data, the posterior probability is expressed as in Eq. (1), where C is a class. In the present work, the class is the age-group of the TV viewer.

$$p(C|F_1, F_2, \ldots, F_N) = \frac{p(C)p(F_1, F_2, \ldots, F_N|C)}{p(F_1, F_2, \ldots, F_N)} \tag{1}$$

Using the chain rule of conditional probability, Eq. (1) can be rewritten as Eq. (2):

$$p(C|F_1, F_2, \ldots, F_N) \propto p(C) \prod_{i=1}^{N} p(F_i|C) \tag{2}$$

where $p(C)$ is the *a priori* probability, and $p(F_i|C)$ is a likelihood. In this paper, the demographic statistics of an age-group are used as $p(C)$. We assume that the likelihood has a multimodal distribution. Therefore, we model the distribution using GMM, as shown in Eq. (3):

$$p(F_i|C = c_a) = \sum_{j=1}^{K_a} w_j \cdot N(F_i|\mu_j, \Sigma_j) \tag{3}$$

where w_j is the prior probability (weight) of the jth Gaussian, $\sum_{j=1}^{K} w_j = 1$ and $0 \leq w_j \leq 1$, and K_a is a peak number for class c_a. To make it possible to select K_a automatically, the Figueiredo–Jain algorithm is used in the training step [4].

The goal of this research is classifying an age-group for each person of family members and the outcome of single classifier based on basic Bayesian model is an existence of respective age-group; therefore, several classifiers are used in this paper to classify age-groups for all members of family.

2.2 Multi-Layered Structure

Our proposed multi-layered classifier assumes that preprocessing of data may be helpful for classifying family members into age-groups. To this end, the system first determines whether anyone in the family is younger than 10 years of age. This is the most important criterion because the composition of a family is highly affected by the presence of this age-group. Therefore, this classifier is at the highest level (C1). If no one is in this age-group, the system next identifies anyone over 60 years of age. This is the second-most important criterion and itself makes up the next highest level of classification (C2). All subsequent classifiers are at the C3 level; they include groups 10–19, 20–29, 30–39, 40–49, and 50–59 years of age (hereafter, the 10s, 20s, 30s, 40s, and 50s age-groups, respectively). Hence, a total of 17 classifiers are represented. Figure 1 shows the proposed system.

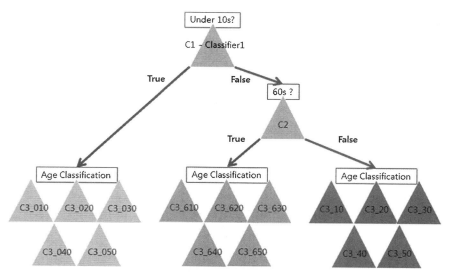

Fig. 1 The proposed multi-layered classifier

2.3 Feature Selection

The feature-selection step verifies the usefulness of each viewing feature (i.e., each combination of genre, day, and time of day) using the training dataset. A score is used to weight each feature based on its estimated usefulness for classification. Here, we assume that each feature is independent of all others.

Equation (4) is a utility function that determines whether a feature can be used to produce a true-positive age-classification result based on the training dataset, where C_1 represents a class of positive data, and C_2 is a class of negative data. Equation (5) does the same for true-negative results. The result of these two utility functions is 1 when a feature always correctly determines true-positive and true-negative results, respectively.

$$r_{i,a} = \begin{cases} 1, & p(C=c_1|f_i \in F_{c_1}) > p(C=c_2|f_i \in F_{c_1}) \\ 0, & p(C=c_1|f_i \in F_{c_1}) < p(C=c_2|f_i \in F_{c_1}) \end{cases} \quad (4)$$

$$s_{i,b} = \begin{cases} 1, & p(C=c_2|f_i \in F_{c_2}) > p(C=c_1|f_i \in F_{c_2}) \\ 0, & p(C=c_2|f_i \in F_{c_2}) < p(C=c_1|f_i \in F_{c_2}) \end{cases} \quad (5)$$

Second, the scoring functions shown in Eqs. (6) and (7) are used to compute the reliability of each feature for correctly classifying the entire training dataset.

$$score(F_i, C=c_1) = \frac{1}{|M|} \sum_{a=1}^{M} r_{i,a} \quad (6)$$

$$score(F_i, C = c_2) = \frac{1}{|M|} \sum_{b=1}^{M} s_{i,b} \tag{7}$$

where M is the number of data points in the training dataset. If a score is above a given threshold, then the feature is considered reliable for classification. We determine this threshold using a heuristic searching process. In other words, through an exhaustive search, the values of $score(F_i, C = c_1)$ and $score(F_i, C = c_2)$ are chosen as the threshold when the accuracy of the posterior probability $p(C|F_{trainning})$ is highest based on the training dataset $F_{training}$.

2.4 A Prior Knowledge for Demographic Data

Next, the age-group classification is processed using the trained classifiers, where $F_{s\in\{S\}}$ is the set of features selected during the previous steps described above. Maximum *a posteriori* (MAP) estimation is applied as shown in Eq. (8):

$$c^* = p(C) \prod_{i=1}^{N} p(F_{s,i}|C) \tag{8}$$

To prevent computational underflow, Eq. (8) is transformed using a logarithmic expression, as shown in Eq. (9):

$$c^* = log\{p(C)\} + \sum_{i=1}^{N} log\{p(F_{s,i}|C)\} \tag{9}$$

where $p(C)$ is the *a prior* probability, and we use the ratio according to age-group for demographic statistics which is analyzed with the training data.

3 Experimental Result

We tested our system on a dataset of the TV watching history of 2060 family members from 689 families. The time period considered was April 1 to June 30, 2012 (13 weeks/91 days). The maximum family size was 6. For the training step, we selected 85 % of the dataset (585 families/1758 members) as the training set. The remaining data points (104 families/302 members) were used as the test set. The experiments were run on a Windows 7 operating system using MATLAB 2010. A modified version of GMMBayes Toolbox [5] was used for GMM training and classification.

Table 1 shows the results of the age-group classification for family members. Method 1, which used unprocessed, raw data, did not work for the 30s and 50s

Table 1 Accuracy of age-classification experiments

Ages (s)	Method 1 (%)	Method 2 (%)	Method 3 (%)
<10	45.19	47.12	50.96
10	64.04	67.31	68.27
20	59.62	77.88	76.92
30	–	44.23	43.27
40	43.27	57.69	59.62
50	–	63.46	63.46
60	65.38	65.50	62.50
Average	**55.50**	**60.03**	**60.71**

age-groups and had a 55.50 % average accuracy. Method 2 used the feature-selection method proposed in this paper and had a 60.03 % average accuracy; additionally, it worked well for the 30s and 50s age-groups. Method 3, which also used *a prior* knowledge estimations, as described above, had an average accuracy of 60.71 %. These results indicate that our proposed feature-selection and *a prior* knowledge steps are meaningful for age-group classification.

The goal of this research is the classification of TV viewer's age-group for family members and even was not known how many people live in a house together. For the ideal watching history data, the viewer should be forced to log in the TV with a remote controller manually. However, in the real situation, almost viewers did rarely log in or change their ID every time. It caused corruption to the viewing history data by other members in the family. Thus, this point is a huge obstacle for an accurate classification of the age-group for family members. We also suffer from this obstacle for classification in this research; however we try to overcome these difficulties with the proposed four approaches. As a result, our proposed method could be reasonably classified the age-group for family members.

4 Conclusions

Our proposed probabilistic model for age-group classification based on complex viewing history data combines Bayesian classifiers with GMM methods. In experiments, our proposed system classified age-groups with reasonable accuracy and well enough to be incorporated into a targeted advertising system.

Acknowledgments This work was supported by the Electronics and Telecommunications Research Institute (ETRI) R&D Program of Korea Communications Commission (KCC), Korea [11921-03001, "Development of Beyond Smart TV Technology"].

References

1. Spangler WE, Gal-Or M, May JH (2003) Using data mining to profile TV viewers. Commun ACM Mob Comput Oppor Chall 46(12):66–72
2. Lee S, Park S, Hong J, Yi C, Jeong S (2012) Inference for the preference of program genre using audience measurement information. In: International conference on information and knowledge, engineering, pp 224–225
3. Wonneberger A, Schoenbach K, Meurs LV (2009) Dynamics of individual television viewing behavior: models, empirical evidence, and a research program. Commun Stud 60(3):235–252
4. Figueiredo MAT, Jain AK (2002) Unsupervised learning of finite mixture models. IEEE Trans Pattern Anal Mach Intell 24(3):381–396
5. GMMBayes Toolbox, Gaussian mixture model learning and Bayesian classification. http://www.it.lut.fi/project/gmmbayes/

Enhancing Utilization of Integer Functional Units for High-Throughput Floating Point Operations on Coarse-Grained Reconfigurable Architecture

Manhwee Jo, Kyuseung Han and Kiyoung Choi

Abstract Supporting floating point operations on coarse-grained reconfigruable architecture (CGRA) becomes essential as the increase of demands on various floating point inclusive applications such as multimedia processing, 3D graphics, augmented reality, or object recognition. However, efficient support for floating point operations on CGRA has not been sufficiently studied yet. This paper proposes a novel technique to enhance utilization of integer units for high-throughput floating point operations on CGRA with experimental results.

Keywords Reconfigurable · CGRA · Floating point · Resource utilization

1 Introduction

Not many years ago, phones are used just for calling, but today they are used for playing audios/videos, surfing web, image processing, and enjoying games. Not only phones but also tablets have been gaining popularity rapidly and are now a part of our daily lives. However, as the funtionality of such mobile devices becomes more diverse and complex, supporting them with limited resources is a big challenge. Multicores are not enough to meet the requirement of compute-intensive programs even though they are suitable for running several control-intensive problems simultanously. ASICs can hardly support various programs since we

M. Jo (✉) · K. Han · K. Choi
Department of EECS, Seoul National University, Seoul, Korea
e-mail: manhwee@dal.snu.ac.kr

K. Han
e-mail: darprin@dal.snu.ac.kr

K. Choi
e-mail: kchoi@dal.snu.ac.kr

J. J. (Jong Hyuk) Park et al. (eds.), *Multimedia and Ubiquitous Engineering*, Lecture Notes in Electrical Engineering 240, DOI: 10.1007/978-94-007-6738-6_143, © Springer Science+Business Media Dordrecht(Outside the USA) 2013

cannot put tens or hundreds of them into a single chip. In addition to that, we have encountered another challenge. While conventional compute-intensive programs such as multimedia applications are based on integer calculation, new ones such as 3D graphics, augmented reality, object recognition, or face recognition require real number operations. Thus, efficient support for both fixed- and floating-point operations with limited resources is also important in future embeded systems.

For the first challenge mentioned above, coarse-grained reconfigruable architecture (CGRA) [1–5] has been proved to be one of viable solutions since it can provide performance and flexibility at the same time. It consists of abundant processing elements (PEs) so it can accelerate the execution of programs by parallel processing. In addition, it can run various applications by changing the functionalilty of hardware dynamically through the reconfiguration of the PEs and the interconnections between them. Thus we focus on CGRA in this paper.

Regarding the second challenge, especially for CGRA, only a few researches have been conducted on CGRA related to floating point operations. Considering that introducing floating-point units in addition to abundant integer units wastes area and power (either of integer units or floating point units are always unused), the approaches in [6–8] share FUs between two types of operations. However, they try only simple approaches resulting in an inefficient use of functional resources. In this paper, we propose a novel architecture that can utilize FUs for both types of operations. In particular, we propose a way to better utilize integer FUs when performing floating point operations on CGRA.

The organization of the paper is as follows. Section 2 presents a base coarse-grained reconfigurable architecture, and Sect. 3 describes a conventional approach to implement floating point operations on the CGRA. Section 4 proposes an approach to enhance the utilization of FUs in PEs. Section 5 compares the base architecture and the proposed architecture in terms of performance and cost. Then Sect. 6 concludes the paper with future directions.

2 Base Architecture

The base coarse-grained reconfigurable architecture (Fig. 1) consists of a general purpose RISC processor, reconfigurable computing module (RCM), hardware for transferring data between local memory and external memory such as direct memory access (DMA) unit. RCM executes compute-intensive code blocks while the RISC processor executes control-intensive code blocks.

RCM has two-dimensional array of PEs, each of which contains functional units such as adder/subtractor, comparator, and shifter, and local register file. Execution control Unit of RCM fetches the configuration for the PE array so that each PE in the PE array executes different operations for its configuration. Each PE has FUs such as adder or shifter, and a local register file. A PE also has an output register connected to the network in the PE array. A PE can communicate with

Fig. 1 A coarse-grained reconfigurable architecture

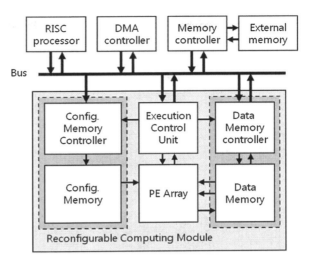

other PEs in the same row and column via the network. Input and output data for the PE array are stored in local data memory which hides the access time to the external global memory.

3 Conventional Approach

Since each PE has only integer FUs inside, a couple of PEs can cooperate in order to execute floating point operations [6]. An floating point operations are divided into a bunch of integer operations and they are executed on a pair of PEs, named an FPU-PE cluster. Specifically, floating point numbers to be calculated are divided into two parts: signed mantissa part and exponent part, then one PE (mantissa PE: PE_M) in a FPU-PE cluster manipulates mantissa parts while the other PE (exponent PE: PE_E) operates exponent parts. PEs in the cluster store the intermediate data in their local registers and communicate with each other via network interconnection between them while running floating point operations. Each PE has a finite state machine (FSM) for configuring the bahaviour of the datapath during multi-cycle floating point operations rather than fetching configuration bits from the configuration memory every cycle. The floating point operations implemented in this work include floating point addition (FADD), subtraction (FSUB) and multiplication (FMUL). The latencies of FADD and FSUB are 8 cycles, while that of FMUL is 4 cycles. In FMUL operation, PE_M multiplies two floating point mantissa parts using integer multiplier module, which is shared by all the PEs in a row for reducing area cost [3].

4 Proposed Approach

Although using integer FUs in order to perform floating point operations increases utility of the architecture, the long latencies of the floating point operations limit the performance of floating point applications (since the base architecture is optimized for running various single cycle integer operations, scheduling integer operations split from a floating point operation requires unnecessary idle cycles due to extra mux and functional unit delays). In order to mitigate it, we propose to share FUs that are temporarily freed during a floating point operation with another floating point operation. Simply, it is overlapping two floating point operations with different operands by sharing FUs between the two floating point operations. That can be done with minimal overhead of additional control logic and registers. Such overlapping of two different floating point operations improves performance by enhancing the utilization of FUs when multiple independent floating point operations are to be executed.

During the execution of floating point operations, each PE employs its FUs to manipulate the input values, a temporal register to keep the intermediate value, and output register to communicate with each other. Since all the FUs in each PE are not used every cycle during execution of a floating point operation, they can be used for another floating point operation. Unlike the FUs, which are frequently freed due to data dependency, two specific registers are utilized almost every cycle so that they cannot be shared with other floating point operations. This problem is solved by inserting an additional register with dedicated communication channel for communication in an FPU-PE cluster for another floating point operation to be overlapped. For intermediate values, there is no need of inserting additional register, but utilizing one of the registers in the local register file in the PE works. Thus just dedicated interconnects to store intermediate value to a register in the register file are added.

For the simplicity of the FSM in each PE, time difference (overlap) between two floating point operations is fixed as a half of the latency of the operation, i.e., 4 cycles for FADD, and 2 cycles for FMUL.

Since the proposed approach cannot be applied to two operations that have between them, we expect it to be mainly applied in unrolled iterations of kernels. In general, however, FADD and FMUL in a kernel program are likely to have data dependency, and thus overlapping FADD and FMUL is not considered in this paper.

Figure 2a shows the usage of FUs of PEs in an FPU-PE cluster for an FADD operation. When another FADD operation starts four cycles later, there are resource conflicts at two different positions shown as black boxes in Fig. 2b. There are two alternative ways of avoiding the conflict: duplication of the conflicted FU, or exetcuting the following floating point operation one cycle later. Fortunately, both of the two conflicts are for incrementing the input value by 1 for rounding. Thus inserting an incrementer instead of another adder/subtractor is good enough for overlapping two floating point operations with 4 cycles of time difference (Fig. 2c).

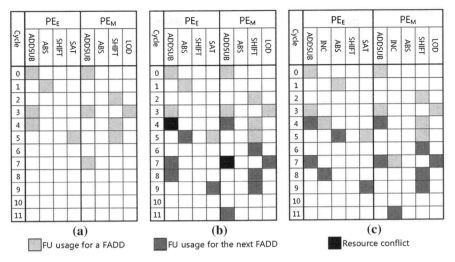

Fig. 2 a Usage of integer FUs in FPU-PE cluster for a FADD, b two overlapping FADDs with conflict, and c two overlapping FADDs with no conflict

On the other hand, there is one conflict on adder/subtractor unit when overlapping two 4-cycle FMUL operations with 2 cycles of time difference. The conflict is also due to a rounding operation, and thus the incrementer inserted for the conflict on FADD is utilized.

The FSM is also changed to support overlapping two floating point operations. Since the cycle difference is fixed, there is no need of parallelizing FSMs but only merging states of the two overlapping floating point operations is needed.

5 Experimental Results

Figure 3 shows the performance of the simple functions running on the existing architecture and the architecture where the proposed approach is applied. Employed benchmark functions are basic arithmetic functions which are frequently used in signal-processing, multimedia applications, 3D graphics, and other mathematical algorithms. Normalized throughput shown in Fig. 3 is throughput ratio of the proposed architecture to the base architecture for each function. By applying the proposed approach of sharing FUs between two floating point operations, we obtain up to 42.9 % better performance and 27.5 % better performance on average compared to the base architecture.

Both of the proposed architecture and the existing architecture are synthesized by Synopsys DesignCompiler with TSMC 130 nm technology library. Each architecture has 8 × 4 array of PEs, one column of shared integer multipliers (i.e., eight multipliers), and configuration and data memory blocks. Synthesis result

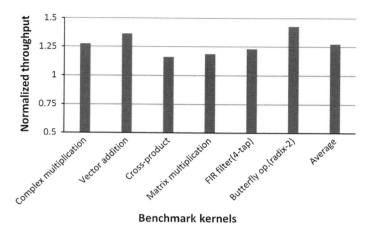

Fig. 3 Normalized throughput of the proposed architecture compared to the base architecture

shows that the area overhead caused by the proposed approach is about 13.8 % at the same clock speed of 330 MHz. It is only about 2.9 % of the whole RCM with configuration and data memory blocks. The area overhead is from FSM and decoder on the control path as well as additional registers and incrementers.

To consider both performance and area cost at the same time, we simply introduce a term, gain, as throughput ratio divided by area ratio. The gain of the proposed approach to the base architecture is 1.12, which indicates about 12 % improvement over the base architecture.

6 Conclusion

This paper proposes a novel mechanism to utilize integer functional units during execution of floating point operations. By sharing functional units, we can support floating point operation with minimal overhead. Especially the proposed overlapping technique enhances the utilization of functional units so that we can accelerate floating point kernels effectively. Experimental results show that our approach has 12 % better throughput-to-area ratio compared to the previous sharing schemes. As a future work, sharing FUs among different types of floating point operations, or operations with data dependency could be researched in order to reach further improvements.

Acknowledgments This work was supported by the National Research Foundation of Korea (NRF) grant funded by the Korean government (MEST) (No. 2012-0006272).

References

1. Hartenstein R (2001) A decade of reconfigurable computing: a visionary retrospective. In: Proceedings of the 4th conference on design, automation and test in Europe. IEEE Press, New York, pp 642–649
2. Singh H, Lee MH, Lu G, Bagherzadeh N, Kurdahi FJ, Filho EMC (2000) Morphosys: an integrated reconfigurable system for data-parallel and computation-intensive applications. IEEE Trans Comput 49(5):465–481
3. Kim Y, Kiemb M, Park C, Jung J, Choi K (2005) Resource sharing and pipelining in coarse-grained reconfigurable architecture for domain-specific optimization. In: Proceedings of the 8th conference on design, automation and test in Europe. IEEE Press, New York, pp 12–17
4. Novo D, Moffat W, Derudder V, Bougard B (2005) Mapping a multiple antenna SDM-OFDM receiver on the ADRES coarse-grained reconfigurable processor. In: Proceedings of the 9th IEEE workshop on signal processing systems design and implementation. IEEE Press, New York, pp 473–478
5. Hasegawa Y, Abe S, Matsutani H, Amano H, Anjo K, Awashima T (2005) An adaptive cryptographic accelerator for IPsec on dynamically reconfigurable processor. In: Proceedings of the IEEE conference on field-programmable technology. IEEE Press, New York, pp 163–170
6. Jo M, Arava VKP, Yang H, Choi K (2007) Implementation of floating-point operations for 3D graphics on a coarse-grained reconfigurable architecture. In: Proceedings of the 20th IEEE international SOC conference. IEEE Press, New York, pp 127–130
7. Syed MA, Schueler E (2006) Reconfigurable parallel computing architecture for on-board data processing. In: Proceedings of the 1st NASA/ESA conference on adaptive hardware and systems. IEEE Press, New York, pp 229–236
8. Brunelli C, Garzia F, Rossi D, Nurmi J (2010) A coarse-grain reconfigurable architecture for multimedia applications supporting subword and floating-point calculations. J Syst Architect 56(1):38–47

An Improved Double Delta Correlator for BOC Signal Tracking in GNSS Receivers

Pham-Viet Hung, Dao-Ngoc Chien and Nguyen-Van Khang

Abstract Multipath is one of main error sources in code tracking in global navigation satellite system. One of the first approaches to mitigate multipath is called Double Delta Correlator (DDC). This technique originally applied to Global Positioning System signals (C/A signals) which are Binary Phase ShiftKeying ones and have a single–peak autocorrelation function. Latter, it also applied to new GNSS signals which are binary offset carrier (BOC) modulated ones. It shows a good code multipath performance for mid-delayed and long-delayed multipath in both cases. For short-delayed multipath, there is still remaining error. This paper presents an Improved DDC that could reduce the error for short-delayed multipath signals. This improved DDC method is proposed for BOC signals. The simulation results show a better performance of proposed method than conventional DDC.

Keywords BOC signal · Double delta correlator · Multipath mitigation

1 Introduction

Currently, binary offset carrier (BOC) modulation has been recommended to some new navigation signals such as Galileo L1OS (OS for Open Service), Global Positioning System (GPS) future L1C (New Civilian L1) [1]. The BOC (m, n) signal is created by modulating a sine wave carrier with the product of a spreading code and a square wave subcarrier. The parameter m stands for the ratio between the subcarrier frequency and the reference frequency $f_0 = 1.023$ MHz, and n stands for the ratio between the code rate and f_0.

P.-V. Hung (✉) · D.-N. Chien · N.-V. Khang
School of Electronics and Telecommunications, Hanoi University of Science
and Technology, Hanoi, Vietnam
e-mail: hung.phamviet@hust.edu.vn

J. J. (Jong Hyuk) Park et al. (eds.), *Multimedia and Ubiquitous Engineering*,
Lecture Notes in Electrical Engineering 240, DOI: 10.1007/978-94-007-6738-6_144,
© Springer Science+Business Media Dordrecht(Outside the USA) 2013

The most common multipath mitigation techniques are based on tracking loop structure which contains carrier phase locked loop (PLL) and code delay locked loop (DLL) [2]. Conventionally, two correlators spaced at one chip from each other are used in delay estimator. It tries to track the delay of the direct signal by correlating the down-converted received signal with replicas of local generating codes in the receiver. However, the classical estimator fails to estimate the multipath error envelope (MEE) accurately. Thus several evolutions have been introduced in the literature in order to mitigate the influence of MP signals, especially in short delayed multipath scenarios. Examples of these are Narrow Correlator [3], DDC [4], early/late slope technique, Early1/Early2 tracker [5]. Initially, these techniques were proposed to GPS signals which are Binary Phase Shift Keying (BPSK) ones and have triangular autocorrelation function.

Among these above mentioned MP mitigation techniques, DDC have the best code multipath performance for medium-to-long delayed MP [4]. However, DDC are still not good enough for short-delayed MP environment such as urban canyons, which is a key motivation for present researchers.

The purpose of this paper is to introduce an improved DDC method suitable for general BOC signals. It is based on conventional DDC. The proposed method will estimate the tracking error of DDC when tracking signal in closely spaced MP environment. Then, the estimate is applied to the DDC as modification in order to reduce tracking error.

The rest of this paper is organized as follows. In Sect. 2, an overview of DDC is presented. In Sect. 3 the improved DDC is proposed. The simulation results of conventional DDC and improved DDC in multipath delay estimation is discussed in Sect. 4, and finally Sect. 5 concludes this paper.

2 Double Delta Correlator Description

As mentioned above, the conventional delay estimator uses 03 correlators so-called: Early (E), Prompt (P) and Late (L). The space between early and late correlator is one (01) chip. For this classical approach, the multipath envelope error could not be estimated accurately. Narrowing chip space is one of the solutions for the improvement of MP error. Whereas the classical discriminator uses 1 chip space, the narrow correlator reduces it to 0.1 chip space [3]. However, band-limitation of the pre-correlation filter rounds the autocorrelation peak so that the space between correlators could not be too small [6].

Also narrowing the chip space, DDC uses 05 correlators in the tracking loop instead of 03 ones (two early, one in prompt and two late) [4]. The mathematical definition of correlators of DDC is shown as

$$
\begin{aligned}
P &= R(\tau); \\
E_1 &= R(\tau - d/2); \quad E_2 = R(\tau - d) \\
L_1 &= R(\tau + d/2); \quad L_2 = R(\tau + d)
\end{aligned}
\tag{1}
$$

An Improved Double Delta Correlator for BOC Signal Tracking 1171

where P relates to the prompt correlator, E_1 and L_1 relate to the inner correlators with a chip spacing of d, whereas E_2 and L_2 relate to the outer correlators with a chip spacing of $2d$. And, R is defined as autocorrelation function (ACF) of BOC signal.

In case of carrier phase locked tracking, the output of the DDC discriminator is equal to [7]

$$D_{DDC} = (E_1 - L_1) - (E_2 - L_2)/2 \qquad (2)$$

However, in multipath environment, the signal reaching the receivers consist one direct path and M − 1 multipath. For theoretical analysis, we set up the dedicated scenario with one direct path and only one multipath and a constant signal attenuation factor. Analytically, the direct signal and multipath signal can be calculated separately and the total correlation function is given as [8]

$$R_{total}(\tau) = R(\tau) + \alpha R(\tau + \delta) \qquad (3)$$

where α is the multipath to direct signal amplitude ratio and δ is the delay of the multipath signal relative to the direct signal.

Consequently, in the scenario of the presence of multipath, the ACF of BOC (1,1) is distorted and discriminator output of DDC is equal to zero at non zero delay ($\tau \neq 0$).

Assuming the chip space is $d = 0.1\,chip$, the ACF of BOC (1,1) signal in multipath environment with $\delta = 0.08$ and $\alpha = 0.5$ is illustrated in Fig. 1. The discriminator output of DDC in that scenario is shown in Fig. 2.

The impact of multipath on code tracking is represented in Fig. 3. It illustrates the maximum error resulting from one multipath with certain delay, amplitude and phase, so-called MEE. In case the MP signal and the LOS signal are in-phase, it is corresponded to upper part of the Fig. 3 [2]. On the other hand, if the MP signal and LOS signal are out of phase, it is corresponded to the lower part of the figure.

As shown in Fig. 3, even though the DDC mitigates perfectly the error for medium-long delayed MP, but for short-delayed MP ($\delta < 0.2\,chip$), it cannot remove it and the error is significant. The proposed method in this paper tries to reduce the short-delayed MP error.

3 The Improved Double Delta Correlator

In order to mitigate the MP in short-delayed MP, the proposed method evaluates the DDC tracking error, then, it is applied to DDC. The tracking error evaluation is based on the geometric analysis of ACF for several MP delays. When DDC code tracking is accuracy, the P correlator is located on the true correlation peak, it means that DDC remove perfectly the error. When DDC code tracking is wrong, the P correlator is located on right side (left side) of the true correlation peak, depending on in-phase (out-phase) MP signals.

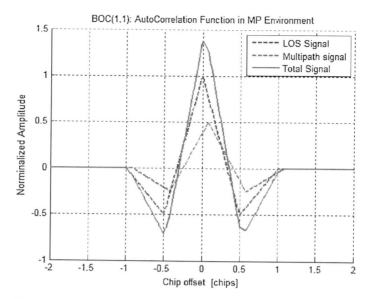

Fig. 1 ACF of BOC (1,1) signal in MP environment

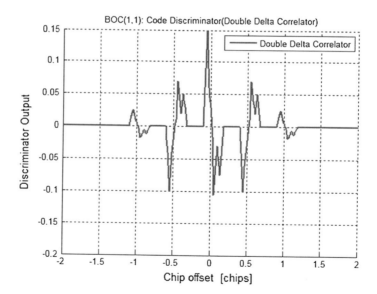

Fig. 2 Discriminator output with DDC for BOC (1,1) signal

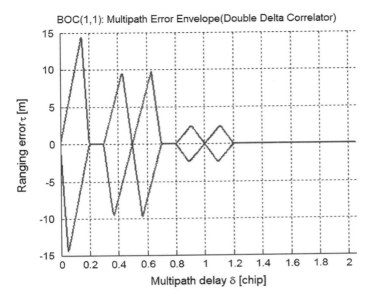

Fig. 3 MEE for BOC (1,1) signal with DDC (d = 0.1 chip)

3.1 In-Phase MP signals

The ACFs of BOC (1,1) in MP environment where MP signal is in-phase with LOS signal is shown in Fig. 4. All parts of the all ACFs (LOS ACF, MP ACF and total signal ACF) are modeled as mathematical equations.

Setting ε is DDC tracking error, a is the maximum amplitude of LOS ACF, b is the maximum amplitude of MP ACF, δ is the MP delay and d is the chip space.

The ACF of the received BOC (1, 1) signal is the sum of LOS ACF and MP ACF. As illustrated in Fig. 4, the shape of the distorted ACF can be separated into 09 lines in order to model each line by an equation as

$$\begin{aligned} d_1 &= y_1; & d_2 &= y_1 + y_5; & d_3 &= y_2 + y_5 \\ d_4 &= y_2 + y_6; & d_5 &= y_3 + y_6; & d_6 &= y_3 + y_7 \\ d_7 &= y_4 + y_7; & d_8 &= y_4 + y_8; & d_9 &= y_8 \end{aligned} \quad (4)$$

In the scenario of the short-delayed MP ($d = 0.1\ chip$, $0.05 < \delta < 0.2$) DDC fails tracking the true correlation peak, so the P correlator is located on the right of the true peak. Consequently, P correlator output is on the line of d_5, E_1 and E_2 are on the line of d_4, L_1 and L_2 are on the d_6.

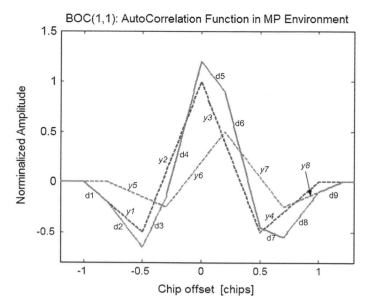

Fig. 4 Geometry of ACF for BOC (1,1) signal with a in-phase MP

Based on Eq. (4), the amplitudes of all correlator outputs are calculated as

$$\begin{aligned} P &= (-a+b)3\varepsilon + a + (1-\delta)b \\ E_1 &= 3(a+b)(\varepsilon - d) + a + (1-\delta)b \\ E_2 &= 3(a+b)(\varepsilon - 2d) + a + (1-\delta)b \\ L_1 &= -3(a+b)(\varepsilon + d) + a + (1+\delta)b \\ L_2 &= -3(a+b)(\varepsilon + 2d) + a + (1+\delta)b \end{aligned} \quad (5)$$

Solving Eq. (5), the DDC tracking error can be expressed as

$$(E_1 - E_2) - (P - E_1) = 6a\varepsilon \quad (6)$$

As seen in Eq. (6), the left side of it could be used as an indicator of DDC tracking error with a scale of $6a$. The left side is the difference between the distance of E_1, E_2 and the distance of E_1, P. If the difference is zero, it means that E_1 is located at the midpoint of E_2 and P, resulting in E_1, E_2 and P are on the same line. Consequently, there is no error in tracking loop ($\varepsilon = 0$) and P correlator output is located on the true correlation peak. When there is tracking error ($\varepsilon > 0$), P correlator output has to move to the right side of true peak due to $E_1 - E_2$ is larger than $E_1 - P$.

However, the tracking error should be calculated without depending on the maximum amplitude of LOS ACF as well as MP ACF. Also derived from Eq. (5), the sum of maximum amplitude of LOS ACF and the one of MP ACF is calculated as

$$a + b = (L_1 - L_2)/3d \tag{7}$$

In almost natural cases, the amplitude of direct component is larger than the one of MP components and thus a $\geq$ b. Therefore, Eq. (7) could be changed as

$$2a \geq (L_1 - L_2)/3d \tag{8}$$

Taking Eq. (8) into Eq. (6), we yield

$$\varepsilon \leq d \frac{(E_1 - E_2) - (P - E_1)}{L_1 - L_2} \tag{9}$$

Therefore, the maximum of DDC tracking error could be chosen as the modification of the output of the DDC discriminator. The output of proposed DDC discriminator is written as

$$D_{\text{Improved}} = D_{DDC} + Modification$$
$$= (E_1 - L_1) - \frac{E_2 - L_2}{2} + d \frac{(E_1 - E_2) - (P - E_1)}{L_1 - L_2} \tag{10}$$

3.2 Out-Phase MP Signals

The geometry of ACF for BOC (1,1) in case of an out-phase MP signal with a half of the LOS amplitude is shown in Fig. 5.

There is no difference between LOS signal in two cases (in-phase MP and out-phase MP.

For MP ACF, there is the difference between two cases. In $[-1 + \delta; 1 + \delta]$ chip offset, ACF is separated into 04 lines, each line is modeled by an equation, respectively. From left side to right side, 04 lines are represented as

$$y_5 = b(x + 1 - \delta); \quad y_6 = -b(3x + 1 - \delta)$$
$$y_7 = b(3x - 1 - \delta); \quad y_8 = -b(x - 1 - \delta) \tag{11}$$

As the same as the scenario of in-phase MP signal, the ACF of the received BOC (1,1) signal is the sum of LOS signal and MP signal. The geometry of ACF is also separated into 09 parts of line and they are mathematical modeled as Eq. (4).

When the short-delayed out-phase MP signal ($d = 0.1$ chip, $0.05 < \delta < 0.2$) is taken into account, the P correlator output is moved to the left of the true correlation peak. Resulting in the P is on the d_4 line, E_1 and E_2 are also on the d_4 line, whereas L_1 is on the d_5 line and L_2 are on the d_6 line. Therefore, based on Eqs. (4) and (11), the amplitudes of all correlator outputs are calculated as

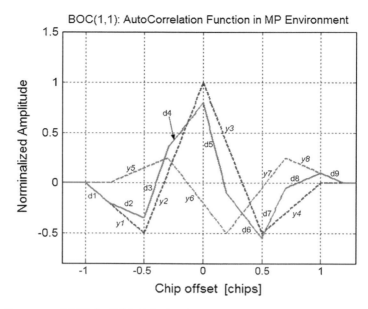

Fig. 5 Geometry of ACF for BOC (1,1) signal with a out-phase MP

$$P = (a - b)3(-\varepsilon) + a - (1 - \delta)b$$
$$E_1 = 3(a - b)(-\varepsilon - d) + a - (1 - \delta)b$$
$$E_2 = 3(a - b)(-\varepsilon - 2d) + a - (1 - \delta)b \quad (12)$$
$$L_1 = -3(a + b)(-\varepsilon + d) + a - (1 - \delta)b$$
$$L_2 = -3(a - b)(-\varepsilon + 2d) + a - (1 + \delta)b$$

As shown in [4], the difference between the distance of L_1, L_2 and the distance of L_1, P may be used as an indicator of DDC tracking error. However, in this case, L_1 and L_2 are on two different lines and causes $(L_1 - L_2) - (P - L_1)$ could not be used. Another Late correlator should be created for this purpose with the chip spacing $d/2$ from P correlator and so called L_{12}. Therefore, L_{12} are on d_5 and the amplitude of L_{12} output could be calculated as

$$L_{12} = -3(a + b)(-\varepsilon + d/2) + a - (1 - \delta)b \quad (13)$$

Hence, the DDC tracking error could be written by solving Eq. (12) and Eq. (13) as

$$(P - L_{12}) - (L_{12} - L_1) = -6a\varepsilon \quad (14)$$

An Improved Double Delta Correlator for BOC Signal Tracking

The sum of maximum amplitude of LOS ACF and the one of MP ACF in this case is also derived from above equation as

$$(P - L_{12}) - (L_{12} - L_1) = -6a\varepsilon \qquad (15)$$

Finally, with the same calculation in in-phase case, the exactly DDC tracking error could be expressed as

$$\varepsilon \leq d\frac{(L_{12} - L_1) - (P - L_{12})}{2(L_{12} - L_1)} \qquad (16)$$

Therefore, the output of improved DDC discriminator is constructed as

$$D_{\text{Improved}} = D_{DDC} + Modification$$
$$= (E_1 - L_1) - \frac{E_2 - L_2}{2} + d\frac{(L_{12} - L_1) - (P - L_{12})}{2(L_{12} - L_1)} \qquad (17)$$

Equations (10) and (17) is the output of improved DDC discriminator. With these modifications, the DDC tracking error could be reduced for MP delays $0.05 < \delta < 0.2$.

4 Simulation Results

The simulations have been carried out for BOC (1,1) modulated signals for an infinite front-end bandwidth. In MEE analysis, several simplifying assumptions have been made in order to ensure the error source is only multipath signals. Such assumptions are: zero Additive White Gaussian Noise (AWGN), ideal infinitive-length PRN codes and zero residual Doppler [9]. Moreover, the scenario is only one direct signal plus one multipath signal so that the analysis expression for MEE is not complicated. The simulations parameters are: number of paths is 02 paths; Chip spacing is $d = 0.125$ *chip*; the ratio between the amplitudes of MP signal and direct signal is $\alpha = 0.5$.

For above configuration, as shown in Fig. 6, the ranging error due to multipath signal is about 15 m in the short-delayed MP scenario (chip-delayed $0.05 < \delta < 0.2$). The improved DDC has reduced ranging error from 15 to 10 m for both cases in-phase (upper part) and out-phase (lower part) MP signals. For medium and long-delayed MP $(0.5 < \delta < 1.0)$, the performance of improved DDC is similar to the conventional DDC. It means that, the improved DDC focuses on improving the performance of DDC in the scenario of short-delayed MP signals. It's very important because the short-delayed MP signal occurs frequently.

If the chip spacing is reduced, the DDC tracking error of the conventional DDC as well as the improved DDC are reduced. Therefore, the ranging error is also reduced. However, the ranging error of the improved DDC is always smaller than the one of the conventional DDC.

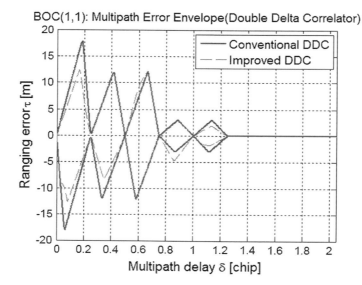

Fig. 6 MEE of improved DDC for BOC (1,1) signal

5 Conclusions

Multipath is one of major sources of ranging error in precise application in GNSS receivers. Many multipath mitigation methods have been researched and proposed in theoretical case. In non- parametric approach, Double Delta Correlator provides the best performance of multipath mitigation. It reduces almost multipath errors in the medium and long delayed multipath environment. For short-delayed multipath, DDC only achieves the multipath mitigation effect same as Narrow Correlator [4]. In this study, the improved DDC was proposed. It has improved the multipath mitigation performance in short-delayed multipath more than 30 % in comparison with the conventional DDC.

References

1. Betz JW (2001) Binary offset carrier modulations for radio navigation. J Inst Navig 48(4):227–246
2. Borre K, Akos DM, Bertelsen N, Rinder P, Jensen SH (2007) A software-defined GPS and Galileo receiver—a single-frequency approach. Birkhäuser, Boston
3. Dierendonck AJV, Fenton P, Ford T (1992) Theory and performance of narrow correlator spacing in a GNSS receiver. J Inst Navig 39(3):265–283
4. Irsigler M, Eissfeller B (2003) Comparison of multipath mitigation techniques with consideration of future signal structures. In: The 16th international technical meeting of the satellite division of the institute of navigation (ION GNSS '03), pp 2584–2592

An Improved Double Delta Correlator for BOC Signal Tracking 1179

5. Dierendonck AJV, Braasch MS (1997) Evaluation of GNSS receiver correlation processing techniques for multipath and noise mitigation. The 1997 national technical meeting of the institute of navigation, pp 207–215
6. Zhu X, Chen X, Chen X (2011) Comparison between strobe correlator and narrow correlator on MBOC DLL tracking loop. In: Instrumentation and measurement technology conference (I2MTC), pp 1–4
7. Tawk Y, Botteron C, Jovanovic A, Farine PA (2010) Performance comparison of different correlation Techniques for the AltBOC modulation in multipath environments. In: 2010 IEEE international conference on communications (ICC), pp 1–6
8. Jovanovic A, Tawk Y, Botteon C, Farine PA (2010) Multipath mitigation techniques for CBOC, TMBOC and AltBOC signals using advanced correlators architectures. Position location and navigation symposium (PLANS), pp 1127–1136
9. Bhuiyan MZH, Lohan ES (2010) Advanced multipath mitigation techniques for satellite— based positioning applications. Int J Navig Obs 1–15 Hindawi Publishing Corporation

Implementation of Automatic Failure Diagnosis for Wind Turbine Monitoring System Based on Neural Network

Ming-Shou An, Sang-June Park, Jin-Sup Shin, Hye-Youn Lim and Dae-Seong Kang

Abstract The global action began to resolve the problem of global warming. Thus, the wind power has been emerged as an alternative energy of existing fossil fuel energy. The existing wind power has limitation of location requirements and noise problems. In case of Korea, the existing wind power has difficulties on limitation of location requirements and the noise problems. The wind power turbine requires bigger capacity to ensure affordability in the market. Therefore, expansion into sea is necessary. But due to the constrained access environment by locating sea, the additional costs are occurred by secondary damage. In this paper, we suggest automatic fault diagnosis system based on CMS (Condition Monitoring System) using neural network and wavelet transform to ensure reliability. In this experiment, the stator current of induction motor was used as the input signal. Because there was constraint about signal analysis of large wind turbine. And failure of the wind turbine is determined through signal analysis based wavelet transform. Also, we propose improved automatic monitoring system through neural network of classified normal and error signal.

Keywords Wavelet transform · Neural network · Automatic failure diagnosis · CMS · Wind turbine monitoring system

1 Introduction

Due to the maturity of wind turbine technology, the unit cost of wind power was decreased. Therefore, the unit cost of wind power is similar to the cost of existing fossil fuels compared to other renewable energy. The global action also began to

---

M.-S. An · S.-J. Park · J.-S. Shin · H.-Y. Lim · D.-S. Kang (✉)
Department of Electronics Engineering, Dong-A University, 840 Hadan 2-Dong, Saha-Gu, Busan, Korea
e-mail: dskang@dau.ac.kr

J. J. (Jong Hyuk) Park et al. (eds.), *Multimedia and Ubiquitous Engineering*, Lecture Notes in Electrical Engineering 240, DOI: 10.1007/978-94-007-6738-6_145, © Crown Copyright 2013

resolve the problem of global warming. Thus, the wind power has been emerged as an alternative energy of existing fossil fuel energy. Because of technological advances over the past 20 years, the wind turbines have become bigger. However, the larger turbines inevitably increase tower height and blade length. Also, the components of wind turbine increase mechanical and electrical permit capacity. Consequently, failure rate of turbine would increase [1]. The existing wind power has limitation of location requirements and noise problems. In case of Korea, the existing wind power has difficulties on limitation of location requirements and the noise problems. The wind power turbine requires bigger capacity to ensure affordability in the market. Therefore, expansion into sea is necessary, but there are some problems for the operation of large wind turbines at sea. The first, mechanical and electrical failures of larger wind turbine are increased. Second, due to the constrained access environment by locating sea, the additional costs are occurred by secondary damage. Therefore, monitoring technology of wind turbine is essential for utilization and reliability [2]. In this paper, failure of the wind turbine is determined through signal analysis based wavelet transform. Also real-time signal analysis was made through wavelet transform. Feature information of classified signal pattern through signal analysis had been learned by using neural network algorithm to implement automatic fault diagnosis system.

2 Condition Monitoring Techniques of Wind Turbine

The surveillance system of wind power is classified into supervisory control and data acquisition (SCADA) system and condition monitoring system (CMS). The SCADA system remotely performs control function of wind turbine in conjunction with turbine controller. It is an essential component to consist of wind power system. However, the current wind turbines are difficult to identify the operating condition because SCADA systems are made differently per the turbine manufacturer.

The other hand, the CMS is prevention system that diagnoses malfunction in advance by closely monitoring, analyzing and predicting component of wind turbine. In the past, the CMS was regarded as an optional component. In case of large and wind turbine located sea, the CMS is recognized as an essential component because of the matter of credibility. In this paper, the stator current of induction motor was used as a input signal. Because the signal analysis of large offshore wind turbines is constrained. Overall, we must ensure the reliability by suggesting automatic fault diagnosis system based on CMS through neural network on the feature information of signal patterns and analysis using the wavelet transform.

2.1 Signal Analysis Using Wavelet Transform

The wavelet analysis appeared by integrating special techniques individually developed to meet special purpose belonging to signal processing system. Basic techniques of computer vision using multi-resolution analysis method, sound and video compression using sub-band coding technique and applied mathematics using wavelet series are developed recently into wavelet theory's special applications. The wavelet transform can understand that input signals are separated into the set of the basis function. The set of basis function used to wavelet transform can be obtained through expansion, reduction and parallel transference of time axis about basis function of wavelet. Basis function of wavelet indicates band-pass filter of special form. And the relative bandwidth invariability of wavelet transform is satisfied by expansion and reduction of temporal axis about wavelet basis function. Thus, the scale is called instead of frequency band in wavelet transform. Unlike Fourier transform, wavelet transform includes high resolvent ability about the scale of signal. Therefore, wavelet transform is time–scale transform.

Wavelet equation about time-domain and frequency-domain is

$$\phi_{ab}(x) = 1/2\phi((x - b)/a), \tag{1}$$

where a means expansion and b which means movement indicates the temporal position. The more a is increased, the more resolution of frequency is increased. The scaling is an expanding or reducing signal. Large-scaling signal is expanded. And small scaling signal responds to the compression [3]. Integration for entire signal of Fourier transform causes high-frequency component by diversion in the finite space, it's difficult to handle flexibly non-stationary function because the analysis about variation of frequency has limit. To overcome these demerits, various analytical methods were devised and wavelet transform among the rest is efficiently used [4]. After studying orthogonal basis functions, discrete wavelet transform (DWT) is developed. DWT which indicates two space of separated frequencies for the signal analysis is defined as follows:

$$x(t) = \sum_{n=-\infty}^{\infty} a_{0,n}(t), g_{0,n}(t) + \sum_{j=0}^{\infty} \sum_{n=-\infty}^{\infty} d_{j,n}(t), h_{j,n}(t), \tag{2}$$

where $a_{0,n}(t)$ is component of the signal, $g_{0,n}(t)$ represents low-pass filter (LPF), $d_{j,n}(t)$ is detail component of the signal and $h_{j,n}(t)$ is high-pass filter (HPF).

In Fig. 1, it is possible to regenerate two-band about the filtering signal. In other words, if separation of 2-level is finished, the original signal is separated into frequency of four bands [5].

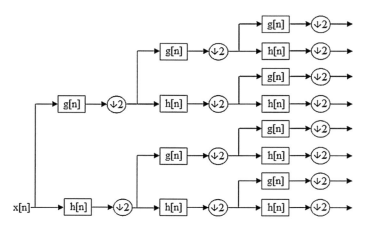

Fig. 1 2-level separation of image signal

2.2 Feature Extraction of Signal Pattern Through Signal Analysis

The wavelet transform is used for noise rejection in signal analysis. In this paper, because of constrained environments, the signal analysis of large wind turbines was substituted for input signal using the stator current of an induction motor. Figure 2 shows the removed noise of input signal using the wavelet transform.

There is very wide variety of wavelet filter according to its purpose. Figure 3 also shows that the coefficients of the filter increase in accordance with N value of Db(N). In this paper, Db(4) is used to analyze input signal of stator current removed noise. Very short basis functions need to suggest discontinuity of input signal, and very long low-frequency basis functions need to analyze frequency

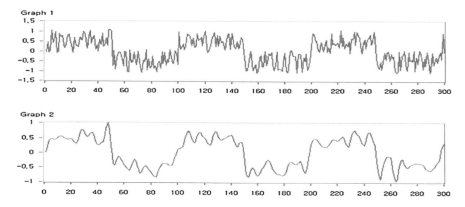

Fig. 2 Noise removal using the wavelet transform

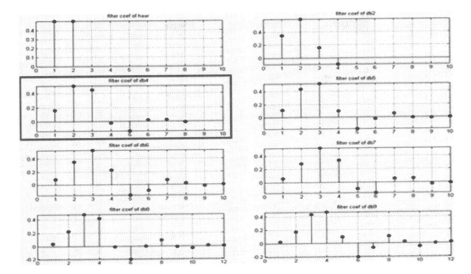

Fig. 3 Db(N) DWT filters

accurately. We simultaneously use two basis functions based on Daubechies wavelet. So, we easily can get the ambiguous feature information on the signal analysis of time–frequency domain.

2.3 Neural Network Modeling

The primary function of neuron calculates input and weighted summing NET of connection strengths in artificial neural network. And the output comes out by the activation function. Therefore, output of neuron is different depending on the activation function.

Step and sigmoid functions are simple activation functions. When the input is over the threshold value, the output is a function that is activated as 1. The sigmoid function for the change of the value has a form to infinitely approach 0 and 1. In other words, the sigmoid function linearly translates disorderly nonlinear values in neural network model.

2.4 Neural Network Learning Algorithm of Back Propagation

The back propagation (BP) algorithm which also called error back propagation algorithm is used to be applied multiple neural network. BP is an universal neural network algorithm used to various fields.

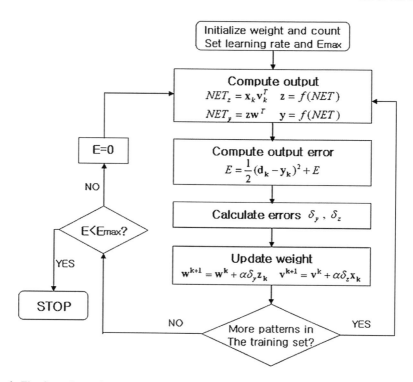

Fig. 4 The flow chart of BP algorithm

The renewal of the connection strength is the most important part in the learning algorithm. The BP algorithm consists of the forward and backward steps. As with other neural network learning algorithm, the learning is made by renewal of the connection strength. Figure 4 is a three-tier structure of BP neural network. The input layer enters classified signal pattern through signal analysis into neural network. And it is multiplied with connection weights connected to the hidden layer. The multiplied values are passed to hidden layer. Through repetition of this process, the end result of output layer is obtained. Error is calculated through the subtraction of the output and the target figure [6].

3 Experimental Results of Automatic Fault Diagnosis System

In this paper, to implement automatic fault diagnosis of offshore wind turbine, we conducted an experiment to improve the reliability by suggesting neural network algorithm and wavelet transform based on CMS for signal analysis.

Implementation of Automatic Failure Diagnosis   1187

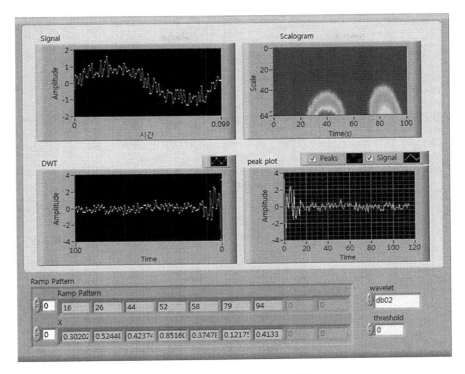

Fig. 5 The automatic failure diagnosis system based on LabView

Figure 5 shows the overall system of automatic failure diagnosis through learning using feature information of classified signal pattern. In this paper, we extracted peak value of classified signal using peak detection function based on LabView for extraction of feature information. And, extracted feature information is used as input to the learning of neural network algorithm to implement automatic failure diagnosis system of offshore wind turbine. In this study, we early diagnose a complex fault signal that occurs over a long time rather than a simple failure. And there are aims to reduce secondary damage. Also, we look forward to applying to a variety of monitoring environments.

Acknowledgments This work was supported by the Human Resources Development of the Korea Institute of Energy Technology Evaluation and planning (KETEP) grant funded by the Ministry of knowledge Economy, Republic of Korea (No. 20114010203060).

References

1. Robi P The wavelet tutorial-fundamental concepts and an overview of the wavelet theory, 2nd edition
2. Park JY (2012) Development of wind power integrated condition monitoring system, Korea Electrical Contractors Association, pp 56–63, Feb 2012
3. Kim CH, Kim H, Ko YH, Byun SH, Aggarwal RK, Allan TJ (2002) A novel fault-detection technique of high-impedance arcing faults in transmission lines using the wavelet transform. IEEE Trans Power Deliv 17(4):921–929
4. Mallat S (1991) Zero crossings of a wavelet transform. IEEE Trans Inf Theory 37(4):1019–1033
5. Wenxian Y, Tavner PJ, Michael W (2008) Wind Turbine condition monitoring and fault diagnosis using both mechanical and electrical signatures. In: Proceedings of the 2008 IEEE/ASME international conference on advanced intelligent mechatronics, pp 1296–1301, July 2008
6. He Q, Du DM (2007) Fault diag -nosis of induction motor using neural networks. In: Proceedings of the 6th international conference on machine learning and cybernetics. vol 2, pp 1090–1095, Aug 2007

Development of Compact Microphone Array for Direction-of-Arrival Estimation

Trình Quốc Võ and Udo Klein

Abstract Direction-of-arrival estimates are required in many applications such as automatic video camera steering and multiparty teleconferencing for beam forming and steering to suppress noise and reverberation and improve speech intelligibility. Ambient noise and multiple reflections of the acoustic source signal significantly degrade the performance of time-difference-of-arrival (TDOA) methods to localize the sound source using only two microphones. In this work, we investigate the performance of a multichannel cross-correlation coefficient (MCCC) algorithm for the estimation of the direction-of-arrival (DOA) of an acoustic source in the presence of significant levels of both noise and reverberation. Simulations and initial experimental results confirm that the DOA estimation robustness and complexity is suitable for a practical micro-phone array using miniature MEMS microphones and an FPGA implementation of the MCCC algorithm.

Keywords Direction of arrival estimation · Microphone arrays · Signal processing algorithms · Time of arrival estimation

1 Introduction

The basic idea to solve the DOA problem is to determine the time difference of a sound source signal arriving at two microphone locations. If the sound source is located in the far-field and the distance b between the two microphones is known,

---

T. QuốcVõ · U. Klein (✉)
School of Electrical Engineering, International University, Vietnam National University, HCMC, Ho Chi Minh City, Vietnam
e-mail: u.klein@ieee.org

J. J. (Jong Hyuk) Park et al. (eds.), *Multimedia and Ubiquitous Engineering*, Lecture Notes in Electrical Engineering 240, DOI: 10.1007/978-94-007-6738-6_146, © Springer Science+Business Media Dordrecht(Outside the USA) 2013

the TDOA $\tau_{12} = (b\cos\theta)/v_a$ is directly related to the DOA angle θ. Here, v_a is the sound velocity in air.

In order to improve the estimate of the DOA in noisy and reverberant environments many algorithms have been proposed using multiple microphones. The linear spatial prediction method [1], the multichannel cross-correlation coefficient algorithm [1], and the broadband multiple signal classification (MUSIC) method [2] all employ the correlation of the aligned microphone signals to estimate the TDOA. The minimum entropy (ME) method [3] uses higher order statistics that could be more suitable for non-Gaussian source signals such as speech. The most reliable TDOA estimation performance in reverberant environments is achieved by adaptive blind multichannel identification (ABMCI) [4], which relies on the blind identification of the real rever-berant impulse response functions of the SIMO system, consisting of the single signal source and multiple microphones. Both ME and ABMCI are considered to give better results than cross-correlation based techniques, although at the cost of higher computational requirements, thereby increasing the hardware complexity and cost.

Because of its relatively low computational complexity the MCCC method has been selected for a low-cost hardware implementation of a microphone array for DOA estimation of an acoustic source in the presence of significant levels of both noise and reverberation. The MCCC method uses the redundancy of the microphone signals by applying an extension of the generalized cross-correlation proposed by Knapp and Carter [5]. In order to evaluate the performance of the MCCC method the algorithm has been implemented in MATLAB$^{®}$ and simulation results are presented in this paper. Following our simulations we have set up a demonstration system using an analog MEMS microphone array and an analog-to-digital (A/D) converter as the interface to the MCCC computer program. First experimental results indicate that the simulated performance can be achieved with compact low-cost hardware using a digital MEMS microphone array and an FPGA implementation of the MCCC algorithm.

2 The MCCC Algorithm

The microphone array consists of L microphones in a linear equidistantly spaced array, from the 1st to the Lth microphone. The delay between the 1st and the lth microphones is then given by $f_l = (l - 1)\tau$ where τ is the time delay between two neighboring microphones. For the application of the MCCC algorithm, we consider the column vector of the aligned signals at the L microphones

$$\mathbf{x}_{1:L}[n - f_L(m)] = [x_1[n - f_L(m) + f_1(m)] \quad x_2[n - f_L(m) + f_2(m)] \cdots x_L[n]]^{\mathrm{T}}$$

with $m/f_s = \hat{\tau}$ as a guess for the delay, where f_s is the sampling frequency. The spatial correlation matrix of the microphone signals can be factored as

Development of Compact Microphone Array

$$R_{m,1:L} = \mathbb{E}\{x_{1:L}[n - f_L(m)] \cdot x_{1:L}^\mathrm{T}[n - f_L(m)]\} = D\tilde{R}_{m,1:L}D$$

with the diagonal matrix D and the symmetric matrix $\tilde{R}_{m,1:L}$ defined as

$$D = \begin{bmatrix} \sqrt{\mathbb{E}\{x_1^2[n]\}} & \cdots & 0 \\ \vdots & \ddots & \vdots \\ 0 & \cdots & \sqrt{\mathbb{E}\{x_L^2[n]\}} \end{bmatrix}, \quad \tilde{R}_{m,1:L} = \begin{bmatrix} 1 & \cdots & \rho_{m,1L} \\ \vdots & \ddots & \vdots \\ \rho_{m,L1} & \cdots & 1 \end{bmatrix},$$

and the cross-correlation coefficients between $x_k[n - f_l(m)]$ and $x_l[n - f_k(m)]$

$$\rho_{m,kl} = \frac{\mathbb{E}\{x_k[n - f_l(m)]x_l[n - f_k(m)]\}}{\sqrt{\mathbb{E}\{x_k^2[n]\}\mathbb{E}\{x_l^2[n]\}}}, \text{ with } k \text{ and } l = 1, 2, \ldots, L.$$

The multichannel cross-correlation coefficient is now defined as [1]

$$\rho_{m,1L}^2 = 1 - \det \tilde{R}_{m,1:L}.$$

The delay estimation is then based on maximizing the cross-correlation coefficient $\rho_{m,1:L}^2$ or by minimizing the cost function, defined as the determinant of the matrix $\tilde{R}_{m,1:L}$, with respect to the guessed delay m.

3 Simulation

Simulation Parameters. In order to simulate the reverberant acoustic environment the image-source method for room acoustics has been employed [6]. Walls, ceiling, and floor of the room are characterized by frequency-independent and incident-angle-independent reflection coefficients. The dimensions of the room are chosen to be 5 m by 5 m by 2.5 m. Reflection coefficients r_i ($i = 1, 2, \ldots, 6$) are varied between 0 and 0.8. The sound source is located at the position (4.0 m, 2.5 m, 1.5 m). For the simulations, up to eight microphones are placed in parallel with the x-axis and with a spacing of 11 cm. The first microphone is located at (2.88 m, 0.5 m, 1.3 m) and the last is at (2.11 m, 0.5 m, 1.3 m). Gaussian noise is added to each microphone signal with all noise signals uncorrelated both with the source signal and with the noise at the other microphones. The signal-to-noise ratio (SNR) has been varied between -5 dB and 20 dB. A 20 s recorded speech signal was sampled at $f_s = 16$ kHz with 16-bit resolution and has been used as the sound source for the simulations. All data processing has been performed on signal frames of 512 samples, corresponding to a sampling length of 32 ms.

The desired resolution of the estimated DOA angle has been chosen to be better than 20 degrees. This means the DOA estimation system needs to be able to detect a maximum delay M between two neighboring microphones of 5 samples. The relation between the minimum spacing $b_{\min}$ between two neighboring microphones, the sampling frequency f_s and the maximum delay M is given by $b_{min} \geq M v_a / f_s$. The resulting minimum distance between two neighboring microphones is $b_{min} = 10.7$ cm, leading to the selected microphone separation of $b = 11$ cm.

Simulation Results. The robustness of the MCCC algorithm for a given source signal depends on the SNR, the degree of reverberation, and the number of microphones in the array. Figure 1a shows the cost function as a function of the guessed delay m in a reverberant-free environment with 0 dB SNR and using two, four, and eight microphones, respectively. Distinctive minima exist at the correct delay estimate of $m = 3$. For a strongly reverberant environment comparable to a typical meeting room, the cost functions in Fig. 1b have much less pronounced minima and for $L = 2$ microphones the minimum at $m = 2$ results in an incorrect delay estimate.

Typical distributions of the time delay estimates for repeatedly applying the MCCC algorithm in a reverberant-free environment and in a strongly reverberant environment with 0 dB SNR are shown in Fig. 2. In a reverberant-free environment and with two microphones, only 80 % of the estimates correspond to the true delay of 3 samples while four- and eight-microphone arrays do not show any incorrect estimates. In a strongly reverberant environment, even an eight-microphone array only achieves 80 % correct delay estimates and a two-microphone array displays erroneous estimates of up to 2 samples off the true delay.

Figure 3 shows the percentage of correct delay estimates in noisy environments with and without reverberation as a function of the SNR. The robustness of the algorithm to estimate the delay correctly increases with the SNR and the number of microphones in the array. The simulation shows that at a SNR of -5 dB in a reverberant-free environment the percentage of successful delay estimates is only about 50 % for two microphones. However, the rate of correct estimates reaches almost 100 % if an array with four or more microphones is used. In a strongly reverberant environment with wall reflection coefficients of $r_i = 0.8$ the

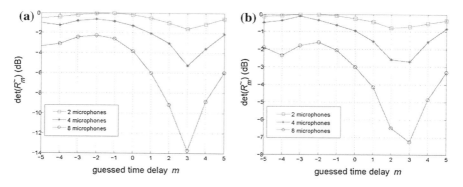

Fig. 1 MCCC cost function $\det \tilde{R}_m$ as a function of the guessed delay m for microphone arrays with $L = 2, 4,$ and 8 microphones and with a SNR of 0 dB (The true delay is 3 samples); **a** reverberant-free and **b** reverberant environment ($r_i = 0.8$)

Development of Compact Microphone Array

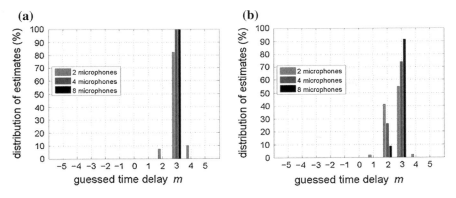

Fig. 2 Distribution of time delay estimates for repeatedly applying the MCCC algorithm 1,000 times with $L = 2$, 4, and 8 microphones and with a SNR of 0 dB (The true delay is 3 samples); **a** reverberant-free and **b** reverberant environment ($r_i = 0.8$).

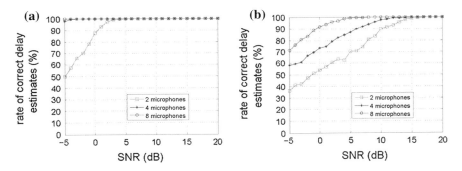

Fig. 3 Robustness of the MCCC algorithm in terms of percentage of correct delay estimates as a function of the SNR for arrays with $L = 2$, 4, and 8 microphones; **a** reverberant-free and **b** reverberant environment ($r_i = 0.8$).

percentage of correct delay estimates is significantly reduced and the eight-microphone array requires a SNR of better than 5 dB to achieve nearly 100 % correct delay estimates.

The simulations confirm the robustness of the MCCC algorithm in estimating the TDOA in both noisy and reverberant environments. A linear equidistantly spaced array of eight microphones performs reliably in strongly reverberant environments with SNRs down to 5 dB and with signal durations as short as 32 ms.

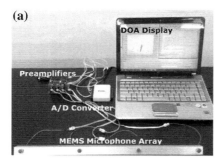 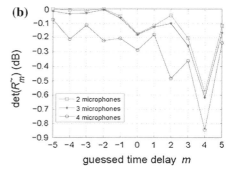

Fig. 4 Demonstration system with 4-channel microphone array; a Analog MEMS microphone array, preamplifiers, and A/D converter interface to the MCCC algorithm in MATLAB® b MCCC cost function det $\tilde{R}_m$ as a function of the guessed delay m for microphone arrays with $L = 2, 3,$ and 4 microphones in a reverberant meeting room environment

4 Demonstration System

We have set up a demonstration system using a microphone array with analog MEMS microphones and an analog-to-digital (A/D) converter as the interface to the MCCC algorithm in MATLAB® (Fig. 4a). The limitations of the A/D converter allow a maximum of four microphone channels at a sample rate of 10.5 kS/s with a resolution of 14 bits. Because of the reduced sample rate compared to the simulation, the distance between the microphones is increased to 17.5 cm. For a sound source at a distance of 2.5 m and at an angle of 35° to the microphone array, and for a SNR of about 10 dB, the cost function is shown in Fig. 4b for microphone arrays with 2, 3, and 4 channels. The estimated delay of $m = 4$ corresponds to a DOA range between 28° and 47°, which includes the correct DOA of 35°.

5 Conclusion

Our investigation confirms that the MCCC algorithm is a suitable candidate for reliable TDOA estimation in real-world environments with a minimum amount of computational cost. The TDOA can be estimated by finding the minimum of the determinant of the cross-correlation coefficient matrix of the aligned microphone array signals.

Results with a 4-channel demonstration system show that the simulated performance is achievable with miniature analog MEMS microphones and an analog-to-digital converter as the interface to the MCCC algorithm in MATLAB®. The full hardware implementation will employ an FPGA and a digital MEMS microphone array. The digital MEMS microphones integrate the microphone, amplifier, and A/D converter in a single component, thereby reducing the system complexity considerably. The system will be smaller, cheaper, and more flexible than conventional analog microphone arrays. In order to evaluate its performance the MEMS microphone arrays will be used to record test data for source localization experiments.

Acknowledgments The authors would like to thank Pirmin Rombach and Armin Schober from EPCOS AG for providing free samples of their MEMS microphones.

References

1. Chen J, Benesty J, Huang Y (2003) Robust time delay estimation exploiting redundancy among multiple microphones. IEEE Speech Audio Process 11:549–557
2. Dmochowski JP, Benesty J, Affes S (2007) Broadband MUSIC: Opportunities and challenges for multiple source localization. In: 2007 IEEE Workshop on the applications of signal processing to audio and acoustics, pp 18–21
3. Benesty J, Huang Y, Chen J (2007) Time delay estimation via minimum entropy. IEEE Signal Process Lett 14:157–160
4. Huang Y, Benesty J (2002) Adaptive multi-channel least mean square and newton algorithms for blind channel identification. Signal Process 82:1127–1138
5. Knapp CH, Carter GC (1976) The generalized correlation method for estimation of time delay. IEEE Trans Acoust Speech Signal Process 24:320–327
6. Lehmann EA (2012) Image-source method: MATLAB code implementation. http://www.eric-lehmann 10 Mar 2012

Design and Implementation of a SoPC System for Speech Recognition

Tran Van Hoang, Nguyen Ly Thien Truong, Hoang Trang and Xuan-Tu Tran

Abstract This paper presents the design of a System on Programmable Chip (SoPC) based on Field Programmable Gate Array (FPGA) for speech recognition in which Mel-Frequency Cepstral Coefficients (MFCC) for speech feature extraction and Vector Quantization (VQ) for recognition are used. The execution speed of the blocks in the speech recognition system is surveyed by calculating the number of clock cycles while executing each block.

Keywords Speech recognition · MFCC · VQ · SoPC · FPGA · Nios

1 Introduction

Speech recognition system is applied in many application fields such as health care, military, human computer interaction, avionics technicians... [1], especially, the applications which support disabled people to communicate with the world in a better way. For that reason, there are many studies on software/hardware implementation of speech recognition systems for many years. However, because of a large number of accents spoken around the world, there are still many challenges that need further research and development, for example, Vietnamese speech recognition.

T. Van Hoang · N. L. T. Truong · H. Trang (✉)
University of Technology, Vietnam National University, HoChiMinh City, Vietnam
e-mail: hoangtrang@hcmut.edu.vn

T. Van Hoang
e-mail: tvhoang@hcmut.edu.vn

N. L. T. Truong
e-mail: nlttruong@hcmut.edu.vn

X.-T. Tran
VNU University of Engineering and Technology, 144 Xuan Thuy, Hanoi, Vietnam
e-mail: tutx@vnu.edu.vn

J. J. (Jong Hyuk) Park et al. (eds.), *Multimedia and Ubiquitous Engineering*,
Lecture Notes in Electrical Engineering 240, DOI: 10.1007/978-94-007-6738-6_147,
© Springer Science+Business Media Dordrecht(Outside the USA) 2013

The research on speech recognition going mainly in two directions, namely: the software runs on Personal Computers (PCs) and embedded systems. For the first direction, many studies and software tools have been developed successfully. In particular, the Hidden Markov Model Toolkit (HTK) is a toolkit for building Hidden Markov Models (HMMs) used in speech recognition successfully [2]. There are also many tools running on the PC or smart phone aimed at the control device via speech. For the second direction, embedded systems have many advantages as high performance, convenience, low cost, and great development potential. However, speech recognition research based on embedded systems is more difficult. This paper will present the implementation of a speech recognition system as an embedded system using FPGA technology.

In fact, the implementation of speech recognition systems has been done using FPGA technology in recent years. In paper [3], speech recognition systems are implemented as hardware/software co-design systems using Hidden Markov Model (HMM). This project use Linear Predictive Coding (LPC) method in feature extraction block. So, the recognition accuracy is not high compared with the MFCC method. In paper [4], the MFCC method is applied, but the optimization was not taken into account yet to increase performance.

Another work, presented in [5] and [6], the author proposed an efficient MFCC hardware implementation for feature extraction in speech recognition. However, this work has been done using ASIC technology and therefore less flexible than FPGA based implementations. Other implementations for speech recognition systems can be found at [7–9]. Among these, the work presented in 8 proposes a hardware/software co-design method to tradeoff between the performance and the flexibility of the recognition system while [7] and [9] present FPGA based implementation of the recognition systems. None of them discuss about the optimization method for MFCC algorithm. In our work, the MFCC method is used with some modifications to increase the performance of the system. The whole system has been implemented using Altera FPGA technology to be more flexible.

The paper is organized as follows. The design and implementation of the proposed speech recognition system as a SoPC (System on Programmable Chip) is mentioned and discussed in Sect. 2. Section 3 will show the achieved experimental results. Finally, conclusions and further discussions will be presented in Sect. 4.

2 Implementation

In the implementation process, some blocks will be adjusted, modified so that the computing speed of the block can be increased. In this section, we will show some improvements in a few blocks to optimize the computing speed of the block. Evaluated results in terms of the number of clock cycles will be presented in the next section.

2.1 Feature Extraction Implementation

Pre-emphasis

In pre-emphasis block, the coefficient "a" has the value from 0.9 to 1. In theory, the normal value of "a" is 0.97. But, when we build the system on SoPC, we must choose the value of "a" so that the program easy to implement. Thus, the pre-emphasis block will run faster. The value $a = 1$, 15/16, 0.97 are surveyed about program performance through assessment of pulse clock.

Transfer function of the filter is described by Eq. 1. In the time domain, the relationship between output and input is shown in Eq. 2.

$$H(z) = 1 - a.z^{-1} \tag{1}$$

$$s'_i = s_i - a.s_{i-1} \tag{2}$$

With $a = 1$, Eq. 2 will be simplified as: $s'_i = s_i - s_{i-1}$.

Advantage of using 15/16 as "a" coefficient is expressed in Eq. 3. $\frac{15}{16} s_{i-1}$ can be realized in binary computation system by shifting s_{i-1} 4 bits to the right. Using this value the multiplication step is simplified to shift and subtract operations.

$$s'_i = s_i - a \cdot s_{i-1}, \quad a = \frac{15}{16} \tag{3}$$

$$s'_i = s_i - \frac{15}{16} s_{i-1} = s_i - \left(s_{i-1} - \frac{1}{16} s_{i-1} \right)$$

Discrete Fourier Transform (DFT)

In general, $X(k)$ and $x(n)$ are the complex numbers. N-point DFT can be calculated as follows:

$$X_R(k) = \sum_{n=0}^{N-1} \left[x_R(n) \cos \frac{2\pi kn}{N} + x_I(n) \sin \frac{2\pi kn}{N} \right], \quad k = 0, 1, 2, \ldots, N - 1. \tag{4}$$

$$X_I(k) = -\sum_{n=0}^{N-1} \left[x_R(n) \sin \frac{2\pi kn}{N} - x_I(n) \cos \frac{2\pi kn}{N} \right], \quad k = 0, 1, 2, \ldots, N - 1. \tag{5}$$

If DFT transformation uses two Eqs. 4 and 5 to calculate, it costs $2N^2$ trigonometric calculations, $4N^2$ real multiplications, and $4N(N - 1)$ additions. This shows that when the direct calculation using the DFT formula above arises large computational cost, it will slow speed program execution. Therefore, in this case we use the Fast Fourier Transform (FFT) algorithm instead. In addition, by using the look-up table of coefficients cosine, sine also increases the computing speed of the program.

Magnitude computation

If using the conventional formula for calculating the complex amplitude as Eq. 6, then the calculation will be very slow speed, thereby reducing the speed of program execution.

$$M = \sqrt{I^2 + Q^2} \tag{6}$$

Therefore, the estimation algorithm is applied. This algorithm calculates very fast amplitude of a complex number almost exact compared to the normal range by taking the square root operation. For complex number $I + jQ$, amplitude estimation algorithm as follows:

$$M \approx \alpha.\max\{|I|, |Q|\} + \beta.\min\{|I|, |Q|\} \tag{7}$$

In this system, we use α as 1 and β as 1/4. This approach reduces the number of calculations with acceptable error.

Mel frequency filter bank

The kth of power coefficient of the nth frame is calculated by the Eq. 8 as

$$S'_{nk} = \sum_j S_{nj}.FC_{kj}, \quad k = 0, 1, \ldots, K \tag{8}$$

where, K is the number of the filters. S_{nj} is the jth point of the nth frame's spectrum, and FC_{kj} is the jth coefficient of the kth filter. When implementing the speech recognition system on SoPC, the rectangular filter bank is used in the new algorithm instead of the triangular filter bank. So, the Eq. 8 becomes

$$S'_{nk} = \sum_j S_{nj}.FC_{kj}, \quad FC_{kj} = 0 \text{ or } 1 \tag{9}$$

The rectangular filters are proposed to be used instead of the triangular filters because the output characteristic of a rectangular filter is either a "1" or a "0", the multiply and sum operations can be simplified to simple "add" and "no add" operations. No multiplication step is required in the proposed approach.

2.2 Training and Recognition Implementation

In this work, training process is done by using Vector Quantization. Codebook size of 128 is considered. K-Mean algorithm is used for training codebook. First, M vectors in the L vectors are randomly chosen for training. The second step, for each training vector v, we find the codeword in the current codebook vectors closest distance this vector and we assign it belongs to the group of the codeword. The third step, for each group, codeword is updated by using the average of all training vectors in this group. Repeat steps 2 and 3 until quantum error smaller than threshold value.

In recognition process, the input speech sample is extracted the feature by the MFCC algorithm first. Then the feature vectors are calculated to find the VQ distortion for each codebook. The word having smallest distortion is the word which needs to be identified.

2.3 SoPC Implementation

The proposed speech recognition system has been intently implemented on Altera FPGAs for high performance. In this system, Nios II Processor is used as the most important component of the system, a processor to execute programs of the system. All compiled C program is stored in the SDRAM. Flash memory is used to store the parameters of the codebook after training. The ADC interface is the part connected to the Audio Codec WM8731 chip. This chip is responsible for data sampling of voices speak into the microphone. LCD is used to show the implementation of the program, the recognition results will be also displayed on LCD. In particular, the Interval Timer is used to calculate the number of the pulse clocks when executing each block.

3 Experimental Results

As mentioned above, we use Interval Timer to survey the program execution speed of each functional block in speech recognition system. The input speech samples are used for system input is 2,400 samples. The clock of the system is 50 MHz.

3.1 Feature Extraction

The number of clock cycles for executing each block in feature extraction are presented in Table 1.

The pre-emphasis block with $a = 1$ is executed fastest. The value $a = 15/16$ in the pre-emphasis block run faster than the pre-emphasis block with $a = 0.97$. In the magnitude computation step, the estimation algorithm calculates amplitude of a complex number much faster than the normal algorithm by taking the square root operation. By using the rectangle filters to replace the triangle filters, the program execution speed of the Mel Frequency Filter Bank block is increased 46 times.

In Fig. 1, the program execution speed of all blocks in MFCC based feature extraction is shown. The FFT block is the slowest, requires 94,874,620 clock cycles to complete the given input samples.

Table 1 Obtained results of program execution speed by coefficient "a"

Block	Algorithm/parameter	Clock cycles (*cycles*)
Pre_emphasis	$a = 1$	2,078,463
	$a = 15/16$	2,155,870
	$a = 0.97$	2,156,018
FFT/DFT	FFT	94,874,620
	DFT	365,586,715
Magnitude_computation	Estimation amplitude	7,463,640
	Accuracy amplitude	80,412,716
Mel-filter-bank	Rectangle filters	418,427
	Triangle filters	19,317,411

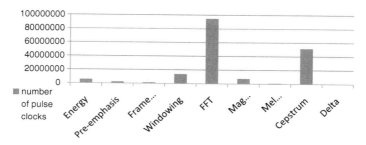

Fig. 1 Program execution speed of the blocks in MFCC based feature extraction

3.2 Vector Quantization

With codebook size of 128, Vector Quantization is used in the recognition step. So, it costs 531,067,721 clock cycles.

3.3 Recognition Accuracy

The whole recognition system with proposed architectures, parameters as stated above has the recognition accuracy of 88 %, in which 7,416 utterances recorded from male and female adults in three regions of the North, Middle, and South of Vietnam are used.

4 Conclusion

In this paper, we propose efficient architectures and design choices for each part in MFCC-HMM-based speech recognition system to improve the processing speed.

Design and Implementation of a SoPC System for Speech Recognition 1203

The determination of design choices are based on the easiness in implementation and experimental results of whole system. The whole system is built on FPGA, verified by testing with 7,416 utterances which are recorded from male and female adults in three regions of North, Middle and South of Vietnam with a recognition accuracy of 88 %.

References

1. Lawrance R, Biing-Hwang J (1993) Fundamentals of speech recognition. Prentice Hall PTR, Upper Saddle River
2. Thomas H, Gunnar E, Dan K, Gareth M, Julian O, Dave O, Dan P, Valtcho V, Phil W, Steve Y (1995–2002) The hidden markov model toolkit (HTK) book (for HTK version 3.2.1). Cambridge University. http://htk.eng.cam.ac.uk/
3. Amudha V, Venkataramani B, Vinoth kumar R, Ravishankar S (2009) Software/Hardware co-design of HMM based isolated digit recognition system. J Compu 4:(2)154–159
4. Zhou H, Han X (2009) Design and implementation of speech recognition system based on field programmable gate array. Mod Appl Sci 3(8):106–111
5. Han W, Chan C-F, Choy C-S, Pun K-P (2006) An efficient MFCC extraction method in speech recognition. In: Proceedings of the 2006 IEEE international symposium on circuits and systems (ISCAS). Greece, pp 145–148
6. Han W (2006) A speech recognition IC with an efficient MFCC extraction algorithm and multi-mixture models. Doctor of philosophy thesis, the Chinese University of Hong Kong
7. Pan S-T, Lai C-C, Tsai B-Y (2011) The implementation of speech recognition systems on FPGA—based embedded systems with SOC architecture. Int J Innovative Comput Inf Control 7(10):5939–5946
8. Cheng O, Abdulla W, Salcic Z (2011) Hardware-Software co-design of automatic speech recognition system for embedded real-time applications. In: Proceedings of the IEEE transactions on industrial electronics, pp 850–859
9. Ge Z, Yin J, Liu, Q, Yang C (2011) A real-time speech recognition system based on the implementation of FPGA. In: Cross strait quad-regional radio science and wireless technology conference (CSQRWC), pp 1375–1378

Index

A
Ahmed, Mohamed A., 841
Ahn, Hyun-Sik, 769, 777
Ahn, Sang-Ho, 367
Aikebaier, Ailixier, 477
Akrout, Belhassen, 43
An, Ming-Shou, 1181
Arpasilp, Dissakan, 879

B
Bae, Changseok, 763
Bae, Nam-Jin, 83, 359, 699
Baek, Mi Ran, 359
Baek, Miran, 83
Baek, Seung Jun, 1047
Bandara, K., 849
Bao, Vo Nguyen Quoc, 919
Bednar, Slavomir, 747
Bien, Franklin, 1065
Bok, Junyeong, 801
Bovet, Gérôme, 93
Byun, Sangmun, 35

C
Chang, Hsuan-pu, 665
Chang, Jae-woo, 277, 345
Chang, Jaewoo, 337
Chang, Wen-Chih, 649
Chao, Yi-Hsiang, 143
Chen, Nan, 887
Chen, Yung Hui, 633
Cheng, H. D., 285
Chien, Dao-Ngoc, 1169
Cho, Do-Eun, 57
Cho, Kyungryong, 691
Cho, Sang Bock, 943

Cho, Seongsoo, 575, 621
Cho, Yong-Yun, 83, 359, 691, 699
Choeysuwan, Patcharaporn, 935
Choi, Dojin, 51
Choi, Hyo Sub, 467, 1137
Choi, Kiyoung, 1161
Choi, KwangHee, 511
Choi, Miae, 35
Chong, Kil To, 1031, 1047
Choomchuay, Somsak, 935
Chul, Son Kwang, 575, 621
Chun, Se-Hak, 353
Chun, Soon-Yong, 1005, 1073, 1081
Chung, Gi-Soo, 545
Chung, Tai-Myoung, 485
Chung, Yeon-ho, 849
Chung, Young-Suk, 127
Claesen, Luc, 177

D
Dang, Hong Quan, 983
Deepunya, Chairak, 871
Dinh, Khanh Quoc, 997
Do, Binh V., 817
Do, Quoc Trinh, 919
Doanh, Nguyen Thi Hong, 895
Dou, Qiang, 449
Duc Dung, Do, 911
Duc, Nguyen Tuan, 895
Duc-Tan, Tran, 911
Duc-Tuyen, Ta, 911

E
Enokido, Tomoya, 477
Eom, Jung-Ho, 485

J. J. (Jong Hyuk) Park et al. (eds.), *Multimedia and Ubiquitous Engineering*,
Lecture Notes in Electrical Engineering 240, DOI: 10.1007/978-94-007-6738-6,
© Springer Science+Business Media Dordrecht(Outside the USA) 2013

F

Feng, Junliang, 193

G

Ganduulga, 1031
Gaspar, Stefan, 713
Gia, Tuan Anh Nguyen, 1013
Giang, Nguyen Linh, 1121, 1129
Gil, Joon-Min, 511, 521, 529
Gu, Bongen, 51
Gu, Junzhong, 193

H

Hai, Nguyen Dai, 1121, 1129
Han, Jong Wook, 293, 377
Han, Kyeong-Soo, 1153
Han, Kyuseung, 1161
Han, Sewon, 827
He, Yi-Jun, 731
Hennebert, Jean, 93
Hiep, Pham Thanh, 857
Hoang, Ngoc-Bach, 1099
Hoang, Nguyen Huy, 857
Hoang, Tran Van, 1197
Hong, Bonghwa, 569, 575
Hong, Dong Pyo, 1031
Hong, Seungtae, 337
Hsu, Victoria, 673
Huang, Fengyun, 683
Hung, Pham-Viet, 1169
HuuTo, Pham, 863
Hyun, DongLim, 415, 423, 459

J

Jang, Min-Ki, 321
Jang, Miyoung, 277
Jang, Mi-Young, 345
Jeon, Byeungwoo, 949, 997
Jeon, Yong-Hee, 505
Jeong, Seungdo, 1153
Jheng, Ming-Ren, 649
Ji, Un-Ho, 1081
Jo, Hyun-ho, 1113
Jo, Manhwee, 1161
Joo, Haejong, 569
Jozef, Zajac, 707
Jun, Kyungkoo, 77
Jung, SeokWon, 1039

K

Kang, Byoung-Doo, 367
Kang, Dae-Seong, 1181
Kang, Dong-oh, 763
Kang, Hee-Jun, 1099
Kang, M. G., 1147
Kao, Bruce C., 633, 641
Keikhosrokiani, Pantea, 785
Khac, Duy Huynh, 1013
Khang, Nguyen-Van, 1169
Kim, ByeongSu, 313, 387
Kim, Byung-Seo, 27, 287
Kim, Chang-Geol, 561
Kim, Chung Hyeok, 621
Kim, Do-Hyeun, 887
Kim, Dong W., 593, 597
Kim, Dong-Hyun, 157, 627
Kim, Dongkyun, 529
Kim, EunGil, 305, 415, 423, 459
Kim, Eun-Young, 353
Kim, Hag-Tae, 1031
Kim, Heung-Shik, 367
Kim, Hong Gean, 359
Kim, HongGeun, 83, 691
Kim, Hoon, 103
Kim, Hyeong-Il, 345
Kim, Hyoung-il, 277
Kim, HyunGon, 1039
Kim, Hyunsung, 203, 253,
    269, 927
Kim, I. K., 1147
Kim, Intaek, 983
Kim, Jae-O, 769
Kim, Jeong Ah, 431
Kim, Jin-Mook, 585, 627
Kim, Jong-Dae, 163, 321
Kim, Jong-Ho, 367
Kim, JongHoon, 305, 313, 387, 415,
    423, 459, 493
Kim, JongJin, 415
Kim, Jong-Seok, 537, 553
Kim, Jun Kyo, 431
Kim, Kisuk, 691
Kim, Mi-Hye, 537, 553, 585
Kim, MinSoo, 1039
Kim, Sang-Chul, 1091
Kim, Sang-Kyoon, 367
Kim, Sangsoo, 569
Kim, Seungcheon, 903
Kim, Seung-Hae, 529
Kim, Si Jung, 57

Index

Kim, Siwan, 227
Kim, Sungho, 975
Kim, Sung-Hwan, 485
Kim, SungWan, 305
Kim, Tae-Eun, 957, 967
Kim, Taeho, 111, 1057
Kim, TaeHun, 313, 387
Kim, Tae Hyung, 359
Kim, Taehyung, 83, 699
Kim, Taejin, 227
Kim, Won-Tae, 1107
Kim, Yongguk, 835
Kim, Yong-Kyun, 377
Kim, Young-Chon, 841
Kim, Yu-Seop, 163, 321
Klein, Udo, 1189
Kohno, Ryuji, 857
Kong, Fei, 211
Kong, Ki-Sik, 511
Koo, Yong Seo, 989
Koonchiang, Krisada, 879
Kresman, Ray, 219
Kun, She, 169
Kwak, Jae Chang, 989
Kwak, Yoonsik, 35, 51
Kweon, In So, 975

L

Lee, Chan-Su, 1005
Lee, DaeWon, 521
Lee, Deok Gyu, 293, 377
Lee, Duck Jin, 1047
Lee, Gangin, 185
Lee, Hankyu, 1153
Lee, Ho-Dong, 597
Lee, Hyeong-Ok, 537, 553
Lee, Hyunjo, 337
Lee, J. S., 1147
Lee, Jae Kyu, 1137
Lee, Jeongsam, 35
Lee, Jeongyong, 35
Lee, JiHwon, 493
Lee, Jong Kwan, 219
Lee, Jooyi, 1107
Lee, Kilhung, 739
Lee, K. T., 1147
Lee, Kwang Yeob, 989
Lee, Kwanyong, 157
Lee, Kyung-Jung, 777
Lee, Myeong Bae, 83, 359
Lee, Sang-Heon, 1005
Lee, Sang Yub, 467, 1137
Lee, Seon-oh, 1113

Lee, Tae-Gyu, 545
Lee, WonBong, 611
Lee, Yunho, 621
Lem, Jeongbin, 35
Li, Gen, 449
Li, Man, 21
Lim, Hye-Youn, 1181
Lim, JungEun, 1073
Lim, KyooSeob, 601, 611
Linh, Mai, 809
Lisowska, Agnieszka, 3
Liu, Jing, 13
Liu, Rui, 285
Lo, Shou-Chih, 329
Lu, Pingjing, 449

M

Ma, Gunil, 227
Maeng, Ji Chan, 1057, 1107
Mahdi, Walid, 43
Marton, David, 747
Masada, Tomonari, 129
Matus, Cuma, 707
Michal, Hatala, 707
Modrak, Vladimir, 747
Moldovyan, Nikolay A., 817
Moon, Chan-Woo, 769
Moon, Jinyoung, 763
Mtonga, Kambombo, 203
Mustaffa, Norlia, 785

N

Nakatsuka, Ryo, 243
Nam, Jung-hak, 1113
Nam, Kyoung-Min, 163
Ngoc Nguyen, T. T., 1065
Nguyen, Bao Ngoc, 809
Nguyen, B. D., 863
Nguyen, Binh Duong, 809
Nguyen, Dinh Uyen, 809
Nguyen, Minh H., 817
Nguyen, Pham Minh Luan, 943
Nguyen, Viet Anh, 949
Ni, Rongrong, 285
Niroopan, P., 849
Noitubtim, Musleemin, 871

O

Oh, JungCheol, 493
Okuda, Kenji, 243
Otsubo, Nobuto, 235

P

Pan, Yun, 177
Park, Chang-Woo, 83
Park, Chan-Young, 163, 321
Park, DongGook, 699
Park, Duck Keun, 1137
Park, Hyeyoung, 157
Park, James J., 521
Park, Jang Woo, 359, 691, 69
Park, Ji-Hoon, 27
Park, Jong-Hyun, 575
Park, Jong-Wook, 593, 597
Park, Koo-Rock, 627
Park, Min-Woo, 485
Park, Sang Hyun, 467, 1137
Park, Sang-June, 1181
Park, Seong Ryoung, 359
Park, Sung-Wook, 593, 597
Park, Tae Ryong, 989
Pasko, Jan, 713
Peyls, Alexander, 177
Pham, Kien T., 863
Pham, Phuong Minh, 997
Phan, Cuong H., 1021
Promwong, Sathaporn, 871, 879
Pyun, Gwangbum, 121

Q

Qiao, Fei, 13

R

Raybourn, Tracey, 219
Ren, Yizhi, 211
Ro, Sunny, 777
Ryang, Heungmo, 113, 135
Ryu, Eun-kyung, 1113
Ryu, Heung-Gyoon, 801, 835
Ryu, Minsoo, 1057, 1107

S

Sakamoto, Syunya, 243
Sarwar, Muhammad Imran, 785
Seo, BoHyeok, 1073
Seo, JaeHyun, 1039
Shin, Chang-Sun, 83, 359, 691, 699
Shin, Jin, 723
Shin, Jin-Sup, 1181
Sim, Dong-gyu, 1113
Skurzok, Dawid, 151
Soh, YoungSung, 187
Song, Byeonghun, 1107

Song, Byung-Seop, 561
Song, Hye-Jeong, 321
Song, Hye-Jung, 163
Song, Seokil, 35
Su, Tran Van, 809
Sun, Caixia, 449
Sung, Mankyu, 405
Syed, Ikram, 103

T

Takasu, Atsuhiro, 129
Takizawa, Makoto, 477
Tang, Dianhua, 269
Taniguchi, Hideo, 235
Tran, Phuoc Vinh, 1013
Tran, Van-Su, 863
Tran, Xuan Nam, 919
Tran, Xuan-Tu, 1197
Trang, Hoang, 1197
Truong, Nguyen Ly Thien, 1197
Truong, Tuyen T. T., 1021

U

Uemura, Shinichiro, 235

V

Vallent, Thokozani Felix, 252
Van, Tram, 863
Võ, Trình Quốc, 1189
Vu, Duc Hiep, 919

W

Wang, Gicheol, 529
Wang, Te-Hua, 657
Wang, Yimu, 177
Wang, Yixing, 261
Wei, Qi, 13
Weng, Martin M., 641
Won, So-Min, 585
Woo, Y. J., 1147
Wu, Shih-Wei, 649

X

Xiaobo, Song, 169
Xiong, Hao, 731
Xu, Jian, 211
Xu, Jinli, 683
Xu, Ming, 211, 261
Xu, Wei, 683

Index

Y

Yamakami, Toshihiko, 69, 395, 439
Yamauchi, Toshihiro, 235, 243
Yan, Xiaolang, 177
Yang, Eun-Suk, 163
Yang, Haomiao, 203, 269
Yang, Hsuan-Che, 649
Yang, Huazhong, 13
Yang, Won-Hyuk, 841
Yang, YoungHoon, 415
Yeo, Sang-Soo, 57
Yi, Chuho, 1153
Yi, Hyunyi, 227
Yi, Jeong Hyun, 227
Yi, Soo-Yeong, 723, 755
Yin, Haibin, 683
Yiu, Siu Ming, 731
Yoon, Eun-Jun, 203
Yoon, Hyunmin, 1057, 1107
Yoon, Min, 277, 345

Yu, Ti-Hsin, 329
Yu, Xiaoqi, 731
Yun, Unil, 113, 121, 135, 185

Z

Zakaria, Nasriah, 785
Zhang, Ganglei, 21
Zhang, Haiping, 211
Zhang, Ying, 449
Zhang, Zheng, 1047
Zhang, Zhitong, 285
Zhao, Yao, 285
Zheng, Ning, 211, 261
Zhiying, Tan, 169
Zhou, Hongwei, 449
Zhou, Zili, 193
Ziólko, Bartosz, 151

Printed by Publishers' Graphics LLC